General data and fundamental constants

Quantity	Symbol	Value	Power of ten	
Speed of light	c	2.997 924 58	10^8	$m\ s^{-1}$
Elementary charge	e	1.602 177	10^{-19}	C
Faraday constant	$F = N_A e$	9.6485	10^4	$C\ mol^{-1}$
Boltzmann constant	k	1.380 66	10^{-23}	$J\ K^{-1}$
Gas constant	$R = N_A k$	8.314 51		$J\ K^{-1}\ mol^{-1}$
		8.314 51	10^{-2}	$L\ bar\ K^{-1}\ mol^{-1}$
		8.205 78	10^{-2}	$L\ atm\ K^{-1}\ mol^{-1}$
		6.2364		$L\ Torr\ K^{-1}\ mol^{-1}$
Planck constant	h	6.626 08	10^{-34}	$J\ s$
	$\hbar = h/2\pi$	1.054 57	10^{-34}	$J\ s$
Avogadro constant	N_A	6.022 14	10^{23}	mol^{-1}
Atomic mass unit	u	1.660 54	10^{-27}	kg
Mass				
electron	m_e	9.109 39	10^{-31}	kg
proton	m_p	1.672 62	10^{-27}	kg
neutron	m_n	1.674 93	10^{-27}	kg
Vacuum permittivity	ε_0	8.854 19	10^{-12}	$J^{-1}\ C^2\ m^{-1}$
	$4\pi\varepsilon_0$	1.112 65	10^{-10}	$J^{-1}\ C^2\ m^{-1}$
Magneton				
Bohr	$\mu_B = e\hbar/2m_e c$	9.274 02	10^{-24}	$J\ T^{-1}$
nuclear	$\mu_N = e\hbar/2m_p c$	5.050 79	10^{-27}	$J\ T^{-1}$
Bohr radius	$a_0 = 4\pi\varepsilon_0 \hbar^2/m_e e^2$	5.291 77	10^{-11}	m
Rydberg constant	$R = m_e e^4/8h^3 c\varepsilon_0^2$	1.097 37		cm^{-1}
Standard acceleration of free fall	g	9.906 65		$m\ s^{-2}$

The Elements of
PHYSICAL
CHEMISTRY

P. W. Atkins

The Elements of
Physical Chemistry

P. W. Atkins

University Lecturer and Fellow
of Lincoln College, Oxford

Oxford • Melbourne • Tokyo
Oxford University Press

Cover image of interactions at the organic-inorganic interface are illustrated by a tetramethylammonium cation trapped within a zeolite cage that has crystallized around it. Modeling and graphics by Catalysis 1.0 software, Biosym Technologies, Inc.

Library of Congress Cataloging-in-Publication Data
Atkins, P. W. (Peter William), 1940–
    The elements of physical chemistry / P. W. Atkins.
        p.   cm.
    Simplified version of the 4th ed. of the author's Physical
chemistry. 1990.
    Includes bibliographical references and index.
    ISBN 0-7167-2364-6
    1. Chemistry, Physical and theoretical.   I. Atkins, P. W. (Peter
William), 1940–   Physical chemistry.   II. Title.
        QD453.2.A87   1992   92-28273
        541.3—dc20   CIP

Printed in the United States of America

Third printing 1995, KP

This edition has been authorized by the Oxford University Press
for sale in the USA and Canada only and not for export therefrom.

Preface

The Elements of Physical Chemistry introduces the basic concepts and techniques of the subject. The concepts are used in many other branches of chemistry—in biochemistry, in chemical engineering, and (increasingly) in the biological and medical sciences—and the subject provides procedures for expressing qualitative ideas in quantitative, testable, terms. Although the text is derived from my *Physical Chemistry* text, and retains that book's general style and approach, it is more than just a crude shortening. I have mined the main text for material, but I have refined the presentation considerably. In many places I have recast arguments, rewritten whole chapters, reorganized the material substantially and have presented material in an entirely different way. The main text provided the theme, and this text has spun variations on that theme.

Apart from the selection of topics, the problem that has given me most concern while writing this book is the level of mathematics to adopt, for a derivation of a simple result can seem daunting and obscure the simplicity of the conclusion. I have therefore used the device of setting off derivations in boxes called *Justifications*: the concept and the conclusion are first presented in the text, then the derivation or supporting argument is presented in the *Justification*. Similarly, many arguments can be presented without calculus, but the calculus-based expression should be shown partly because calculus is an intrinsic part of the apparatus of physical chemistry and partly because the resulting expressions are often of considerably greater power. I have therefore included a number of *Calculus boxes* which re-express a preceding idea in terms of differentials and integrals. I hope that in this way the student will be able to use the text at a variety of levels: achieving one level of understanding on first reading a chapter, and then moving to a deeper level on second reading by taking in the supporting material in the *Justifications* and the *Calculus boxes*.

I am very conscious that students will come to this text from a wide variety of backgrounds. I have therefore provided a number of sections called *Further information* that serve two functions. In one, they provide some necessary background information (such as on the concepts of classical mechanics). In another, they provide information that enriches the text (such as an account of the Boltzmann distribution). As in the parent text, there are many worked *Examples* throughout the text, each with an accompanying exercise. An innovation is the addition of brief *Exercises* in the text itself. These exercises are designed to give the reader practice with a simple numerical manipulation of an expression, and in particular to become familiar with typical orders of magnitudes of quantities that have been introduced in the immediately preceding paragraphs.

I have held in mind the type of reader and their likely interests. Thus, much of the illustrative material and applications make use of biochemically or biologically inclined material, because I see one type of reader as being someone who is anticipating a career in these disciplines. I have also drawn a number of examples from industrial chemistry, because another class of users is likely to be students moving in that direction. It has been a challenge to position the text so that readers from a diverse background of educational traditions will find the material useful. I have tried to write it in such a way that readers who have completed a general chemistry course in their first year and readers who have had no such course will both find the material rewarding. To do so, I have ensured that material that will have been covered in an introductory course is treated in a somewhat more sophisticated style, yet in such a way that sophistication leaves the topic accessible to the newcomer. Overall, though, I have aimed to produce a text which will give any college student an introduction to the main principles of physical chemistry, either as the only course in the subject that they will ever take or as a reasonably gentle introduction to more detailed study in a later course.

In selecting the themes to develop, I have been greatly helped by a number of people who are familiar with the level and aspirations of the intended audience. I would particularly like to thank: G. G. Cameron, University of Aberdeen; T. Cosgrove, University of Bristol; R. DePoy, Metropolitan State College of Denver; J. Hutchinson, Rice University; K. G. McKendrick, University of Edinburgh; C. Metz, College of Charlston; K. Morden, Louisiana State University; M. J. Pilling, University of Leeds; B. Robinson, University of East Anglia; R. Schwenz, University of North Colorado; D. Tildesley, University of Southampton. Professor H. Shanfield, University of Houston, read the entire proofs, and Professor D. Goss, Hunter College, spotted a number of points while she was preparing the *Solutions Manual*. As always I have received excellent support from my publishers, and wish to thank them publicly yet again for their advice, help, and attention to detail.

Oxford
February *1992*

P.W.A.

Contents

Contents

The properties of gases

1

Chemistry is the science of matter and the changes it can undergo; **physical chemistry** is concerned with the physical principles that underlie chemistry, and seeks to account for the structure of matter and the changes it undergoes in terms of fundamental concepts such as atoms, electrons, and energy.

The broadest classification of matter is into one of three **physical states**, namely gas, liquid, or solid. Later we shall see how this classification can be refined, but these three broad classes are a good starting point. The three states of matter can be recognized by their behaviour when they are enclosed in a container:

A **gas** is a fluid state of matter that fills the container it occupies.

A **liquid** is a fluid state of matter that possesses a well-defined surface and (in a gravitational field) fills the lower part of the container it occupies.

A **solid** retains its shape independent of the shape of the container it occupies.

One of the roles of physical chemistry is to establish the link between the properties of bulk matter and the behaviour of the particles— atoms, ions, or molecules—of which it is composed. The existence of the three states is a first illustration of this procedure, for each state is composed of particles with different degrees of freedom:

A **gas** is composed of particles that are in continuous rapid, chaotic motion. A particle travels several (often many) diameters before colliding with another particle, and for most of the time the particles are so far apart that they interact with each other only very weakly.

A **liquid** consists of particles that are in contact with each other, but are able to move past each other. The particles are in a continuous state of motion, but travel only a fraction of a diameter before colliding with a neighbour.

A **solid** consists of particles that are in contact with each other and unable to move past each other. Although the particles oscillate around an average location, they are essentially trapped in their initial positions.

The essential difference between the three states of matter comes down to the freedom of the particles to move past each other; if the particles are, on average, widely separated, then there is hardly any restriction on their motion, and the substance is a gas. If the particles interact so strongly with each other that they are locked together, then the substance is a solid. If the particles have an intermediate mobility between these extremes, then the substance is a liquid. The melting of a solid and the vaporization of a liquid can be understood in terms of the progressive increase in the liberty of the particles as a sample is heated and the particles are able to move more vigorously.

The description of states of matter

The term 'state' has many different meanings in chemistry, and it is important to keep them all in mind. We have already met one meaning in the expression 'the states of matter' and specifically 'the gaseous state'. Now we meet a second: by **state** we shall mean *a description of a sample of matter in terms of its volume, pressure, temperature, and amount of substance present in it.* (The precise meaning of these terms is described below.) If a certain sample of a substance has the same volume, pressure, and temperature, and contains the same amount of atoms or molecules as another sample of the same substance, then the two samples are in the same state. If any of the properties differ between samples, then the substances are in different states.

1.1 The description of states

To see more precisely what is involved in identifying the state of a substance, we need to define the terms we have used. One property—the volume—can be dealt with without any fuss: the **volume** V of a sample is a measure of the space it occupies. Thus, we write $V = 100\,cm^3$ if the sample occupies $100\,cm^3$ of space. Some of the units used to express volume (including cubic metres, m^3; litres, L; millilitres, mL) are reviewed in *Further information* 1 (next page). The other properties we have mentioned (pressure, temperature, and amount of substance) need more introduction, for even though they may be familiar from everyday life, they need to be defined carefully for use in science.

Pressure

Pressure p is force per unit area. When you are standing on ice, you generate a pressure on the ice as a result of the gravitational force pulling you towards the centre of the Earth, but the pressure will be quite low because the force is spread over an area equal to that of the soles of your shoes. However, if you stand on skates, then the area of the blades in contact with the ice is much smaller, so although the force is the same, the pressure is much greater (Fig. 1.1). The pressure may be so great, in fact, that it modifies the arrangement of water molecules at the surface of the ice, and hence allows you to slide smoothly over the surface.

Although the gravitational pull of the Earth on an object can result in a pressure, gravity is not the only origin of pressure. For example, the impact of gas molecules on a surface gives rise to a force, and hence to a pressure. If an object is immersed in the gas, it experiences a pressure over its entire surface because molecules collide with it from all directions. In this way, the atmosphere exerts a pressure on all the objects in it, and the pressure is greatest at sea level because the density of air, and hence the number of colliding molecules, is greatest there. The atmospheric pressure is very considerable: it is the same as would be exerted by placing 1 kg of lead (or any other material) on to a surface of area $1\,cm^2$: we go through our lives under this heavy burden pressing on every square centimetre of our bodies. Some deep-sea creatures are built to withstand even greater pressures: at 1000 m below sea level the pressure is 100 times greater than at the surface, and creatures and submarines must withstand the equivalent of 100 kg of lead loaded on to each square centimetre of their surfaces.

If a gas is confined to a cylinder fitted with a movable piston that separates it from the atmosphere, the position of the piston adjusts until the pressure of the gas inside the cylinder is equal to that exerted by the atmosphere. This equality of pressures is the condition of **mechanical equilibrium**. The pressure of the confined gas arises from the incessant impact of the particles: they batter the inside surface of the piston and oppose the battering of the molecules in the atmosphere that is pressing on the outside surface of the piston (Fig. 1.2). So long as the piston is weightless (that is, so long as we can neglect any gravitational pull on it), the gas is in mechanical equilibrium with the atmosphere whatever the orientation of the piston and cylinder, because the external battering is the same in all directions.

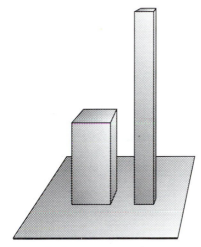

Fig. 1.1 These two blocks of matter have the same mass. They exert the same force on the surface on which they are standing, but the narrow column exerts a stronger pressure because it exerts the force on a smaller area than the fatter block.

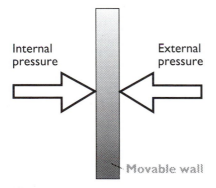

Fig. 1.2 A system is in mechanical equilibrium with its surroundings if it is separated from them by a movable wall and the external pressure is equal to the pressure of the gas in the system.

Pressure is measured in a unit called the **pascal**, Pa:

$$1\,\text{Pa} = 1\,\text{kg}\,\text{m}^{-1}\,\text{s}^{-2}$$

The pressure of the atmosphere is about 10^5 Pa (100 kPa) at sea level. This fact lets us imagine the magnitude of 1 Pa, for we have just seen that 1 kg of lead covering $1\,\text{cm}^2$ exerts the equivalent of an atmosphere's pressure, so $1/100\,000$ of that mass, or 0.01 g, will exert about 1 Pa (which shows that the pascal is rather a small unit of pressure). Some of the other units that are used to report pressures are summarized in *Further information 1*. They include the **bar**, which is

FURTHER INFORMATION I:
Quantities and units

The result of a measurement is a **physical quantity** (such as mass or density) that is reported as a numerical multiple of agreed **units**:

physical quantity = numerical value × unit

For example, the mass of an object may be reported as 2.5 g and its density as $1.01\,\text{g}\,\text{cm}^{-3}$ where the units are respectively grams (g) and grams per cubic centimetre ($\text{g}\,\text{cm}^{-3}$). Units are treated like algebraic quantities, and may be multiplied, divided, and cancelled. Thus the expression (physical quantity)/unit is simply the numerical value of the measurement in the specified units, and hence is a dimensionless quantity. For instance, the mass reported above could be denoted $m/\text{g} = 2.5$ and the density as $\rho/\text{g}\,\text{cm}^{-3} = 1.01$.

In the **International System** of units (SI, from the French *Système International*), the units are formed from seven **base units**:

Physical quantity	Symbol for quantity	Base unit
Length	l	metre, m
Mass	m	kilogram, kg
Time	t	second, s
Electric current	I	ampere, A
Thermodynamic temperature	T	kelvin, K
Amount of substance	n	mole, mol
Luminous intensity	I_v	candela, cd

All other physical quantities may be expressed as combinations of these physical quantities and reported in terms of **derived units**. Thus, volume is $(\text{length})^3$ and may be reported in cubic metres (m^3), and density, which is mass/volume, may be reported in kilograms per cubic metre ($\text{kg}\,\text{m}^{-3}$). A number of derived quantities have special names and symbols. Among the most important for our purposes are the following:

Derived quantity	Derived unit	Name of derived unit
Force	$\text{kg}\,\text{m}\,\text{s}^{-2}$	newton, N
Pressure	$\text{kg}\,\text{m}^{-1}\,\text{s}^{-2}$	pascal, Pa
Energy	$\text{kg}\,\text{m}^2\,\text{s}^{-2}$	joule, J
Power	$\text{kg}\,\text{m}^2\,\text{s}^{-3}$	watt, W

defined as exactly 10^5 Pa, or exactly 100 kPa (so atmospheric pressure is close to 1 bar), and the **atmosphere** (atm), which is exactly 101.325 kPa. The origin of the latter unit will be explained in a moment.

Exercise E1.1 Express a pressure of 1.000 bar in atmospheres.

[0.9869 atm]

In all cases (both for base and derived quantities), the units may be modified by a prefix that denotes a factor of a power of 10. Among the most common prefixes are the following:

Prefix	f	p	n	μ	m	c	d
Name	femto	pico	nano	micro	milli	centi	deci
Factor	10^{-15}	10^{-12}	10^{-9}	10^{-6}	10^{-3}	10^{-2}	10^{-1}
Prefix	k	M	G				
Name	kilo	mega	giga				
Factor	10^3	10^6	10^9				

Examples of the use of prefixes are

$$1\,\text{nm} = 10^{-9}\,\text{m}$$
$$1\,\text{pA} = 10^{-12}\,\text{A}$$
$$1\,\mu\text{mol} = 10^{-6}\,\text{mol}$$

The kilogram (kg) is anomalous: although it is a base unit, it is interpreted as 10^3 g, and prefixes are attached to the gram (as in $1\,\text{mg} = 10^{-3}\,\text{g}$). Powers of units apply to the prefix as well as the unit they modify:

$$1\,\text{cm}^3 = 1\,(\text{cm})^3 = 1\,(10^{-2}\,\text{m})^3 = 10^{-6}\,\text{m}^3$$

There are a number of units that are in wide use but are not a part of the International System. Some are *exactly* equal to multiples of SI units. These include the **litre** (L), which is

exactly 10^3 cm^3 and the **atmosphere** (atm), which is exactly 101.325 kPa. Others rely on the values of fundamental constants, and hence are liable to change when the values of the fundamental constants are modified by more accurate or more precise measurements. Thus, the size of the energy unit **electronvolt** (eV), the energy acquired by an electron that is accelerated through a potential difference of exactly 1 V, depends on the value of the charge of the electron, and the current (1992) conversion factor is

$$1\,\text{eV} \approx 1.602\,18 \times 10^{-19}\,\text{J}$$

The following table gives the conversion factors for a number of these 'convenient' units:

Physical quantity	Name of unit	Symbol for unit	Value
Time	minute	min	60 s
	hour	h	3600 s
	day	d	86 400 s
Length	angstrom	Å	10^{-10} m
Volume	litre	L	1 dm^3
Mass	tonne	t	10^3 kg
Pressure	bar	bar	10^5 Pa
Energy	electron-volt	eV	$\approx 1.602 \times 10^{-19}$ J

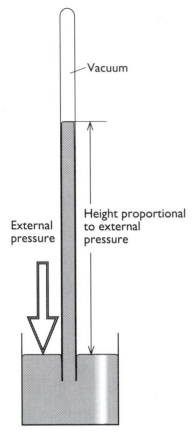

Fig. 1.3 The operation of a mercury barometer. The space above the mercury in the vertical tube is a vacuum, so no pressure is exerted on the top of the mercury column. However, the atmosphere exerts a pressure on the mercury in the reservoir, and pushes the column up the tube until the pressure exerted by the mercury column is equal to that exerted by the atmosphere. The height reached by the column is proportional to the external pressure, so the height can be used as a measure of this external pressure.

Atmospheric pressure (which varies with altitude and the weather) is measured with a **barometer**, an instrument invented by Torricelli, a student of Galileo. A barometer consists of an inverted tube of mercury that is sealed at its upper end and stands with its lower end in a bath of mercury. The height of the mercury in the tube is proportional to the atmosphere pressure (Fig. 1.3). It is found that on a normal day at sea level the atmosphere can support a column of mercury that is 760 mm high (which is reported as a pressure of 760 mmHg, where mmHg denotes 'millimetre of mercury'), but different states of the weather give slightly different heights. The use of the mercury barometer has inspired the introduction of the unit **Torr**, which is defined so that 1 atm is exactly equal to 760 Torr (see *Further information 1*).

The pressure at the foot of an incompressible liquid (a good approximation for any liquid) is proportional to the height h of the column and the density d of the liquid:

$$p = gdh \qquad (1)$$

where g is the acceleration of the free fall, a measure of the Earth's gravitational pull on the liquid ($g = 9.81 \text{ m s}^{-2}$ at sea level). For example, a 760 mm high column of mercury, which has a density of 13.6 g cm^{-3} ($13.6 \times 10^3 \text{ kg m}^{-3}$) at room temperature, exerts a pressure of

$$p = (9.81 \text{ m s}^{-2}) \times (13.6 \times 10^3 \text{ kg m}^{-3}) \times (0.760 \text{ m})$$

$$= 1.01 \times 10^5 \text{ kg m}^{-1} \text{s}^{-2}$$

or 101 kPa.

Exercise E1.2 Determine the relation between 1 mmHg and 1 Torr, and express 1.000 Torr in pascals.
[1 mmHg = 1 Torr; 1.000 Torr = 133.3 Pa]

The pressure of a gas inside a container can be measured by a variety of different pressure gauges. The simplest is a **manometer**, which is a U-tube containing liquid (water is sometimes used), with one limb connected to the container and the other open to the atmosphere (Fig. 1.4). The difference in heights of the liquid in the two arms of the U-tube is proportional to the *difference* in pressure between the gas in the container and the external atmosphere. So, if the atmospheric pressure is 765 Torr and the column of mercury on the apparatus side of the manometer is 140 mm higher than the column on the atmosphere side, we would conclude that the pressure in the apparatus is 765 Torr − 140 Torr = 625 Torr.

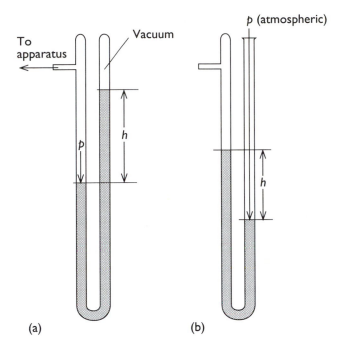

p (atmospheric)

Fig. 1.4 Two versions of a manometer used to measure the pressure of a sample of gas. (a) The height difference h of the two columns in the sealed-tube manometer is directly proportional to the pressure of the sample. (b) The difference in heights of the columns in the open-tube manometer is proportional to the difference in pressure between the sample and the atmosphere. In this case, the pressure of the sample is lower than that of the atmosphere.

(a)   (b)

Temperature

The **temperature** T of a sample is a familiar concept in everyday life—as a measure of how 'hot' or 'cold' something is—but it is quite difficult to give the concept a precise definition. The temperature of an object is a property that determines the direction in which energy will flow when it is in contact with another object: energy ('heat') flows from the higher temperature object to the lower temperature object. When the two objects have the same temperature, there is no net flow of heat between them, and we say that they are in **thermal equilibrium** (Fig. 1.5).

In science, temperatures are measured on either the Celsius or the Kelvin scales. On the **Celsius scale**, the freezing point of water is set at 0 °C and the boiling point is set at 100 °C. This scale is now in widespread everyday use. Temperatures on the Celsius scale will be denoted by θ throughout this text. However, it turns out to be much more convenient in many scientific applications to adopt the **Kelvin scale** and to express the temperature in kelvins, K. Temperatures on the kelvin scale are denoted T. The two scales are related by

$$T \text{ (in kelvin)} = \theta \text{ (in degrees Celsius)} + 273.15$$

That is, to obtain the temperature in kelvins, add 273.15 to the temperature in degrees Celsius. Thus, water freezes at 273 K and boils at 373 K; a warm day (25 °C) corresponds to 298 K. The Kelvin scale is much more fundamental than the Celsius scale (as will become clear), and in this text, *whenever we write T, we shall mean a temperature on the Kelvin scale.*

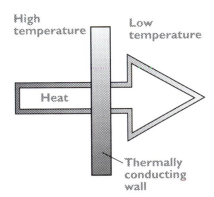

Fig. 1.5 The temperatures of two objects act as a signpost showing the direction in which energy will flow as heat through a thermally conducting wall: heat always flows from high temperature to low temperature.

A more sophisticated way of expressing the relation between T and θ, and one that we shall use in other contexts, is to regard the value of T as the product of a number and a unit (K), so that T/K is a pure number (for example, $T/K = 298$ if $T = 298$ K), and likewise for $\theta/°C$ (for example, $\theta/°C = 25$ if $\theta = 25\ °C$). Then we can write

$$T/K = \theta/°C + 273.15 \tag{2}$$

This expression is a relation between pure numbers.

> **Exercise E1.3** Use eqn (2) to express body temperature, 37 °C, in kelvins.
>
> [310 K]

Amount of substance

The **mass** m of an object is a measure of the quantity of matter it contains. Thus, 1 cm³ of lead contains more matter than the same volume of lithium, and on average a man contains more matter than a woman. Mass is measured in **kilograms** (kg), with 1 kg being defined as the mass of a certain block of platinum–iridium alloy which is preserved at Sèvres, just outside Paris. For typical laboratory-sized samples it is usually more convenient to use a smaller unit and to express mass in **grams** (g), where $1\ \text{kg} = 10^3\ \text{g}$.

In chemistry it is usually more useful to know the number of atoms, molecules, or ions in a sample rather than the mass itself. Because even 10 g of water consists of about 10^{23} H_2O molecules it is clearly appropriate to define a new unit that can be used to express such large numbers in a much simpler fashion. Chemists have therefore introduced the **mole** (mol; the name is derived, ironically, from the Latin word meaning 'massive heap'):

> 1 mol of particles is equal to the number of atoms in exactly 12 g of carbon-12.

In practice,

$$1\ \text{mol} = 6.022 \times 10^{23}\ \text{particles}$$

For example, a sample of hydrogen gas that contains 6.022×10^{23} hydrogen molecules consists of 1.000 mol H_2, and a sample of water that contains 1.2×10^{24} $(=2.0 \times 6.022 \times 10^{23})$ water molecules consists of 2.0 mol H_2O. A number of awe-inspiring illustrations of the number of particles in a mole have been devised; for instance, 1 mol of soft-drink cans would bury the surface of the earth to a depth of 200 miles.

Note that we must always specify the particles when using the unit mole, for that avoids any ambiguity. If, improperly, we said that a sample consisted of 1 mol of hydrogen, it would not be clear whether it consisted of 6×10^{23} hydrogen atoms (1 mol H) or 6×10^{23} hydrogen molecules (1 mol H_2). The important feature of the mole is that if we know that a sample of gas contains 1 mol O_2 and another sample contains 1 mol N_2, then we know that both samples contain the same number of molecules. The two samples would have different masses (different quantities of matter), but the numbers of molecules would be the same in each case.

Just as a kilogram is the unit used to report a certain physical property (the mass), the mole is the unit used when reporting the value of another physical property. In its case the property is called the **amount of substance** n of a sample. Thus, we can write $n = 1$ mol H_2, and say that the amount of hydrogen molecules in a sample is 1 mol. The term 'amount of substance', however, has not yet found wide acceptance among chemists, and in casual conversation you will often hear them speaking of 'the number of moles' in a sample.

There are various useful concepts that stem from the introduction of the concept of the amount of substance and its unit, the mole. One is **Avogadro's constant**, N_A. This constant is the number of particles (of any kind) per mole of substance:

$$N_A = 6.022 \times 10^{23} \, \text{mol}^{-1}$$

Avogadro's constant makes it very simple to convert from the number of particles N in a sample to the amount of substance n (in moles) it contains:

$$n = \frac{\text{Number of particles}}{\text{Number of particles per mole}} = \frac{N}{N_A} \tag{3}$$

For example, 8.8×10^{22} Cu atoms correspond to

$$n = \frac{8.8 \times 10^{22} \, \text{Cu}}{6.022 \times 10^{23} \, \text{mol}^{-1}} = 0.15 \, \text{mol Cu}$$

Notice how much easier it is to report the amount of copper atoms present rather than their actual number.

Exercise E1.4 How many Xe atoms are present in a sample that contains 1.8 mol Xe?

$$[1.1 \times 10^{24} \, \text{Xe atoms}]$$

The second very important concept is that of **molar mass**, M, the mass per mole of substance. For example, the molar mass of C atoms is $12.01 \, \text{g mol}^{-1}$, and that of H_2O molecules is $18.02 \, \text{g mol}^{-1}$. (The former names for molar mass were atomic weight, for the mass per mole of atoms, and molecular weight, for the mass per mole of molecules; both terms are still far from dead). In principle, the molar mass of a substance can be determined by measuring the mass m of a known amount of substance, but in practice other methods (especially mass spectrometry) are used instead.

The molar mass is used to convert from the mass of a sample (which we can measure) to the amount of substance present (which, in chemistry, we often need to know):

$$n = \frac{\text{Mass of atoms or molecules}}{\text{Mass per mole of atoms or molecules}} = \frac{m}{M} \qquad (4)$$

For example, to find the amount of C atoms present in $21.5 \, \text{g}$ of carbon, given that the molar mass of C is $12.01 \, \text{g mol}^{-1}$, we write

$$n = \frac{21.5 \, \text{g C}}{12.01 \, \text{g mol}^{-1}} = 1.79 \, \text{mol C}$$

Exercise E1.5 What amount of H_2O molecules is present in $100 \, \text{g}$ of water, given the molar mass of H_2O as $18.02 \, \text{g mol}^{-1}$.
[$5.55 \, \text{mol } H_2O$]

The molar masses of atoms are given in the periodic table printed inside the front cover of this book; the molar masses of molecules (such as H_2O) and of formula units of ionic compounds (such as $CaCO_3$) are found by adding together the molar masses of all the atoms of which the molecule or formula unit is composed.

Equations of state

Now we are at the stage where we can specify the state of any sample of substance by giving the values of the following properties:

V, the volume (in m^3, cm^3, etc.) it occupies.
p, the pressure (in Pa) to which it is subjected.
T, its temperature (in K).
n, the amount of substance (in mol) it contains.

However, an astonishing experimental fact of nature is that *these four quantities are not independent*. For instance, we cannot arbitrarily choose to have a sample of 5.55 mol H_2O in a volume of 100 cm^3 under a pressure of 100 kPa at 500 K: it is found *experimentally* that the state simply does not exist. If we select the amount, the volume, and the temperature, then we find that we have to accept a particular pressure. The same is true of all substances. This experimental generalization is summarized by saying that the substance obeys a particular **equation of state**, an equation that relates one of the four properties to the other three.

The equation of state of most substances are not known, so in general we cannot write down the mathematical relation between the four properties that define a state. However, certain equations of state are known. In particular, the equation of state of low-pressure gases is known, and proves to be very simple and very useful: it is used to describe the behaviour of gases taking part in reactions, the properties of the atmosphere, and as a starting point for problems in chemical engineering. It even occurs in astrophysics in the description of the structures of stars.

1.2 The perfect gas equation of state

The equations of states of gases were among the first results to be established in physical chemistry. The original experiments were carried out by Robert Boyle in the seventeenth century, and there was a resurgence in interest in the late seventeenth century when people began to fly in balloons and technological progress demanded more knowledge about the response of gases to changes of pressure and temperature.

Boyle's law: the response to pressure

Robert Boyle measured the variation of the volume of a gas as the pressure on it was changed while the temperature was held constant. Boyle found that, for a given sample of gas, *the volume is inversely proportional to the applied pressure*:

$$\textbf{Boyle's law:} \quad V \propto \frac{1}{p} \quad \text{(at constant } n, T\text{)} \tag{5}$$

(In science, a 'law' is a summary of experimental observations—a distillation of experience, not something laid down that substances must obey.) It follows from Boyle's law that doubling the pressure on a gas reduces its volume to one half of its original value. We can understand the ease with which a gas can be compressed by noting that, because there is so much space between the chaotically moving particles, they can be confined into a smaller region relatively easily.

If Boyle's law is rearranged to

$$p \propto \frac{1}{V} \quad \text{(at constant } n, T\text{)}$$

Fig. 1.6 The volume of a gas decreases as the pressure on it is increased. For a sample that obeys Boyle's law and that is kept at constant temperature, the graph showing the dependence is a hyperbola, as shown here. Each curve corresponds to a single temperature, and hence is an isotherm.

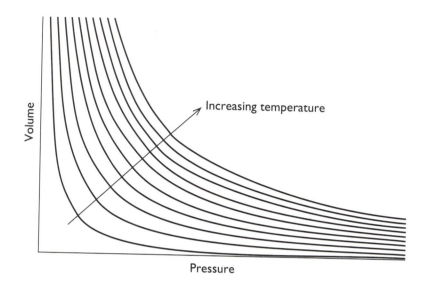

we see that the increase in the pressure of a gas as it is confined to smaller and smaller volumes should follow a curve like that shown in Fig. 1.6. Such a curve is called an **isotherm** because it shows the variation of a property (the pressure) at constant temperature. ('*Iso*' is Greek for 'equal' and '*therm*' is derived from the Greek for 'heat'.) Different temperatures lead to different curves, but each isotherm has the form $p \propto 1/V$.

Example Using Boyle's law

A sample of helium gas occupies 1.50 L when it is subjected to a pressure of 98.0 kPa; what pressure does it exert when it is confined to 1.00 L at the same temperature?

Answer We should anticipate that the pressure of the gas will be greater when it is confined to a smaller volume because the number of collisions the atoms make with the walls will be greater. Specifically, because pressure is inversely proportional to volume, it follows that for two volumes V_1 and V_2,

$$\frac{p_2}{p_1} = \frac{V_1}{V_2}$$

and therefore that

$$p_2 = p_1 \times \frac{V_1}{V_2} = 98.0 \, \text{kPa} \times \frac{1.50 \, \text{L}}{1.00 \, \text{L}} = 147 \, \text{kPa}$$

Exercise E1.6 To what volume must the same sample of gas be confined to achieve a pressure of 33.2 kPa?

[4.43 L]

The best way of testing Boyle's law is to measure the volume of a sample of gas held at constant temperature (by immersion in a constant-temperature bath) as the pressure acting on it is changed, and then to plot $1/V$ against p. Boyle's law would be confirmed if the graph is a straight line. In practice, the graph is not perfectly straight (Fig. 1.7), but it becomes progressively straighter as the pressure is reduced (or samples containing smaller amounts of substance are used). When there is so little substance present that the pressure of the gas is very low over the entire range of volume studied, then a very good straight line is obtained. We can expect that if the amount of substance was so slight that the pressure was close to zero over the entire volume range, then we would get a perfectly straight line, and Boyle's law would be obeyed exactly. This is summarized by saying that Boyle's law is a **limiting law**: it is not obeyed exactly by any actual gas, but describes a gas in the limit of zero pressure.

A hypothetical gas that obeys Boyle's law at *any* pressure is called a **perfect gas**. From what has just been said, an actual gas, which is termed a **real gas**, behaves more and more like a perfect gas as its pressure is reduced, and behaves exactly like a perfect gas when the pressure has been reduced to zero. In practice, normal atmospheric pressure at sea level ($p \approx 100$ kPa) is already low enough for most real gases to behave almost perfectly, and we shall always assume in this text that the gases we encounter behave like a perfect gas (and therefore that Boyle's law can be used to describe the relation between their volume and pressure when the temperature is constant). The reason why a real gas behaves differently from a perfect gas can be traced to the attractions and repulsions that exist between actual molecules and which are absent in a perfect gas; the origin of these interactions is described in Chapter 10.

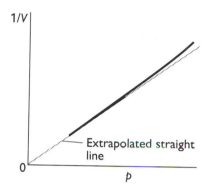

Fig. 1.7 A good test of Boyle's law is to plot $1/V$ against p (at constant temperature), when a straight line should be obtained. This diagram shows that the observed volumes (the heavy lines) approach a straight line as the pressure is reduced. A perfect gas would follow the straight line at all pressures; real gases obey Boyle's law in the limit of low pressures.

Charles's law: the response to temperature

The French scientist Jacques Charles, a keen balloonist, had an interest in the response of gases to the conditions. In particular, he established that *the volume of a gas is proportional to the temperature*.†

Charles's law: $V \propto T$ (at constant n, p) (6)

Therefore, doubling the temperature (on the Kelvin scale, such as from 300 K to 600 K, corresponding to an increase from 27 °C to 327 °C) will double the volume.

---

† Charles's results preceded a more accurate series of measurements made by Gay-Lussac, and Charles's law is often referred to as Gay-Lussac's law.

Example Using Charles's law

A sample of air occupies 150 mL at 20 °C. What volume will it occupy when the sample is immersed in a sterilizing bath at 100 °C, supposing that it is free to expand against a constant atmospheric pressure?

Answer We anticipate that, because the temperature is raised, the gas will occupy a greater volume at 100 °C than at 20 °C. Specifically, because the volume is proportional to the temperature (on the Kelvin scale), we can write

$$\frac{V_2}{V_1} = \frac{T_2}{T_1}$$

Then, with $T_1 = 293$ K (corresponding to 20 °C) and $T_2 = 373$ K (corresponding to 100 °C), we find that

$$V_2 = V_1 \times \frac{T_2}{T_1} = 150 \text{ mL} \times \frac{373 \text{ K}}{293 \text{ K}} = 191 \text{ mL}$$

Exercise E1.7 To what temperature must a sample of air of volume 125 mL at 30 °C be cooled to reduce its volume to 100 mL at the same pressure?

[−31 °C]

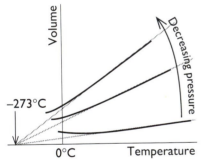

Fig. 1.8 This diagram illustrates the content and implications of Charles's law, which asserts that the volume occupied by a gas (at constant pressure) is proportional to the temperature. The dark lines are schematic versions of the experimental results for a series of temperatures. When plotted against Celsius temperatures (as here), all gases give straight lines that extrapolate to $V = 0$ at −273 °C. This extrapolation suggests that −273 °C is the lowest attainable temperature.

Charles's experiments were the first indication that there exists an **absolute zero** of temperature, a temperature below which it is impossible to cool an object, and which on the Kelvin scale occurs at $T = 0$. Thus, if the volume of a sample of gas is plotted against the *Celsius* temperature (Fig. 1.8), we find that the graphs are all approximately straight lines, and that samples at different pressures or containing different amounts of substance give straight lines of different slopes. However, all the straight lines, regardless of the identity or amount of gas, extrapolate to zero volume at −273 °C. At temperatures below −273 °C the extrapolations suggest that the volume of the sample should be negative, which is absurd. This observation suggests that −273 °C *is the lowest possible temperature for any substance.* It is therefore natural to set $T = 0$ at this temperature, for in that way a more fundamental temperature scale is obtained than one that is based on the properties of a particular substance (such as water, and the use of its freezing and boiling points to define the Celsius scale).

Avogadro's principle: the variation with amount of substance

The Italian chemist Amedeo Avogadro (whose constant we have already encountered) proposed that, *at a given temperature and pressure, equal volumes of gas contain the same number of molecules.*

That is, if a sample of air at 1 bar occupies 1.00 L at 300 K, and a sample of pure carbon dioxide occupies the same volume under the same conditions, then we can infer that both samples contain the same number of molecules. Moreover, doubling the number of molecules results in a doubling of the volume of the sample, as long as its pressure and temperature remain constant. We can therefore write

Avogadro's principle: $V \propto n$ (at constant p, T)   (7)

Avogadro's suggestion is a principle rather than a law, because it is not a direct summary of experience (it is based on the model of a gas as a collection of molecules).

The **molar volume** V_m of any substance is the volume it occupies per mole of molecules present in the sample:

$$V_m = \frac{V}{n} \qquad (8)$$

Avogadro's principle implies that the molar volume of a gas should be the same for all gases at the same temperature and pressure. That this conclusion is almost true under normal conditions (normal atmospheric pressure of about 1 bar and room temperature, close to 25 °C) can be confirmed by measuring out 1 mol of molecules of various gases and determining the volume they occupy. Some experimental data are given in Table 1.1: they show that the molar volumes of several gases are nearly the same under normal conditions, and close to 25 L mol^{-1}.

Table 1.1 The molar volumes of gases at 25 °C and 1 bar

Gas	V_m/L mol^{-1}
Perfect gas	24.79
Ammonia	24.8
Argon	24.4
Carbon dioxide	24.6
Nitrogen	24.8
Oxygen	24.8
Hydrogen	24.8
Helium	24.8

The perfect gas law

At this stage we have identified three properties of a gas

Boyle's law: $p \propto \dfrac{1}{V}$ or $pV = $ constant at constant n, T

Charles's law: $V \propto T$ at constant n, p

Avogadro's principle: $V \propto n$ at constant p, T

All three relations can be combined into a single expression:

$$pV \propto \text{constant} \times nT$$

We can verify that this expression combines all the three written above:

If n and T are constant, $pV = $ constant, which is Boyle's law.

If n and p are constant, constant $\times V = $ constant $\times T$, so $V \propto T$, which is Charles's law.

If p and T are constant, constant $\times V = $ constant $\times n$, so $V \propto n$, which is Avogadro's principle.

The combined expression is called the **perfect gas equation of state**; the constant is called the **gas constant** and is denoted by the letter R:

Perfect gas equation of state: $pV = nRT$ (9)

This expression is one of the most important equations in physical chemistry, and we shall use it many times in the following pages. The equation is an exact description of a perfect gas (under all conditions), is approximately valid for gases under normal conditions, and becomes progressively more accurate as the pressure of the gas decreases. It is an equation of state in the sense defined previously: it establishes a relation between the four variables that define the state of a gas. Thus, if we know the amount of substance in the sample, the volume it occupies, and the temperature, then the pressure of the gas must be

$$p = \frac{nRT}{V}$$

The value of the constant R may be obtained once and for all by measuring the pressure of a given amount of gas in a container of known volume and temperature, and substituting the values in

$$R = \frac{pV}{nT}$$

In practice, the values of R should be plotted against the pressure of the sample, and the graph extrapolated to $p = 0$ to ensure that the real gas is behaving perfectly. The result is that

$$R = 8.3145 \, \text{kPa L K}^{-1} \, \text{mol}^{-1}$$

Values of R in other units of pressure and volume are given in Table 1.2.

Example Expressing R in alternative units
Express R in the units joules per kelvin per mole ($\text{J K}^{-1} \, \text{mol}^{-1}$).
Answer We need to note (from *Further information 1*) that

$$1 \, \text{Pa} = 1 \, \text{N m}^{-2}$$
$$1 \, \text{J} = 1 \, \text{N m}$$
$$1 \, \text{L} = 1 \, \text{dm}^3 = 10^{-3} \, \text{m}^3$$

then it follows that

$$R = 8.3145 \times 10^3 \, \text{Pa L K}^{-1} \, \text{mol}^{-1}$$
$$= 8.3145 \times 10^3 \times 10^{-3} \, (\text{N m}^{-2}) \times \text{m}^3 \, \text{K}^{-1} \, \text{mol}^{-1}$$
$$= 8.3145 \, \text{N m K}^{-1} \, \text{mol}^{-1} = 8.3145 \, \text{J K}^{-1} \, \text{mol}^{-1}$$

Exercise E1.8 Express R in litre-atmospheres per kelvin per mole ($\text{L atm K}^{-1} \, \text{mol}^{-1}$).
$$[8.2058 \times 10^{-2} \, \text{L atm K}^{-1} \, \text{mol}^{-1}]$$

The perfect gas equation of state is very useful for calculating the pressure of a gas under stated conditions and the final state of a gas that has undergone more than one change (for example, a gas that has been heated and compressed). The following examples illustrate the procedures.

Example Using the perfect gas equation of state 1
What pressure is exerted by 1.25 g of nitrogen gas in a flask of volume 250 mL at 20 °C?

Answer The perfect gas equation of state can be arranged into a form that gives the pressure in terms of the other variables:

$$p = \frac{nRT}{V}$$

To use this formula, we need to know the amount of N_2 molecules in the sample, but that can easily be obtained from the molar mass of N_2, which is 28.02 g mol^{-1}:

$$n = \frac{\text{Mass}}{\text{Molar mass}} = \frac{1.25 \text{ g N}_2}{28.02 \text{ g mol}^{-1}} = 4.46 \times 10^{-2} \text{ mol N}_2$$

Then, substitution of the data gives

$$p = \frac{(4.46 \times 10^{-2} \text{ mol}) \times (8.3145 \text{ kPa L K}^{-1} \text{mol}^{-1}) \times (293 \text{ K})}{250 \times 10^{-3} \text{ L}}$$

$$= 435 \text{ kPa}$$

Exercise E1.9 Calculate the pressure exerted by 1.22 g of carbon dioxide confined to a 500 mL flask at 37 °C.

[143 kPa]

Example Using the perfect gas equation of state 2
What is the pressure of a sample of gas that has been compressed from 500 mL to 200 mL and heated from 25 °C to 250 °C given that its initial pressure is 57 kPa?

Answer The perfect gas law states that the pressure of the gas is inversely proportional to the volume and directly proportional to the temperature, so we can write

$$p_2 = \frac{T_2}{T_1} \times \frac{V_1}{V_2} \times p_1$$

Substitution of the data (after conversion of the temperatures to the Kelvin scale) then gives

$$p_2 = \frac{523 \text{ K}}{298 \text{ K}} \times \frac{500 \text{ mL}}{200 \text{ mL}} \times 57 \text{ kPa} = 250 \text{ kPa}$$

Table 1.2 The gas constant, R, in various units

$R =$ 8.314 51 J K^{-1} mol^{-1}

8.205 78 $\times$ 10^{-2} L atm K^{-1} mol^{-1}

62.364 L Torr K^{-1} mol^{-1}

1.987 22 cal K^{-1} mol^{-1}

Exercise E1.10 What is the final volume of a sample of gas that has been heated from 25 °C to 1000 °C and its pressure increased from 10.0 kPa to 150.0 kPa, given that its initial volume was 15 mL?

[4.3 mL]

The perfect gas law can also be used to calculate the molar volume V_m of a perfect gas at any temperature and pressure:

$$V_m = \frac{V}{n} = \frac{RT}{p}$$

Chemists have found it convenient to report much of their data at a particular set of standard conditions. By **standard ambient temperature and pressure** (SATP) they mean a temperature of 25 °C (more precisely, 298.15 K) and a pressure of 100 kPa (that is, 1 bar).† The **conventional temperature** of 25 °C is often denoted $\mathcal{T}$, so $\mathcal{T} = 298.15$ K exactly. The **standard pressure** is normally denoted $p^{\ominus}$, with $p^{\ominus} = 1$ bar exactly. At SATP, the molar volume of a perfect gas is 24.79 L mol^{-1}, as can be verified by substituting the values of the temperature and pressure into eqn (9).

1.3 Mixtures of gases: partial pressures

Chemists are often concerned with mixtures of gases, such as when they are considering the properties of the atmosphere (in meteorology), the composition of exhaled air (in medicine), or the mixtures of hydrogen and nitrogen used in the industrial synthesis of ammonia. The kind of question we need to be able to answer is: what is the contribution that each component of a gaseous mixture makes to the total pressure?

In the early nineteenth century, John Dalton carried out a series of experiments that led him to formulate the following law:

Dalton's law: the pressure exerted by a mixture of perfect gases is the sum of the pressures exerted by the individual gases occupying the same volume alone.

The contribution that a gas makes to the total pressure is the **partial**

---

† An earlier set of standard conditions, which is still encountered, is 'standard temperature and pressure' (STP), namely 0 °C and 1 atm.

pressure of that gas, and for a gas A is denoted p_A. It follows that:

1. The total pressure of a mixture of gases is the sum of the partial pressures of all the gases present.

2. The partial pressure of each gas is equal to the pressure it would exert if it occupied the container alone.

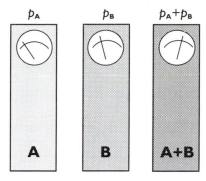

Fig. 1.9 The partial pressure p_A of a perfect gas A is the pressure that it would exert if it occupied a container alone; similarly, the partial pressure p_B of a perfect gas B is the pressure that it would exert if it occupied the same container alone. The total pressure p, when both gases simultaneously occupy the container, is the sum of their partial pressures.

Dalton's law is strictly valid only for mixtures of perfect gases (or for real gases at such low pressures that they are behaving perfectly). Under most of the conditions we shall encounter, the gases we shall deal with can be considered to be almost perfect.

As an example, suppose that we were interested in the composition of inhaled and exhaled air, and we knew that a certain mass of carbon dioxide exerts a pressure of 5 kPa when present alone in a container, and that a certain mass of oxygen exerts 20 kPa when present alone in the same container at the same temperature. Then, when both gases are present in the container, the partial pressure of the carbon dioxide in the mixture is 5 kPa, that of the oxygen is 20 kPa, and the total pressure of the mixture is 25 kPa (Fig. 1.9).

The partial pressure of each component in a mixture can be calculated from the perfect gas equation of state once we know the amount of molecules of each gas. For example, suppose the mixture consisted of an amount n_A of A molecules and n_B of B molecules, then the partial pressures of A and B are given by

$$p_A = \frac{n_A RT}{V} \qquad p_B = \frac{n_B RT}{V} \tag{10}$$

According to Dalton's law, the total pressure of the mixture is the sum of these two partial pressures:

$$p = p_A + p_B$$

A similar calculation can be carried out for mixtures of three or more gases, with each partial pressure being given by a term like those in eqn (10).

The easiest way of setting up partial-pressure calculations is to introduce the concept of mole fraction (we shall find this concept useful in several other contexts too):

The **mole fraction**, x_A, of a species A is the amount of A expressed as a fraction of the total amount of molecules present in the sample.

That is, in a sample that consists of an amount n_A of a species A, an amount n_B of a species B, and so on, the mole fraction of A present in the mixture is

$$x_A = \frac{n_A}{n_A + n_B + \cdots} \qquad (11)$$

with similar expressions (with n_B, etc. in the numerator) for the mole fractions of B, C, etc. For a **binary mixture** (one that consists of two species)

$$x_A = \frac{n_A}{n_A + n_B}, \qquad x_B = \frac{n_B}{n_A + n_B}, \qquad \text{and} \quad x_A + x_B = 1 \qquad (12)$$

When only A is present, $x_A = 1$ and $x_B = 0$; when only B is present, $x_A = 0$ and $x_B = 1$, and when both are present in the same amounts, $x_A = \frac{1}{2}$ and $x_B = \frac{1}{2}$.

> **Exercise E1.11** Calculate the mole fractions of N_2, O_2, and Ar in dry air at sea level, given that 100.0 g of air consists of 75.5 g of N_2, 23.2 g of O_2, and 1.3 g of Ar.
> [*Hint:* begin by converting each mass to an amount in moles.]
> [0.780, 0.210, 0.009]

The advantage of introducing the mole fractions of the species present in a mixture is that each partial pressure is related to the total pressure by

$$p_A = x_A \times p \qquad (13)$$

(and likewise for any other gases present). Therefore, if we know the mole fraction of the species and the total pressure of the mixture, we can immediately write down its partial pressure.

JUSTIFICATION

The total pressure p exerted by a sample that consists of an amount n of molecules (of any kind, including a mixture of molecules) is related to the volume and temperature by the perfect gas equation of state, $pV = nRT$. It follows that

$$\frac{RT}{V} = \frac{p}{n}$$

Substitution of this expression into eqn (10) results

in

$$p_A = n_A \times \frac{p}{n}$$

However, n_A/n is the mole fraction of A in the mixture (because n is the total amount of molecules present, $n = n_A + n_B + \cdots$). Therefore,

$$p_A = x_A \times p$$

as written above.

Example Calculating partial pressures
Many environmental and biophysical arguments depend on a knowledge of the composition of the atmosphere, and in particular on the partial pressures of its principal components. Given that the composition of a 100.0 g sample of dry air at sea level is 75.5 g N_2, 23.2 g O_2, and 1.3 g Ar, what is the partial pressure of each component at 100 kPa total pressure?
Answer We need to know the mole fractions of the components; these were calculated in Exercise 1.11 as

$$x(N_2) = 0.780, \qquad x(O_2) = 0.210, \qquad x(Ar) = 0.009$$

It then follows from eqn (13) that

$$p(N_2) = x(N_2) \times p = 0.780 \times 100 \text{ kPa} = 78.0 \text{ kPa}$$

$$p(O_2) = x(O_2) \times p = 0.210 \times 100 \text{ kPa} = 21.0 \text{ kPa}$$

$$p(Ar) = x(Ar) \times p = 0.009 \times 100 \text{ kPa} = 0.9 \text{ kPa}$$

The partial pressure of oxygen in air plays an important role in the aeration of water, which enables aquatic life to thrive, and in the absorption of oxygen by blood in our lungs (see Section 4.3).

Exercise E1.12 Calculate the partial pressures of a sample of gas consisting of 2.50 g O_2 and 6.43 g CO_2 with a total pressure of 88 kPa.
[31 kPa, 57 kPa]

The kinetic theory of gases

We have remarked that a gas may be pictured as a collection of particles in continuous, chaotic motion (Fig. 1.10). Now we shall develop this model of the gaseous state of matter to see how it accounts for the perfect gas laws. One of the most important functions of physical chemistry is to convert qualitative notions into quantitative statements that can be tested experimentally by making measurements and comparing the results with predictions. Indeed, an important component of science as a whole is its technique of proposing a qualitative model and then expressing that model mathematically. The kinetic theory of gases is an excellent example of this procedure: the

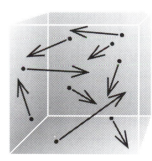

Fig. 1.10 The model used for discussing the molecular basis of the physical properties of a perfect gas. The molecules move chaotically with a range of speeds and directions, both of which change when they collide with the walls or with other molecules.

model is very simple, and the quantitative prediction (the perfect gas law) is experimentally verifiable.

The **kinetic theory of gases** is based on three assumptions:

1. A gas consists of molecules of mass m and diameter d in ceaseless random motion.
2. The size of the molecules is negligible in the sense that their diameters are much smaller than the average distance travelled between collisions.
3. The molecules do not interact, except during collisions.

The assumption that the molecules do not interact implies that there is no potential energy of interaction between them (that is, their energy is independent of their separation). Because the total energy is the sum of the kinetic energy (the energy arising from motion) and the potential energy (the energy arising from position), and the latter is zero, the energy of the sample is the sum of the kinetic energies of all the molecules present in it, and the faster they travel the greater the energy of the gas. (The various contributions to the energy of a system are reviewed in *Further information 2* on p. 24.)

1.4 The pressure of a gas

The kinetic theory accounts for the steady pressure exerted by a gas in terms of the collisions the molecules make with the walls of the container. These collisions are so numerous that the walls experience a virtually constant force, and hence a steady pressure (Fig. 1.11).

The calculation of the pressure exerted by the gas need not be done in detail (but see *Further information 3* on p. 25). Briefly, to

Fig. 1.11 The wall of a vessel containing a gas experiences collisions from the gas molecules. The impacts have different strengths (depending on the velocities of the molecules that strike the surface). This diagram illustrates the strengths of the impulses that a wall experiences in a very brief period (the entire interval shown could be as brief as 10^{-15} s, or even less, depending on the pressure and temperature of the gas). The impulses vary so rapidly that the wall experiences an almost perfectly steady force, which is the mean of the individual impacts.

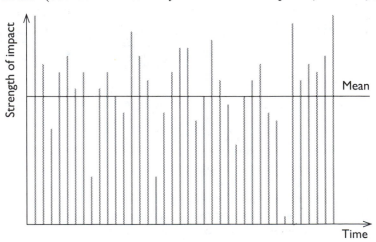

calculate the pressure of the gas, we need to calculate the force exerted by the molecules per unit area of wall as they collide with it. That force is calculated using Newton's second law of motion, which states that force is equal to the rate at which the linear momentum (the product of mass and velocity) of a particle changes: if a heavy particle collides with a wall and undergoes a reversal of direction, then we know that a strong force has acted. So the calculation boils down to finding the average change in linear momentum which occurs when molecules collide with the wall and change their direction of travel. The final outcome of the calculation (see *Further information 3*) is

$$p = \frac{nMc^2}{3V} \tag{14}$$

where M is the molar mass and c is the **root-mean-square speed** (r.m.s. speed) of the molecules. The latter is the square root of the average of the square of the speeds of all the molecules in the sample; for a sample consisting of N molecules with speeds $s_1, s_2, \ldots, s_N$,

$$c = \sqrt{\frac{s_1^2 + s_2^2 + \cdots + s_N^2}{N}} \tag{15}$$

The root-mean-square speed is quite close in value to the mean speed of the molecules (the mean speed is smaller by a factor of 0.92).

Exercise E1.13 Cars pass a point travelling at 45.00 (5), 47.00 (7), 50.00 (9), 53.00 (4), 57.00 (1) km h^{-1}, where the number of cars is given in parentheses. Calculate (a) the r.m.s. speed and (b) the mean speed of the cars.

[(a) 49.06 km h^{-1}, (b) 48.96 km h^{-1}]

We can see at once that the kinetic theory gives an expression that already resembles the perfect gas equation of state, for eqn (14) can be rearranged to

$$pV = \tfrac{1}{3}nMc^2 \tag{16}$$

which is very similar in form to $pV = nRT$. If we suppose that the r.m.s. speed of the molecules depends only on the temperature, then at constant temperature and for a fixed amount of molecules, the kinetic theory leads to

$$pV = \text{constant}$$

which is Boyle's law. This conclusion is a major success of the kinetic model, for from the model we have successfully derived a result which is valid experimentally.

FURTHER INFORMATION 2: Energy and force

Matter can store energy in two ways, as kinetic energy and as potential energy.

Kinetic energy is the energy that a body (a block of matter, an atom, or an electron) possesses by virtue of its motion. The formula for calculating the kinetic energy of a body of mass m that is travelling at a speed v is

$$\text{kinetic energy} = \tfrac{1}{2}mv^2$$

This expression shows that a body may have a high kinetic energy if it is heavy (m large) and is travelling rapidly (v large). A stationary body ($v = 0$) has zero kinetic energy, whatever its mass. The energy of a sample of perfect gas is entirely due to the kinetic energy of its molecules: they travel more rapidly (on average) at high temperatures than at low, and so raising the temperature of a gas increases the kinetic energy of its molecules.

Potential energy is the energy that a body has by virtue of its position. A body on the surface of the Earth has a potential energy on account of the gravitational force it experiences: if the body is raised, then its potential energy is increased. There is no general formula for calculating the potential energy of a body because there are several kinds of force. For a body of mass m at a height h above the surface of the Earth, the **gravitational potential energy** is

$$\text{potential energy} = mgh$$

where g is the acceleration of free fall ($g = 9.81 \text{ m s}^{-2}$). A heavy object at a certain height has a greater potential energy than a light object at the same height. One very important contribution to the potential energy is encountered when a charged particle is brought up to another charge. In this case the potential energy is inversely proportional to the distance between the charges (see *Further Information 5* on p. 206):

$$\text{potential energy} \propto \frac{1}{r}$$

This **Coulomb potential energy** decreases with distance, and two infinitely widely separated charged particles have zero potential energy of interaction. The Coulomb potential plays a central role in the structures of atoms, molecules, and solids, as we see in Chapters 8–11.

The **total energy** of a body is the sum of its kinetic and potential energies. It is a central feature of physics that *the total energy of a body that is free from external influences is constant*. Thus, a stationary ball at a height h above the surface of the Earth has a potential energy of magnitude mgh; if it is released and begins to fall to the ground, it loses potential energy (as it loses height), but gains the same amount of kinetic energy (and therefore accelerates). Just before it hits the surface, it has lost all its potential energy, and all its energy is then kinetic.

The state of motion of a body is changed by a **force**. According to Newton's second law of motion, a force changes the state of motion of a body such that the **acceleration** of the body (its rate of change of velocity) is proportional to the strength of the force:

$$\text{force} = \text{mass} \times \text{acceleration}$$

Thus, to accelerate a heavy particle by a given amount requires a stronger force than to accelerate a light particle by the same amount. A force can be used to change the kinetic energy of a body by accelerating the body to a higher speed. It may also be used to change the potential energy of a body by moving it to another position (for example, by raising it near the surface of the Earth).

The units of energy and force are explained in *Further information 1* on p. 4.

FURTHER INFORMATION 3:
The kinetic theory of gases

Consider the system in Fig. 1. When a particle of mass m collides with the wall on the right its component of linear momentum (the product of its mass and its velocity) parallel to the x axis changes from mv_x (when it is travelling to the right) to $-mv_x$ (when it is travelling to the left). Its momentum therefore changes by $2mv_x$ on each collision. The number of collisions in an interval Δt is equal to the number of particles able to reach the wall in that interval. Because a particle with speed v_x can travel a distance $v_x \Delta t$ in an interval Δt, all the particles within a distance $v_x \Delta t$ of the wall will strike it if they are travelling towards it. Therefore, if the wall has area A, then all the particles in a volume $Av_x \Delta t$ will reach the wall (if they are travelling towards it). If the number density, the number of particles per unit volume, is N, the number in the volume $Av_x \Delta t$ is $NAv_x \Delta t$.

On average, half the particles are moving to the right, and half are moving to the left. Therefore, the average number of collisions with the wall during the interval Δt is $\frac{1}{2}NAv_x \Delta t$. The total momentum change in that interval is the product of this number and the change $2mv_x$ that an individual molecule experiences:

$$\text{momentum change} = \tfrac{1}{2}NAv_x \Delta t \times 2mv_x$$
$$= mNAv_x^2 \Delta t$$

The rate of change of momentum is this change of momentum divided by the interval Δt during which it occurs:

$$\text{rate of change of momentum} = mNAv_x^2$$

The rate of change of momentum is equal to the force (by Newton's second law of motion), and so the force exerted by the gas on the wall is also $mNAv_x^2$. It follows that the pressure, the force divided by the area A on which the force acts, is

$$\text{pressure} = mNv_x^2$$

The detected pressure, p, is the average (denoted $\langle \cdots \rangle$) of the quantity just calculated:

$$p = mN\langle v_x^2 \rangle$$

The r.m.s. speed, c, of the particles is

$$c = \langle v^2 \rangle^{1/2} = (\langle v_x^2 \rangle + \langle v_y^2 \rangle + \langle v_z^2 \rangle)^{1/2}$$

However, because the particles are moving randomly, the average of v_x^2 is the same as the average of the analogous quantities in the y and z directions. Because $\langle v_x^2 \rangle$, $\langle v_y^2 \rangle$, and $\langle v_z^2 \rangle$ are all equal,

$$c = (3\langle v_x^2 \rangle)^{1/2}, \quad \text{implying that } \langle v_x^2 \rangle = \tfrac{1}{3}c^2$$

Therefore,

$$p = \tfrac{1}{3}Nmc^2$$

The number density of molecules is the product of the amount (the number of moles, n) and Avogadro's constant N_A divided by the volume, V, that the molecules occupy, so the last equation becomes

$$pV = \tfrac{1}{3}nN_A mc^2 = \tfrac{1}{3}nMc^2$$

where $M = m \times N_A$ is the molar mass of the molecules. This expression is the equation used in the text.

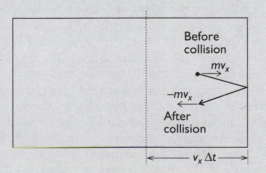

Fig. 1

1.5 The speeds of gas molecules

The ability of the kinetic theory to account for Boyle's law suggests that it is a valid model of perfect gas behaviour. That being so, we can take another major step. We shall suppose that the expression for pV derived from kinetic theory (eqn (16)) is indeed the equation of state of a perfect gas, which means that we can equate the expression on the right of $pV = \frac{1}{3}nMc^2$ to nRT, and hence obtain a formula for calculating the r.m.s. speed of the gas molecules at any temperature:

$$c = \sqrt{\frac{3RT}{M}} \qquad (17)$$

The r.m.s. speed of O_2 molecules (of molar mass 32 g mol^{-1}) at 25 °C works out as 482 m s^{-1}, and that of N_2 molecules is 515 m s^{-1}. Both these values are not far off the speed of sound in air (346 m s^{-1} at 25 °C): that is reasonable, because sound is a wave of pressure variation that is transmitted by the movement of molecules, and so the speed of propagation of a wave should be approximately the same as the average speed at which molecules can adjust their locations.

The important conclusion that we can now draw is that *the r.m.s. speed of molecules in a gas is proportional to the square root of the temperature.* Hence, doubling the temperature (on the Kelvin scale) increases the r.m.s. speed of the molecules by a factor of $\sqrt{2} = 1.4$. As another illustration, cooling a sample of air from 25 °C (298 K) to 0 °C (273 K) reduces the r.m.s. speed of the molecules by a factor of

$$\sqrt{\frac{273}{298}} = 0.957$$

So, on a cold day, the molecules of the air are moving on average at about 4 per cent more slowly than on a warm day.

The Maxwell distribution of speeds

So far, we have dealt only with the *mean* speeds of molecules. Not all molecules, however, travel at the same speed: some are travelling more slowly than the average (until they collide, and get accelerated to a high speed, like the effect of the impact of a bat on a ball), and others may, briefly, be travelling at much higher speeds than the average. There is continual interchange of speeds among molecules as they undergo collisions, and at one moment a molecule may be travelling much faster than the average, but at the next moment it might be brought to an almost complete standstill when it collides with another. A molecule collides about once every nanosecond (1 ns = 10^{-9} s) or so in a gas under normal conditions.

The fraction of molecules that has a particular speed is called the **distribution of molecular speeds**. Thus, the distribution might tell us that at 20 °C a fraction 1.9×10^{-2} O_2 molecules (19 in 1000) have a speed in the range between 300 and 310 m s^{-1}, that 2.1×10^{-2} molecules (21 in 1000) have a speed in the range 400 to 410 m s^{-1}, and so on. The precise form of the distribution was worked out by James Clerk Maxwell towards the end of the nineteenth century, and is known as the **Maxwell distribution of speeds**. According to Maxwell, the fraction f of molecules that has a speed in a narrow range s to $s + \Delta s$ is

$$f = 4\pi \left(\frac{M}{2\pi RT}\right)^{3/2} s^2 e^{-Ms^2/2RT} \Delta s \tag{18}$$

This is the formula used to calculate the figures quoted above. Although eqn (18) looks quite complicated, its features can be picked out quite readily. First, the fraction in the range Δs increases in proportion to the width of the range: if at a given speed we increase the range (but still ensure that it is narrow), then the fraction in that range increases. Second, the fact that the expression includes a decaying exponential (e^{-x}, with x proportional to s^2) implies that the fraction of molecules with very high speeds will be very small. Third, the factor s^2 that multiplies the exponential goes to zero as s goes to zero, so the fraction with very low speeds will also be very small.

The graph of the Maxwell distribution is very important, and is shown in Fig. 1.12. As we should expect from the form of eqn (18), we see that only a very few molecules in the sample have speeds much smaller than the average speed and, similarly, only very few molecules have speeds much greater than the average. However, the fraction with very high speeds increases sharply as the temperature is raised, for the tail of the distribution then reaches up to higher speeds. This feature plays an important role in the rates of chemical reactions, for (as we shall see in Section 7.5), the rate of a reaction in the gas phase depends on the energy with which two molecules crash together, which in turn depends on their speeds.

Another feature of the Maxwell distribution is shown in Fig. 1.13, where it is drawn for molecules with different molar masses. As can be

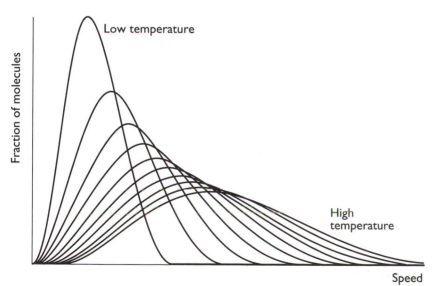

Fig. 1.12 The Maxwell distribution of speeds and its variation with the temperature. Note the broadening of the distribution and the shift of the r.m.s. speed to higher values as the temperature is increased.

Fig. 1.13 The Maxwell distribution of speeds also depends on the molar mass of the molecules: molecules of low molar mass have a broad spread of speeds, and a significant fraction may be found travelling much faster than the r.m.s. speed of the sample. The distribution is much narrower for heavy molecules, and most of them travel with speeds close to the r.m.s. value.

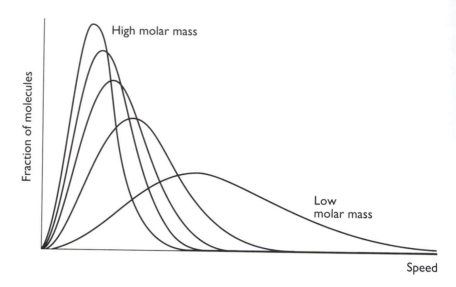

seen, not only do heavy molecules have lower average speeds at a given temperature, but they also have a significantly narrower spread of speeds: most molecules will be found with speeds close to the average. In contrast, light molecules (such as H_2) have high average speeds and a wide spread of speeds: many molecules will be found travelling either much more slowly or much more quickly than the average. This feature plays an important role in determining the composition of planetary atmospheres, because it means that a high fraction of light molecules travel at sufficiently high speeds for them to escape from the planet's gravitational attraction. The ability of light molecules to escape is one reason why hydrogen (molar mass $2.02 \, \text{g mol}^{-1}$) and helium ($4.00 \, \text{g mol}^{-1}$) are very rare in the Earth's atmosphere.

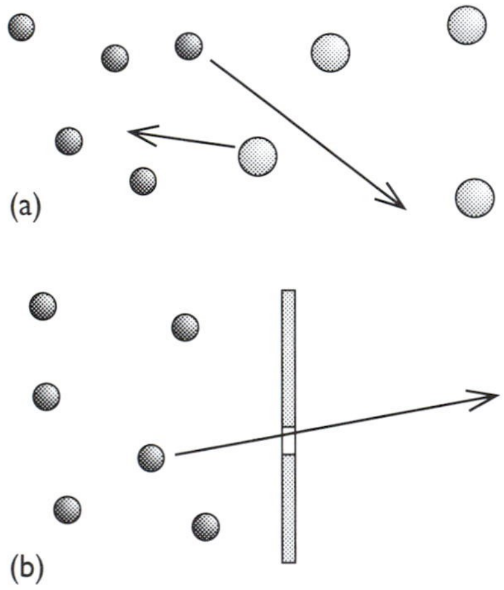

Diffusion and effusion

The process by which the molecules of different substances mingle with each other is called **diffusion**. The atoms of two solids diffuse into each other when the two solids are in contact, but the process is very slow. The diffusion of a solid through a liquid solvent is much faster, but mixing normally needs to be encouraged by stirring or shaking the solid in the liquid (the process is then no longer pure diffusion). Much faster than either is the diffusion of gases into each other. Gaseous diffusion accounts for the largely uniform composition of the atmosphere, for if a gas is produced by a localized source (such as carbon dioxide from the respiration of animals, oxygen from photosynthesis by green plants, and pollutants from cars and industrial sources), then the molecules of gas will diffuse from the source and in due course result in a uniform composition throughout the atmosphere. (In practice, the process of mixing is accelerated by winds.) The process of **effusion** is the escape of a gas through a small hole, as in a puncture in an inflated balloon or tyre (Fig. 1.14).

Fig. 1.14 (a) Diffusion is the spreading of the molecules of one substance into the region initially occupied by another species. Note that molecules of both substances move, and each substance diffuses into the other. (b) Effusion is the escape of molecules through a small hole in a confining wall.

The rates of diffusion and effusion increase with increasing temperature, for they both depend on the motion of molecules, and molecular speeds increase with temperature. The rates also decrease with increasing molar mass, for molecular speeds decrease with molar mass. The dependence on molar mass, however, is simple only in the case of effusion, for then only a single substance is in motion, not the several gases that may be involved in diffusion.

The experimental observations on the dependence of the rate of effusion of a gas on its molar mass are summarized by a law proposed by Thomas Graham in 1833:

Graham's law of effusion: at a given pressure and temperature, the rate of effusion of a gas is inversely proportional to the square root of its molar mass.

For example, the rates at which hydrogen $(2.02 \text{ g mol}^{-1})$ and carbon dioxide $(44.01 \text{ g mol}^{-1})$ effuse under the same conditions of pressure and temperature are in the ratio

$$\frac{\text{Rate of effusion of } H_2}{\text{Rate of effusion of } CO_2} = \sqrt{\frac{44.01 \text{ g mol}^{-1}}{2.02 \text{ g mol}^{-1}}} = 4.67$$

The difference in rates of effusion through a porous barrier is employed in the separation of uranium-235 from the more abundant uranium-238 in the processing of nuclear fuel. The process depends on the formation of uranium hexafluoride, a volatile solid. However, because the ratio of the molar masses of $^{238}UF_6$ and $^{235}UF_6$ is only 1.008, the ratio of the rates of effusion is only $\sqrt{1.008} = 1.004$; hence thousands of successive effusion stages are required to achieve a significant separation. The rates of effusion of gases have been used to determine the molar masses of gases and volatile liquids, for comparison of the rate of effusion of the gas with that of a gas of known molar mass can be interpreted using Graham's law in terms of the relative molar masses of the two gases. However, there are now much more precise procedures (mass spectrometry, for instance).

The form of Graham's law can be explained by referring to the expression for the r.m.s. speed of molecules in a gas, eqn (17), which is inversely proportional to the square root of the molar mass. Because the rate of effusion through a hole in a container is proportional to the rate at which molecules pass through the hole, it follows that the rate should be inversely proportional to $\sqrt{M}$, which is in accord with Graham's law.

1.6 Molecular collisions

The average distance that a molecule travels between collisions is called its **mean free path**, λ (Fig. 1.15). The mean free path in a liquid

Fig. 1.15 (a) In a gas, a molecule traces a chaotic path—a random walk—that changes direction on each collision; each step is over a different distance. (b) The *average* of the distances that any molecule travels between collisions is the mean free path.

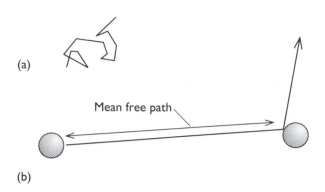

(a)

Mean free path

(b)

is smaller than the diameter of the molecules, for they encounter a neighbour after they move only a fraction of a diameter. However, in gases the mean free paths of molecules can be several hundred molecular diameters. The **collision frequency**, z, is the rate at which a single molecule collides with other molecules (the number of collisions per second). It follows that the inverse of the collision frequency, $1/z$, is the average time that a molecule spends in flight between two collisions: as we shall see, this average time is typically about 1 ns under normal conditions.

The collision frequency and the mean free path are related, because the r.m.s. speed c can be identified with the average length of the flight of a molecule between collisions (that is, with λ) divided by the average length of time needed for the flight (that is, $1/z$). Therefore, the mean free path and the collision frequency are related by

$$c = \lambda z \qquad (19)$$

It follows that if we can calculate either λ or z, then we can find the other from the eqn (17) for c.

To find expressions for λ and z we need a slightly more elaborate version of the kinetic theory of gases. The kinetic theory supposes that the molecules are effectively point-like; however, to obtain collisions, we need to assume that two 'points' score a hit whenever they come within a certain range, d, of each other (Fig. 1.16). When this feature

Fig. 1.16 To calculate features of the gas that are related to collisions (such as the mean free path and the collision frequency), a point is regarded as being surrounded by a sphere which, if entered by another point, counts as a collision. The radius of the sphere is d.

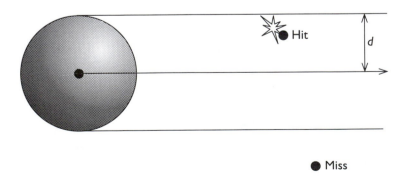

Hit

d

Miss

is added to the model, it turns out that the mean free path and the collision frequency depend on the quantity $\sigma = \pi d^2$, which is called the **collision cross-section** of the molecules:

$$\lambda = \frac{RT}{\sqrt{2}\, N_A \sigma p} \qquad z = \frac{\sqrt{2}\, N_A \sigma c p}{RT} \qquad (20)$$

Some values of the collision cross-sections of common species are given in Table 1.3. For example, the mean free path of O_2 molecules in a gas at 100 kPa and 25 °C (298 K) is

$$\lambda = \frac{(8.3145\ \mathrm{J\,K^{-1}\,mol^{-1}}) \times (298\ \mathrm{K})}{\sqrt{2} \times (6.022 \times 10^{23}\ \mathrm{mol^{-1}}) \times (0.40 \times 10^{-18}\ \mathrm{m^2}) \times (1.00 \times 10^5\ \mathrm{Pa})}$$

$$= 7.3 \times 10^{-8}\ \mathrm{m}$$

or 73 nm. (In the calculation, we used $1\,\mathrm{J} = 1\,\mathrm{Pa\,m^3}$, see *Further information 1*.) Under the same conditions, the collision frequency is $6.2 \times 10^9\ \mathrm{s^{-1}}$ (6.2 billion collisions per second).

The important features of eqn (20) are as follows. First, we see that the *mean free path decreases as the pressure increases* ($\lambda \propto 1/p$). This decrease is a result of the greater number of molecules present in a given volume as the pressure is increased, so each molecule will travel a shorter distance before it collides with a neighbour. For example, the mean free path of an O_2 molecule in a sample of oxygen decreases from 73 nm to 36 nm when the pressure is increased from 1.0 bar to 2.0 bar. Second, *the mean free path is shorter for molecules with large collision cross-sections* ($\lambda \propto 1/\sigma$). For instance, at the same pressure and temperature, the collision cross-section of a benzene molecule, 0.88 nm², is about four times greater than that of a helium atom (0.21 nm²), and its mean free path will be four times shorter. Thirdly, *the collision frequency increases with the pressure of the gas* ($z \propto p$), because (provided the temperature is the same), a molecule takes less time to travel to its neighbour. For example, although the collision frequency for an O_2 molecule in oxygen gas at 25 °C and 1.0 bar is $6.2 \times 10^9\ \mathrm{s^{-1}}$ (6.2 billion collisions per second, corresponding to an average time between collisions of 0.16 ns), at 2.0 bar the collision frequency is doubled, to $1.2 \times 10^{10}\ \mathrm{s^{-1}}$. Finally, *the collision frequency depends on the average speed of the molecules in the sample* ($z \propto c$), so—providing their collision cross-sections are almost the same— heavy molecules have lower collision frequencies than light molecules.

Table 1.3 Collision cross-sections of atoms and molecules

Species	$\sigma/\mathrm{nm^2}$
Ar	0.36
C_2H_4	0.64
C_6H_6	0.88
CH_4	0.46
Cl_2	0.93
CO_2	0.52
H_2	0.27
He	0.21
N_2	0.43
Ne	0.24
O_2	0.40
SO_2	0.58

$1\ \mathrm{nm^2} = 10^{-18}\ \mathrm{m^2}.$

Real gases

So far, everything we have said applies to perfect gases, ones in which the molecules are so far apart on average that they move completely

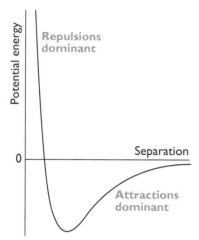

Fig. 1.17 The variation of the potential energy of two molecules on their separation. High positive potential energy (at very small separations) indicates that the interactions between them are strongly repulsive at these distances. At intermediate separations, where the potential energy is negative, the attractive interactions dominate. At large separations (on the right) the potential energy is zero and there is no interaction between the molecules.

independently of each other. In terms of the quantities introduced in the previous section, a perfect gas is a gas for which the mean free path, λ, of the molecules in the sample (the distance over which they travel freely between collisions) is much greater than d (the separation at which they come into contact): $\lambda \gg d$. We can conclude that a perfect gas is a gas in which the only contribution to the energy comes from the kinetic energy of the motion of the molecules and there is no contribution from the potential energy arising from the interaction of the molecules with one another.

1.7 Intermolecular forces

All real gases are imperfect, and when their molecules are within range of each other they interact with each other. There are two contributions to the interaction between molecules. At relatively long distances (a few molecular diameters), two molecules attract each other, and their potential energy decreases as they approach. The **intermolecular attraction** that results in this lowering of the potential energy is responsible for the cohesion of molecules into liquids when the temperature is reduced so much that the molecules have insufficient kinetic energy to escape from each other's attraction. The same intermolecular attraction has to be overcome when a liquid is vaporized: that is why heat must be supplied to a liquid in order to vaporize it. Second, molecules repel each other when they are in contact. This **intermolecular repulsion** is responsible for the fact that liquids and solids have a definite bulk and do not collapse to an infinitesimal point. The general form of the variation of the intermolecular interaction is illustrated in Fig. 1.17, which shows that initially (at large separations) the interactions are attractive (energy-lowering), but that at short distances the repulsions (the energy-increasing interactions) dominate.

The effect of the intermolecular interactions shows up in the bulk properties of a gas. For example, the isotherms of real gases have shapes that differ from those given by Boyle's law, particularly at high pressures. A set of experimental isotherms for carbon dioxide are shown in Fig. 1.18 and should be compared with the perfect-gas isotherms shown in Fig. 1.6. Although the experimental isotherms resemble the perfect-gas isotherms at high temperatures (and at low pressures, off the scale on the right of the graph), there are very striking differences between the two at temperatures below 50 °C and at pressures above about 1 bar.

1.8 The critical temperature

The sharp curves in the graph at 20 °C (and at other temperatures) show the formation of a liquid by the application of pressure. Thus, at point A the sample is a gas. As the volume of the sample is reduced to B by pressing in a piston, the pressure increases broadly in agreement

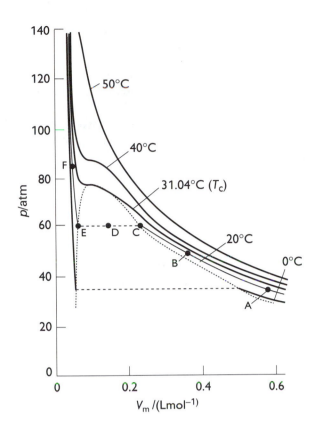

Fig. 1.18 The experimental isotherms of carbon dioxide at several temperatures. The critical isotherm is at 31.04 °C.

with Boyle's law, and the increase continues until point C is reached. Then the piston can be pushed in without any further increase in pressure, through D to E. After E has been reached, the reduction in volume results in a very steep increase in pressure. In fact, inspection of the sample shows that it has condensed to a liquid; therefore, the collapse in volume from C to E represents the conversion of the sample from a gas to a highly compact liquid. The subsequent reduction in volume, from E to F, corresponds to the very high pressure needed to compress a liquid into a smaller volume. In terms of intermolecular interactions, the step from C to E corresponds to the molecules being forced to be so close on average that they attract each other and cohere into a liquid. The step from E to F represents the effect of trying to force the molecules even closer together when they are in contact, and hence trying to overcome the strong repulsive interactions between them.

If we looked inside the container at point D, we would see a liquid separated from the remaining gas by a sharp surface (Fig. 1.19). At a higher temperature (30 °C, for instance), a liquid forms, but it needs a higher pressure. It might be difficult to make out the surface because the remaining gas is at such a high pressure that its density is similar to that of the liquid. At the special temperature of 304.19 K (31.04 °C) the density of the gas is the same as that of the liquid, and no surface is formed. At this temperature, which is called the **critical temperature**

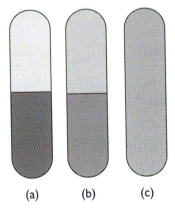

(a)　　　(b)　　　(c)

Fig. 1.19 When a liquid is heated in a sealed container, the density of the vapour phase increases and that of the liquid phase decreases. There comes a stage at which the two densities are equal and the interface between the two fluids disappears. This occurs at the critical temperature. The container needs to be strong: the critical temperature of water is at 373 °C and the vapour pressure is then 218 atm.

Table 1.4 The critical temperatures of gases

Gas	Critical temperature/°C
Noble gases	
He	−268 (5.2 K)
Ne	−229
Ar	−123
Kr	−64
Xe	17
Halogens	
Cl_2	144
Br_2	311
Small inorganic species	
H_2	−240
O_2	−118
H_2O	374
N_2	−147
NH_3	132
CO_2	31
Organic compounds	
CH_4	−83
CCl_4	283
C_6H_6	289

T_c, and at all higher temperatures, a single phase fills the container at all stages of the compression and there is no separation of a liquid from the gas. We have to conclude that *a gas cannot be condensed to a liquid by the application of pressure unless the temperature is lower than the critical temperature*. Some critical temperatures are listed in Table 1.4. We see, for example, that no degree of compression of nitrogen gas will result in the formation of a liquid (which we would detect by the appearance of a condensed phase separated from the gas by a well defined surface) unless the temperature is below 126 K (−147 °C).

1.9 Real equations of state

According to the perfect gas equation of state,

$$\frac{pV}{nRT} = \frac{pV_m}{RT} = 1 \tag{21}$$

where V_m is the molar volume of the gas. When the value of the **compression factor** $Z = pV_m/RT$ is measured for actual gases, however, it is found to differ from 1 and its value varies with pressure as shown in Fig. 1.20. The deviation of pV_m/RT from 1 arises from the intermolecular interactions in the gas. At low pressures, some gases (methane, CH_4; ethane, C_2H_6; and ammonia, NH_3 in the illustration) show a decrease below 1, which means that for a given temperature and pressure, the molar volume of the gas is less than that expected for a perfect gas. The reason for the reduction in molar volume can be traced to the attractive interactions between the molecules, which tend to draw the molecules together and hence reduce the space they occupy.

The experimental value of pV_m/RT increases above 1 at high pressures whatever the identity of the gas, and for some gases (hydrogen in the illustration) its value is greater than 1 at all pressures. When $pV_m/RT > 1$, the molar volume is greater than expected for a perfect gas of the same temperature and pressure. This behaviour can be traced to repulsive forces, which tend to drive the molecules apart when they are forced to be close together at high pressures (the case with methane, ethane, and ammonia) or that have such weak attractive interactions (as is the case with hydrogen) that the repulsive interactions dominate even at quite low pressures.

The virial equation of state

The departure of the value of pV_m/RT from 1 can be used to construct an *empirical* (observation based) equation of state by supposing that the perfect gas equation of state is only the first term of a lengthier expression, and writing

$$\frac{pV_m}{RT} = 1 + \frac{B}{V_m} + \frac{C}{V_m^2} + \cdots \tag{22}$$

This expression is called the **virial equation of state** (the word 'virial' comes from the Latin word for 'force', and it reflects the fact that intermolecular forces are now significant). The empirical coefficients B, C, etc., which vary from gas to gas, are called the **virial coefficients**. From the graphs in Fig. 1.20 it follows that, for the temperature to which the data apply, B must be positive for hydrogen (so that $pV_m/RT > 1$) but negative for methane, ethane, and ammonia (so that $pV_m/RT < 1$ until the term C/V_m^2 becomes large at high pressures—when V_m^2 is very small). The values of virial coefficients for many gases are known from measurements of pV_m/RT over a range of pressures, and eqn (22) is widely used to discuss the properties of real gases. We can see that at very low pressures, when the molar volume is very large, the terms B/V_m and C/V_m^2 are very small, and to a good approximation the gas is described by the perfect gas equation of state:

$$\frac{pV_m}{RT} \to 1 \quad \text{as} \quad V_m \to \infty \quad \text{when} \quad p \to 0 \qquad (23)$$

The van der Waals equation of state

Although it is the most reliable equation of state (because it is a fit of observed pressure and molar volumes), the virial equation does not give us much immediate insight into the behaviour of gases and their condensation to liquids. The **van der Waals equation**, which was proposed in 1873 by the Dutch physicist Johannes van der Waals, is only an approximate equation of state, but it has the advantage of showing how the intermolecular interactions contribute to the deviations of a gas from perfect gas behaviour. We can view the van der Waals equation as another example of taking a soundly based physical idea and expressing it mathematically. To do so, we recognize that a real gas differs from a perfect gas in that the molecules interact with each other. The interaction is attractive when the molecules are within a few molecular diameters of each other, but is strongly repulsive as soon as they touch.

The presence of the repulsive interaction implies that two molecules cannot come closer than a certain distance of each other. Therefore, instead of being free to travel anywhere in a volume V, the actual volume in which the molecules can travel is reduced by an amount that is proportional to the number of molecules present and the volume they each exclude. We can therefore model the effect of the repulsive forces (the volume-excluding forces) by changing V in the perfect gas equation to $V - nb$, where b is the proportionality constant between the reduction in volume and the amount of molecules present in the container. With this modification, the perfect gas equation of state changes to

$$p = \frac{nRT}{V - nb}$$

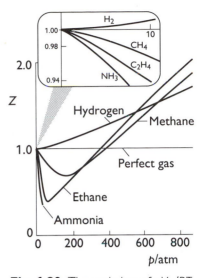

Fig. 1.20 The variation of pV_m/RT with pressure for several gases at 0 °C. A perfect gas has $pV_m/RT = 1$ at all pressures. Of the gases shown, hydrogen shows positive deviations at all pressures (at this temperature); all the other gases show negative deviations initially but positive deviations at high pressures. The negative deviations are a result of the attractive interactions between molecules and the positive deviations are a result of the repulsive interactions.

This equation of state (it is not yet the full van der Waals equation) should describe a gas in which repulsions are important. Note that when the pressure is low, so the molar volume is large compared with the volume excluded by the molecules ($V_m \gg b$), the nb term in the denominator can be ignored, and the equation reduces to the perfect gas equation of state.

The presence of attractive interactions between the molecules is to reduce the pressure that the gas exerts. We can estimate how the reduction in pressure depends on n and V by noting that the attraction experienced by a given molecule is proportional to the concentration, n/V, of molecules in the container. Because the attractions slow the molecules down, they strike the walls less frequently and with a lower impact. Because the pressure is determined by the impact of molecules, and they are fewer and weaker, the reduction in pressure is proportional to the *square* of the molar concentration:

reduction in pressure $\propto$ (rate of impacts) $\times$ (average strength of impact)

$$\propto \frac{n}{V} \times \frac{n}{V}$$

If the constant of proportionality is written a, the last equation becomes

$$\text{reduction in pressure} = a \times \left(\frac{n}{V}\right)^2$$

It follows that the equation of state allowing for both repulsions and attractions is

$$p = \frac{nRT}{V - nb} - a\left(\frac{n}{V}\right)^2 \tag{24}$$

This expression is the **van der Waals equation of state**.

We have built the van der Waals equation using physical arguments about the volumes of molecules and the effects of forces between them. It can be derived in other ways, but the present method has the advantage that it shows how to derive the form of an equation out of general ideas. The derivation also has the advantage of keeping imprecise the significance of the coefficients called the **van der Waals parameters** a and b: they are much better regarded as empirical parameters than as precisely defined molecular properties.

We can judge the reliability of the van der Waals equation by comparing the isotherms it predicts with the experimental isotherms shown in Fig. 1.18. Some calculated isotherms are shown in Fig. 1.21, and apart from the oscillations below the critical temperature they do resemble experimental isotherms quite well. The oscillations, which are called **van der Waals's loops**, are unrealistic because they suggest that under some conditions an increase of pressure results in an increase of volume. They are therefore replaced by horizontal lines. The van der Waals coefficients are found by fitting the calculated

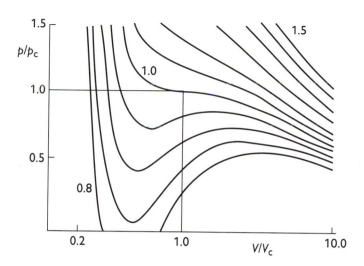

Fig. 1.21 Isotherms calculated using the van der Waals equation of state. The axes are labelled with the 'reduced' pressure p/p_c and volume V/V_c, where $p_c = a/27b^2$ and $V_c = 3b$, and the isotherms are labelled with the 'reduced' temperature T/T_c, where $T_c = 8a/27Rb$. The isotherm labelled 1.0 is the critical isotherm (the isotherm at the critical temperature).

curves to the experimental, and the values for some gases are listed in Table 1.5.

There are two notable features of the van der Waals equation.

> 1. Perfect-gas isotherms are obtained at high temperatures and large molar volumes.

To confirm this remark, we should note that when the temperature is high, RT may be so large that the first term in eqn (24) greatly exceeds the second. Furthermore, as we have already seen, if the molar volume is large (in the sense $V_m \gg b$, so that $V \gg nb$), then the denominator $V - nb \approx V$. Hence, under these conditions (of high temperature and large molar volume) the equation reduces to $p = nRT/V$, the perfect gas equation.

> 2. Liquids and gases coexist when cohesive and dispersing effects are in balance.

The van der Waals loops, which mark values of pressure and volume at which liquid and gas coexist, occur when both terms in eqn (24) have similar magnitudes. The first term arises from the kinetic energy of the molecules and their repulsive interactions, both of which tend to disperse the molecules; the second represents the effect of the

Table 1.5 Van der Waals parameters of gases

	a/L^2 atm mol^{-2}	b/L mol^{-1}
Air	1.4	0.039
Ammonia	4.17	0.037
Argon	1.35	0.032
Carbon dioxide	3.59	0.043
Ethane	5.49	0.064
Ethene	4.47	0.057
Helium	0.034	0.024
Hydrogen	0.244	0.027
Nitrogen	1.39	0.039
Oxygen	1.36	0.032
Xenon	4.19	0.051

attractive interactions, which cause the molecules to cohere together, and so opposes their dispersal.

The van der Waals equation of state is an interesting demonstration of the conversion of physical concepts into a quantitatively testable theory. The equation also provides yardsticks to judge when a gas can be treated as perfect (when V_m is large compared with a/RT and b). It is also useful in practice for estimating the pressure of a gas that cannot be treated as perfect.

1.10 The liquefaction of gases

A gas may be liquefied by cooling it below its boiling point at the pressure of the experiment. For example, chlorine can be liquefied by cooling it to below $-34\,°C$ in a bath cooled with 'dry ice' (solid carbon dioxide). For gases with very low boiling points (such as oxygen and nitrogen, at $-183\,°C$ and $-196\,°C$, respectively), such a simple technique is not practicable unless another very cold substance is available. One alternative and widely used commercial technique makes use of the forces that act between molecules. We saw earlier that the average speed of molecules in a gas is proportional to the square root of the temperature (eqn (17)). It follows that if there is a way of reducing the average speed of the molecules in a gas, then that is equivalent to cooling the gas. If the speed of the molecules can be reduced to the point at which neighbours can capture each other by their intermolecular attractions, then the cooled gas will condense to a liquid.

To slow the gas molecules, we make use of an effect like the one that happens when a ball is thrown into the air: as it rises it slows on account of the gravitational attraction of the Earth. Molecules attract each other, as we have seen (the attraction is not gravitational, but the effect is the same), and if we can cause them to move apart from each other, like a ball rising from a planet, then they should slow. It is very easy to move molecules apart from each other: we allow the gas to expand, which increases the average separation of the molecules. To cool a gas, therefore, we allow it to expand without allowing any heat to enter from outside. As it does so, the molecules climb apart, struggling against the attraction of all the other molecules, and as they move apart, they travel more slowly. Because the molecules now travel more slowly on average, the gas is cooler than before the expansion. This process of cooling by expansion is called the **Joule–Thomson effect**, because it was first observed by James Joule (whose name is commemorated in the unit of energy) and William Thomson (who later became Lord Kelvin).

In practice, the gas is allowed to expand several times by recirculating it through a device called a **Linde refrigerator** (Fig. 1.22). On each successive expansion the gas becomes cooler, and as it flows past the incoming gas, the latter is cooled further. After several successive expansions, the gas becomes so cold that it condenses to a liquid.

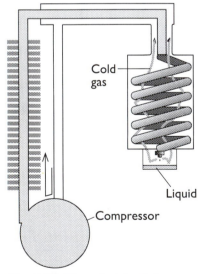

Fig. 1.22 The principle of the Linde refrigerator is shown in this diagram. The gas is recirculated and cools the gas that is about to undergo expansion through the throttle. The expanding gas cools still further. Eventually liquefied gas drips from the throttle.

EXERCISES

1.1 A sample of air occupies 1.0 L at 25 °C and 1.00 atm. What pressure is needed to compress it to 100 cm^3 at this temperature?

1.2 (a) Could 131 g of xenon in a vessel of volume 1.0 L exert a pressure of 20 atm at 25 °C if it behaved as a perfect gas? If not, what pressure would it exert? (b) What pressure would it exert if it behaved as a van der Waals gas?

1.3 A perfect gas undergoes isothermal compression, which reduces its volume by 2.20 L. The final pressure and volume of the gas are 3.78×10^3 Torr and 4.65 L respectively. Calculate the original pressure of the gas in (a) Torr, (b) atm.

1.4 To what temperature must a 1.0 L sample of a perfect gas be cooled from room temperature to reduce its volume to 100 cm^3 at constant pressure?

1.5 A perfect gas at 340 K is heated at constant pressure until its volume has increased by 14 per cent. What is the final temperature of the gas?

1.6 At sea level, where the pressure is 104 kPa, the gas in a balloon occupies 2.0 m^3. To what volume will the balloon expand when it has risen to an altitude where the pressure is (a) 52 kPa, (b) 880 Pa? Assume that the material of the balloon is infinitely extensible and assume constant temperature.

1.7 A sample of 255 mg of neon occupies 3.00 L at 122 K. Use the perfect gas law to calculate the pressure of the gas.

1.8 A gas mixture consists of 320 mg of methane, 175 mg of argon, and 225 mg of neon. The partial pressure of neon at 300 K is 15.2 kPa. Calculate (a) the volume and (b) the total pressure of the mixture.

1.9 The density of a gaseous compound was found to be 1.23 g L^{-1} at 330 K and 25.5 kPa. What is the molar mass of the compound?

1.10 In an experiment to measure the molar mass of a gas, 250 cm^3 of the gas was confined in a glass vessel. The pressure was 152 Torr at 298 K, and after correcting for buoyancy effects, the mass of the gas was 33.5 mg. What is the molar mass of the gas?

1.11 Calculate the pressure exerted by 1.0 mol C_2H_6 behaving as (a) a perfect gas, (b) a van der Waals gas, when it is confined under the following conditions: (i) at 273.15 K in 22.414 L, (ii) at 1000 K in 100 cm^3. Use the data in Table 1.5.

1.12 A vessel of volume 22.4 L contains 2.0 mol H_2 and 1.0 mol N_2 at 273.15 K. Calculate (a) their partial pressures and (b) the total pressure.

1.13 A diving bell has an air space of 3 m^3 when on the deck of a boat. What is the volume of the air space when the bell has been lowered to a depth of 50 m? Take the mean density of sea water to be 1.025 g cm^{-3} and assume that the temperature is the same as on the surface.

1.14 What pressure difference must be generated across the length of a 15 cm vertical drinking straw in order to drink a water-like liquid of density 1.0 g cm^{-3}?

1.15 A meteorological balloon had a radius of 1.0 m when released at sea level at 20 °C and expanded to a radius of 3.0 m when it has risen to its maximum altitude where the temperature was −20 °C. What is the pressure inside the balloon at that altitude?

1.16 Express the van der Waals equation of state as a virial expansion in powers of $1/V_m$ and obtain expressions for B and C in terms of the parameters a and b. The expansion you will need is

$$\frac{1}{1-x} = 1 + x + x^2 + \cdots$$

Measurements on argon gave $B = -21.7$ cm^3 mol^{-1} and $C = 1200$ cm^6 mol^{-2} for the virial coefficients at 273 K. What are the values of a and b in the corresponding van der Waals equation of state?

1.17 Calculate the mean speed of (a) He atoms, (b) CH_4 molecules at (i) 77 K, (ii) 298 K, (iii) 1000 K.

1.18 At what pressure does the mean free path of argon at 25 °C become comparable to the size of a 1 L vessel that contains it? Take $\sigma = 0.36$ nm^2.

1.19 At what pressure does the mean free path of argon at 25 °C become comparable to the diameters of the atoms themselves? ($\sigma = 0.36$ nm^2.)

1.20 At an altitude of 20 km the temperature is 217 K and the pressure 0.05 atm. What is the mean free path of N_2 molecules? ($\sigma = 0.43$ nm^2.)

1.21 How many collisions does a single Ar atom make in 1.0 s when the temperature is 25 °C and the pressure is (a) 10 bar, (b) 100 kPa, (c) 1.0 Pa?

1.22 Calculate the total number of collisions per second in 1.0 L of argon under the same conditions as in Exercise 1.21.

1.23 How many collisions per second does an N_2 molecule make at an altitude of 20 km? (See Exercise 1.20 for data.)

1.24 Calculate the mean free path of diatomic molecules in air using $\sigma = 0.43$ nm^2 at 25 °C and (a) 10 bar, (b) 103 kPa, (c) 1 Pa.

1.25 Use the Maxwell distribution of speeds to estimate the fraction of N_2 molecules at 500 K that have speeds in the range 290 to 300 m s^{-1}.

1.26 How does the mean free path in a sample of a gas vary with temperature in a constant-volume container?

2

Thermodynamics: the first law

The branch of physical chemistry known as **thermodynamics** is concerned with the study of the transformations of energy, and in particular its transformation from heat into work and vice versa. That concern might seem remote from chemistry; indeed, thermodynamics was originally formulated by physicists and engineers interested in the efficiency of steam engines. However, thermodynamics has proved to be of immense importance in chemistry, for not only does it deal with the energy output of chemical reactions (as when a fuel burns and the heat is converted into work in an engine) but it also deals with questions that lie right at the heart of everyday chemistry, such as the properties of reactions at equilibrium and the generation of electricity in electrochemical (and biological) cells.

Almost every argument and explanation in chemistry boils down to a consideration of some aspect of a single property: the **energy**. We shall find that energy determines what molecules may form, what reactions may occur, how fast they may occur, and (with a refinement in our conception of energy), in which direction a reaction has a tendency to occur. Energy is absolutely central to chemistry, yet it is extraordinarily difficult to give a satisfactory account of what energy *is*. It is easy to give a bland definition:

Energy is the capacity to do work.

(We shall explain the scientific meaning of work shortly.) This definition implies that a raised weight has more energy than one on the ground (because the former can do work as it falls to the level of the latter). It also implies that a sample of gas at a high temperature has more energy than the same sample at a low temperature, because the former has a higher pressure and can do more work in driving out a piston: this effect is the principle of the internal combustion engine.

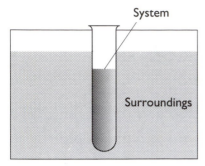

Fig. 2.1 The sample is the system of interest; the rest of the world is its surroundings. The surroundings are where observations are made on the system. They can often be modelled (as here) by a large water bath.

The contributions to energy that arise from a particle's position and speed (its potential and kinetic energies, respectively) are reviewed in *Further information 2* (p. 24). It is always possible to relate the energy changes that we shall meet in this chapter to changes in potential energy and kinetic energy of the atoms and molecules that constitute the sample of interest. Thus, a raised weight has a high energy because the potential energy of all its atoms is raised when they are all simultaneously lifted to a height above the surface of the Earth. A hot gas has a higher energy than when it is cool because the individual molecules have a higher kinetic energy.

The conservation of energy

People struggled for centuries to create energy from nothing, for they believed that if they could create energy, then they could produce work (and wealth) endlessly. However, without exception, despite strenuous efforts (many of which degenerated into deceit), they failed, and we have come to recognize that energy can be neither created nor destroyed. This property of energy, that it can be neither created nor destroyed, is called the **conservation of energy**. The conservation of energy is of great importance in chemistry because most chemical reactions release energy or absorb it as they occur, and so we can be confident that all such changes must involve only the conversion of energy from one form to another, and neither its creation nor its annihilation.

In this first section we shall see how to take into account the various forms in which energy can be transferred into or out of a sample of matter. To do so, we must make a careful distinction between a **system**, the part of the world in which we have a special interest, and the **surroundings**, where we make our observations. Thus, a system may be a stoppered flask containing a reaction mixture; the surroundings may be a constant-temperature bath in which the flask is immersed (Fig. 2.1). We also need to distinguish three types of system (Fig. 2.2):

An **open system** is a system that can exchange matter and energy with the surroundings.

A **closed system** is a system that cannot exchange matter with its surroundings.

An **isolated system** is a system that can exchange neither matter nor energy with its surroundings.

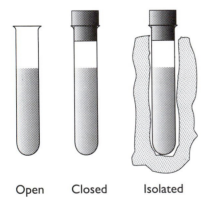

Open   Closed   Isolated

Fig. 2.2 A system is open if it can exchange matter and energy with its surroundings, closed if it can exchange energy but not matter, and isolated if it can exchange neither energy nor matter.

An example of an open system is a flask that is not stoppered and to which various substances can be added; a biochemical cell is an open system because nutrients and waste can pass through the cell wall. An example of a closed system is a stoppered flask: energy can be exchanged with the contents of the flask because the walls may be able to conduct heat, or there may be a pair of electrodes sealed into the flask and through which an electric current can flow. An example of an isolated system is a sealed flask that is thermally, mechanically, and electrically insulated from its surroundings. A system that is thermally insulated from its surroundings (like water in a vacuum flask) is called an **adiabatic system** (from the Greek words for 'not passing through'). For the rest of this chapter we consider closed and isolated systems. Such systems are very good approximations to many of the actual systems that we meet in practice, and by focusing on them the discussion is greatly simplified. For example, a sealed flask and the water bath in which it is immersed jointly make up a very good approximation to an isolated system.

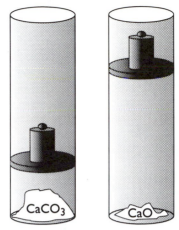

Fig. 2.3 When calcium carbonate decomposes, the carbon dioxide produced must push back the surrounding atmosphere (represented by the weight resting on the piston), and hence must do work on its surroundings. This is an example of energy leaving a system as work.

2.1 Work and heat

There are only two ways in which the energy of a closed system can be changed: by transferring energy as work and by transferring energy as heat.

> **Work** is a transfer of energy that can be used to change the height of a weight somewhere in the surroundings.
>
> **Heat** is a transfer of energy as a result of a temperature difference between the system and the surroundings.

For instance, consider a chemical reaction that produces gases, such as the thermal decomposition of calcium carbonate

$$CaCO_3(s) \xrightarrow{\Delta} CaO(s) + CO_2(g)$$

which takes place inside a cylinder fitted with a piston. (Δ signifies an elevated temperature; for this reaction, of about $800\,°C$.) The gas produced in the reaction can drive out the piston and raise a weight in the surroundings (Fig. 2.3). In this case, the system has transferred energy to the surroundings because a weight has been raised in the surroundings and they (the surroundings) are capable of doing more work than they were before the weight was raised. Alternatively, suppose we carry out a typical acid–base neutralization, such as

$$HCl(aq) + NaOH(aq) \rightarrow NaCl(aq) + H_2O(l)$$

in a flask immersed in an ice bath (Fig. 2.4). No weight is raised in the surroundings (so the reaction does no work), but some of the ice melts. We have to conclude that energy has left the system (because ice must be supplied with energy if it is to melt) and has travelled from the system to the ice bath; because the latter has a lower temperature than the system, we conclude that energy has left the system as heat.

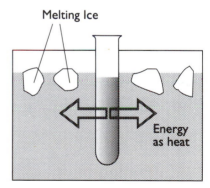

Fig. 2.4 Energy may also leave a system as heat. The loss into the surroundings can be detected by noting whether ice melts as the process proceeds. Note that in this case and in Fig. 2.3 the distinction between energy transfer by work or by heat is made by observations in the surroundings (the raising of a weight or the melting of ice).

The molecular nature of heat and work

Thermodynamics is a collection of relations between *observations,* and does not employ any information about the behaviour of atoms and molecules. Nevertheless, it is interesting, and enriching, to look behind the definitions to find a deeper understanding of the nature of work and heat

The clue to the nature of work comes from an understanding of the motion of a weight in terms of its component atoms. When a weight is raised, all its atoms move in the same direction (upwards). This observation is the clue we need to the nature of work: *work is the transfer of energy that achieves or utilizes uniform motion in the surroundings* (Fig. 2.5(a)).

Now consider heat. When energy is transferred to an ice bath and some of the ice melts, the H_2O molecules oscillate more rapidly around their positions in the ice, and molecules at the surface may escape into the surroundings liquid. The key point is that the motion stimulated by the arrival of energy from the system is chaotic, not uniform as in the case of work. This observation reveals the nature of heat: *heat is the transfer of energy that achieves or utilizes chaotic motion in the surroundings* (Fig. 2.5(b)).

Fig. 2.5 (a) Work is transfer of energy that causes or utilizes *uniform* motion of atoms in the surroundings. For example, when a weight (the solid cylinder, a part of the surroundings) is raised, all the atoms of the weight (shown magnified) move in unison in the same direction. (b) Heat is the transfer of energy that causes or utilizes *chaotic* motion in the surroundings. When energy leaves the system (the shaded region on the left), it generates chaotic motion in the surroundings (shown magnified).

Surroundings

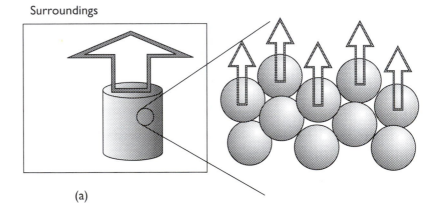

(a)

Surroundings

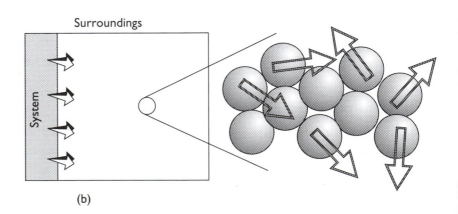

(b)

It is an interesting social point that the molecular difference between work and heat correlates with the chronological order of their application. Thus, the liberation of heat by fire is a relatively unsophisticated procedure (because the energy emerges in a chaotic fashion from the burning fuel), and was developed—stumbled upon—early in the history of civilization. The generation of work by a burning fuel relies on a carefully controlled transfer of energy so that myriads of molecules move in unison: it came thousands of years later, with the development of the steam engine.

The measurement of work

The general definition of work is that it is equal to the product of distance and the force opposing the motion:

$$\text{work} = \text{distance} \times \text{opposing force}$$

It follows that moving through a long distance against a strong opposing force requires a lot of work. If the force is the gravitational attraction of the Earth on a mass m, then the force opposing the raising of the mass vertically is mg, where g is the acceleration of free fall (the acceleration due to gravity, $9.81 \, \text{m s}^{-2}$). Therefore, the work needed to raise a mass m through a height h on the surface of the Earth is

$$\text{work} = h \times mg$$

For example, raising a book like this one (of mass about 1.0 kg) from the floor to the table 75 cm above requires

$$w = 0.75 \, \text{m} \times 1.0 \, \text{kg} \times 9.81 \, \text{m s}^{-2} = 7.4 \, \text{kg m}^2 \, \text{s}^{-2}$$

The unit used to report energy is the joule (J), which is named after James Joule, the Manchester brewer who made a detailed study of heat and work in the nineteenth century:

$$1 \, \text{J} = 1 \, \text{kg m}^2 \, \text{s}^{-2}$$

Therefore, the work we have just calculated would be reported as 7.4 J. Each beat of the human heart does work equal to about 1 J, so about 100 kJ of energy is expended each day driving the blood around one's body.

Exercise E2.1 A useful relation between joules and pascals is $1 \, \text{J} = 1 \, \text{Pa} \times 1 \, \text{m}^3$. Confirm this relation from the definition of pascals in *Further information* 1 (p. 4).

$$[1 \, \text{Pa} \times 1 \, \text{m}^3 = 1 \, \text{kg m}^{-1} \, \text{s}^{-2} \times 1 \, \text{m}^3 = 1 \, \text{kg m}^2 \, \text{s}^{-2} = 1 \, \text{J}]$$

In chemistry, a very important type of work is **expansion work**, the work done when a system expands against an opposing pressure. The thermal decomposition of calcium carbonate illustrated in Fig. 2.3 is an example of expansion work. It is quite easy to calculate the work that the system does as it expands through a volume ΔV by considering a piston of area A moving out through a distance h (Fig. 2.6). The force opposing the expansion is the constant external

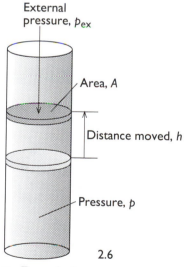

External pressure, p_{ex}

Area, A

Distance moved, h

Pressure, p

2.6

Fig. 2.6 When a piston of area A moves out through a distance h, it sweeps out a volume $\Delta V = Ah$. The external pressure p_{ex} opposes the expansion with a force $p_{ex}A$.

pressure p_{ex} times the area of the piston (because pressure is force per unit area), so the work done is

$$\text{work done by system} = \text{distance} \times \text{opposing force}$$
$$= h \times (p_{ex} \times A) = p_{ex} \times (h \times A)$$
$$= p_{ex} \times \Delta V \tag{1}$$

The last line follows from the fact that hA is the volume of the cylinder swept out by the piston as the gas expands ($hA = \Delta V$).

Exercise E2.2 Calculate the work done by a system in which a reaction results in the formation of 1.0 mol of gas at 25 °C and 100 kPa.

Hint: the increase in volume will be 25 L under these conditions (if the gas is treated as perfect); note the relation between pascals and joules established in Exercise E2.1.

[2.5 kJ]

A very important type of expansion work is a special case of the one we have just described. Suppose we are interested in the *maximum* work that we can extract from a gas as it expands isothermally (at constant temperature). According to the expression above, the maximum work is obtained when the external pressure has its maximum value. However, that external pressure cannot be greater than the pressure of the gas inside the system, for otherwise the external pressure would compress the gas instead of allowing it to expand. Therefore, *maximum work is obtained when the external pressure is only infinitesimally smaller than the pressure of the gas in the system.* In effect, the two pressures are the same. In Chapter 1 we called this balance of pressures a state of mechanical equilibrium; therefore, we can conclude that *a system in mechanical equilibrium does maximum expansion work.*

There is another way of expressing the condition for obtaining maximum work. Because the external pressure is infinitesimally smaller than the pressure of the gas, the piston moves outwards. However, suppose we increased the pressure so that it becomes infinitesimally greater than the pressure of the gas: now the piston moves inwards. That is, *when a system is in a state of mechanical equilibrium, an infinitesimal change in the pressure can result in opposite directions of change.* We say that a state in equilibrium is **reversible**. In everyday life 'reversible' means a process that can be reversed; in thermodynamics it has a stronger meaning—it means that a process can be reversed by an *infinitesimal* modification in some property (such as the pressure).

We can summarize this discussion by the following remarks.

1. A system does maximum expansion work when the external pressure is equal to that of the system, $p_{ex} = p$.

2. A system does maximum expansion work when it is in mechanical equilibrium with its surroundings.

3. A system does maximum expansion work when it is changing reversibly.

All three statements are equivalent, but they reflect different degrees of sophistication in the way the point is expressed.

The maximum work obtainable from the isothermal expansion from an initial volume V_i to a final volume V_f of a gas cannot be written down simply by replacing p_{ex} in eqn (1) by p because, as the piston moves out and the volume of the gas increases, the pressure inside the system falls (by Boyle's law). Therefore, at each stage of the expansion, the external pressure must be reduced slightly to match the slightly lower pressure of the gas. The calculation of the work available from the system under such circumstances can be done using calculus, and the result is

$$\text{work done by system} = nRT \ln \frac{V_f}{V_i} \qquad (2)$$

where n is the amount of substance of gas and T is the temperature.

JUSTIFICATION

The derivation of the expression for the maximum expansion work is carried out by considering the work done during an infinitesimal expansion dV against a pressure p_{ex} that, at every stage of the expansion, has been adjusted to be equal to the pressure p of the gas:

$$\text{work done by system} = p \, dV$$

According to the perfect gas equation of state, as the expansion continues, the pressure is given by

$$p = \frac{nRT}{V}$$

Therefore, the *total* work of expansion as the system expands from an initial volume V_i to a final volume V_f is the integral

$$\text{work done by system} = \int_{V_i}^{V_f} p \, dV = nRT \int_{V_i}^{V_f} \frac{dV}{V}$$

$$= nRT \ln \frac{V_f}{V_i} \left(\text{because} \int \frac{dx}{x} = \ln x \right)$$

Exercise E2.3 Calculate the work done when 1.0 mol Ar that is confined in a cylinder of volume 1.0 L at 25 °C expands isothermally and reversibly to 2.0 L.

[1.7 kJ]

Measuring the energy transferred as heat

The energy that is transferred as heat can be measured in two ways, both of which involve the use of a **calorimeter**, a device for measuring the heat transferred during a process (the name comes from 'calor', the Latin word for heat). A typical calorimeter consists of a container, in which the reaction occurs, and a surrounding water bath. The entire assembly is insulated from the rest of the surroundings, so overall it forms an isolated system.

If the surrounding bath contains melting ice in water at $0\,°C$, the apparatus is known as an **ice calorimeter**. In this arrangement, the mass of ice that melts when the reaction occurs is proportional to the heat released. It is known from separate measurements that to melt $1.00\,g$ of ice at $0\,°C$ requires $0.333\,kJ$ of heat. Therefore, if in a certain experiment we observed that $22.5\,g$ of ice melted, then we could conclude that the reaction released $22.5 \times 0.333\,kJ = 7.50\,kJ$ of energy as heat. Conversely, if an additional $22.5\,g$ of ice were *formed,* then we would conclude that the system had *absorbed* $7.50\,kJ$ of heat.

A more reliable procedure is to observe the change in temperature of the water in a bath surrounding the reaction vessel and to **calibrate** the calorimeter (the entire assembly—the reaction vessel and the water bath) by comparing the observed change in temperature with a change in temperature brought about by a given quantity of heat. One procedure is to heat the calorimeter electrically by passing a known current for a specific time through a heater, and observing the increase in temperature. The heat supplied by the heater is given by

$$\text{heat supplied to system} = IVt \tag{3}$$

where I is the current in amperes (A), V is the potential difference of the supply in volts (V), and t is the time for which the current flows (in seconds). For example, if a $1.23\,A$ current from a $12.0\,V$ source flows for $123\,s$, the heat supplied is

$$\text{heat supplied to system} = 1.23\,A \times 12.0\,V \times 123\,s = 1.82\,kJ$$

(We have used the relation $1\,A\,V\,s = 1\,J$, which follows from the definition of the electrical units.) If, furthermore, we observe that the temperature of the calorimeter rises by $4.47\,°C$, then the **calorimeter constant**, the increase in temperature per unit energy supplied, is

$$C = \frac{\text{heat supplied}}{\text{increase in temperature}} = \frac{1.82\,kJ}{4.47\,°C} = 0.407\,kJ\,(°C)^{-1}$$

Then, if in an experiment on an unknown reaction we measure a temperature rise of $3.22\,°C$ in the same calorimeter, the heat output would be

$$\text{heat output} = C \times \text{increase in temperature}$$
$$= 0.407\,kJ\,(°C)^{-1} \times 3.32\,°C = 1.31\,kJ$$

Similar procedures apply when a reaction releases heat or absorbs heat: the temperature of a calorimeter rises if heat is released and it falls if heat is absorbed. A process that releases heat into the surroundings is called **exothermic**; one that absorbs heat from the surroundings is called **endothermic**. All combustions are exothermic. Endothermic reactions are much less common, but one spectacular example is the reaction that takes place between ammonium thiocyanate, NH_4SCN, and barium hydroxide octahydrate, $Ba(OH)_2 \cdot 8H_2O$, when the two solids are ground together:

$$2NH_4SCN(s) + Ba(OH)_2 \cdot 8H_2O(s) \rightarrow Ba(SCN)_2(aq)$$
$$+ 10H_2O(l) + 2NH_3(g)$$

The reaction releases the water of hydration from the barium hydroxide, and is so endothermic that the resulting solution freezes, as does any water that happens to be on the outside of the flask.

The internal energy

When a system releases 10 kJ of energy to the surroundings as work (that is, when 10 kJ of work is done in the surroundings by raising a weight), we say that the **internal energy** U of the system has decreased by 10 kJ, and write $\Delta U = -10$ kJ, the minus sign signifying the reduction in internal energy that has occurred. If the system releases 20 kJ of energy as heat, then the internal energy decreases by 20 kJ, and we write $\Delta U = -20$ kJ. If the system releases 10 kJ as work *and* 20 kJ as heat (as in an inefficient internal combustion engine), the internal energy falls by 30 kJ, and we write $\Delta U = -30$ kJ. On the other hand, if we do 10 kJ of work *on* the system, for instance, by winding a spring it contains, or pushing in a piston to compress a gas (Fig. 2.7), then the internal energy of the system increases by 10 kJ, and we write $\Delta U = +10$ kJ. Likewise, if we supply 20 kJ of energy as heat, then the internal energy increases by 20 kJ, and we write $\Delta U = +20$ kJ. (Notice that ΔU *always* carries a sign explicitly, even if it is positive: we never write $\Delta U = 20$ kJ but always $+20$ kJ.)

The internal energy is the sum of all the kinetic and potential contributions to the energy of all the atoms, ions, and molecules in the system: it is the grand total energy of the system. In practice, we do not know this total energy, because it includes the kinetic and potential energies of all the electrons and all the components of the atomic nuclei. Nevertheless, there is no problem with dealing with the *changes* in internal energy, because we can determine those by monitoring the energy supplied or lost as heat or work. All practical applications of thermodynamics deal only with ΔU, not with U itself.

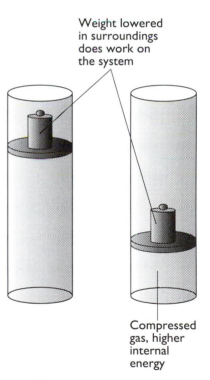

Weight lowered in surroundings does work on the system

Compressed gas, higher internal energy

Fig. 2.7 Work is done *on* a system ($w > 0$) if a weight is lowered in the surroundings and compresses the material of the system. As a result, the internal energy of the system increases ($\Delta U > 0$).

2.2 Notation

The preceding remarks can be summarized by introducing the following notation:

> w is the energy supplied *to* the system as work;
>
> q is the energy supplied *to* the system as heat.

These conventions mean that

$w = +10\,\text{kJ}$ if 10 kJ of energy is supplied to the system as work;

$w = -10\,\text{kJ}$ if 10 kJ of energy is lost from the system as work;

$q = +10\,\text{kJ}$ if 10 kJ of energy is supplied to the system as heat;

$q = -10\,\text{kJ}$ if 10 kJ of energy is lost from the system as heat.

Then, in terms of this notation, the comments about the change in internal energy can be expressed succinctly as

$$\Delta U = w + q \tag{4}$$

The previous expressions we have written for expansion work can now be written in terms of w and the appropriate sign.

Expansion against a constant external pressure:

$$w = -p_{\text{ex}}\,\Delta V, \quad \text{with} \quad \Delta V = V_f - V_i \tag{5}$$

Reversible, isothermal expansion:

$$w = -nRT \ln \frac{V_f}{V_i} \tag{6}$$

In each case, the negative sign signifies that energy leaves the system as work when it expands. Thus, when the system expands against a constant pressure, ΔV is positive and $-p_{\text{ex}}\,\Delta V$ is negative: the system has *lost* energy as work. Likewise, in the reversible expansion, the ratio V_f/V_i is greater than 1 (because V_f is larger than V_i), the logarithm of a number larger than 1 is positive, and so the value of w is negative. Once again, energy has left the system as work.

> **Example** Calculating the change in internal energy
> Nutritionists are interested in the use of energy by the human body, and we can consider our own body as a thermodynamic

'system'. Calorimeters have been constructed which can accommodate a person to measure (nondestructively) their net energy output. Suppose in the course of an experiment a person does 622 kJ of work on an exercise bicycle and loses 82 kJ of energy as heat. What is the change in internal energy of the person?

Answer This example is an exercise in keeping track of signs correctly. When energy is lost from the system, w or q is negative; when energy is gained by the system, w or q is positive. It follows that $w = -622$ kJ and $q = -82$ kJ, so

$$\Delta U = w + q = -622 \text{ kJ} - 82 \text{ kJ} = -704 \text{ kJ}$$

and the person's internal energy falls by 704 kJ.

Exercise E2.4 An electric battery is charged by the supply of 250 kJ of energy as electrical work, but in the process it loses 25 kJ of energy as heat to the surroundings. What is the change in internal energy of the battery?

[+225 kJ]

CALCULUS

In some of the calculations that we shall need to do, it will be necessary to consider infinitesimally small changes, such as the supply of an infinitesimal quantity of heat, dq, or work, dw. Then the change in internal energy will also be infinitesimally small, and we denote it dU. When the changes are infinitesimal, the internal energy changes by

$$dU = dw + dq$$

The same sign conventions apply as before: when energy is supplied to the system, dw and dq are positive; when energy is lost, they are negative. An example of an infinitesimal quantity of work is the expansion considered earlier: for an infinitesimal expansion of the system through a volume dV against a pressure p_{ex}, the energy lost as work is

$$dw = -p_{ex}\, dV$$

Equation (4) expresses the fact that *work and heat are equivalent ways of changing the internal energy of a system*. That is, it is immaterial whether we supply energy to a system as heat, work, or a mixture of the two: the change in internal energy is the same in each case, and internal energy that has been supplied as work (for instance) can be withdrawn as heat. The system is like a bank that can accept deposits and make payments in either of two currencies (as work or heat), but stores its reserves as the thermodynamic equivalent of gold (that is, as internal energy).

An important characteristic of the internal energy is that it is a state

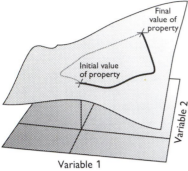

Fig. 2.8 The curved sheet shows how a property (for example, the altitude) varies as two coordinates (for example, latitude and longitude) are changed. The altitude is a state property, because it depends only on the current state of the system. The change in the value of a state property is independent of the path between the two states. For example, the difference in altitude between the initial and final states shown in the diagram is the same whatever path (as depicted by the dark and light lines) is used to travel between them. The internal energy is a state property in the same sense, but now the variables include the pressure and temperature of the system.

property:

> A **state property** is a physical property that depends only on the current state of the system and is independent of the path by which that state was prepared.

A state property is very much like altitude: each point on the surface of the Earth can be specified by quoting its latitude and longitude, and (on land areas, at least) there is a unique property, the altitude, that has a fixed value at that point. The altitude at any location is independent of the path that we may have taken to arrive there, so altitude is a state property that is determined by the two variables latitude and longitude (Fig. 2.8).

The internal energy, being the sum of all the kinetic and potential energies that the atoms in a system happen to have at a particular instant, is a property of the state of the system. If we were to change the temperature of the system, then change the pressure, then adjust the temperature and pressure back to their original values, the internal energy would return to its original value too.

More generally, the fact that U is a state property implies that *a change in the internal energy between two states of a system is independent of the path between them.* For example, if we had a sample of gas which we compressed to a certain pressure and then cooled to a certain temperature, then the internal energy change would have a particular value. If, on the other hand, we changed the temperature and then the pressure, but ensured that the two final values were the same as in the first experiment, then the overall change in internal energy would be exactly the same as before. This path independence of ΔU is of the greatest importance in chemistry, as we shall soon see.

2.3 The first law

Finally, we can move to the climax of this part of the chapter. Suppose we now consider an isolated system, one that is insulated mechanically and thermally from its surroundings. Then since the system can neither do work nor supply heat, it follows that the internal energy cannot change. That is,

> The internal energy of an isolated system is constant.

This statement is the **first law of thermodynamics**. It is obviously closely related to the law of conservation of energy, but it should be remembered that the internal energy is expressed in terms of work and heat and their equivalence; so the first law is a statement that concerns *thermodynamics,* the transformation of energy into different forms (including heat), not just mechanics (which does not deal with the concept of heat).

Just as many people have sought to find exceptions to the law of the conservation of energy, so people have also looked for exceptions to the first law of thermodynamics. They have tried to invent **perpetual motion machines**, which are devices for producing work without consuming energy. All such attempts have failed, and no exception has ever been found to the first law of thermodynamics.

The measurement of ΔU

The definition of ΔU in terms of w and q points to a very simple method for measuring the change in internal energy of a system when a reaction occurs. We have seen already that the work done by a system when it pushes against an external pressure p_{ex} and expands through a volume ΔV is given by eqn (5) $(w = -p_{ex} \Delta V)$. It follows that, if expansion work is the only type of work that a system can do, then

$$\Delta U = -p_{ex} \Delta V + q \tag{7}$$

However, suppose we seal the reaction into a container that cannot change its volume, then $\Delta V = 0$ and this expression simplifies to

$$\Delta U = q \text{ at constant volume} \tag{8}$$

It follows that, to measure a change in internal energy, we need only measure the heat supplied to or absorbed by a system that cannot change its volume. The apparatus used in practice is called a **bomb calorimeter** (Fig. 2.9). It consists of a central, sturdy steel container (the bomb) where the reaction takes place at constant volume. The calorimeter is used like the other calorimeters we have described: it is calibrated (either electrically or using a reaction of known heat output), and then the temperature rise caused by the reaction of interest is measured and interpreted in terms of a heat output. The value of q obtained in this way is identified with the change in internal energy of the system.

The heat capacity

The internal energy of a system can be increased by raising the temperature. For example, to raise the temperature of 100 g of water requires the input of heat, and the rise in temperature is proportional to the quantity of heat supplied. This proportionality is written

$$\Delta T = \frac{q}{C} \tag{9}$$

where the constant C is called the **heat capacity**; its units are joules per kelvin, $J\,K^{-1}$. Because ΔT is inversely proportional to the heat

Fig. 2.9 A constant-volume bomb calorimeter. The 'bomb' is the central, sturdy vessel, which is strong enough to withstand high pressures. The calorimeter is the entire assembly shown here. In order to ensure that no heat escapes into the surroundings, the calorimeter may be immersed in a water bath with a temperature that is continuously adjusted to that of the calorimeter at each stage of the combustion.

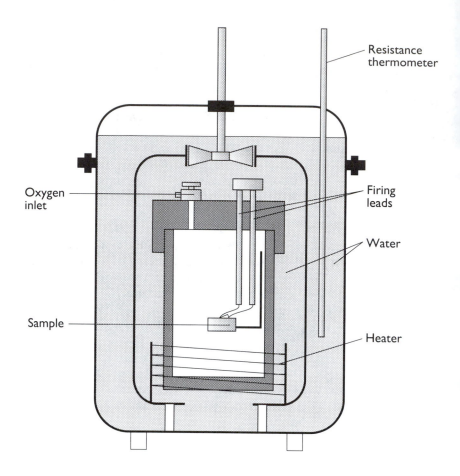

capacity, a system with a large heat capacity undergoes only a small increase in temperature for a given input of heat. The heat capacity of a substance depends on the size of the sample and it is common to report values as the **specific heat capacity**, the heat capacity per unit mass of substance (typically, in joules per kelvin per gram, $J\,K^{-1}\,g^{-1}$), and the **molar heat capacity**, the heat capacity per unit amount of substance (in joules per kelvin per mole, $J\,K^{-1}\,mol^{-1}$). The value for water, $75\,J\,K^{-1}\,mol^{-1}$, for example, shows that the increase in temperature of 100 g of water (5.55 mol H_2O) when 1.0 kJ of heat is supplied is

$$\Delta T = \frac{1.0 \times 10^3\,J\,mol^{-1}}{5.55\,mol \times 75\,J\,K^{-1}\,mol^{-1}} = 2.4\,K$$

It turns out that the heat capacity of a substance depends on whether the substance is maintained at the same volume (like a gas contained in a rigid vessel) or is free to expand against a constant external pressure (as when a solid or liquid is heated in the atmosphere). The two heat capacities are called, respectively, the **constant-volume heat capacity**, C_V (for a system that cannot expand) and the **constant-pressure heat capacity**, C_p (for a system that is free to expand). The numerical values of the two types of heat capacity are very similar for solids and liquids (neither of which change volume very much with temperature, even if they are free to), but they differ appreciably for gases. For perfect gases, the two heat capacities are related by

$$C_p = C_V + nR \qquad (10)$$

The constant-pressure heat capacity of a gas is larger than the constant-volume heat capacity. The difference arises because some of the energy supplied as heat at constant volume is used to drive back the atmosphere as the system expands, and so a given input of energy as heat brings about a smaller rise in temperature, corresponding to a larger heat capacity. Some typical values are given in Table 2.1.

Exercise E2.5 The molar constant-pressure heat capacity of water vapour at 25 °C is 33.58 J K^{-1} mol^{-1}. What is its molar constant-volume heat capacity if it behaves as a perfect gas?
[25.27 J K^{-1} mol^{-1}]

Table 2.1   Heat capacities of some materials

Substance	Specific heat capacity C_p/J K^{-1} g^{-1}	Molar heat capacity† C_p/J K^{-1} mol^{-1}
Air	1.01	29
Benzene, C_6H_6	1.05	136.1
Brass	0.37	
Copper, Cu	0.38	24.4
Ethanol, C_2H_5OH	2.42	111.5
Glass (Pyrex)	0.78	
Granite	0.80	
Marble	0.84	
Polyethylene	2.3	
Stainless steel	0.51	
Water, H_2O	2.03 (solid)	37
	4.18 (liquid)	75.29
	2.01 (vapour)	33.58

† Molar heat capacities are given only for air and well defined, pure substances.

The internal energy of a system increases as heat is supplied to it, and we have already seen that at constant volume $\Delta U = q$. This relation can be combined with the expression defining the heat capacity, $q = C\,\Delta T$, to obtain

$$\Delta U = C_V\,\Delta T \quad \text{at constant } V$$

For example, when 100 g of liquid water (5.55 mol H_2O, $C_V = 75\,\text{J K}^{-1}\,\text{mol}^{-1}$) is heated through 5.0 K, its internal energy increases by

$$\Delta U = 5.55\,\text{mol} \times 75\,\text{J K}^{-1}\,\text{mol}^{-1} \times 5.0\,\text{K} = +2.1\,\text{kJ}$$

The enthalpy

Much of chemistry takes place in vessels that are open to the atmosphere and subjected to constant pressure as distinct from constant volume. In general, when a change takes place in a system, its volume changes. For example, the thermal decomposition of 1.0 mol $CaCO_3$ at 1 bar results in an increase in volume of 89 L at 800 °C on account of the carbon dioxide gas produced. To create this large volume for the carbon dioxide to occupy, the surrounding atmosphere must be pushed back. That means that the system must perform expansion work. Therefore, although a certain amount of heat may be supplied to bring about the decomposition, the increase in internal energy of the system is not equal to the energy supplied as heat because some of that energy has been used to do work of expansion (Fig. 2.10). Because the volume has changed, some of the heat has leaked back into the surroundings as work. Another example is the oxidation of a fat, such as tristearin, to carbon dioxide in the body. The overall reaction is

$$2C_{57}H_{110}O_6(aq) + 163O_2(g) \rightarrow 114CO_2(g) + 110H_2O(g)$$

In this reaction there is a net increase in volume equivalent to the formation of 61 mol of gas molecules (if water is produced as vapour), which corresponds to an increase in volume of over 800 mL at 25 °C for the consumption of 1 g of the fat. In this example, not the whole of the change in internal energy is released as heat because some energy must be used to push back the atmosphere to make space for the products. Thus, although we might know that the internal energy of the system has decreased by 100 kJ (for the combustion of an appropriate mass of fat), only 90 kJ might be released as heat, the other 10 kJ being used to make room for the products.

Although we could always estimate the work that a system had to do, and hence determine the energy released as heat, it is more convenient to introduce a property that is equal to the heat supplied at constant pressure. Then we need not keep track of the work that has to be done as the volume of the system changes. This property is

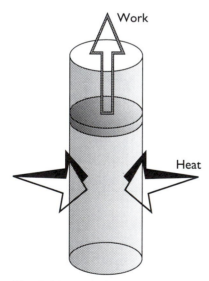

Fig. 2.10 The change in internal energy of a system that is free to expand or contract is not equal to the energy supplied as heat because some energy may escape back into the surroundings as work. However, the change in *enthalpy* of the system under these conditions is equal to the energy supplied as heat.

called the 'enthalpy': it will be at the centre of our attention throughout the rest of the chapter and will recur throughout the book.

2.4 The definition of enthalpy

The **enthalpy**, H, is defined so that

$$\Delta H = q \quad \text{at constant } p \tag{11}$$

That is, if 10 kJ of heat is supplied to a system that is free to change its volume at constant pressure, then the enthalpy increases by 10 kJ and we write $\Delta H = +10$ kJ. On the other hand, if the reaction is exothermic and releases 10 kJ of heat when it occurs, then we write $\Delta H = -10$ kJ. For the particular case of the combustion of tristearin mentioned above, in which 90 kJ of energy is released as heat, we would write $\Delta H = -90$ kJ. Note that an *endothermic reaction corresponds to an increase in enthalpy* whereas *an exothermic process corresponds to a decrease in enthalpy*. All combustion reactions (including the controlled combustions that contribute to respiration), for example, release heat, are exothermic, and are accompanied by a reduction in the enthalpy of the reaction system. These relations are consistent with the name 'enthalpy', which is derived from the Greek words meaning 'heat inside': the 'heat inside' the system is increased if the process is endothermic and absorbs heat from the surroundings; it is decreased if the process is exothermic and releases heat into the surroundings.

The enthalpy of a substance increases when its temperature is raised. For example, the enthalpy of 100 g of water is greater at 80 °C than at 20 °C. The change can be measured simply by monitoring the energy that must be supplied as heat to raise the temperature through 60 °C when the sample is open to the atmosphere (or subjected to some other constant pressure), and it is found that $\Delta H \approx +25$ kJ in this instance. The change can be calculated by combining eqn (11) ($\Delta H = q$ at constant pressure) with the definition of the constant pressure heat capacity $C_p(q = C_p \Delta T)$, which gives

$$\Delta H = C_p \Delta T \quad \text{at constant } p \tag{12}$$

For example, when the temperature of 100 g of water (corresponding to 5.55 mol H_2O) is raised from 20 °C to 80 °C (so $\Delta T = +60$ K), the enthalpy of the sample changes by

$$\Delta H = 5.55 \text{ mol} \times 75.29 \text{ J K}^{-1} \text{mol}^{-1} \times (+60 \text{ K}) = +25.1 \text{ kJ}$$

in good agreement with the experimental value. (Any discrepancy between the experimental and calculated values can be traced to a combination of experimental error and the fact that the heat capacity changes with temperature.)

There is a very simple relation between the enthalpy and the internal energy of a system. The formal *definition* of enthalpy is

$$H = U + pV \tag{13}$$

Therefore, to calculate the enthalpy of a system, we simply add the product of its pressure and its volume to the value of its internal energy.

JUSTIFICATION

The consistency of eqns (11) and (13) can be verified quite easily by considering a system that is open to the atmosphere, so that its pressure p is constant. Initially, the enthalpy is

$$H_i = U_i + pV_i$$

and after a reaction has occurred it is

$$H_f = U_f + pV_f$$

The change in enthalpy is therefore the difference between these two quantities, and is

$$H_f - H_i = U_f - U_i + p(V_f - V_i) \quad \text{at constant } p$$

or

$$\Delta H = \Delta U + p\,\Delta V \quad \text{at constant } p$$

However, we know that the change in internal energy is given by eqn (7) (which states that $\Delta U = -p_{ex}\Delta V + q$), and substituting that expression into this one, with $p_{ex} = p$, gives

$$\Delta H = (-p\,\Delta V + q) + p\,\Delta V = q \quad \text{at constant } p$$

which is the same result as in eqn (11). We can therefore conclude that if the enthalpy is defined as in eqn (13), then the change in enthalpy is equal to the heat absorbed at constant pressure.

The importance of the relation between H and U is that it shows that *the enthalpy is a state property*. This conclusion follows from the fact that U, p, and V are all state properties, and so H must be too. That the enthalpy is a state property, and hence has a value that is independent of how the state was prepared, is of the greatest importance in chemistry, as we shall see. The rest of this chapter is in fact an extended illustration of the role of enthalpy in chemistry; throughout it, keep in mind the origin of the name: the *heat inside*.

2.5 The enthalpy of physical change

In this section, we consider two types of physical change: the conversion of one bulk phase (such as a liquid) to another (such as a vapour), and the conversion of individual atoms and molecules to their ions or other fragments.

Bulk change

The vaporization of a liquid, such as the conversion of liquid water to water vapour when a pool of water evaporates at 20 °C or a kettle boils at 100 °C, is an endothermic process because heat must be supplied to bring about the change. At a molecular level, molecules are being driven apart from the grip exerted on them by their attractive interaction with their neighbours, and this process requires an input of energy. One of the body's strategies for maintaining its temperature at about 37 °C is to use the endothermic character of the vaporization of water, for the evaporation of perspiration requires heat and withdraws it from the skin.

Table 2.2 Standard enthalpies of physical change†

Substance	Formula	Freezing point T_f/K	$\Delta H_{fus}^{\ominus}/$ kJ mol^{-1}	Boiling point T_b/K	$\Delta H_{vap}^{\ominus}/$ kJ mol^{-1}
Acetone	CH$_3$COCH$_3$	177.8	5.72	329.4	29.1
Ammonia	NH$_3$	195.3	5.65	239.7	23.4
Argon	Ar	83.8	1.2	87.3	6.5
Benzene	C$_6$H$_6$	278.7	9.87	353.3	30.8
Ethanol	C$_2$H$_5$OH	158.7	4.60	351.5	43.5
Helium	He	3.5	0.02	4.22	0.08
Mercury	Hg	234.3	2.292	629.7	59.30
Methane	CH$_4$	90.7	0.94	111.7	8.2
Methanol	CH$_3$OH	175.5	3.16	337.2	35.3
Water	H$_2$O	273.2	6.01	373.2	40.7

† Values correspond to the transition temperature. For values at 25 °C, use the data in Appendix 1.

The heat that must be supplied at constant pressure per mole of molecules that are vaporized is called the **enthalpy of vaporization** of the liquid, and is denoted ΔH_{vap} (Table 2.2). For example, the heat required to vaporize 1 mol of H$_2$O molecules from the liquid at 25 °C is 44 kJ, so $\Delta H_{vap} = +44$ kJ mol^{-1}. Alternatively, we can report the same information by writing the **thermochemical equation**

$$H_2O(l) \rightarrow H_2O(g) \qquad \Delta H = +44 \text{ kJ}$$

A thermochemical equation shows the enthalpy change that accompanies the conversion of an amount equal to the stoichiometric coefficients in the accompanying chemical equation (1 mol H$_2$O, in this case). If the stoichiometric coefficients in the chemical equation are multiplied through by 2, then the thermochemical equation would be written

$$2H_2O(l) \rightarrow 2H_2O(g) \quad \Delta H = +88 \text{ kJ}$$

This equation signifies that 88 kJ of heat is required to vaporize 2 mol H$_2$O.

Example Determining the enthalpy of vaporization of a liquid
Ethanol, C$_2$H$_5$OH, is brought to the boil at a pressure of 1 atm. When an electric current of 0.682 A from a 12.0 V

supply is passed for 500 s, it is found that 4.33 g of ethanol is vaporized. What is the enthalpy of vaporization of ethanol at its boiling point?

Answer The heat supplied is

$$q = IVt \quad \text{(see eqn (3))}$$

$$= 0.682 \text{ A} \times 12.0 \text{ V} \times 500 \text{ s} = 4.09 \times 10^3 \text{ A V s}$$

$$= 4.09 \text{ kJ} \quad \text{(because } 1 \text{ J} = 1 \text{ A V s)}$$

Because the heat is supplied at constant pressure, q can be identified with the change in enthalpy of the ethanol when it vaporizes:

$$\Delta H = +4.09 \text{ kJ} \quad \text{(an endothermic process)}$$

The enthalpy of vaporization is the change in enthalpy per mole of ethanol molecules vaporized, and the latter can be obtained from the mass vaporized and the molar mass of ethanol (46.07 g mol^{-1}):

$$n = \frac{4.33 \text{ g}}{46.07 \text{ g mol}^{-1}} = 0.0940 \text{ mol}$$

It follows that

$$\Delta H_{vap} = \frac{4.09 \text{ kJ}}{0.0940 \text{ mol}} = +43.5 \text{ kJ mol}^{-1}$$

Exercise 2.6 In a similar experiment, it was found that 1.36 g of boiling benzene, C_6H_6, was vaporized when a current of 0.835 A from a 12.0 V source was passed for 53.5 s. What is the enthalpy of vaporization of benzene at its boiling point?
$$[+30.8 \text{ kJ mol}^{-1}]$$

There are some striking differences: although the value for water is $+44 \text{ kJ mol}^{-1}$, that for methane, CH_4, at its boiling point is only $+8 \text{ kJ mol}^{-1}$. Even allowing for the fact that vaporization is taking place at different temperatures, the difference between the enthalpies of vaporization signifies that water molecules are held together in the bulk liquid much more tightly than methane molecules are in liquid methane. (We shall see in Chapter 10 that the interaction responsible for the low volatility of water is the hydrogen bond.) The high enthalpy of vaporization of water has profound ecological consequences, for it is partly responsible for the survival of the oceans and the generally low humidity of the atmosphere: if only a small amount of heat had to be supplied to vaporize the oceans, the atmosphere would be much more heavily saturated with water vapour than is in fact the case.

Another common phase transition is **fusion**, or melting, as when ice melts to water or iron becomes molten. The enthalpy per mole of molecules that accompanies fusion is called the **enthalpy of fusion**, ΔH_{fus}. Its value for water at 0 °C is +6.01 kJ mol^{-1}, which signifies that 6.01 kJ of energy is needed to melt 1 mol $H_2O(s)$ at 0 °C. Notice that the enthalpy of fusion of water is much less than its enthalpy of vaporization. In the latter transition, the molecules become completely separated from each other, whereas when a solid melts, the molecules are merely loosened without separating completely (Fig. 2.11), so the change in enthalpy is much smaller.

The reverse of vaporization is **condensation** and the reverse of fusion is **freezing**. The enthalpy changes are, respectively, the negative of the enthalpies of vaporization and fusion, because the heat that is supplied to bring about vaporization or melting is released when the substance condenses or freezes. In general *the enthalpy change of a reverse transition is the negative of the enthalpy change of the forward transition* (under the same conditions of temperature and pressure):

$$H_2O(s) \rightarrow H_2O(l) \qquad \Delta H = +6.01 \text{ kJ}$$

$$H_2O(l) \rightarrow H_2O(s) \qquad \Delta H = -6.01 \text{ kJ}$$

This relation follows from the fact that H is a state property, and so it must return to the same value if a forward change is followed by the reverse of that change. The high enthalpy of vaporization of water (+44 kJ mol^{-1}), signifying a strongly endothermic process, implies that the condensation of water ($\Delta H = -44$ kJ mol^{-1}) is a strongly exothermic process. That exothermicity is the origin of the ability of steam to scald severely.

The direct conversion of a solid to a vapour is called **sublimation**. Sublimation can be observed on a cold, frosty morning, when frost vanishes as vapour without first melting. The vaporization of solid carbon dioxide ('dry ice') is another example of sublimation. The molar enthalpy change accompanying sublimation is called the **enthalpy of sublimation**, ΔH_{sub}. Because enthalpy is a state property, the same change in enthalpy must be obtained both in the *direct* conversion of solid to vapour and in the *indirect* conversion, in which the solid first melts to the liquid and then that liquid vaporizes (**1**):

$$\Delta H_{sub} = \Delta H_{fus} + \Delta H_{vap} \tag{14}$$

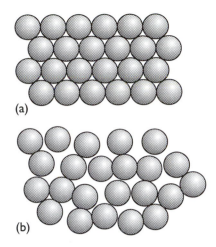

(a)

(b)

Fig. 2.11 When a solid (a) melts to a liquid (b), the molecules separate from one another only slightly, the intermolecular interactions are reduced only slightly, and there is only a small change in enthalpy. When a liquid vaporizes (not shown), the molecules are separated by a considerable distance, the intermolecular forces are reduced almost to zero, and the change in enthalpy is much greater.

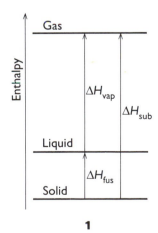

1

Exercise E2.7 Calculate the enthalpy of sublimation of ice at 0 °C from its enthalpy of fusion at 0 °C (+6.01 kJ mol^{-1}) and the enthalpy of vaporization of water at 0 °C (+45.07 kJ mol^{-1}). Note that we must use the values of the enthalpy changes at the same temperature.

[+51.08 kJ mol^{-1}]

So far we have used the fact that enthalpy is a state property in two different ways:

1. Because enthalpy is a state property, the enthalpy change in a reverse process is the negative of the enthalpy change in the forward process between the same two states:

$$\Delta H(\text{reverse}) = -\Delta H(\text{forward})$$

2. Because enthalpy is a state function, the enthalpy change in a direct route between two states is equal to the sum of the enthalpy change for a sequence of changes between the same two sets:

$$\Delta H(\text{indirect route}) = \Delta H(\text{direct route})$$

Atomic and molecular change

Another group of enthalpy changes that we shall employ quite often in the following pages are those accompanying changes to individual atoms and molecules. Among the most important of these properties is the **enthalpy of ionization**, ΔH_{ion}, the molar enthalpy change accompanying the removal of an electron from a gas-phase atom (or ion). For example, because

$$H(g) \rightarrow H^+(g) + e^-(g) \qquad \Delta H = +1312 \text{ kJ}$$

the enthalpy of ionization of hydrogen atoms is reported as $+1312 \text{ kJ mol}^{-1}$. This value signifies that 1312 kJ of heat must be supplied to ionize 1 mol of H atoms. Ionization of neutral atoms is endothermic in all cases, so all enthalpies of ionization of neutral atoms are positive (Table 2.3).

It is often convenient to consider a succession of ionizations, such as the conversion of magnesium atoms to Mg^+ ions, and then ionization of Mg^+ ions to Mg^{2+} ions, and so on. The successive molar enthalpy changes are called, respectively, the **first ionization enthalpy**, $\Delta H_{\text{ion},1}$, the **second ionization enthalpy**, $\Delta H_{\text{ion},2}$, and so on. For magnesium, these enthalpies refer to the processes

$$Mg(g) \rightarrow Mg^+(g) + e^-(g) \qquad \Delta H = +738 \text{ kJ}$$
$$Mg^+(g) \rightarrow Mg^{2+}(g) + e^-(g) \qquad \Delta H = +1451 \text{ kJ}$$

Note that the second ionization enthalpy is larger than the first: it takes more energy to separate an electron from a positively charged ion than from the neutral atom. Note also that enthalpies of ionization refer to the ionization of the *gas-phase atom or ion,* not to the

Table 2.3   First and second (and some higher) ionization enthalpies of the elements in kilojoules per mole (kJ mol^{-1})

				H			He
				1310			2370
							5250
Li	**Be**	**B**	**C**	**N**	**O**	**F**	**Ne**
519	900	799	1090	1400	1310	1680	2080
7300	1760	2420	2350	2860	3390	3370	3950
		14 800	3660				
			25 000				
Na	**Mg**	**Al**	**Si**	**P**	**S**	**Cl**	**Ar**
494	736	557	786	1060	1000	1260	1520
4560	1450	1820					
	7740	2740					
		11 600					
K	**Ca**	**Ga**	**Ge**	**As**	**Se**	**Br**	**Kr**
418	590	577	762	966	941	1140	1350
3070	1150						
	4940						
Rb	**Sr**	**In**	**Sn**	**Sb**	**Te**	**I**	**Xe**
402	548	556	707	833	870	1010	1170
2650	1060						
	4120						
Cs	**Ba**	**Tl**	**Pb**	**Bi**	**Po**	**At**	**Rn**
376	502	812	920	1040	812	920	1040
2420	966						
3300	3390						

ionization of an atom or ion in a solid.

Example Combining enthalpy changes

The enthalpy of sublimation of magnesium at 25 °C is $+148\ \text{kJ mol}^{-1}$. What quantity of heat must be supplied to 1.00 g of solid magnesium metal to produce a gas of Mg^{2+} ions and electrons?

Answer The overall strategy is to calculate the enthalpy change for the process

$$Mg(s) \rightarrow Mg^{2+}(g) + 2e^{-}(g)$$

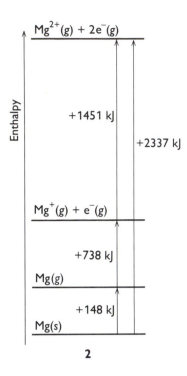

Enthalpy

$Mg^{2+}(g) + 2e^-(g)$

$+1451$ kJ

$+2337$ kJ

$Mg^+(g) + e^-(g)$

$+738$ kJ

$Mg(g)$

$+148$ kJ

$Mg(s)$

2

by combining the enthalpies of the steps into which this overall process may be divided:

		$\Delta H/kJ$
Sublimation	$Mg(s) \rightarrow Mg(g)$	$+148$
First ionization	$Mg(g) \rightarrow Mg^+(g) + e^-(g)$	$+738$
Second ionization	$Mg^+(g) \rightarrow Mg^{2+}(g) + e^-(g)$	$+1451$
Overall (sum)	$Mg(s) \rightarrow Mg^{2+}(g) + 2e^-(g)$	$+2337$

These processes are illustrated diagrammatically in **2** (in which enthalpy increases upwards). Because the molar mass of Mg is $24.31\ g\ mol^{-1}$, 1.0 g of magnesium corresponds to

$$n = \frac{1.00\ g}{24.31\ g\ mol^{-1}} = 0.0411\ mol$$

Therefore, the heat that must be supplied is

$$q = 0.0411\ mol \times (+2337\ kJ\ mol^{-1}) = +96.1\ kJ$$

This quantity of heat is approximately the same as that needed to vaporize about 43 g of boiling water.

Exercise E2.8 The enthalpy of sublimation of aluminium is $+326\ kJ\ mol^{-1}$. Use this information and the ionization enthalpies given in Table 2.3 to calculate the quantity of heat that must be supplied to convert 1.00 g of solid aluminium metal to a gas of Al^{3+} ions and electrons at 25 °C.

[+203 kJ]

The reverse of ionization is **electron gain**, and the corresponding molar enthalpy change is called the **electron-gain enthalpy**, ΔH_{eg}. For example, because experiments show that

$$Cl(g) + e^-(g) \rightarrow Cl^-(g) \qquad \Delta H = -349\ kJ$$

it follows that the electron-gain enthalpy of Cl atoms is $-349\ kJ\ mol^{-1}$. Notice that electron gain by Cl is an *exo*thermic process, so heat is released when a Cl atom captures an electron and forms a gas-phase Cl^- ion. It can be seen from Table 2.4, which lists a number of electron-gain enthalpies, that some electron gains are exothermic, others are endothermic. For example, electron gain by an O^- ion is strongly endothermic because it takes energy to push an electron on to an already negatively charged species:

$$O^-(g) + e^-(g) \rightarrow O^{2-}(g) \qquad \Delta H = +844\ kJ$$

The **electron affinity** E_A of a species is the negative of its electron-gain enthalpy. (The convention is based on the view that an exothermic

Table 2.4 *Electron-gain enthalpies of the main-group elements, $\Delta H_{eg}/kJ\ mol^{-1}$)†*

Li	Be	B	C	H	N	O	F	He
				-72				$+21$
Li	Be	B	C	N	O	F	Ne	
-60	$+60$	-28	-122	$+7$	-141	-328	$+29$	
					$+844$			
Na	Mg	Al	Si	P	S	Cl	Ar	
-53	$+232$	-44	-120	-72	-200	-349	$+35$	
					$+532$			
K	Ca	Ga	Ge	As	Se	Br	Kr	
-48	$+156$	-29	-117	-77	-195	-325	$+39$	
Rb	Sr	In	Sn	Sb	Te	I	Xe	
-47	$+52$	-29	-121	-101	-190	-295	$+41$	

† Where two values are given, the first refers to the formation of the ion X^- from the neutral atom X; the second refers to the formation of X^{2-} from X^-.

electron gain, corresponding to a positive electron affinity, represents an 'affinity' of the species for the incoming electron.) Thus, the electron affinity of Cl is $+349\ kJ\ mol^{-1}$. It is sometimes convenient to express discussions in terms of electron affinities, but we shall employ the electron-gain enthalpy in most cases.

The final atomic and molecular process that we need consider is the **dissociation** of a chemical bond, as in the process

$$H—Cl(g) \rightarrow H(g) + Cl(g) \qquad \Delta H = +431\ kJ$$

The corresponding molar enthalpy change is called the **bond enthalpy**, so we would report the bond enthalpy of H—Cl as $431\ kJ\ mol^{-1}$ (because all bond enthalpies are positive, they are normally reported without the + sign). Some values are given in Table 2.5. Note that the nitrogen–nitrogen bond in molecular nitrogen, N_2, is very strong, at $945\ kJ\ mol^{-1}$, which helps to account for the chemical inertness of nitrogen and the fact that it dilutes the reactive oxygen of the atmosphere. In contrast, the fluorine–fluorine bond in molecular fluorine, F_2, is very weak, at $155\ kJ\ mol^{-1}$; the weakness of this bond contributes to the high reactivity of elemental fluorine. However, bond enthalpy is not the full reason, because although the bond in molecular iodine is even weaker, I_2 is less reactive than F_2. Most other bonds lie in the range between these two (but the bond in carbon monoxide is stronger, and as we have remarked there are some bonds that are even weaker than that in molecular fluorine).

**Table 2.5 Selected bond enthalpies,
$\Delta H(A—B)/kJ\ mol^{-1}$**

Diatomic molecules

H—H	436	O=O	497	F—F	155	H—F	565
		N≡N	945	Cl—Cl	242	H—Cl	431
		O—H	428	Br—Br	193	H—Br	366
		C=O	1074	I—I	151	H—I	299

Polyatomic molecules

H—CH$_3$	435	H—NH$_2$	431	H—OH	492
H—C$_6$H$_6$	469	O$_2$N—NO$_2$	57	HO—OH	213
H$_3$C—CH$_3$	368	O=CO	531	HO—CH$_3$	377
H$_2$C=CH$_2$	699			Cl—CH$_3$	452
HC=CH	962			Br—CH$_3$	293
				I—CH$_3$	234

A complication when dealing with bond enthalpies is that their value depends on the molecule in which the two linked atoms occur. For instance, the total enthalpy change for the process

$$H_2O(g) \rightarrow 2H(g) + O(g) \qquad \Delta H = +927\ kJ$$

is not twice an OH bond enthalpy even though two OH bonds are dissociated. There are in fact two *different* dissociation steps. In the first step, an OH bond is broken in an H$_2$O molecule:

$$H_2O(g) \rightarrow HO(g) + H(g) \qquad \Delta H = +499\ kJ$$

In the second step, the OH bond is broken in an OH radical:

$$HO(g) \rightarrow H(g) + O(g) \qquad \Delta H = +428\ kJ$$

The sum of the two steps is the complete atomization of the molecule. As can be seen from this example, the O—H bonds in H$_2$O and HO have similar but not identical bond enthalpies. Although accurate calculations must use bond enthalpies for the molecule in question, it is sometimes possible to make estimates by using **mean bond enthalpies**, B, which are the averages of bond enthalpies over a related series of compounds. For example, the mean HO bond enthalpy, $B(H—O) = 463\ kJ\ mol^{-1}$, is the mean of the two HO bond enthalpies in H$_2$O and several other similar compounds, including methanol, CH$_3$OH.

Example Using mean bond enthalpies
Estimate the enthalpy change that accompanies the reaction

$$C(s, graphite) + 2H_2(g) + \tfrac{1}{2}O_2(g) \rightarrow CH_3OH(l)$$

Table 2.6 Mean bond enthalpies, $B/kJ\ mol^{-1}$†

	H	C	N	O	F	Cl	Br	I	S	P	Si
H	436										
C	412	348 (1)									
		612 (2)									
		518 (a)									
N	388	305 (1)	163 (1)								
		613 (2)	409 (2)								
		890 (3)	945 (3)								
O	463	360 (1)	157	146 (1)							
		743 (2)		497 (2)							
F	565	484	270	185	155						
Cl	431	338	200	203	254	242					
Br	366	276				219	193				
I	299	238				210	178	151			
S	338	259			496	250	212		264		
P	322									200	
Si	318			466							226

† Values are for single bonds except where otherwise stated (in parentheses).
(a) Denotes aromatic.

in which liquid methanol is formed from its elements at 25 °C. Use information from Appendix 1 and bond enthalpy data from Tables 2.5 and 2.6.

Answer This calculation requires data from several sources, as can be seen by breaking the reaction into three steps. The first step is the atomization of all the reactants

$$C(s, graphite) + 2H_2(g) + \tfrac{1}{2}O_2(g) \rightarrow C(g) + 4H(g) + O(g)$$

The next step is the formation of gaseous methanol from the atoms:

$$C(g) + 4H(g) + O(g) \rightarrow CH_3OH(g)$$

The final step is the condensation of the vapor to a liquid:

$$CH_3OH(g) \rightarrow CH_3OH(l)$$

which is the reverse of its vaporization at 25 °C. For the first step we need the enthalpy of sublimation of carbon (graphite) and the bond enthalpies of H_2 and O_2:

	$\Delta H/\text{kJ}$	
Atomization of graphite,		
$\quad$ C(s, graphite) $\rightarrow$ C(g)	+716.68	(from Appendix 1)
Dissociation of $2H_2$,		
$\quad$ $2H_2(g) \rightarrow 4H(g)$	+871.88	(from Table 2.5)
Dissociation of $\frac{1}{2}O_2$,		
$\quad$ $\frac{1}{2}O_2(g) \rightarrow O(g)$	+249.17	(from Table 2.5)
Overall	+1837.73	

In the second step, three CH bonds, one CO bond, and one OH bond are formed, and we can estimate their enthalpies from the mean values of such bonds. The enthalpy change for bond formation (the reverse of dissociation) is the negative of the mean bond enthalpy B (obtained from Table 2.6):

Formation of three C—H bonds	−1236 kJ
Formation of one C—O bond	−360 kJ
Formation of one O—H bond	−463 kJ
Overall	−2059 kJ

The final stage of the reaction is

$\quad$ condensation of methanol,

$\qquad$ $CH_3OH(g) \rightarrow CH_3OH(l)$ $\qquad$ −38.00 kJ

(from Table 2.2). The sum of the enthalpy changes is

$$\Delta H = +1837.73 \text{ kJ} + (-2059 \text{ kJ}) + (-38.00 \text{ kJ})$$

$$= -259 \text{ kJ}$$

The experimental value is −239.00 kJ.

Exercise E2.9 Estimate the enthalpy change for the combustion of methanol in the reaction

$$2CH_3OH(l) + 3O_2(g) \rightarrow 2CO_2(g) + 4H_2O(g)$$

$\qquad$ [−988 kJ; the experimental value is −1452 kJ]

2.6 The enthalpy of chemical change

In the rest of this chapter we shall be considering **thermochemistry**, the study of the heat required or absorbed by chemical reactions, such as the heat output of a combustion or respiration reaction. The central property in all the following discussions will be the 'standard enthalpy of reaction'. We shall introduce this concept in two stages, first as the 'enthalpy of reaction' and then as its 'standard' value.

> The **enthalpy of reaction** is the enthalpy change accompanying a reaction as specified in the accompanying chemical equation.

For example, for the hydrogenation in which hydrogen is added to ethene,

$$CH_2{=}CH_2(g) + H_2(g) \rightarrow CH_3CH_3(g) \qquad \Delta H = -137 \text{ kJ at } 25\,°C$$

the reaction enthalpy of -137 kJ signifies that the enthalpy of the system decreases by 137 kJ (and, if the reaction takes place at constant pressure, that 137 kJ of heat is released into the surroundings) when 1 mol $CH_2{=}CH_2$ combines with 1 mol H_2 at $25\,°C$. As in the case of all thermochemical equations, if the stoichiometric coefficients are all multiplied by a factor, so too is the enthalpy of reaction:

$$3CH_2{=}CH_2(g) + 3H_2(g) \rightarrow 3CH_3CH_3(g) \qquad \Delta H = -411 \text{ kJ at } 25\,°C$$

This thermochemical equation tells us that 411 kJ of heat will be released into the surroundings when 3 mol $CH_2{=}CH_2$ is hydrogenated.

The numerical value of the enthalpy of reaction depends on the conditions—the state of purity of the reactants and products, the pressure, and the temperature—under which the reaction takes place. Chemists have therefore found it convenient to report their data for a set of standard conditions at the temperature of their choice:

> **Standard conditions** are pure, unmixed substances at a pressure of exactly 1 bar.

(Solutions are a special case, and are dealt with in Section 4.3.) For example, the standard state of hydrogen is the pure gas at 1 bar, the standard state of calcium carbonate is the pure solid at 1 bar. The physical state needs to be specified because we can speak of the standard states of the solid, liquid, or vapour phases of methanol, which are the pure solid, the pure liquid, or the pure vapour, respectively, at 1 bar in each case. The temperature is not a part of the definition of a standard state, and it is possible to speak of the standard state of hydrogen gas at 100 K, 273.15 K, or any other temperature. It is conventional, though, for data to be reported at 298.15 K (25.00 °C, the conventional temperature $\mathcal{F}$), and from now on, unless specified otherwise, all data will be for substances at that temperature.

With the concept of standard state in mind, we can combine it with the definition of enthalpy of reaction:

The **standard enthalpy of reaction**, $\Delta H^{\ominus}$, is the enthalpy of reaction for the conversion of the reactants in their standard states into products in their standard states.

For example, for the reaction

$$2H_2(g) + O_2(g) \rightarrow 2H_2O(l) \qquad \Delta H^{\ominus} = -572 \text{ kJ at } 25°C$$

The value quoted shows that when 2 mol H_2 as pure hydrogen gas at 1 bar combines with 1 mol O_2 as pure oxygen gas at 1 bar to form 2 mol H_2O as pure liquid water at 1 bar, the initial and final states being at 25 °C, then the enthalpy decreases by 572 kJ, and (at constant pressure) 572 kJ of heat is released into the surroundings.

One commonly encountered reaction is **combustion**, the reaction of a substance with oxygen, as in the combustion of magnesium

$$2Mg(s) + O_2(g) \rightarrow 2MgO(s) \qquad \Delta H^{\ominus} = -1204 \text{ kJ}$$

or the combustion of methane, as in a natural-gas flame

$$CH_4(g) + 2O_2(g) \rightarrow CO_2(g) + 2H_2O(l) \qquad \Delta H^{\ominus} = -890 \text{ kJ}$$

The **standard enthalpy of combustion** $\Delta H^{\ominus}_{com}$ is the standard reaction enthalpy per mole of combustible substance. In these two examples, we would write $\Delta H^{\ominus}_{com}(Mg, s) = -602 \text{ kJ mol}^{-1}$ for the standard enthalpy of combustion of solid magnesium and $\Delta H^{\ominus}_{com}(CH_4, g) = -890 \text{ kJ mol}^{-1}$ for methane gas. Some typical values are given in Table 2.7.

Enthalpies of combustion are widely used in thermochemistry, as we shall see in the following section. One immediate application is to judge the suitability of a fuel. For example, from the value of the standard enthalpy of combustion for methane, we know that for each mole of CH_4 supplied to a furnace, 890 kJ of heat can be released, whereas for each mole of iso-octane (C_8H_{18}, 2,2,4–trimethylpentane **3**, a typical component of gasoline) supplied to an internal combustion engine, 5461 kJ of heat is released, because $\Delta H^{\ominus}_{com}(C_8H_{18}, l) = -5461 \text{ kJ mol}^{-1}$. The much larger value for iso-octane is a consequence of each molecule having eight carbon atoms to contribute to the formation of carbon dioxide, whereas methane has only one.

Table 2.7 Standard enthalpies of combustion

Substance†	Formula	$\Delta H^{\ominus}_{com}/$ kJ mol^{-1}
Acetylene	$C_2H_2(g)$	−1300
Benzene	$C_6H_6(l)$	−3268
Benzoic acid	$C_6H_5COOH(s)$	−3227
Carbon	$C(s, \text{graphite})$	−394
Carbon monoxide	$CO(g)$	−394
Ethanol	$C_2H_5OH(l)$	−1368
Glucose	$C_6H_{12}O_6(s)$	−2808
Hydrogen	$H_2(g)$	−286
Methane	$CH_4(g)$	−890
Methanol	$CH_3OH(l)$	−726
Octane	$C_8H_{18}(l)$	−5471
Phenol	$C_6H_5OH(s)$	−3054
Propane	$C_3H_8(g)$	−2220
Sucrose	$C_{12}H_{22}O_{11}(s)$	−5645
Toluene	$C_6H_5CH_3(l)$	−3910
Urea	$CO(NH_2)_2(s)$	−632

† C is converted to $CO_2(g)$, H to $H_2O(l)$, and N to $N_2(g)$.

3 2,2,4-trimethylpentane

Table 2.8 *Thermochemical properties of some fuels*

Fuel	Combustion equation	$\Delta H_c^{\ominus}/$ kJ mol^{-1}	Specific enthalpy/ kJ g^{-1}	Enthalpy density†/ kJ L^{-1}
Hydrogen	$2H_2(g) + O_2(g)$ $\rightarrow 2H_2O(l)$	-286	142	13
Methane	$CH_4(g) + 2O_2(g)$ $\rightarrow CO_2(g) + 2H_2O(l)$	-890	55	40
Octane	$2C_8H_{18}(l) + 25O_2(g)$ $\rightarrow 16CO_2(g) + 18H_2O(l)$	-5471	48	3.8×10^4
Methanol	$2CH_3OH(l) + 3O_2(g)$ $\rightarrow 2CO_2(g) + 4H_2O(l)$	-726	23	1.8×10^4

† At atmospheric pressures and room temperature.

The heating power of fuels is often expressed in terms of the **specific enthalpy**, the heat released per unit mass of compound (typically in kilojoules per gram, kJ g^{-1}, Table 2.8). Because 1 mol CH_4 releases 890 kJ of heat and the mass of 1 mol CH_4 is 16.04 g, the specific enthalpy of methane is

$$\frac{890 \text{ kJ}}{16.04 \text{ g}} = 55.5 \text{ kJ g}^{-1}$$

(All specific enthalpies are positive, and are typically written without the + sign.) Likewise, because 1 mol C_8H_{18} releases 5461 kJ and the mass of 1 mol C_8H_{18} is 114.23 g, the specific enthalpy of iso-octane is 47.8 kJ g^{-1}. Therefore, on a mass basis, it is more economical to carry fuel as methane than as iso-octane. The specific enthalpy of hydrogen is even more superior: $\Delta H_{com}^{\ominus}(H_2, g) = -286$ kJ mol^{-1} and the molar mass of H_2, 2.016 g mol^{-1}, give a value of 142 kJ g^{-1}, a much larger value than for methane. This high value is one of the reasons why liquid hydrogen is adopted as a rocket fuel, for then mass is of critical importance.

The combination of reaction enthalpies

The standard enthalpies of reactions can be combined together to obtain the enthalpy of another reaction. The basis of this application is widely called 'Hess's law':

Hess's law: The standard enthalpy of a reaction is the sum of the standard enthalpies of the reactions into which the overall reaction may be divided.

The individual steps need not be actual reactions that can be carried out in the laboratory—they may be entirely hypothetical reactions, the only requirement being that they should balance. Hess's law is not really another law of thermochemistry: it is simply a recognition of the fact that enthalpy is a state property, and therefore the value of $\Delta H^{\ominus}$ for the direct path between reactants and products must be equal to the value of $\Delta H^{\ominus}$ for an indirect path in which a series of intermediate substances are first formed and then used in subsequent steps.

Example Using Hess's law
Given the thermochemical equations

$$C_3H_6(g) + H_2(g) \rightarrow C_3H_8(g) \qquad \Delta H^{\ominus} = -124 \text{ kJ}$$

$$C_3H_8(g) + 5O_2(g) \rightarrow 3CO_2(g) + 4H_2O(l) \qquad \Delta H^{\ominus} = -2220 \text{ kJ}$$

where C_3H_6 denotes propene ($CH_2{=}CHCH_3$) and C_3H_8 denotes propane ($CH_3CH_2CH_3$), calculate the standard enthalpy of combustion of propene.
Answer We need to add or subtract the thermochemical equations (including the values of $\Delta H^{\ominus}$) together with any others that are needed (from Appendix 1) so as to reproduce the thermochemical equation for the reaction required, which is

$$C_3H_6(g) + \tfrac{9}{2}O_2(g) \rightarrow 3CO_2(g) + 3H_2O(l) \qquad \Delta H^{\ominus} = ?$$

This equation can be recreated from the following sum:

	$\Delta H^{\ominus}/\text{kJ}$
$C_3H_6(g) + H_2(g) \rightarrow C_3H_8(g)$	-124
$C_3H_8(g) + 5O_2(g) \rightarrow 3CO_2(g) + 4H_2O(l)$	-2220
$H_2O(l) \rightarrow H_2(g) + \tfrac{1}{2}O_2(g)$	$+286$
$C_3H_6(g) + \tfrac{9}{2}O_2(g) \rightarrow 3CO_2(g) + 3H_2O(l)$	-2058

It follows that the standard enthalpy of combustion of propene is $-2058 \text{ kJ mol}^{-1}$.

Exercise E2.10 Calculate the standard enthalpy of the reaction

$$C_6H_6(l) + 3H_2(g) \rightarrow C_6H_{12}(l)$$

from the standard enthalpies of combustion of benzene (C_6H_6) and cyclohexane (C_6H_{12}).

$$[-205 \text{ kJ}]$$

Standard enthalpies of formation

Whereas it would be possible (but lengthy) to list the standard enthalpies of all reactions, it is much more economical to list quantities relating to individual species that can be used to calculate the standard enthalpies of the reactions in which they participate. Consequently, we introduce the 'standard enthalpy of formation' of a substance:

The **standard enthalpy of formation**, $\Delta H_f^{\ominus}$, of a substance is the standard enthalpy of the reaction in which it is formed from its elements in their reference states, expressed as an enthalpy change per mole of substance.

The standard enthalpies of formation of elements in their reference states are zero by definition (because their formation is the null reaction: element $\rightarrow$ element). By the **reference state** of an element is meant the most stable form of the element under the prevailing conditions. The reference states of some common elements at 25 °C are shown in the following table:

Hydrogen, nitrogen, oxygen, fluorine, chlorine	Gas
Bromine, mercury	Liquid
All metallic elements other than mercury	Solid
Carbon	Graphite
Sulfur	Rhombic sulfur
Phosphorus	White phosphorus†
Tin	White tin

† White phosphorus is not the most stable form of phosphorus, but it is more reproducible than the other allotropes, and so is used instead of one of them.

73

For example, the standard enthalpy of formation of liquid water (at 25 °C, as always in this text) is obtained from the thermochemical equation

$$H_2(g) + \tfrac{1}{2}O_2(g) \rightarrow H_2O(l) \qquad \Delta H^{\ominus} = -286\ kJ$$

by expressing the enthalpy reaction as a quantity per mole of H_2O:

$$\Delta H_f^{\ominus}(H_2O, l) = -286\ kJ\ mol^{-1}$$

Similarly, the standard enthalpy of formation of liquid carbon disulfide is obtained from the experimentally determined thermochemical equation

$$C(s,\ graphite) + 2S(s,\ rhombic) \rightarrow CS_2(l) \qquad \Delta H^{\ominus} = +90\ kJ$$

and is

$$\Delta H_f^{\ominus}(CS_2, l) = +90\ kJ\ mol^{-1}$$

The values of some standard enthalpies of formation at 25 °C are given in Table 2.9, and a longer list is given in Appendix 1 at the end of the book.

The reference states of the elements define a thermochemical 'sea level' and the enthalpies of formation are thermochemical 'altitudes' above or below sea level. Compounds that have negative standard enthalpies of formation (such as water) are classified as **exothermic compounds** for they lie at a lower enthalpy than their component elements (they lie below 'sea level'). Compounds that have positive standard enthalpies of formation (such as carbon disulfide) are classified as **endothermic compounds**, and possess a higher enthalpy than their component elements (they lie above 'sea level').

The great usefulness of standard enthalpies of formation is that they can be combined together to give the standard enthalpy of any reaction. To do so, we regard a reaction as proceeding by 'unforming' the reactants into their elements and then forming those elements into the products. The value of $\Delta H^{\ominus}$ for the overall reaction is the sum of these 'unforming' and forming enthalpies. Because 'unforming' is the reverse of forming, the enthalpy of an unforming step is the negative of the enthalpy of formation. For example, the enthalpy of the reaction between hydrazoic acid, HN_3, and nitrogen monoxide, NO (nitric oxide)

$$2HN_3(l) + 2NO(g) \rightarrow H_2O_2(l) + 4N_2(g) \qquad \Delta H^{\ominus} = ?$$

is the sum of the contributions given in the following table.

Unform 2 mol $HN_3(l)$:

$\quad 2HN_3(l) \rightarrow N_2(g) + \tfrac{3}{2}H_2(g) \quad 2\,mol \times \{-\Delta H_f^{\ominus}(HN_3, l)\} = -528.0\ kJ$

Unform 2 mol $NO(g)$:

$\quad 2NO(g) \rightarrow N_2(g) + O_2(g) \quad 2\,mol \times \{-\Delta H_f^{\ominus}(NO, g)\} = -180.50\ kJ$

Form 1 mol $H_2O_2(l)$:

$\quad H_2(g) + O_2(g) \rightarrow H_2O_2(l) \quad 1\,mol \times \Delta H_f^{\ominus}(H_2O_2, l) = -187.78\ kJ$

Table 2.9 Some standard enthalpies of formation at 25 °C

Substance†	Formula	$\Delta H_f^{\ominus}/\text{kJ mol}^{-1}$
Inorganic compounds		
Ammonia	$NH_3(g)$	−46.11
Ammonium nitrate	$NH_4NO_3(s)$	−365.56
Carbon monoxide	$CO(g)$	−110.53
Carbon disulfide	$CS_2(l)$	+89.70
Carbon dioxide	$CO_2(g)$	−393.51
Dinitrogen tetroxide	$N_2O_4(g)$	+9.16
Dinitrogen oxide	$N_2O(g)$	+82.05
Hydrogen chloride	$HCl(g)$	−92.31
Hydrogen fluoride	$HF(g)$	−271.1
Hydrogen sulfide	$H_2S(g)$	−20.63
Nitric acid	$HNO_3(l)$	−174.10
Nitric oxide	$NO(g)$	+90.25
Nitrogen dioxide	$NO_2(g)$	+33.18
Sodium chloride	$NaCl(s)$	−411.15
Sulfur dioxide	$SO_2(g)$	−296.83
Sulfur trioxide	$SO_3(g)$	−395.72
Sulfuric acid	$H_2SO_4(l)$	−813.99
Water	$H_2O(l)$	−285.83
	$H_2O(g)$	−241.82
Organic compounds		
Acetylene	$C_2H_2(g)$	+226.73
Benzene	$C_6H_6(l)$	+49.0
Ethane	$C_2H_6(g)$	−84.68
Ethanol	$C_2H_5OH(l)$	−277.69
Ethylene	$C_2H_4(g)$	+52.26
Glucose	$C_6H_{12}O_6(s)$	−1268
Methane	$CH_4(g)$	−74.81
Methanol	$CH_3OH(l)$	−238.86
Sucrose	$C_{12}H_{22}O_{11}(s)$	−2222

† A longer list is given in Appendix 1 at the end of the book.

The sum of the thermochemical equations is

$$2HN_3(l) + 2NO(g) \rightarrow H_2O_2(l) + 4N_2(g)$$

$$\Delta H^{\ominus} = -528.0\,\text{kJ} + (-180.50\,\text{kJ}) + (-187.78\,\text{kJ}) = -896.3\,\text{kJ}$$

Therefore, even if it is impossible to measure the enthalpy of the reaction, it may be possible to calculate it from data that have been compiled in other measurements. Indeed, one of the great strengths of thermodynamics is its ability to provide methods for calculating a desired property from information compiled from other, apparently unrelated, experiments.

The calculation we have just illustrated has the form

$$\Delta H^{\ominus} = \sum_{\text{Products}} n\,\Delta H_f^{\ominus}(\text{products}) - \sum_{\text{Reactants}} n\,\Delta H_f^{\ominus}(\text{reactants}) \quad (15)$$

where the n are the stoichiometric coefficients in the thermochemical equation expressed in moles. For example, the calculation illustrated above would be written

$$\Delta H^{\ominus} = \{1\,\text{mol} \times \Delta H_f^{\ominus}(H_2O_2, l) + 4\,\text{mol} \times \Delta H_f^{\ominus}(N_2, g)\}$$
$$- \{2\,\text{mol} \times \Delta H_f^{\ominus}(HN_3, l) + 2\,\text{mol} \times \Delta H_f^{\ominus}(NO, g)\}$$

which gives the same result as before once we note that $\Delta H_f^{\ominus}(N_2, g) = 0$.

Example Using standard enthalpies of formation
Calculate the standard enthalpy of combustion of liquid benzene from the standard enthalpies of formation of the components of the reaction.

Answer The thermochemical reaction is

$$C_6H_6(l) + \tfrac{15}{2}O_2(g) \rightarrow 6CO_2(g) + 3H_2O(l) \qquad \Delta H_f^{\ominus} = ?$$

Then, from eqn (15),

$$\Delta H^{\ominus} = \{6\,\text{mol} \times \Delta H_f^{\ominus}(CO_2, g) + 3\,\text{mol} \times \Delta H_f^{\ominus}(H_2O, l)\}$$
$$- \{1\,\text{mol} \times \Delta H_f^{\ominus}(C_6H_6, l) + \tfrac{15}{2}\,\text{mol} \times \Delta H_f^{\ominus}(O_2, g)\}$$
$$= \{6 \times (-393.51\,\text{kJ}) + 3 \times (-285.83\,\text{kJ})\}$$
$$- \{1 \times (+49.0\,\text{kJ}) + 0\}$$
$$= -3268\,\text{kJ}$$

It follows that the standard enthalpy of combustion of liquid benzene is $-3268\,\text{kJ mol}^{-1}$

Exercise E2.11 Use standard enthalpies of formation to calculate the enthalpy of combustion of propane gas (C_3H_8) to carbon dioxide and water vapour.

$$[-2220\,\text{kJ mol}^{-1}]$$

2.7 The variation of enthalpy with temperature

We have enough information now to be able to predict the enthalpy at any temperature from its value at another temperature. For example, we might want to know the enthalpy of a particular reaction at body temperature, 37 °C, but may have data available for 25 °C; similarly, we might want to determine whether the oxidation of glucose is more or less exothermic when it takes place inside an arctic fish (which inhabits water at 0 °C) than at mammalian body temperatures.

Consider, as a simple example of this type of calculation, the reaction

$$2H_2(g) + O_2(g) \rightarrow 2H_2O(l) \quad \Delta H^{\ominus}$$

where the standard enthalpy of reaction is known at one temperature (for example, at 25 °C, from tables). The enthalpy of reaction can be expressed as the difference in the enthalpy (not, in this case, the enthalpy of formation) of each substance:

$$\Delta H^{\ominus} = 2\,\text{mol} \times H^{\ominus}(H_2O, l) - \{2\,\text{mol} \times H^{\ominus}(H_2, g)$$
$$+ 1\,\text{mol} \times H^{\ominus}(O_2, g)\}$$

Then, if the reaction takes place at a higher temperature, the enthalpy of each substance would be increased by $C_p \Delta T$, where C_p is the heat capacity of the substance. For example, the enthalpy of water changes to

$$H^{\ominus\prime}(H_2O, l) = H^{\ominus}(H_2O, l) + C_p(H_2O, l)\,\Delta T$$

The standard enthalpy of reaction then changes to

$$\Delta H^{\ominus\prime} = \Delta H^{\ominus} + \Delta C_p\,\Delta T \tag{16}$$

where

$$\Delta C_p = 2C_p(H_2O, l) - \{2C_p(H_2, g) + C_p(O_2, g)\}$$

is the difference between the heat capacities of the products and the reactants. Equation (16) is known as **Kirchhoff's law**. In general,

$$\Delta C_p = \sum_{\text{Products}} nC_p(\text{products}) - \sum_{\text{Reactants}} nC_p(\text{reactants}) \tag{17}$$

and the enthalpy of reaction at any temperature can be calculated from the enthalpy of reaction at one temperature and a knowledge of the heat capacities of all the substances. The derivation of Kirchhoff's law supposes that the heat capacities are constant over the range of temperature of interest, so the law is best restricted to small temperature differences (of not more than 100 K or so).

Example Using Kirchhoff's law
The standard enthalpy of formation of gaseous water at 25 °C is $-241.82\,\text{kJ}\,\text{mol}^{-1}$. Estimate its value at 100 °C given the

molar heat capacities of $H_2O(g)$, $H_2(g)$, and $O_2(g)$ as $33.58\,J\,K^{-1}\,mol^{-1}$, $28.84\,J\,K^{-1}\,mol^{-1}$, and $29.37\,J\,K^{-1}\,mol^{-1}$, respectively.

Answer The formation equation is

$$H_2(g) + \tfrac{1}{2}O_2(g) \rightarrow H_2O(g) \quad \Delta H^{\ominus}$$

The difference in heat capacities that we require is

$$\begin{aligned}
\Delta C_p &= 1\,mol \times C_p(H_2O, g) \\
&\quad - \{1\,mol \times C_p(H_2, g) + \tfrac{1}{2}\,mol \times C_p(O_2, g)\} \\
&= 33.58\,J\,K^{-1} - \{28.84\,J\,K^{-1} + \tfrac{1}{2} \times 29.37\,J\,K^{-1}\} \\
&= -9.95\,J\,K^{-1}
\end{aligned}$$

Then, since $\Delta T = +75\,K$,

$$\Delta H^{\ominus\prime} = -241.82\,kJ + (-9.95\,J\,K^{-1}) \times (75\,K) = -242.57\,kJ$$

It follows that the enthalpy of formation of $H_2O(g)$ at $100\,°C$ is $-242.57\,kJ\,mol^{-1}$.

Exercise E2.12 Estimate the standard enthalpy of formation of $NH_3(g)$ at $400\,K$ given the molar heat capacities of NH_3, H_2, and N_2 as 36.00, 28.82, and $29.13\,J\,K^{-1}\,mol^{-1}$, respectively.
[$-48.4\,kJ\,mol^{-1}$]

The calculation in the example shows that the reaction enthalpy at $100\,°C$ is only slightly different from that at $25\,°C$. The reason is that the change in the reaction enthalpy is proportional to the *difference* between the heat capacities of the products and the reactants, which is not very large. It is generally the case that enthalpies of reactions vary only slightly with temperature (over small temperature ranges) because the heat capacities of the products differ only a little from those of the reactants. A reasonable first approximation in thermochemistry is that the enthalpy of reaction may be assumed to be independent of temperature.

EXERCISES

Assume all gases are perfect unless stated otherwise. To two significant figures, 1.0 atm is the same as 1.0 bar. Unless otherwise stated, thermochemical data are for 298 K.

2.1 Calculate the work that a person must do to raise a mass of 1.0 kg through 10 m on the surface of (a) the Earth ($g = 9.81\,m\,s^{-2}$) and (b) the Moon ($g = 1.60\,m\,s^{-2}$).

2.2 Calculate the work needed for a person of mass 65 kg to climb through 4.0 m on the surface of the Earth.

2.3 A chemical reaction takes place in a container of cross-sectional area $100\,cm^2$; the container has a loosely fitted piston at one end. As a result of the reaction, the piston is pushed out through 10 cm against an external pressure of 1.0 atm. Calculate the work done by the system.

2.4 A sample of 4.50 g of methane occupies 12.7 L at 310 K. (a) Calculate the work done when the gas expands isothermally against a constant external pressure of 200 Torr until its volume has increased by 3.3 L. (b) Calculate the work that would be done if the same expansion occurred reversibly.

2.5 In the isothermal reversible compression of 52.0 mmol of a perfect gas at 260 K, the volume of the gas is reduced to one-third its initial value. Calculate w for this process.

2.6 A certain liquid sample occupies 0.450 L at 0 °C and 1 bar, and shrinks by 0.67 per cent when subjected to compression under constant external pressure of 95 bar. Calculate w.

2.7 A strip of magnesium of mass 15 g is dropped into a beaker of dilute hydrochloric acid. Given that the magnesium is the limiting reactant, calculate the work done by the system as a result of the reaction. The atmospheric pressure is 1.0 atm and the temperature 25 °C.

2.8 Calculate the heat required to melt 750 kg of sodium metal at 370.95 K.

2.9 When 229 J of energy is supplied as heat to 3.0 mol Ar(g), the temperature of the sample increases by 2.55 K. Calculate the molar heat capacities at constant volume and constant pressure of the gas.

2.10 The heat capacity of air at room temperature and pressure is approximately 21 J K^{-1} mol^{-1}. How much heat is required to raise the temperature of a 5.0 m × 5.0 m × 3.0 m room by 10 °C? If losses are neglected, how long will it take a 1.0 kW heater to achieve that increase given that 1 W = 1 J s^{-1}?

2.11 A sample of liquid of mass 25 g is cooled from 290 K to 275 K at constant pressure by the extraction of 1.2 kJ of energy as heat. Calculate q and ΔH and estimate the heat capacity of the sample.

2.12 When 3.0 mol of O_2 is heated at a constant pressure of 3.25 atm, its temperature increases from 260 K to 285 K. Given that the molar heat capacity of O_2 at constant pressure is 29.4 J K^{-1} mol^{-1}, calculate q, ΔH, and ΔU.

2.13 A certain liquid has $\Delta H_{vap}^{\ominus} = +26.0$ kJ mol^{-1}. Calculate q, w, ΔH, and ΔU when 0.50 mol is vaporized at 250 K and 750 Torr.

2.14 The standard enthalpy of formation of ethylbenzene is −12.5 kJ mol^{-1}. Calculate its standard enthalpy of combustion.

2.15 Calculate the standard enthalpy of hydrogenation of 1-hexene to hexane given that the standard enthalpy of combustion of 1-hexene is −4003 kJ mol^{-1}.

2.16 Calculate the standard internal energy of formation of liquid methyl acetate from its standard enthalpy of formation, which is −442 kJ mol^{-1}.

2.17 The standard enthalpy of combustion of naphthalene is −5157 kJ mol^{-1}. Calculate its standard enthalpy of formation.

2.18 The temperature of a bomb calorimeter rose by 1.617 K when a current of 3.20 A was passed for 27.0 s from a 12.0 V source. Calculate the heat capacity of the calorimeter.

2.19 When 120 mg of naphthalene, $C_{10}H_8(s)$, was burned in a bomb calorimeter, the temperature rose by 3.05 K. Calculate the heat capacity of the calorimeter. By how much will the temperature rise when 100 mg of phenol, $C_6H_5OH(s)$, is burned in the calorimeter under the same conditions?

2.20 When 0.3212 g of glucose was burned in a bomb calorimeter with a calorimeter constant of 641 J K^{-1} the temperature rose by 7.793 K. Calculate (a) the standard molar enthalpy of combustion, (b) the standard internal energy of combustion, and (c) the standard enthalpy of formation of glucose.

2.21 Calculate the standard enthalpy of solution of AgCl(s) in water from the enthalpies of formation of the solid and the aqueous ions.

2.22 The standard enthalpy of decomposition of the yellow complex NH_3—SO_2 into NH_3 and SO_2 is +40 kJ mol^{-1}. Calculate the standard enthalpy of formation of NH_3—SO_2.

2.23 Given that the enthalpy of combustion of graphite is −393.5 kJ mol^{-1} and that of diamond is −395.41 kJ mol^{-1}, calculate the enthalpy of the graphite → diamond transition.

2.24 The mass of a typical sugar cube is 1.5 g. Calculate the energy released as heat when a cube is burned in air. To what height could a 65-kg person climb on the energy a cube provides assuming 25 per cent of the energy is available for work?

2.25 The standard enthalpy of combustion of propane gas is $-2220\,kJ\,mol^{-1}$ and the standard enthalpy of vaporization of the liquid is $+15\,kJ\,mol^{-1}$. Calculate (a) the standard enthalpy and (b) the standard internal energy of combustion of the liquid.

2.26 Classify as endothermic or exothermic the following reactions:

(a) $CH_4(g) + 2O_2(g) \rightarrow CO_2(g) + 2H_2O(l)$

$$\Delta H^{\ominus} = -890\,kJ\,mol^{-1}$$

(b) $2C(s) + H_2(g) \rightarrow C_2H_2(g)$

$$\Delta H^{\ominus} = +227\,kJ\,mol^{-1}$$

(c) $NaCl(s) \rightarrow NaCl(aq)$

$$\Delta H^{\ominus} = +3.9\,kJ\,mol^{-1}$$

2.27 Use standard enthalpies of formation to calculate the standard enthalpies of the following reactions:

(a) $2NO_2(g) \rightarrow N_2O_4(g)$

(b) $NH_3(g) + HCl(g) \rightarrow NH_4Cl(s)$

(c) $Cyclopropane(g) \rightarrow propene(g)$

(d) $HCl(aq) + NaOH(aq) \rightarrow NaCl(aq) + H_2O(l)$

2.28 Calculate the standard enthalpy of formation of N_2O_5 from the following data:

$2NO(g) + O_2(g) \rightarrow 2NO_2(g)$

$$\Delta H^{\ominus} = -114.1\,kJ\,mol^{-1}$$

$4NO_2(g) + O_2(g) \rightarrow 2N_2O_5(g)$

$$\Delta H^{\ominus} = -110.2\,kJ\,mol^{-1}$$

$N_2(g) + O_2(g) \rightarrow 2NO(g)$

$$\Delta H^{\ominus} = +180.5\,kJ\,mol^{-1}$$

2.29 Use the information in Appendix 1 to predict the standard reaction enthalpy of $2NO_2(g) \rightarrow N_2O_4(g)$ at $100\,°C$ from its value at $25\,°C$. The molar heat capacities of NO_2 and N_2O_4 are 37.20 and $77.28\,J\,K^{-1}\,mol^{-1}$, respectively.

3

Thermodynamics: the second law

CONTENTS

Some things happen; some things don't. A gas expands to fill the vessel it occupies; a gas that already fills a vessel does not suddenly contract into a smaller volume. A hot object cools to the temperature of its surroundings; a cool object does not suddenly become hotter than its surroundings. Hydrogen and oxygen combine explosively (once their potential to do so has been liberated by a spark) and form water; water does not decompose into hydrogen and oxygen. That is, changes can be divided into **spontaneous changes**, changes that have a natural tendency to occur, and **nonspontaneous changes**, changes that have no natural tendency to occur. Nonspontaneous changes can be *made* to occur: gas can be compressed into a smaller volume by pushing in a piston, a cool object can be heated by forcing an electric current through a heater that is attached to it, and water can be decomposed by the passage of an electric current. However, in each case we need to do work on the system to bring the nonspontaneous change about.

There must be some feature of the world that accounts for the distinction between the two types of change. Once we have identified it, we shall be able to discuss why some reactions are spontaneous but others are not, and hence be able to understand why some reactions are successful but others are not. In particular, we shall be able to express the condition for **chemical equilibrium**, which is when the tendency of a reaction to occur appears to be exhausted and it has no tendency to form more products, and no tendency to reverse and re-form reactants.

Throughout the chapter we shall use the terms 'spontaneous' and 'nonspontaneous' to signify, respectively, that a change does or does not have a natural tendency to occur. The term spontaneous is not a synonym for fast, and a spontaneous change may be so slow that it takes millions of years to come about. *A spontaneous change is one that is poised for change, and is not necessarily in the process of change*: substances poised to undergo a spontaneous change are like water in a reservoir held back by a dam. For example, although the reaction between hydrogen and oxygen is spontaneous, it does not actually occur until the mixture is ignited by a spark: without the spark, the mixture can survive indefinitely, like water behind a dam. Thermodynamics deals only with *tendencies* to change: it is silent on the rate at which changes occur.

Entropy

A few moments thought is all that is needed to identify the reason why some changes are spontaneous and others are not. That reason is *not* the tendency of the system to move towards lower energy. This point is easily established, because a perfect gas expands spontaneously into a vacuum, but its total potential energy does not change in the process, because the separation of the molecules does not affect their energy; and the total kinetic energy does not change either, because (at constant temperature) the molecules continue to travel at the same average speed. Indeed, even in a process in which the energy does decrease, the first law requires the total energy to be constant, so the energy of another region of the world must increase. For instance, a hot block of metal in contact with a cool block cools and loses energy; however, the second block becomes warmer, and increases in energy. It is equally valid to say that the second block moves spontaneously to higher energy as it is to say that the first block has a tendency to go to lower energy!

3.1 The direction of spontaneous change

We shall now show that *the driving force of spontaneous change is the tendency of energy and matter to disperse chaotically*. For example, the molecules of a gas may all be in one region of a container initially, but their continuous chaotic motion rapidly ensures that they spread rapidly throughout the entire volume of the container (Fig. 3.1). Because their motion is so chaotic, there is an extremely low probability that all the molecules will be found back in the initial region of the container, and such a phenomenon has never been observed. In this instance, there has been no change in the energy of the system (because, in a perfect gas, the energy of the molecules is independent of their separation), and the natural direction of change is in the direction in which matter has dispersed.

A similar explanation accounts for spontaneous cooling, but now we need to consider the dispersal of energy. Thus, in a hot block of metal, the atoms are oscillating vigorously, the hotter the block the more vigorous their motion. The cooler surroundings also consist of

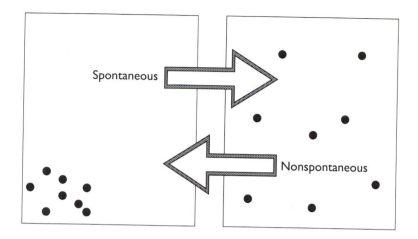

Fig. 3.1 One fundamental type of spontaneous process is the chaotic dispersal of matter. This tendency accounts for the spontaneous tendency of a gas to spread into and fill the container it occupies. It is extremely unlikely that all the particles will collect into one small region of the container. (In practice, the number of particles is of the order of 10^{23}.)

oscillating atoms (or chaotically moving atoms if they are a gas), but their motion is less vigorous. The vigorously oscillating atoms of the hot block jostle their neighbours in the surroundings, and their energy is handed on to those neighbours (Fig. 3.2). The process continues, until the vigour with which the atoms in the system are oscillating has fallen to that of the surroundings. It is extremely improbable that a block with atoms that were oscillating as vigorously as those in the surroundings would suddenly acquire enough energy from the random jostlings of the surroundings to appear to have become suddenly significantly hotter than them. That is, *the natural direction of change corresponds to the chaotic dispersal of energy*.

In summary, we have identified two basic types of spontaneous process:

1. Matter tends to disperse chaotically.

2. Energy tends to disperse chaotically.

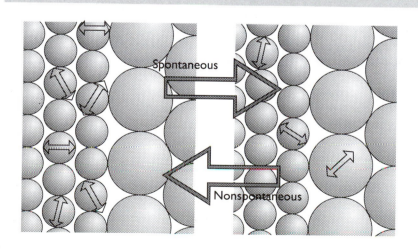

Fig. 3.2 Another fundamental type of spontaneous process is the chaotic dispersal of energy (represented by the open arrows). In these diagrams, the small spheres represent the system and the large spheres represent the surroundings. The double headed arrows represent the thermal motion of the atoms.

We must now trace the consequences of these primitive types of physical change into chemistry, and see why some chemical reactions are spontaneous and others are not. Moreover, we must also see how to make the discussion quantitative, so that we can make numerical predictions about reactions.

3.2 Entropy and the second law

The measure of the dispersal of matter and energy used in thermodynamics is called the **entropy**, S. Initially we can take entropy to be a synonym for the extent of chaotic dispersal of matter and energy, but we shall define it precisely and quantitatively shortly. When energy disperses chaotically, the entropy increases; when matter disperses chaotically, the entropy increases. It then follows that we can express the tendency of energy and matter to disperse chaotically in a single statement:

> The **second law of thermodynamics:** the entropy of an isolated system increases in a spontaneous change.

The remarkable feature of this law is that it accounts for change in all its forms: for precipitation reactions, acid–base reactions, and redox reactions, as well as the physical changes we have already considered.

The definition of an entropy change

To make progress and turn the second law into a quantitatively useful statement, we need to define entropy precisely. The following definition turns out (as we shall see) to be appropriate for isothermal processes:

$$\Delta S = \frac{q_{\text{rev}}}{T} \tag{1a}$$

That is, the change in entropy of a system (ΔS) is equal to the heat transferred to it *reversibly* (we shall explain that in a moment) divided by the temperature (on the Kelvin scale) at which the transfer of heat takes place.

First, we consider the meaning of 'reversible' in this context. We met the concept of mechanical reversibility in Section 2.1, where we saw that it involved matching the external pressure acting on a system to the pressure of the system itself. To achieve a reversible transfer of heat, the *temperature* of the surroundings is matched to that of the system almost exactly. (The temperature needs to be infinitesimally higher to ensure that heat flows into the system, or infinitesimally lower to ensure that heat flows out from the system.) By making the transfer reversible we ensure that there are no hot spots generated in the object which later disperse spontaneously and add to the entropy.

Reversible transfer is a smooth, careful, restrained transfer. If there were hot spots generated, then their dispersal would add to the entropy change and ΔS would be larger than the value calculated by dividing q by T. That is, for an *irreversible* transfer of heat to a body at a temperature T,

$$\Delta S > \frac{q_{\text{irrev}}}{T} \qquad (1b)$$

Equation (1a) can be justified formally, but it should be quite easy to accept that it is plausible. From what we know about entropy, it should be plausible that the extent to which disorder is stirred up is proportional to the energy transferred as heat: the more energy that is transferred as heat, the greater the chaotic motion that is stimulated, and so the greater the increase in entropy. Transfers of energy as work do not result in an entropy change: recall from Fig. 2.5 that work stimulates the *orderly* motion of atoms in the surroundings, so their degree of disorder is left unchanged when work is done. The presence of the temperature in the denominator of eqn (1) takes into account the chaos that is already present. If a given quantity of heat is transferred to a hot object (one in which there is already a lot of chaotic thermal motion), then the additional chaos that is generated is less significant than if the same quantity of heat is transferred to a cold object (in which the atoms have less chaotic motion). The difference is like sneezing in a busy street (heat transferred to a hot body that already has a lot of chaotic motion) and sneezing in a quiet library (heat transfer to a cold body in which there is only little chaotic motion initially). For example, transferring 100 kJ of heat to a large sample of water (large, so that its temperature does not change significantly) at 0 °C results in a change in entropy of

$$\Delta S = \frac{100 \times 10^3 \text{ J}}{273 \text{ K}} = +366 \text{ J K}^{-1}$$

whereas the same transfer at 100 °C results in

$$\Delta S = \frac{100 \times 10^3 \text{ J}}{373 \text{ K}} = +268 \text{ J K}^{-1}$$

Notice the units of entropy: they are joules per kelvin, J K^{-1}; later, when we deal with molar entropies, their units will be joules per kelvin per mole, $\text{J K}^{-1} \text{mol}^{-1}$.

We shall not confirm it explicitly, but *entropy is a state property*. That should also be plausible, for the entropy is a measure of the current state of disorder of the system, and how that disorder was achieved is not relevant to its current value. That is, the entropy of a system depends only on its present state, not on how that system was prepared or what has happened to it previously: 100 g of liquid water at 60 °C and under a pressure of 98 kPa has exactly the same entropy irrespective of what has happened to it in the past. Moreover, because entropy is a state property, *the change in entropy between any two*

Fig. 3.3 Entropy is a state property, and so the change in entropy when a system changes from an initial state to a specified final state is independent of the path (as specified here by the values of pressure and temperature that the system takes as it migrates from one state to the other).

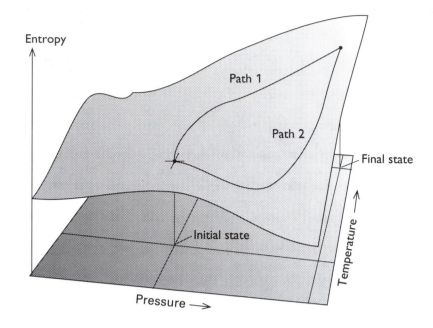

states of a system is the same however the change is brought about (Fig. 3.3).

The determination of entropy

The entropy of a gas increases as its volume is increased because the greater the volume in which the molecules can move, the greater the disorder of the sample. The change in entropy when a perfect gas expands isothermally from a volume V_i to a volume V_f is given by the expression

$$\Delta S = nR \ln \frac{V_f}{V_i} \qquad (2)$$

We can check the plausibility of this expression by noting that if $V_f > V_i$, as in an expansion, $V_f/V_i > 1$ and the logarithm is positive, then ΔS is positive, corresponding to an increase in entropy. The increase in entropy with increasing volume is illustrated in Fig. 3.4. So long as there is no change in the entropy of the surroundings when the expansion takes place, the total entropy increases, and the expansion is spontaneous.

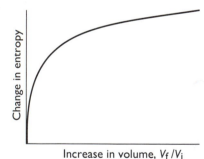

Fig. 3.4 The entropy of a sample of fixed composition and temperature increases logarithmically as its volume is increased. The graph shown here is a plot of eqn (2); the increase in volume is the ratio of the final volume to the initial volume.

Exercise E3.1 Calculate the change in entropy when 1.0 mol of hydrogen gas doubles its volume isothermally. [*Hint*: set $V_f = 2V_i$.]

[$+5.8\,\mathrm{J\,K^{-1}}$]

JUSTIFICATION

The proof of eqn (2) for the isothermal expansion of a perfect gas makes use of the first law expressed in the form

$$\Delta U = w + q$$

When a perfect gas expands isothermally, its internal energy remains unchanged because the molecules continue to travel at the same average speed, and therefore with the same total kinetic energy. Therefore, $\Delta U = 0$, which implies that $q = -w$ for the isothermal expansion of a perfect gas. To calculate the entropy change, we need to know the heat absorbed by the gas during a reversible expansion. When the expansion is reversible, the gas does maximum work, which from eqn (6) in Chapter 2, is

$$w_{max} = -nRT \ln \frac{V_f}{V_i}$$

(The negative sign indicates that the system is losing energy on account of the work that it is doing.) It follows that

$$\Delta S = \frac{q_{rev}}{T} = \frac{(-w_{max})}{T} = nR \ln \frac{V_f}{V_i}$$

The second type of change we consider is the effect on the entropy of increasing the temperature of a system.† We should expect the entropy to increase, because the motion of the molecules is more vigorous and the thermal chaos of the system is greater. If we assume that the heat capacity of a substance is independent of temperature in the range of interest (this assumption is strictly valid only for monatomic gases), the change in entropy of a sample when its temperature is increased from T_i to T_f at constant volume (the latter condition ensures that there is no change in entropy arising from the greater spread in position of the particles) is

$$\Delta S = C_V \ln \frac{T_f}{T_i} \tag{3}$$

As in the analogous expression for the increase in entropy with volume, when $T_f > T_i$, the value of T_f/T_i is greater than 1, so the logarithm is positive, which implies that ΔS is positive, and that the entropy has increased. The increase in entropy with temperature predicted by this expression is illustrated in Fig. 3.5.

> **Exercise E3.2** Calculate the change in entropy when the temperature of 1.0 mol H_2 (which has a molar heat capacity at constant volume of 20.44 J K^{-1} mol^{-1}) is raised from 20 °C (293 K) to 30 °C (303 K).
>
> [+0.69 J K^{-1}]

† It may be puzzling that we can develop an expression for the change in entropy of a system when heated, given that eqn (1) applies to an *isothermal* process. However, the entropy change may be divided into a sequence of infinitesimal steps, each of which occurs at a different but fixed temperature. So we can use eqn (1) for each one of them, and sum the values to obtain the overall change in entropy.

Fig. 3.5 The entropy of a sample of fixed composition and volume and with a heat capacity that is independent of temperature (such as a monatomic perfect gas) increases logarithmically as the temperature is increased. The increase is proportional to the heat capacity of the sample. The graphs shown here are plots of eqn (3).

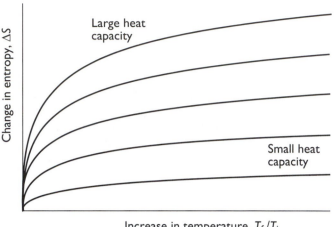

When the heat capacity cannot be considered to be independent of temperature (when the range of temperatures is wide), the change in entropy is determined by measuring the heat capacity C_V, plotting a graph of C_V/T against T, and then determining the area under the curve that lies between the two temperatures (Fig. 3.6):

ΔS = area under graph of C_V/T versus T between T_i and T_f

The heat capacities of all substances decrease as the temperature is lowered (see Section 8.2), so the graph of C_V/T resembles that in the illustration.

JUSTIFICATION

When an infinitesimal quantity of heat dq_{rev} is added reversibly to a substance of heat capacity C_V, the temperature of the substance increases by dT, where dq_{rev} and dT are related by $dq_{rev} = C_V \, dT$. Therefore, the entropy of the sample changes by

$$dS = \frac{dq_{rev}}{T} = \frac{C_V \, dT}{T}$$

The total change in entropy for a change in temperature from T_i to T_f is the sum (integral) of all such terms:

$$\Delta S = \int_{T_i}^{T_f} \frac{C_V}{T} \, dT$$

The value of the integral, and hence the value of ΔS, is the area under the curve of C_V/T against T between T_i and T_f. If C_V is constant in the range of interest, then it may be taken outside the integral; the latter may then be evaluated using

$$\int \frac{1}{x} \, dx = \ln x$$

to give

$$\Delta S = C_V \int_{T_i}^{T_f} \frac{1}{T} \, dT = C_V \ln T \big|_{T_i}^{T_f} = C_V \ln \frac{T_f}{T_i}$$

as in eqn (3).

The entropy of a substance changes when it undergoes a phase transition. Suppose a solid is at its melting temperature, then any addition of heat causes melting (fusion) and any removal of heat causes freezing. These two opposite changes can be achieved by raising the temperature of the surroundings infinitesimally (which results in melting) or by lowering the temperature infinitesimally (which results in freezing). Therefore, at the melting temperature, the phase transition is reversible in the thermodynamic sense. Furthermore, because the transition occurs at constant pressure, we can identify the heat transferred with the enthalpy of the transition (the enthalpy of melting, in this case). Therefore, the change in entropy of a sample at its melting temperature T_f is

$$\Delta S = \frac{q_{rev}}{T_f} = \frac{\Delta H_{fus}}{T_f} \qquad (4)$$

The melting of ice, for example, is accompanied by an increase in entropy, because the orderly structure of the solid collapses into the more disorderly liquid (Fig. 3.7).

Exercise E3.3 Calculate the change in entropy when 1.0 mol $H_2O(s)$ melts at $0\,°C$.

$$[+22\,J\,K^{-1}]$$

The entropy change accompanying other transitions may be discussed similarly. Thus, the change in entropy at the boiling temperature T_b of a liquid is related to its enthalpy of vaporization at that temperature:

$$\Delta S = \frac{\Delta H_{vap}}{T_b} \qquad (5)$$

The increase in entropy when water vaporizes at $100\,°C$ is greater than the increase when ice melts at $0\,°C$ despite the fact that the heat transfer occurs at a higher temperature, because there is a much greater change in disorder on going from a liquid to a gas than from a solid to a liquid.

Exercise E3.4 Calculate the change in entropy when 1.0 mol $H_2O(l)$ vaporizes at $100\,°C$.

$$[+109\,J\,K^{-1}]$$

When any liquid vaporizes, the compact condensed phase changes into a widely dispersed gas that occupies approximately the same volume in each case. To a good approximation, therefore, we can expect the change in entropy that accompanies vaporization, which is equal to $\Delta H_{vap}/T_b$, to be almost the same for all liquids at their boiling temperatures. This argument is the rationalization of the empirical

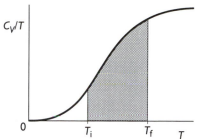

Fig. 3.6 The experimental determination of the change in entropy of a sample which has a heat capacity that varies with temperature involves measuring the heat capacity over the range of temperatures of interest, then plotting C_V/T against T and determining the area under the curve (the shaded area shown here). The heat capacity of all solids decreases toward zero as the temperature is reduced.

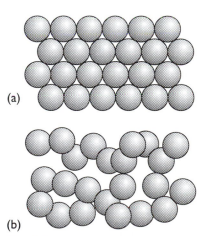

Fig. 3.7 When a solid (depicted by the orderly array of spheres, (a)) melts, the molecules form a more chaotic liquid (the disorderly array of spheres, (b)). As a result, the entropy of the sample increases.

Table 3.1 Entropies of vaporization at 1 atm and the normal boiling point

	$\Delta S_{vap}/$ $J\,K^{-1}\,mol^{-1}$
Bromine, Br_2	+88.6
Benzene, C_6H_6	+87.2
Carbon tetrachloride, CCl_4	+85.9
Cyclohexane, C_6H_{12}	+85.1
Hydrogen sulfide, H_2S	+87.9
Ammonia, NH_3	+97.4
Water, H_2O	+109.1

observation known as **Trouton's rule**, that $\Delta H_{vap}/T_b$ is approximately the same ($85\,J\,K^{-1}\,mol^{-1}$) for almost all liquids. Some values of Trouton's constant (that is, the entropy of vaporization) are listed in Table 3.1. As we see, they do indeed have very similar values except in a few cases. The exceptions to Trouton's rule are liquids in which the interactions between molecules result in the liquid being less disorderly. For example, the high value for water implies that the H_2O molecules are linked together in a loose structure (by hydrogen bonding, Section 10.2), with the result that the entropy change is greater when the relatively orderly liquid forms the disorderly gas.

Trouton's rule is valid for liquids in which there is no hydrogen bonding nor other special type of bonding (such as the metallic bonding in liquid mercury). For liquids for which it is valid, Trouton's rule provides a convenient way of estimating the boiling temperature from the enthalpy of vaporization or vice versa.

Example Using Trouton's rule
Estimate the enthalpy of vaporization of liquid bromine from its boiling temperature, 59.2 °C.
Answer Because there is no hydrogen bonding in liquid bromine, we can use Trouton's rule in the form

$$\Delta H_{vap} = T_b \times 85\,J\,K^{-1}\,mol^{-1}$$
$$= 332.4\,K \times 85\,J\,K^{-1}\,mol^{-1} = 28\,kJ\,mol^{-1}$$

The experimental value is $29\,kJ\,mol^{-1}$.

Exercise E3.5 Predict the enthalpy of vaporization of ethane from its boiling temperature, -88.6 °C.

[$16\,kJ\,mol^{-1}$]

Absolute entropies and the third law of thermodynamics

The graphical procedure summarized by Fig. 3.6 and eqn (3) for the determination of the difference in entropy of a substance at two temperatures has a very important application. If $T_i = 0$ (that is, the absolute zero of temperature), then eqn (3) can be used to write the entropy of a substance at a temperature T in terms of its entropy at $T = 0$, $S(0)$:

$$S = S(0) + \Delta S(\text{from 0 to } T)$$
$$= S(0) + \text{area under graph of } C_V/T \text{ versus } T \text{ between 0 and } T$$

If there are any phase transitions (for example, melting) in the temperature range of interest, then the entropy of each transition, calculated from eqns (4) or (5), must be added to the result obtained in this way.

At $T = 0$, all the motion of the atoms has been eliminated, and there is no thermal disorder. Moreover, if the substance is perfectly crystalline, with every atom in a well-defined location, then there is no spatial disorder either. That is, *at $T = 0$, the entropy is zero*. For example, the entropy of a crystal of sucrose at $T = 0$ is zero, as all the atoms lie in a perfectly orderly array and there is no thermal motion; the same is true of a perfect crystal of sodium chloride and of solid carbon dioxide. This conclusion is generalized by the 'third law' of thermodynamics:

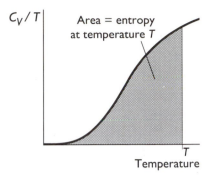

Fig. 3.8 The absolute entropy (or third-law entropy) of a substance is calculated by extending the measurement of heat capacities down to $T = 0$ (or as close to that value as possible), and then determining the area of the graph of C_V/T against T up to the temperature of interest. The area is equal to the absolute entropy at the temperature T.

> **The third law of thermodynamics:** the entropy of a perfectly crystalline material is zero at $T = 0$.

According to the third law, $S(0) = 0$ for all perfectly ordered crystalline materials. It therefore follows that the **absolute entropy** at any temperature is given by

$$S(T) = \text{area under graph of } C_V/T \text{ versus } T \text{ between 0 and } T \quad (6)$$

This result is summarized in Fig. 3.8. Once again, this conclusion is valid only if there are no phase transitions below the temperature T; if there are any phase changes, then the entropy of each transition, calculated from eqn (4) or eqn (5), must be included.

The value of the absolute entropy of a substance at any given temperature depends on the pressure (a high pressure, for instance, would confine a gas to a smaller volume and so reduce its entropy). It is therefore common to select a standard pressure and to report the **standard molar entropy**, $S^{\ominus}$, the entropy per unit amount of substance under a pressure of 1 bar (the standard pressure $p^{\ominus}$) at the temperature of interest. Some values at 25 °C (the conventional temperature $T = 298.15$ K) are given in Table 3.2.

It is worth spending a moment to look at the values in Table 3.2, for they are consistent with our understanding of entropy. All standard molar entropies are positive, because raising the temperature of a sample above $T = 0$ invariably increases its entropy above the value $S(0) = 0$. Another feature is that the standard molar entropy of diamond ($2.4 \, \text{J K}^{-1} \, \text{mol}^{-1}$) is lower than that of graphite ($5.7 \, \text{J K}^{-1} \, \text{mol}^{-1}$) because the atoms are linked less rigidly in graphite than in diamond, and its thermal disorder is greater. The standard molar entropy of ice, water, and water vapour at 25 °C are, respectively, $45 \, \text{J K}^{-1} \, \text{mol}^{-1}$, $70 \, \text{J K}^{-1} \, \text{mol}^{-1}$, and $189 \, \text{J K}^{-1} \, \text{mol}^{-1}$, and the increase in values corresponds to the increasing disorder on going from a solid to a liquid to a gas.

Table 3.2 Standard molar entropies of some substances at 25 °C

Substance	$S^{\ominus}/$ $J\,K^{-1}\,mol^{-1}$
Gases	
Ammonia, NH_3	192.5
Carbon dioxide, CO_2	213.7
Helium, He	126.2
Hydrogen, H_2	130.7
Neon, Ne	146.3
Nitrogen, N_2	191.6
Oxygen, O_2	205.1
Water vapour, H_2O	188.8
Liquids	
Benzene, C_6H_6	173.3
Ethanol, CH_3CH_2OH	160.7
Water, H_2O	69.9
Solids	
Calcium oxide, CaO	39.8
Calcium carbonate, $CaCO_3$	92.9
Copper, Cu	33.2
Diamond, C	2.4
Graphite, C	5.7
Lead, Pb	64.8
Magnesium carbonate, $MgCO_3$	65.7
Magnesium oxide, MgO	26.9
Sodium chloride, NaCl	72.1
Sucrose, $C_{12}H_{22}O_{11}$	306.2
Tin, Sn (white)	51.6
Sn (grey)	44.1

See Appendix 1 for more values.

The entropy of reaction

When a reaction takes place, reactants change into products and there is an accompanying change in entropy of the reaction system. This change can be expressed in terms of the absolute entropies of the substances:

> The **standard entropy of reaction**, $\Delta S^{\ominus}$, is the difference in the standard entropies of the reactants and products.

As for the enthalpy of reaction, we write

$$\Delta S^{\ominus} = \sum_{Products} nS^{\ominus}(products) - \sum_{Reactants} nS^{\ominus}(reactants) \qquad (7)$$

For example, the standard entropy of the reaction

$$2H_2(g) + O_2(g) \rightarrow 2H_2O(l)$$

is

$$\Delta S^{\ominus} = (2\ mol) \times S^{\ominus}(H_2O, l) - \{(2\ mol) \times S^{\ominus}(H_2, g)$$
$$+ (1\ mol) \times S^{\ominus}(O_2, g)\}$$
$$= 2 \times 70\ J\,K^{-1} - \{(2 \times 131\ J\,K^{-1}) + (205\ J\,K^{-1})\}$$
$$= -327\ J\,K^{-1}$$

(Do not make the mistake of setting the entropies of elements equal to zero: the $S^{\ominus}$ are *absolute* entropies, not entropies of formation, and all elements have nonzero entropies for $T > 0$.) We see that the formation of the liquid product from two gases has resulted in a decrease in entropy of nearly $327\ J\,K^{-1}$ per mole of O_2 consumed. This is a huge decrease in entropy, and it corresponds to the formation of a compact liquid from two gases.

> **Exercise E3.6** Calculate the standard entropy of the reaction $N_2(g) + 3H_2(g) \rightarrow 2NH_3(g)$ at 25 °C.
>
> $[-198.76\ J\,K^{-1}]$

The spontaneity of reactions

The result of the H_2O calculation should be rather surprising at first sight. We know that the reaction between hydrogen and oxygen is spontaneous (once initiated, it proceeds with explosive violence), yet the entropy change that accompanies it is negative: the reaction results in less disorder, yet it is spontaneous!

The resolution of this apparent paradox underscores a feature of entropy that recurs throughout chemistry: it is always essential to consider the entropy of the system *and its surroundings* when determining whether a process is spontaneous or not. The reduction in entropy by $327\,\mathrm{J\,K^{-1}}$ relates to the *system,* not to the system plus its surroundings. To apply the second law we need to calculate the *total* entropy change, ΔS_{tot}, the sum of the changes in the system and the surroundings.

To calculate the total entropy change, we note that when a reaction occurs, heat may enter or leave the system. For example, for the water formation reaction written above, $\Delta H^{\ominus} = -572\,\mathrm{kJ}$, and $572\,\mathrm{kJ}$ of heat enters the surroundings if the reaction takes place at constant pressure. The change in entropy of the surroundings (which are maintained at $25\,°\mathrm{C}$) is therefore

$$\Delta S_{sur} = \frac{572\,\mathrm{kJ}}{298\,\mathrm{K}} = +1.92 \times 10^3\,\mathrm{J\,K^{-1}}$$

This is a huge increase, and far outweighs the reduction in entropy of the system: the total change in entropy is

$$\Delta S_{tot} = \Delta S^{\ominus} + \Delta S_{sur} = -327\,\mathrm{J\,K^{-1}} + 1.92 \times 10^3\,\mathrm{J\,K^{-1}}$$
$$= +1.59 \times 10^3\,\mathrm{J\,K^{-1}}$$

The total entropy change is positive, indicating (as we know from experience) that the reaction is strongly spontaneous. In this case, the spontaneity is a result of the considerable disorder that the reaction generates in the surroundings: water is dragged into existence, even though H_2O has a lower entropy than the reactants, by the tendency of energy to disperse into the surroundings.

In general, when the enthalpy of reaction is ΔH, the heat transferred into the surroundings is $q_{sur} = -\Delta H$ (note the change of sign: what leaves the system enters the surroundings, so a decrease in the enthalpy of the system by $10\,\mathrm{kJ}$ corresponds to the addition of $10\,\mathrm{kJ}$ of heat to the surroundings), and the change in entropy of the surroundings is

$$\Delta S_{sur} = -\frac{\Delta H}{T} \tag{8}$$

This expression is applicable when the temperature of the surroundings is the same as that of the system, so that the heat released in the reaction is transferred to them reversibly.

Example Estimating the entropy change in the surroundings
A typical resting person generates about $100\,\mathrm{W}$ of heat. Estimate the resulting entropy generated in the surroundings in the course of a day at $20\,°\mathrm{C}$.
Answer Although the heat transfer is not reversible (the person's body is about $17\,°\mathrm{C}$ warmer than their surroundings)

we can estimate the approximate change in entropy from eqn (8), and be confident (from the argument that led to eqn (1b)) that the actual entropy change will not be less than we calculate. Because $1\,W = 1\,J\,s^{-1}$, and there are $86\,400\,s$ in a day,

heat released into surroundings

$$= 86\,400\,s \times 100\,J\,s^{-1} = 8.64 \times 10^6\,J$$

Because the transfer occurs to surroundings at $20\,^\circ C$ ($293\,K$), the increase in entropy is

$$\Delta S_{sur} = \frac{8.64 \times 10^6\,J}{293\,K} = +2.95 \times 10^4\,J\,K^{-1}$$

That is, the entropy production is about $30\,kJ\,K^{-1}$. To stay alive, each person on the planet contributes about $30\,kJ\,K^{-1}$ per day to the ever-increasing entropy of their surroundings. The use of transport, machinery, and communications generates far more in addition.

Exercise E3.7 Suppose a small reptile, which takes on the temperature of its surroundings, operates at $0.50\,W$. What entropy does it generate in the course of a day in the water of the lake that it inhabits, where the water is at $15\,^\circ C$?

$$[150\,J\,K^{-1}]$$

Free energy

One of the problems with entropy calculations is already apparent: it is necessary to calculate two entropy changes: the change in the system and the change in the surroundings. The great American theoretician J. W. Gibbs, who laid the foundations of chemical thermodynamics towards the end of the nineteenth century, found that the two calculations can be combined into one. The combination of the two procedures in fact turns out to be of much greater relevance than just saving a little labour, and throughout this text we shall see consequences of the procedure he developed.

3.3 Focusing on the system

The total entropy change that accompanies a process is

$$\Delta S_{tot} = \Delta S + \Delta S_{sur} \quad (\Delta S_{tot} > 0 \text{ for a spontaneous process}) \quad (9)$$

where ΔS is the entropy change of the system. However, we have also seen that, if the process occurs at constant pressure and at the same temperature as the surroundings, then the change in entropy of the surroundings can be expressed in terms of the enthalpy change of the system by using eqn (8). When that expression is inserted into this one, we obtain

$$\Delta S_{tot} = \Delta S - \frac{\Delta H}{T}$$

The great advantage of this formula is that it expresses the total entropy change of the system and its surroundings in terms of properties of the system alone.

Two further steps can now be taken. First, we multiply through by $-T$, to obtain

$$-T\Delta S_{tot} = -T\Delta S + \Delta H$$

Then we define the **Gibbs free energy**,† G:

$$G = H - TS \qquad (10)$$

A change in the value of G at constant temperature arises from changes in the enthalpy and entropy, and is

$$\Delta G = \Delta H - T\,\Delta S \qquad (11)$$

It follows that, at constant temperature and pressure,

$$\Delta G = -T\,\Delta S_{tot} \qquad (12)$$

Thus, we see that the change in free energy is simply another way of expressing the change in the overall entropy when the process occurs at constant temperature and pressure.

Properties of the free energy

The difference in sign between ΔG and ΔS_{tot} implies that the condition for a process being spontaneous changes from $\Delta S_{tot} > 0$ to $\Delta G < 0$. That is, *in a spontaneous change, the free energy decreases*. It may seem more natural to think of a system as falling to a lower value of some property; however, it must never be forgotten that to say that a system tends to fall towards lower free energy is only a modified way of saying that a system and its surroundings jointly tend towards a greater total entropy. *The only criterion for spontaneous change is the value of the total entropy of the system and its surroundings: the free energy merely contrives a way of expressing that total change in terms of the properties of the system alone.* Every chemical reaction that is spontaneous under conditions of constant temperature and pressure, including those that drive the processes of growth, learning, and reproduction, are reactions that change in the direction of lower free energy, or—equivalently—result in the overall entropy of the system and its surroundings becoming greater.

† The Gibbs free energy—commonly referred to as simply the 'free energy' is also called the Gibbs function. There is another related property, the Helmholtz free energy (or Helmholtz function), denoted A, which is useful for discussing processes that take place at constant volume rather than (as for the Gibbs free energy) at constant pressure.

A second feature of the free energy is that *its value gives the maximum amount of nonexpansion work that can be extracted from a system that is undergoing a change at constant temperature and pressure*. By **nonexpansion work** is meant any work other than that arising from the expansion of the system, so it may include electrical work (if the process takes place inside an electrochemical or biological cell) or other kinds of mechanical work, such as the winding of a spring or the contraction of a muscle. If we denote nonexpansion work by w', then the *total* work of a system that can do both types of work is

$$w = w' - p_{ex} \Delta V \tag{13}$$

That is, an important property of the free energy is that

$$\Delta G = w'(\text{maximum}) \quad \text{at constant temperature and pressure} \tag{14}$$

For example, we shall shortly see that for the formation of liquid water at 25 °C and 1 bar,

$$H_2(g) + \tfrac{1}{2}O_2(g) \rightarrow H_2O(l) \qquad \Delta G = -237\,\text{kJ}$$

It follows that, in a system in which hydrogen and oxygen react to form $1\,\text{mol}\ H_2O(l)$ at 25 °C and 1 bar, the maximum nonexpansion work is

$$w'(\text{maximum}) = -237\,\text{kJ}$$

JUSTIFICATION

We shall first consider infinitesimal changes, and derive the relation between dG and dw'. To do so, we begin by substituting the definitions of dH and dU into the expression for dG:

$dG = dH - T\,dS$ (at constant T)

$\quad = dU + p\,dV - T\,dS$ (because $dH = dU + p\,dV$

at constant p)

$\quad = dq + dw + p\,dV - T\,dS$ (because $dU = dq + dw$)

$\quad = dq + dw' - p_{ex}\,dV$ (because $dw = $

$\quad\quad + p\,dV - T\,dS$ $dw' - p_{ex}\,dV$)

The derivation is valid, as the notes in parentheses indicated, for a process that is occurring at constant temperature and constant pressure. Now we recall from Section 2.1 that maximum work is obtained when a process occurs reversibly. For the *expansion work* to be done reversibly, the external pressure p_{ex} must match the pressure of the system; for a process that is producing maximum work we can set $dw' = dw'(\text{maximum})$ and $p_{ex}\,dV = p\,dV$, whereupon the third and fourth terms on the right cancel. Because the *heat transfer* is also reversible, we can also set $dq = dq_{rev}$, which, from the definition of entropy (eqn (1)), is equal to $T\,dS$. As a result, the first and fifth terms cancel, leaving

$dG = dw'(\text{maximum})$

      for a reversible change at constant T, p.

This relation holds for each infinitesimal step between specified initial and final states of the system, and so it applies to the overall change too, and we can write

$\Delta G = w'(\text{maximum})$

      for a reversible change at constant T, p.

The negative sign indicates that 237 kJ of work can be done by the system (that is, 237 kJ of energy leaves the system as work). If the reaction takes place in a fuel cell like those used on the space shuttle, then up to 237 kJ of electrical energy can be generated for each mole of H_2O produced.

Some insight into the relationship between ΔG and maximum work can be obtained by noting that for the same reaction under the same conditions, $\Delta H = -286$ kJ. This enthalpy of reaction shows that, for each mole of H_2O produced, 286 kJ of heat can be produced. In summary:

• If the reaction is run under conditions of constant pressure and temperature, and no attempt is made to extract the energy as work, then 286 kJ (in general, ΔH) of heat will be produced.

• Alternatively, if the reaction is run under conditions such that some of the energy released is used to do work, then up to 237 kJ (in general, ΔG) of nonexpansion work will be produced.

A question we need to explore is why not *all* the 286 kJ that can be released as heat be captured as nonexpansion (for example, electrical) work.

The answer lies in the entropy change that accompanies the reaction, which, under the same conditions, is

$$H_2(g) + \tfrac{1}{2}O_2(g) \rightarrow H_2O(l) \qquad \Delta S = -163 \text{ J K}^{-1}$$

That is, the entropy of the reaction system decreases (as a compact liquid is formed from two gases). It follows that the reaction will be spontaneous only if some heat is released into the surroundings to increase their entropy by at least 163 J K^{-1} so that, overall, ΔS_{tot} is positive. To generate 163 J K^{-1} of entropy in the surroundings at 25 °C (which is enough to cancel the reduction of entropy in the system) we must release

$$q_{rev} = T \, \Delta S_{sur} = 298 \text{ K} \times 163 \text{ J K}^{-1} = 49 \text{ kJ}$$

which is exactly the difference between the energy we can extract as heat and the maximum energy we can extract as nonexpansion work.

In summary:

> The difference between ΔG and ΔH is the energy that must be discarded into the surroundings as heat to ensure that the process is spontaneous and hence can produce work.

It is as though nature exerts a 'heat tax' on energy changes that are released as work: to ensure that a process can generate work, it may be necessary to pay the heat tax to the surroundings to ensure that, overall, there is an increase in disorder. The fact that ΔG represents the energy net of heat tax, the energy that we are free to use for nonexpansion work of various kinds, is the origin of the name *free energy* for G (Fig. 3.9).

Fig. 3.9 The arrows entering or leaving the system represent the flow of energy as heat in exothermic ((a) and (b)) and endothermic (c) reactions and the vertical arrows represent the change in entropy of the system and the surroundings. (a) A strongly exothermic reaction results in a large increase in entropy of the surroundings, and even though the system might undergo a reduction in entropy, the reaction is spontaneous. (b) In a less exothermic reaction, the entropy of the surroundings increases a small amount but the reaction may still be spontaneous, particularly (but not necessarily) if the entropy of the system increases too. (c) If the reaction is endothermic, then the entropy of the surroundings falls, and the reaction can be spontaneous *only* if the entropy of the system increases (and by a more than compensating amount).

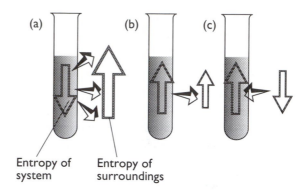

Entropy of system

Entropy of surroundings

Example Estimating a change in free energy

Suppose a certain small bird has a mass of 30 g. What is the minimum mass of glucose that it must consume to fly to a branch 10 m above the ground? The free energy change that accompanies the oxidation of 1.0 mol $C_6H_{12}O_6$ to carbon dioxide and water vapour at 25 °C is −2828 kJ.

Answer The work needed to raise a mass m through a height h on the surface of the Earth is equal to mgh, where g is the acceleration of free fall (9.81 m s^{-2}). The bird must do at least the following:

$$\text{work done by bird} = (30 \times 10^{-3}\,\text{kg}) \times (9.81\,\text{m s}^{-2}) \times (10\,\text{m})$$
$$= 2.9\,\text{J}$$

Therefore, enough glucose must be consumed to release 2.9 J of free energy. Because 1 mol releases 2828 kJ, 2.9 J can be obtained from 1.0×10^{-6} mol. The molar mass of $C_6H_{12}O_6$ is 180 g mol^{-1}, so the mass of glucose that must be consumed is 180 g mol$^{-1} \times 1.0 \times 10^{-6}$ mol $= 1.8 \times 10^{-4}$ g, or about 0.18 mg.

Exercise E3.8 A hard-working human brain, perhaps one that is grappling with physical chemistry, operates at about 25 W. What mass of glucose must be consumed to sustain that power output for an hour?

[5.7 g]

The great importance of the free energy in chemistry should be beginning to become apparent. At this stage, we see that it is a measure of the useful driving power of chemical reactions. In some cases, the nonexpansion work is extracted as electrical energy: that is the case when the reaction takes place in an electrochemical cell, as we shall see in Chapter 6. In other cases, the reaction may be used to build other molecules. That is the case in biological cells, where the free energy stored in ATP (adenosine triphosphate) is used to build proteins from amino acids, to power muscular contraction, and to derive the neural circuits in our senses and our brains.

The reaction free energy

The change in Gibbs free energy when a reaction takes place is called the **reaction free energy**, ΔG. However, we need to distinguish between the reaction free energy under standard conditions and the reaction free energy under other conditions. The former is the easier concept, and we deal with it first.

The standard reaction free energy is defined as follows:

> The **standard reaction free energy**, $\Delta G^{\ominus}$, is the change in Gibbs free energy that occurs when reactants in their standard states form products in their standard states.

For example, the standard free energy of the reaction

$$2H_2(g) + O_2(g) \rightarrow 2H_2O(l)$$

is the change in free energy when 2 mol H_2 as pure hydrogen gas at 1 bar reacts with 1 mol O_2 as pure oxygen gas at 1 bar to produce pure liquid water at the same pressure.

The standard reaction free energy can be calculated from the standard enthalpy and entropy of the reaction by using

$$\Delta G^{\ominus} = \Delta H^{\ominus} - T\Delta S^{\ominus} \tag{15}$$

in conjunction with the value of $\Delta H^{\ominus}$ calculated from tables of enthalpies of formation and $\Delta S^{\ominus}$ calculated from tables of absolute entropies.

Example Determining the Gibbs free energy of a reaction
What is the standard Gibbs free energy of the water-formation reaction specified above?
Answer We need two items of information. One is the standard enthalpy of reaction, which is obtained from tables of enthalpies of formation:

$$\Delta H^{\ominus} = (2\,\text{mol}) \times \Delta H_f^{\ominus}(H_2O, l)$$

$$- \{(2\,\text{mol}) \times \Delta H_f^{\ominus}(H_2, g) + (1\,\text{mol}) \times \Delta H_f^{\ominus}(O_2, g)\}$$

$$= 2 \times (-285.83\,\text{kJ}) - \{0 + 0\} = -571.66\,\text{kJ}$$

The second is the standard entropy of reaction, which is obtained from the standard entropies of the reactants and products:

$$\Delta S^{\ominus} = (2\,\text{mol}) \times S^{\ominus}(\text{H}_2\text{O}, l)$$
$$- \{(2\,\text{mol}) \times S^{\ominus}(\text{H}_2, g) + (1\,\text{mol}) \times S^{\ominus}(\text{O}_2, g)\}$$
$$= 2 \times 69.91\,\text{J K}^{-1} - \{2 \times 130.68\,\text{J K}^{-1} + 205.14\,\text{J K}^{-1}\}$$
$$= -326.68\,\text{J K}^{-1}, \text{ or } -0.32668\,\text{kJ K}^{-1}$$

It follows that

$$\Delta G^{\ominus} = -571.66\,\text{kJ} - (298.15\,\text{K}) \times (-0.32668\,\text{kJ K}^{-1})$$
$$= -474.26\,\text{kJ}$$

Exercise E3.9 Use the information in Appendix 1 to determine the standard Gibbs free energy of reaction of $3\text{O}_2(g) \rightarrow 2\text{O}_3(g)$ from standard enthalpies of formation and absolute entropies.

[+326.4 kJ]

The value of $\Delta G^{\ominus}$ is negative if $\Delta H^{\ominus}$ is negative (an exothermic reaction) and $\Delta S^{\ominus}$ is positive (a reaction system that becomes more disorderly, such as by forming a gas). The value of $\Delta G^{\ominus}$ may also be negative if the reaction is endothermic ($\Delta H^{\ominus}$ is positive) and $T\Delta S^{\ominus}$ is sufficiently large (and positive): this is so if the increase in disorder of the system is so great that it dominates the reduction in disorder of the surroundings (as measured by $\Delta H^{\ominus}$). Thus, in certain cases, and so long as the temperature is not too low (so that $T\Delta S^{\ominus}$ is large enough to overcome $\Delta H^{\ominus}$), endothermic reactions may be spontaneous (this point was illustrated in Fig. 3.9(c)). For an endothermic reaction to have a negative reaction free energy, its entropy of reaction *must* be positive. However, that alone is not sufficient: the temperature must also be high enough for $T\Delta S^{\ominus}$ to overcome the positive value of $\Delta H^{\ominus}$. The switch of $\Delta G^{\ominus}$ from positive to negative occurs at

$$T = \frac{\Delta H^{\ominus}}{\Delta S^{\ominus}} \tag{16}$$

As an example, consider the thermal decomposition of a solid, such as

$$\text{CaCO}_3(s) \xrightarrow{\Delta} \text{CaO}(s) + \text{CO}_2(g)$$
$$\Delta H^{\ominus} = +178\,\text{kJ}, \Delta S^{\ominus} = +161\,\text{J K}^{-1}$$

This reaction (and others like it) is endothermic because bonds are broken as the solid decomposes; it also has a positive entropy of reaction because a compact solid decomposes into a dispersed gas. Because the entropy of decompositon is similar for all such reactions (they all involve the decomposition of a solid into a gas), we can conclude that *the decomposition temperature of a solid increases with its enthalpy of decomposition.* For calcium carbonate, this decomposition temperature is

$$T = \frac{178 \times 10^3 \, \text{J}}{161 \, \text{J K}^{-1}} = 1.11 \times 10^3 \, \text{K}$$

or about 830 °C. It should be noted that an endothermic reaction can be made to produce products by heating *not* because the heating supplies the necessary energy but because the higher temperature reduces the adverse change in entropy of the surroundings to the point that the reaction becomes spontaneous. (The increased temperature also increases the rate of the reaction, but that is an entirely different matter.)

Just as the standard enthalpy of reaction can be expressed in terms of standard enthalpies of formation, so the standard free energy of reaction can be expressed in terms of the 'standard free energy of formation' of each species taking part in the reaction.

> The **standard free energy of formation** $\Delta G_f^{\ominus}$, of a compound is the standard free energy of formation of the compound from its elements in their reference states.

(The concept of reference state was introduced in Section 2.6; the temperature is arbitrary, but we shall almost always take it as 25 °C.) For example, the standard free energy of formation of liquid water is obtained from the thermochemical equation

$$H_2(g) + \tfrac{1}{2}O_2(g) \rightarrow H_2O(l) \qquad \Delta G^{\ominus} = -237 \, \text{kJ}$$

and is reported in kilojoules per mole of formula units of the compound:

$$\Delta G_f^{\ominus}(H_2O, l) = -237 \, \text{kJ mol}^{-1}$$

Some standard free energies of formation are listed in Table 3.3 (this table is a brief version of a longer one in Appendix 1). Note that it follows from the definition that the values for the formation of elements in their reference states are zero. Many of the values have been compiled by combining the enthalpy of formation of the compounds with the entropies of the compound and the elements, as illustrated above, but there are other sources of data, and we shall encounter some of them later.

The standard free energies of formation of compounds are like a measure of the thermodynamic altitude of a compound above or below a 'sea level' of stability represented by the elements (Fig. 3.10). If the free energy of formation is positive, so that the compound lies above sea level (like ozone), the compound has a spontaneous tendency to sink towards thermodynamic sea level and decompose into the elements: we say that the compound is **thermodynamically unstable**

Table 3.3 Some standard free energies of formation at 25 °C

Substance	$\Delta G_f^{\ominus}/$ kJ mol^{-1}
Gases	
Ammonia, NH_3	−16.5
Carbon dioxide, CO_2	−394.4
Dinitrogen tetroxide, N_2O_4	+97.9
Hydrogen iodide, HI	+1.7
Nitrogen dioxide, NO_2	+51.3
Sulfur dioxide, SO_2	−300.2
Water, H_2O	−228.6
Liquids	
Benzene, C_6H_6	+124.3
Ethanol, CH_3CH_2OH	−174.8
Water, H_2O	−237.1
Solids	
Calcium carbonate, $CaCO_3$	−1128.8
Iron(III) oxide, Fe_2O_3	−742.2
Silver bromide, AgBr	−96.9
Silver chloride, AgCl	−109.8

Additional values are given in Appendix 1.

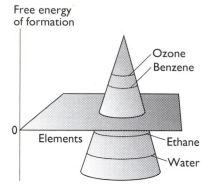

Fig. 3.10 The free energy of formation of compounds is like a measure of a compound's 'altitude' above (or below) 'sea level': compounds that lie above sea level have a spontaneous tendency to decompose into the elements (and to revert to sea level). Compounds that lie below sea level are stable with respect to decomposition into the elements.

with respect to its elements. Thus, ozone has a spontaneous tendency to decompose into oxygen under standard conditions at 25 °C: because $\Delta G_f^{\ominus}(O_3, g) = +163 \text{ kJ mol}^{-1}$, we know that

$$2O_3(g) \rightarrow 3O_2(g) \qquad \Delta G^{\ominus} = -326 \text{ kJ}$$

Although ozone is thermodynamically unstable, it can survive if the reactions that convert it into oxygen are slow: that is the case in the upper atmosphere, and the O_3 molecules in the ozone layer (which help to protect us from the ultraviolet radiation of the sun) survive for long periods. Benzene is also thermodynamically unstable with respect to its elements, with $\Delta G_f^{\ominus}(C_6H_6, l) = +124 \text{ kJ mol}^{-1}$, and it follows that

$$C_6H_6(l) \rightarrow 6C(s, \text{graphite}) + 3H_2(g) \qquad \Delta G^{\ominus} = -124 \text{ kJ}$$

However, the fact that bottles of benzene are everyday laboratory commodities also reminds us that *spontaneity is a thermodynamic tendency that might not be realized at a significant rate in practice*.

There is no point in searching for *direct* syntheses of thermodynamically unstable compounds from their elements (under standard conditions, at the temperature to which the determination applies), because the decomposition reaction, not the formation reaction, will occur once a small amount of the compound is formed. Such compounds must be synthesized by alternative routes or under conditions (such as those of temperature or pressure) for which their free energy of formation is negative and they lie beneath thermodynamic sea level.

Compounds with negative free energies of formation are said to be **thermodynamically stable** with respect to their elements: such compounds lie below the thermodynamic sea level of the elements (under standard conditions). An example is ethane gas, with $\Delta G_f^{\ominus}(C_2H_6, g) = -33 \text{ kJ mol}^{-1}$, so that its formation is spontaneous:

$$2C(s, \text{graphite}) + 3H_2(g) \rightarrow C_2H_6(g) \qquad \Delta G^{\ominus} = -33 \text{ kJ}$$

and its decomposition (under standard conditions at 25 °C) is nonspontaneous:

$$C_2H_6(g) \rightarrow 2C(s, \text{graphite}) + 3H_2(g) \qquad \Delta G^{\ominus} = +33 \text{ kJ}$$

Note that we have to specify the species with respect to which the compound is stable or unstable: although ethane is stable with respect to its elements, its combustion is spontaneous:

$$2C_2H_6(g) + 7O_2(g) \rightarrow 4CO_2(g) + 6H_2O(l) \qquad \Delta G^{\ominus} = -2934 \text{ kJ}$$

and so we could say that ethane is unstable with respect to carbon dioxide and water.

Once standard free energies of formation are available, it is easy to combine them to obtain the standard free energies of almost any reaction. We use a version of the now familiar expression

$$\Delta G^{\ominus} = \sum_{\text{Products}} n \, \Delta G_f^{\ominus}(\text{products}) - \sum_{\text{Reactants}} n \, \Delta G_f^{\ominus}(\text{reactants}) \quad (17)$$

For example, to determine the standard free energy of the reaction

$$2CO(g) + O_2(g) \rightarrow 2CO_2(g)$$

we carry out the following calculation:

$$\Delta G^{\ominus} = (2 \text{ mol}) \times \Delta G_f^{\ominus}(CO_2, g) - \{(2 \text{ mol}) \times \Delta G_f^{\ominus}(CO, g)$$
$$+ (1 \text{ mol}) \times \Delta G_f^{\ominus}(O_2, g)\}$$
$$= 2 \times (-394 \text{ kJ}) - \{2 \times (-137 \text{ kJ}) + 0\} = -514 \text{ kJ}$$

Calculations of this kind are of considerable use, for we shall now go on to see that from them we can predict the equilibrium constant of the corresponding reaction.

Exercise E3.10 Calculate the standard free energy of the partial oxidation of ammonia to nitric oxide in the reaction used for the production of nitric oxide: $4NH_3(g) + 5O_2(g) \rightarrow 4NO(g) + 6H_2O(g)$.

$$[-959.42 \text{ kJ mol}^{-1}]$$

Reaction free energies at arbitrary concentrations

The standard free energy of reaction relates to a very specific type of change: from the pure, unmixed reactants in their standard states to the pure, unmixed products in their standard states. It then follows that if $\Delta G^{\ominus}$ is negative, then change from pure, separated reactants to pure, separated products is spontaneous: the overall entropy increases. However, in chemistry we may be more interested in knowing whether a particular *mixture* of reactants and products has a tendency to form more products, and $\Delta G^{\ominus}$ would not appear to be relevant to that question. For example, consider the synthesis of hydrogen iodide at 25 °C:

$$H_2(g) + I_2(s) \rightarrow 2HI(g) \qquad \Delta G^{\ominus} = +3 \text{ kJ}$$

We would conclude from the value of $\Delta G^{\ominus}$ that the formation of hydrogen iodide gas is not spontaneous at 25 °C. However, in practice, hydrogen and iodine do react, and produce a mixture in which, although hydrogen and iodine are still present, some hydrogen iodide gas is also present. Apparently, although the free energy of pure HI is higher than that of pure H_2 and I_2, there is an intermediate composition at which all three substances are present and which has a lower free energy than either the pure reactants or the pure products.

To identify the composition that corresponds to the lowest free energy, we introduce the **reaction free energy** ΔG. This quantity is defined as the change in free energy that occurs when a reaction takes place in such a huge volume of material that the composition of the reaction mixture remains virtually constant. For example, consider the reaction

$$H_2(g) + I_2(s) \rightarrow 2HI(g)$$

that has reached a state with a certain composition. Suppose, for example, that all three substances are present in equal abundance, with 10^{10} mol H_2, 10^{10} mol I_2, and 10^{10} mol HI (or some similarly vast but equal amounts). Then, when 1 mol H_2 reacts with 1 mol I_2, the free energy of the system changes by ΔG, but the composition remains virtually constant. We could choose a system with a different composition, such as one in which there are 0.5×10^{10} mol H_2, 0.5×10^{10} mol I_2, and 2×10^{10} mol HI (or some other vast amounts in the ratio $1:1:2$), and once again consider the reaction in which 1 mol H_2 reacts with 1 mol I_2: this time a *different* value of ΔG will be found.

The same procedure may be carried out for any composition of the reaction mixture: in each case we consider a vastly magnified system (so that the production or consumption of any substance has virtually no effect on the composition), and allow the reaction to proceed according to the stoichiometric coefficients in the thermochemical equation. The change in free energy under these circumstances is the reaction free energy at that specified composition. It is very important to keep the distinction clear between ΔG and $\Delta G^{\ominus}$:

The value of $\Delta G^{\ominus}$ is the change in Gibbs function that occurs when the pure reactants change into the pure products, with each species in its standard state.

The value of ΔG is the change in Gibbs function for the reaction as written when the reaction occurs under conditions of constant composition of the reaction mixture.

We have seen that (at constant temperature and pressure) a process tends to move in the direction corresponding to a decrease in free energy. Therefore, if ΔG is negative at a certain composition, then the reactants will have a spontaneous tendency to form more products. Conversely, if ΔG is positive, then the *reverse* reaction is spontaneous, and the products that are already present tend to decompose into reactants. If, however, ΔG happens to be zero at a particular composition, then the reaction system will have no tendency to form either products or reactants. In other words, the system will be at equilibrium. We can conclude that *the criterion for chemical equilibrium is that, at constant temperature and pressure,*

$$\Delta G = 0 \tag{18}$$

All that remains to be done is to determine how ΔG varies with the composition of the system, for then we can identify the composition corresponding to this criterion. This important problem is solved in the following section.

3.4 The relation between ΔG and $\Delta G^{\ominus}$

First, we shall review some empirical results relating to chemical equilibria. Experiments during the nineteenth century showed that the composition of a system at equilibrium could be described by an **equilibrium constant** K that is characteristic of the reaction and which depends only on the temperature.

Definition and properties of the equilibrium constant

The expression for the equilibrium constant of a particular reaction can be written down by inspection of the balanced chemical equation. First, we formulate the **reaction quotient** Q by writing the partial pressures or molar concentrations of the products in the numerator and those of the reactants in the denominator, with each one raised to a power equal to the corresponding stoichiometric coefficient. An example is

$$N_2(g) + 3H_2(g) \rightarrow 2NH_3(g) \qquad Q = \frac{(p_{NH_3}/p^{\ominus})^2}{(p_{N_2}/p^{\ominus})(p_{H_2}/p^{\ominus})^3}$$

where p_X is the partial pressure of the species X and $p^{\ominus}$ is the standard pressure ($p^{\ominus} = 1$ bar). Then, when the system has its equilibrium composition, the reaction quotient is equal to a characteristic constant, the equilibrium constant K, for the reaction:

$$N_2(g) + 3H_2(g) \rightleftharpoons 2NH_3(g) \qquad Q_{eq} = K$$

where Q_{eq} is evaluated using the partial pressures of the gases in the equilibrium mixture. The sign $\rightleftharpoons$ signifies a state of **dynamic equilibrium** in which the forward and reverse reactions are continuing but as their rates are equal there is no net change. The appearance of the last equation but one can be simplified by writing $a_X = p_X/p^{\ominus}$ for each species, for then

$$Q = \frac{a_{NH_3}^2}{a_{N_2}a_{H_2}^3}$$

There are several advantages to this change of notation: one is the simplicity of the expressions that result; the second is the ease with which this expression is generalized to equilibria that involve other phases; and the third is the ease with which it is possible to make the transition to discussing real gases (and other nonideal systems). All we need note at this stage, though, is that a is a dimensionless quantity.

The first development of these remarks is to generalize them so that we can discuss equilibria in all types of system. All chemical equilibria are dynamic, and at equilibrium the forward and reverse rates are equal. This equality is true of the equilibrium between oxygen and haemoglobin (Hb) in blood that exists in our lungs:

$$Hb(aq) + O_2(g) \rightleftharpoons HbO_2(aq)$$

At equilibrium, the rate at which O_2 attaches to haemoglobin molecules is equal to the rate at which it escapes. It is true of the equilibrium between an acid, such as carbonic acid (H_2CO_3), and its anion, which exists in water, and helps to maintain the acidity of natural waters in the environment:

$$H_2CO_3(aq) + H_2O(l) \rightleftharpoons H_3O^+(aq) + HCO_3^-(aq)$$

At equilibrium, the rate at which hydrogen ions migrate from H_2CO_3 molecules to H_2O molecules is equal to the rate at which they transfer from H_3O^+ ions back to HCO_3^- ions. It is also true of reactions of industrial importance (when such reactions are allowed to reach equilibrium, which is not always the case), including the oxidation of nitrogen oxide (NO) to nitrogen dioxide (NO_2) in the manufacture of nitric acid:

$$2NO(g) + O_2(g) \rightleftharpoons 2NO_2(g)$$

At equilibrium, the rate at which NO reacts with O_2 molecules is equal to the rate at which NO_2 molecules break apart into NO and O_2 molecules.

In general, a chemical reaction and its reaction quotient have the form

$$aA + bB \rightarrow cC + dD \qquad Q = \frac{a_C^c a_D^d}{a_A^a a_B^b} \tag{19}$$

At equilibrium

$$aA + bB \rightleftharpoons cC + dD \qquad Q_{eq} = K \tag{20}$$

where the as are **activities**. For our purposes, it is sufficient to use the following values of a:

for perfect gases: $a_X = p_X/p^{\ominus}$

for pure liquids and solids: $a_X = 1$

for solutes at low concentration: $a_X = [X]/(1 \text{ mol L}^{-1})$

In more advanced work, the activities take into account the effects of interactions between species (such as intermolecular interactions in reactions of real gases and interionic forces in electrolyte solutions). An example of this general formula is

$$H_2(g) + I_2(s) \rightarrow 2HI(g)$$

$$Q = \frac{a_{HI}^2}{a_{H_2} a_{I_2}} = \frac{(p_{HI}/p^{\ominus})^2}{(p_{H_2}/p^{\ominus})} \qquad Q_{eq} = K$$

because I_2 is present as a solid reactant and $a_{I_2} = 1$.

Example Expressing a reaction quotient and an equilibrium constant

An amino acid in solution is in equilibrium with its *zwitterionic* form, in which a proton is lost from the carboxyl group and gained by the amino group. Express the reaction quotient and equilibrium constant for the equilibrium in the case of glycine, NH_2CH_2COOH.

Answer The equilibrium in solution is

$$NH_2CH_2COOH(aq) \rightleftharpoons {}^+NH_3CH_2CO_2^-(aq)$$

(The proton acquired by the amino group is not necessarily supplied by the carboxyl group: it might come from an H_2O molecule.) The reaction quotient at any stage of the attainment of equilibrium is

$$Q = \frac{a_{{}^+NH_3CH_2CO_2^-}}{a_{NH_2CH_2COOH}} = \frac{[{}^+NH_3CH_2CO_2^-]}{[NH_2CH_2COOH]}$$

where the second equality follows from the cancellation of the $1\ mol\ L^{-1}$ that appears in the definition of each activity. The second equality is also valid only when the solution is so dilute that ion–ion interactions can be ignored. At equilibrium, Q is equal to a constant K, the equilibrium constant for zwitterion formation.

Exercise E3.11 Write the reaction quotient and equilibrium constant for a liquid-phase esterification reaction of the form

$$CH_3COOH + C_2H_5OH \rightleftharpoons CH_3(CO)OC_2H_5 + H_2O$$

$$\left[Q = \frac{[CH_3(CO)OC_2H_5][H_2O]}{[CH_3COOH][C_2H_5OH]}, \quad K = Q_{eq} \right]$$

Physical transitions

The concept of equilibrium constant applies to physical transformations as well as to chemical reactions. For example, for the vaporization of water

$$H_2O(l) \rightarrow H_2O(g) \qquad Q = a_{H_2O(g)} = \frac{p_{H_2O}}{p^{\ominus}}$$

and at equilibrium

$$H_2O(l) \rightleftharpoons H_2O(g) \qquad Q_{eq} = K$$

from which it follows that

$$p_{H_2O} = K \times p^{\ominus}$$

These relations show that K (after multiplication by 1 bar) is essentially the **vapour pressure** of water, the pressure of the vapour in equilibrium with the liquid at the temperature of the experiment.

The general significance of K

It is established in introductory chemistry courses that an equilibrium constant is a guide to the feasibility of a reaction. If $K \gg 1$ (typically, if K is larger than about 10^3), then at equilibrium the products dominate the reaction mixture. If $K \ll 1$ (typically, less than about 10^{-3}), then at equilibrium the reactants dominate the mixture and only a small proportion of products are formed. If $K \approx 1$, then both reactants and products are present in similar abundance.

The variation of ΔG with composition

The justification for the use of a reaction quotient for expressing the composition of a general reaction mixture stems from the relation between ΔG and $\Delta G^{\ominus}$, which we shall now establish. The existence of an equilibrium constant is justified, and its value is obtained, by applying the criterion $\Delta G = 0$ to the relation between ΔG and $\Delta G^{\ominus}$.

The values of ΔG and $\Delta G^{\ominus}$ are related by a simple expression which can be derived in two stages. First, we need to know that the free energy of a perfect gas is related to its partial pressure in a mixture by

$$G = G^{\ominus} + nRT \ln \frac{p}{p^{\ominus}} \tag{21}$$

where $p^{\ominus} = 1$ bar is the standard pressure. A simpler version of this expression is

$$G = G^{\ominus} + nRT \ln a \tag{22}$$

where, as usual, $a = p/p^{\ominus}$. Thus, if the gas is at a pressure of 1 bar, then $a = p/p^{\ominus} = 1$, $\ln 1 = 0$, and $G = G^{\ominus}$. The logarithmic term takes into account the possibility that the gas is present at a partial pressure of other than 1 bar.

Exercise E3.12 Calculate the free energy of 1 mol of gas when its pressure falls from 1.00 bar to 0.50 bar as a result of a reaction at 25 °C.

$$[G = G^{\ominus} - 1.7 \text{ kJ}]$$

JUSTIFICATION

The proof of eqn (21) starts at the definition of dG for a change at constant temperature

$$dG = dH - T\,dS$$

We know that $H = U + pV$, so a change in H can be expressed in terms of changes in U, p, and V:

$dH = dU + p\,dV + V\,dp$

$\quad = dq + dw + p\,dV + V\,dp$ (because dU

$\qquad\qquad\qquad\qquad\qquad\quad = dq + dw$)

$\quad = T\,dS - p\,dV + p\,dV + V\,dp$ (because

$\qquad\qquad\qquad\qquad\qquad\quad dq = T\,dS$ and

$\qquad\qquad\qquad\qquad\qquad\quad dw = -p\,dV$

$\qquad\qquad\qquad\qquad\qquad\quad$ for reversible

$\qquad\qquad\qquad\qquad\qquad\quad$ changes)

$\quad = T\,dS + V\,dp$

When this expression is substituted into the one for dG we obtain

$$dG = V\,dp$$

If the sample is incompressible, so that V is independent of p, this expression integrates to

$$\Delta G = V\,\Delta p$$

Therefore, to find the change in G when the pressure is increased isothermally, we multiply the volume of the sample by the change in pressure.

If the substance is a gas, the overall change in Gibbs function when the pressure of the gas changes from p_i to p_f must allow for the fact that V changes with pressure. If we treat the gas as perfect, we can express the volume using the perfect gas equation of state, $V = nRT/p$, and obtain

$$\Delta G = \int_{p_i}^{p_f} V\,dp = nRT \int_{p_i}^{p_f} \frac{1}{p}\,dp = nRT \ln\frac{p_f}{p_i}$$

If we take $p_i = p^{\ominus}$ (at which pressure $G = G^{\ominus}$) and $p_f = p$ (when the free energy has the value G, so that $\Delta G = G - G^{\ominus}$), we obtain

$$G - G^{\ominus} = nRT \ln\frac{p}{p^{\ominus}}$$

which rearranges to the expression used in the text.

In the next step we consider a reaction of the form

$$A(g) + 2B(g) \rightarrow 3C(g)$$

The free energy of 3 mol C at a stage of the reaction when its partial pressure is p_C is

$$(3\text{ mol}) \times \{G^{\ominus}(C) + RT \ln a_C\}$$

$$= (3\text{ mol}) \times G^{\ominus}(C) + (1\text{ mol}) \times RT \ln a_C^3$$

where $a_C = p_C/p^{\ominus}$, and the total free energy of 1 mol A and 2 mol B when their partial pressures are p_A and p_B is

$$(1\text{ mol}) \times \{G^{\ominus}(A) + RT \ln a_A\} + (2\text{ mol}) \times \{G^{\ominus}(B) + RT \ln a_B\}$$

$$= \{(1\text{ mol}) \times G^{\ominus}(A) + (2\text{ mol}) \times G^{\ominus}(B)\} + (1\text{ mol}) \times RT \ln a_A a_B^2$$

The reaction free energy at the stated composition is the difference between these two quantities. To simplify the result, we note that the difference of standard free energies is the standard reaction free energy:

$$\Delta G^{\ominus} = (3\text{ mol}) \times G^{\ominus}(C) - \{(1\text{ mol}) \times G^{\ominus}(A) + (2\text{ mol}) \times G^{\ominus}(B)\}$$

Then,

$$\Delta G = \Delta G^{\ominus} + (1\text{ mol}) \times RT \times \{\ln a_C^3 - \ln a_A a_B^2\}$$

which rearranges to

$$\Delta G = \Delta G^{\ominus} + (1 \text{ mol}) \times RT \ln \frac{a_C^3}{a_A a_B^2}$$

To simplify this result we introduce the **molar free energy of reaction**,

$$\Delta G_m = \frac{\Delta G}{1 \text{ mol}} \tag{23}$$

and the reaction quotient

$$Q = \frac{a_C^3}{a_A a_B^2}$$

Then the expression becomes

$$\Delta G_m = \Delta G_m^{\ominus} + RT \ln Q \tag{24}$$

Equation (24) applies to all reactions, the only difference being that Q is the relevant reaction quotient. At this stage, we see that the reaction quotient appears naturally in the expression for the difference between the reaction free energy at the specified composition and the standard reaction free energy.

Exercise E3.13 Calculate the free energy of the reaction $N_2(g) + 3H_2(g) \rightarrow 2NH_3(g)$ at 25 °C when the partial pressures of nitrogen, hydrogen, and ammonia are 0.20 bar, 0.42 bar, and 0.61 bar, respectively. In which direction is the reaction spontaneous under these conditions?

[−22.91 kJ, forward]

3.5 The condition of equilibrium

The term $RT \ln Q$ in eqn (24) either adds to (when $Q > 1$ and $\ln Q > 0$) or subtracts from (when $Q < 1$ and $\ln Q < 0$) the value of $\Delta G_m^{\ominus}$, and so ΔG_m changes in the course of the reaction. This brings us to a very important point.

Suppose the reaction has reached equilibrium: it has no further tendency to change and $\Delta G = 0$. At equilibrium, the reaction quotient has a particular value K. It follows that, at equilibrium,

$$0 = \Delta G_m^{\ominus} + RT \ln K$$

and hence that

$$\Delta G_m^{\ominus} = -RT \ln K \tag{25}$$

This is one of the most important equations in the whole of chemical thermodynamics. Its principal use is to predict the value of the equilibrium constant of any reaction from tables of thermodynamic data, like those in Appendix 1. Alternatively, it can be used to measure $\Delta G^{\ominus}$ of a reaction following an experimental determination of the equilibrium constant of the reaction.

Example Predicting an equilibrium constant
Use the information in Table 3.3 to calculate the equilibrium constant of the reaction

$$H_2(g) + I_2(s) \rightleftharpoons 2HI(g)$$

at 25 °C.

Answer The reaction quotient is

$$Q = \frac{a_{HI}^2}{a_{H_2}a_{I_2}} = \frac{(p_{HI}/p^{\ominus})^2}{(p_{H_2}/p^{\ominus})} = \frac{p_{HI}^2}{p_{H_2}p^{\ominus}}$$

because $a_{I_2} = 1$ for a pure solid. We need the standard Gibbs function of the reaction at 25 °C (298 K), which is

$$\Delta G^{\ominus} = (2 \text{ mol}) \times \Delta G_f^{\ominus}(HI, g) = +3.40 \text{ kJ}$$

and hence

$$\Delta G_m^{\ominus} = \frac{+3.40 \text{ kJ}}{1 \text{ mol}} = +3.40 \text{ kJ mol}^{-1}$$

It follows that

$$\ln K = -\frac{\Delta G_m^{\ominus}}{RT} = -\frac{3.40 \times 10^3 \text{ J mol}^{-1}}{(8.3145 \text{ J K}^{-1} \text{ mol}^{-1}) \times (298 \text{ K})} = -1.37$$

Hence,

$$K = e^{-1.37} = 0.25$$

Therefore, at equilibrium, the partial pressures of hydrogen and hydrogen iodide satisfy the relation

$$\frac{p_{HI}^2}{p_{H_2}p^{\ominus}} = 0.25$$

That is, since $p^{\ominus} = 1$ bar, at equilibrium

$$p_{HI}^2 = p_{H_2} \times p^{\ominus} \times 0.25 = 0.25 p_{H_2} \text{ bar}$$

It follows that if the partial pressure of hydrogen happens to be 0.20 bar at equilibrium in a certain experiment, then the partial pressure of hydrogen iodide in the mixture will be

$$p_{HI} = \sqrt{(0.25 \text{ bar} \times 0.20 \text{ bar})} = 0.22 \text{ bar}$$

Exercise E3.14 Calculate the equilibrium constant of the reaction $N_2(g) + 3H_2(g) \rightleftharpoons 2NH_3(g)$ at 25 °C.

$$[6.0 \times 10^5]$$

The relation between K and $\Delta G^{\ominus}$ enables us to judge the feasibility of a chemical reaction from tables of standard free energies of formation. First, we note that the equation implies that $K > 1$ if $\Delta G^{\ominus}$ is negative. Broadly speaking, an equilibrium constant greater than 1 implies that the products will be dominant at equilibrium, so we can conclude that *a reaction is thermodynamically feasible if $\Delta G^{\ominus}$ is negative*. Conversely, because eqn (25) implies that $K < 1$ when $\Delta G^{\ominus}$ is positive, then we know that the *reactants* will be dominant in a reaction mixture at equilibrium if, for that reaction, $\Delta G^{\ominus}$ is positive. Some care must be exercised with these rules, however, because the products will be significantly more abundant than reactants only if K is *much* larger than 1 (more than about 10^3, as remarked earlier).

3.6 The response of equilibria to the conditions

The presence of a catalyst does not change the equilibrium constant of a reaction. The justification of this statement relies on the recognition of the fact that the value of K is determined by $\Delta G^{\ominus}$, which is the difference between the standard free energies of the products and reactants, and has the same value however the reaction is brought about in practice. Therefore, even though a catalyst may speed a reaction by providing a different reaction pathway between the reactants and the products, it has no effect on $\Delta G^{\ominus}$ and therefore no effect on K: *the equilibrium composition of a reaction mixture is independent of the presence of a catalyst.*

The effect of temperature

The equilibrium constant of a reaction changes when the temperature is changed. The easiest way of deriving the magnitude and direction of the effect is to consider the effect of temperature on $\Delta G_{m}^{\ominus}$, and to do so we return to its expression in terms of $\Delta H_{m}^{\ominus} - T\,\Delta S_{m}^{\ominus}$ (where $\Delta H_{m}^{\ominus} = \Delta H^{\ominus}/(1\text{ mol})$, and likewise for $\Delta S_{m}^{\ominus}$). We shall make the approximation that the standard reaction enthalpy and entropy are independent of temperature over the range of interest (this approximation was justified for ΔH in Section 2.7), so the entire temperature dependence of $\Delta G^{\ominus}$ stems from the T in

$$\Delta G^{\ominus} = \Delta H^{\ominus} - T\,\Delta S^{\ominus}$$

At a temperature T,

$$\ln K = -\frac{\Delta G_{m}^{\ominus}}{RT} = -\frac{\Delta H_{m}^{\ominus}}{RT} + \frac{\Delta S_{m}^{\ominus}}{R}$$

At another temperature T', when

$$\Delta G^{\ominus} = \Delta H^{\ominus} - T'\,\Delta S^{\ominus}$$

a similar expression holds:

$$\ln K' = -\frac{\Delta H_{m}^{\ominus}}{RT'} + \frac{\Delta S_{m}^{\ominus}}{R}$$

The difference between the two is

$$\ln K - \ln K' = -\frac{\Delta H_m^{\ominus}}{R}\left(\frac{1}{T}-\frac{1}{T'}\right)$$

which can be rearranged to

$$\ln K' = \ln K + \frac{\Delta H_m^{\ominus}}{R}\left(\frac{1}{T}-\frac{1}{T'}\right) \qquad (26)$$

This formula is a version of the **van't Hoff equation**.

Consider the case when T' is higher than T. Then the term in parentheses in eqn (26) is positive. If $\Delta H^{\ominus}$ is also positive, corresponding to an endothermic reaction, then the second term *adds* to $\ln K$, and so $\ln K'$ is larger than $\ln K$. That being so, we conclude that K' is larger than K for an endothermic reaction. In general, *the equilibrium constant of an endothermic reaction increases with temperature*. The opposite will be true when $\Delta H^{\ominus}$ is negative, so we can conclude that *the equilibrium constants of an exothermic reaction decreases with temperature*.

These conclusions are of considerable commercial and environmental significance. For example, the synthesis of ammonia

$$N_2(g) + 3H_2(g) \rightarrow 2NH_3(g) \qquad \Delta H^{\ominus} = -92\,kJ$$

is exothermic, so its equilibrium constant decreases as the temperature is increased; in fact, it falls below 1 at about 200 °C. Unfortunately, the reaction is slow at low temperatures, and is commercially feasible only if the temperature exceeds about 750 °C even in the presence of a catalyst. We shall see shortly how Fritz Haber, the inventor of the Haber process for the commercial synthesis of ammonia, was able to resolve this dilemma. Another example is the oxidation of nitrogen:

$$N_2(g) + O_2(g) \rightarrow 2NO(g) \qquad \Delta H^{\ominus} = +180\,kJ$$

This reaction is endothermic (largely as a consequence of the very high bond enthalpy of N_2), and its equilibrium constant increases with temperature. It is for this reason that nitrogen monoxide (nitric oxide) is formed in significant quantities in the hot exhausts of jet engines and in the hot exhaust manifolds of internal combustion engines, and then goes on to contribute to the problems of acid rain.

The effect of pressure

Because $\ln K$ is proportional to $\Delta G^{\ominus}$, and $\Delta G^{\ominus}$ is defined as the difference between the free energies of substances in their standard states (and therefore at 1 bar pressure), $\Delta G^{\ominus}$ has the same value whatever the actual pressure used for the reaction. Hence K *is independent of pressure*. Thus, if the pressure of a reaction vessel in which ammonia is being synthesized is increased, the equilibrium constant remains unchanged.

This rather startling conclusion should not be misinterpreted. The value of K is independent of the pressure to which the system is subjected, but that does not mean that the *individual* partial pressures or concentrations are unchanged. Suppose, for example, the volume of the reaction vessel in which the hydrogen iodide reaction has reached equilibrium is reduced by a factor of 2 and the system is allowed to reach equilibrium again. If the partial pressures were simply to double (that is, there is no adjustment of composition by further reaction), the reaction quotient would change from

$$Q_1 = \frac{(p_{HI}/p^{\ominus})^2}{(p_{H_2}/p^{\ominus})} = \frac{p_{HI}^2}{p_{H_2}p^{\ominus}}$$

to

$$Q_2 = \frac{(2 \times p_{HI})^2}{(2 \times p_{H_2})p^{\ominus}} = 2Q_1$$

However, the two expressions clearly cannot both be equal to the equilibrium constant K. Therefore the two partial pressures must adjust by different amounts. In this instance, the two reaction quotients remain equal if the partial pressure of HI changes by a smaller amount than a factor of 2 and the partial pressure of H_2 increases by more than a factor of 2, for then the numerator in Q_2 decreases and the denominator increases, with the result that Q_2 is smaller than we have just calculated (and would in fact have to be equal to Q_1). In other words, the equilibrium composition must shift in the direction of the reactants in order to preserve the equilibrium constant.

The general rule of thumb for predicting the effect of increased pressure on a gas-phase reaction at equilibrium is as follows:

> When the pressure is increased, the composition of a gas-phase equilibrium adjusts so as to reduce the number of molecules in the gas phase.

In the hydrogen iodide reaction, one H_2 molecule forms two HI molecules, so an increase in pressure favours the decomposition of HI into hydrogen (and solid iodine). In the synthesis of ammonia,

$$N_2(g) + 3H_2(g) \rightleftharpoons 2NH_3(g)$$

from each four reactant molecules, only two product molecules are formed, so an increase in pressure favours the formation of ammonia. Indeed, this is the key to resolving Haber's dilemma, for by working at high pressure he was able to increase the yield of ammonia even though the equilibrium constant for the reaction remained at its low, high-temperature value.

Exercise E3.15 Is the formation of products in the reaction $4NH_3(g) + 5O_2(g) \rightarrow 4NO(g) + 6H_2O(g)$ favoured by an increase or a decrease in pressure?

[decrease]

Care should be taken to note that under certain circumstances an increase in pressure has *no* effect on the equilibrium composition. A simple example is the effect of pressure on a reaction in which the number of gas phase molecules is the same in the reactants as in the products. An example is the oxidation of nitrogen mentioned above; another is the gas phase synthesis of hydrogen iodide:

$$H_2(g) + I_2(g) \rightleftharpoons 2HI(g)$$

A more subtle example is the effect of the addition of an inert gas to a reaction mixture that is contained inside a vessel of constant volume. The overall pressure increases as the gas (such as argon) is added, but the addition of a foreign gas does not affect the partial pressures of the other gases present. (The partial pressure, recall, is the pressure a gas would exert if it alone occupied the vessel, so it is independent of the presence or absence of any other gases.) Therefore, under these circumstances, not only does the equilibrium constant remain unchanged, but the partial pressures of the reactants and products remain the same whatever the stoichiometry of the reaction.

EXERCISES

3.1 A 1.75 kg sample of aluminium is cooled at constant pressure from 300 K to 265 K. Calculate the energy that must be removed as heat and the change in entropy of the sample. Use $C_p = 24\,J\,K^{-1}\,mol^{-1}$.

3.2 A 25 g sample of methane gas at 250 K and 18.5 atm expands isothermally until its pressure is 2.5 atm. Calculate the change in entropy of the gas.

3.3 A sample of perfect gas that initially occupies 15.0 L at 250 K and 1.00 atm is compressed isothermally. To what volume must the gas be decreased in order to reduce its entropy by $5.0\,J\,K^{-1}$?

3.4 Calculate the change in entropy when 50 g of water at 80 °C is poured into 100 g of water at 10 °C in an insulated vessel given that $C_p = 75.5\,J\,K^{-1}\,mol^{-1}$.

3.5 The enthalpy of vaporization of chloroform ($CHCl_3$) is $29.4\,kJ\,mol^{-1}$ at its normal boiling point

of 334.88 K. Calculate the entropy of vaporization of chloroform at this temperature. What is the entropy change in the surroundings?

3.6 Calculate the standard reaction entropy at 298 K of
(a) $2CH_3CHO(g) + O_2(g) \rightarrow 2CH_3COOH(l)$
(b) $2AgCl(s) + Br_2(l) \rightarrow 2AgBr(s) + Cl_2(g)$
(c) $Hg(l) + Cl_2(g) \rightarrow HgCl_2(s)$
(d) $Zn(s) + Cu^{2+}(aq) \rightarrow Zn^{2+}(aq) + Cu(s)$
(e) $C_{12}H_{22}O_{11}(s) + 12O_2(g)$
    $\rightarrow 12CO_2(g) + 11H_2O(l)$

3.7 Combine the reaction entropies calculated in Exercise 3.6 with the reaction enthalpies and calculate the standard reaction free energies of the reactions at 298 K.

3.8 Use standard free energies of formation to calculate the standard reaction free energy at 298 K of the reactions in Exercise 3.6.

3.9 The standard enthalpy of combustion of solid phenol (C_6H_5OH) is $-3054\,kJ\,mol^{-1}$ at 298 K and its standard molar entropy is $144.0\,J\,K^{-1}\,mol^{-1}$. Calculate the standard free energy of formation of phenol at 298 K.

3.10 Calculate the change in entropy when a perfect gas is compressed to half its volume and simultaneously heated to twice its initial temperature.

3.11 Calculate the maximum nonexpansion work per mole that may be obtained from a fuel cell in which the chemical reaction is the combustion of methane at 298 K.

3.12 The enthalpy of the graphite $\rightarrow$ diamond phase transition, which under 100 kbar occurs at 2000 K, is $+1.9\,kJ\,mol^{-1}$. Calculate the entropy change of the transition.

3.13 The equilibrium constant for the isomerization of *cis*-2-butene to *trans*-2-butene is $K = 2.07$ at 400 K. Calculate the standard reaction free energy.

3.14 The standard reaction free energy of the isomerization of *cis*-2-pentene to *trans*-2-pentene at 400 K is $-3.67\,kJ\,mol^{-1}$. Calculate the equilibrium constant of the isomerization.

3.15 The standard reaction enthalpy of $Zn(s) + H_2O(g) \rightarrow ZnO(s) + H_2(g)$ is approximately constant at $+224\,kJ\,mol^{-1}$ from 920 K up to 1280 K. The standard reaction free energy is $+33\,kJ\,mol^{-1}$ at 1280 K. Assuming that both quantities remain constant, estimate the temperature at which the equilibrium constant becomes greater than 1.

3.16 The equilibrium constant of the reaction

$$2C_3H_6(g) \rightleftharpoons C_2H_4(g) + C_4H_8(g)$$

is found to fit the expression

$$\ln K = -1.04 - \frac{1088\,K}{T} + \frac{1.51 \times 10^5\,K^2}{T^2}$$

between 300 K and 600 K. Calculate the standard reaction enthalpy and standard reaction entropy at 400 K. *Hint*: Begin by calculating ln K at 390 K and 410 K; then use eqn (26).

3.17 The standard reaction free energy of the isomerization of borneol to isoborneol in the gas phase at 503 K is $+9.4\,kJ\,mol^{-1}$. Calculate the reaction free energy in a mixture consisting of 0.15 mol of borneol and 0.30 mol of isoborneol when the total pressure is 600 Torr.

3.18 The equilibrium pressure of H_2 over a mixture of solid uranium and solid uranium hydride at 500 K is 1.04 Torr. Calculate the standard free energy of formation of $UH_3(s)$ at 500 K.

3.19 The equilibrium constant for the gas-phase isomerization of borneol ($C_{10}H_{17}OH$) to isoborneol at 503 K is 0.106. A mixture consisting of 7.50 g of borneol and 14.0 g of isoborneol in a 5.0 L container is heated to 503 K and allowed to come to equilibrium. Calculate the mole fractions of the two substances at equilibrium.

3.20 Use the data in Appendix 1 to decide which of the following reactions have $K > 1$ at 298 K.
(a) $HCl(g) + NH_3(g) \rightleftharpoons NH_4Cl(s)$
(b) $2Al_2O_3(s) + 3Si(s) \rightleftharpoons 3SiO_2(s) + 4Al(s)$
(c) $Fe(s) + H_2S(g) \rightleftharpoons FeS(s) + H_2(g)$
(d) $FeS_2(s) + 2H_2(g) \rightleftharpoons Fe(s) + 2H_2S(g)$
(e) $2H_2O_2(l) + H_2S(g) \rightleftharpoons H_2SO_4(l) + 2H_2(g)$

3.21 Which of the products in Exercise 3.20 are favoured by a rise in temperature at constant pressure (in the sense that K increases)?

3.22 What is the standard enthalpy of a reaction for which the equilibrium constant is (a) doubled, (b) halved when the temperature is increased by 10 K at 298 K?

3.23 The standard free energy of formation of $NH_3(g)$ is $-16.5\,kJ\,mol^{-1}$ at 298 K. What is the reaction free energy when the partial pressure of the N_2, H_2, and NH_3 (treated as perfect gases) are 3.0 bar, 1.0 bar and 4.0 bar respectively? What is the spontaneous direction of the reaction in this case?

3.24 The dissociation vapour pressure (the pressure of gaseous products in equilibrium with the solid reactant) of NH_4Cl at 427 °C is 608 kPa but at 459 °C it has risen to 1115 kPa. Calculate (a) the equilibrium constant, (b) the standard reaction free energy, (c) the standard enthalpy, (d) the standard entropy of dissociation, all at 427 °C. Assume that the vapour behaves as a perfect gas and that $\Delta H^{\ominus}$ and $\Delta S^{\ominus}$ are independent of temperature in the range given.

3.25 Estimate the temperature at which (a) $CaCO_3$ decomposes and (b) $CuSO_4 \cdot 5H_2O$ undergoes dehydration to the anhydrous solid.

Phase equilibria

Boiling, freezing, and the conversion of graphite to diamond are all examples of **phase transitions**, or changes of phase without change of chemical composition. Many phase changes are common everyday phenomena, and their description is an important part of physical chemistry. They occur whenever a solid changes into a liquid (as in the melting of ice) or a liquid changes into a vapour (as in the vaporization of water in our lungs). They also occur when one solid phase changes into another, as in the conversion of graphite into diamond under high pressure, or one phase of iron into another as it is heated in the process of steel making. Phase changes are important geologically too, for calcium carbonate is typically deposited as aragonite, but then gradually changes into another crystal form, calcite.

We shall see that the presence of a solute in a solvent can affect the properties of the phases, including the freezing and boiling temperatures of liquids. A related effect is of immense importance in living systems too, for the same mechanism that gives rise to changes of freezing and boiling points also gives rise to the phenomenon of osmosis, which is responsible for the transport of nutrients in living things, and also helps maintain their structures.

In this chapter we describe the properties of phases of pure substances and simple mixtures, and see how the presence of a solute can affect the properties of a solvent. Throughout, we use as our guiding principle the tendency of systems at constant temperature and pressure to adjust their free energy to a lower value. Thus, if the solid phase has a lower free energy than the liquid phase, then the liquid will solidify spontaneously; if the opposite is true, then the solid will melt spontaneously to a liquid. By the **thermodynamically stable phase** of a substance we shall mean the phase that, under the specified conditions (of pressure, temperature, and composition), has the lowest free energy. The thermodynamically stable phase is therefore the phase to which all other phases of the same substance under the same conditions have a spontaneous tendency to convert. For example, if ice is the thermodynamically stable phase of water at a particular pressure and temperature (which it is if the pressure is 1 atm and the temperature below 0 °C), then water vapour and liquid water will both tend to convert to ice. At 1 atm and below 0 °C, ice has a lower free energy than liquid water, but at 1 atm and above 0 °C, the opposite is true:

$$H_2O(l) \rightarrow H_2O(s) \qquad \Delta G < 0 \text{ below } 0\,°C,$$

$$\Delta G > 0 \text{ above } 0\,°C \text{ (at 1 atm)}$$

Exercise E4.1 The standard free energy of formation of metallic white tin (α-Sn) is 0 at 25 °C, and that of the nonmetallic grey tin (β-Sn) is $+0.13\,\text{kJ mol}^{-1}$ at the same temperature. Which is the thermodynamically stable phase at 25 °C?

[white tin]

It is found empirically that each phase transition of a substance occurs at a characteristic temperature for a given pressure. For instance, at 1 atm, the phase transition between ice and water occurs at 0 °C and the phase transition between grey and white tin occurs at 13 °C. The **transition temperature** is the temperature at which the two phases are in equilibrium (and at which $\Delta G = 0$):

$$H_2O(l) \rightarrow H_2O(s) \qquad \Delta G = 0 \text{ at } 0\,°C \text{ (at 1 atm)}$$

Transition temperatures, particularly melting points, of substances are useful diagnostic tests for their identification and purity.

As always when using thermodynamic arguments, it is important to keep in mind the distinction between the *spontaneity* of a phase transition and its *rate*: a phase transition that is predicted from thermodynamics to be spontaneous may in practice occur so slowly as to be unimportant. For instance, at normal temperatures and pressures the free energy of graphite is lower than that of diamond, and so there is a thermodynamic tendency for diamonds to convert into graphite:

$$C(s, \text{ diamond}) \rightarrow C(s, \text{ graphite}) \quad \Delta G^{\ominus} = -3 \text{ kJ (at 25 °C and 1 atm)}$$

However, for this transition to take place, the C atoms of diamond must change their locations, and this process is unmeasurably slow except at high temperatures because the bonds between the atoms are so strong and large numbers of bonds must change simultaneously. The rate of attainment of equilibrium is a *kinetic* problem (one to do with rates), not a thermodynamic problem (which deals only with tendencies). In gases and liquids the mobilities of the molecules allow phase transitions to occur rapidly, but in solids thermodynamic instability may be 'frozen in' and the thermodynamically unstable phase may persist for thousands of years.

Phase diagrams of single substances

The various possible phases of a substance become thermodynamically stable at different temperatures and pressures. The **phase diagram** of a substance is a map that shows the conditions at which its various phases are thermodynamically stable (Fig. 4.1). For example, at point A in the phase diagram in the illustration, the gaseous phase of the substance is thermodynamically the most stable, but at B the liquid phase is the most stable. The boundaries between regions, which are called **phase boundaries**, show the values of p and T at which the two phases are in equilibrium. For example, if the system is arranged to have a pressure and temperature represented by point C, then the liquid and its vapour are in equilibrium. If the temperature is reduced at constant pressure, the system moves to point B, and only liquid is present. If the temperature is reduced still further to D, the solid and the liquid phases can coexist in equilibrium. A further reduction in temperature to E takes the system into the region where the solid alone is present.

4.1 Phase boundaries

The liquid–gas phase boundary can be determined by measuring the pressure of a vapour in the presence of its liquid phase for a series of temperatures and plotting the graph of the pressure against temperature. For example, if a liquid is introduced into the vacuum at the top of a barometer, then it will depress the mercury and the difference in height is proportional to the pressure its vapour exerts (Fig. 4.2). If enough liquid is added so that some remains after the vapour forms, then the liquid and vapour phases coexist in equilibrium, and the pressure exerted by the vapour enables a point to be located on the liquid–gas phase boundary curve. The temperature can then be changed and another point on the curve determined, and so on. The pressure measured under these circumstances (with some liquid still present, so that vapour and liquid coexist in equilibrium) is called the **vapour pressure** of the substance. A plot of vapour pressure against temperature, the liquid–gas phase boundary, is called the **vapour pressure curve** of the substance (Fig. 4.3). The vapour pressure of a substance invariably increases with increasing temperature because as the temperature is raised the molecules in the liquid acquire more energy, and so can escape more readily from the attractive forces that cause them to cohere as a mobile liquid or even a solid. (The molecular factors affecting the relative magnitudes of the vapour pressures of substances are discussed in Chapter 10.)

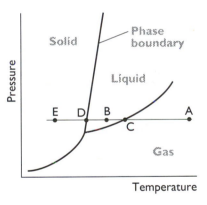

Fig. 4.1 A typical phase diagram, showing the regions of pressure and temperature at which each phase is the most stable. The phase boundaries (three are shown here) show the values of pressure and temperature at which the two phases separated by the line are in equilibrium. The significance of the letters A, B, C, D and E is explained in the text and is referred to in Fig. 4.4.

Fig. 4.2 When a small volume of water is introduced into the vacuum above the mercury in a barometer (a), the mercury is depressed (b) by an amount that is proportional to the vapour pressure of the liquid. (c) The same pressure is observed however much liquid is present (so long as some is present).

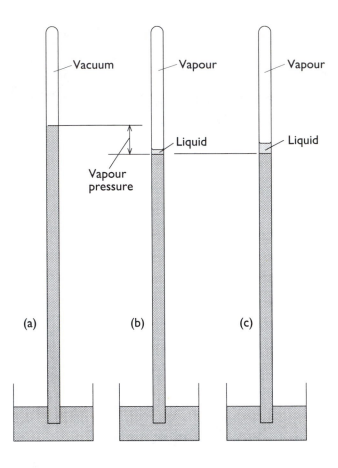

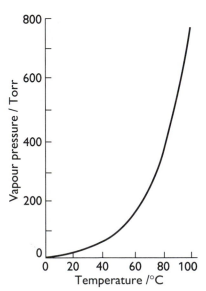

Fig. 4.3 The variation of the vapour pressure of water with temperature.

Exercise E4.2 What would be observed if a pressure of 50 Torr were applied to a sample of water in equilibrium with its vapour at 25 °C, when its vapour pressure is 23.8 Torr?

[condense to liquid]

The same approach can be used to determine the solid–vapour boundary. However, the vapour pressures of solids (the pressure of the vapour in equilibrium with a solid at a particular temperature) is usually much lower than the vapour pressure of a liquid. Special techniques have been developed for solids with very low vapour pressures: they make use of the relation between the vapour pressure and the rate of effusion of the vapour.

A more sophisticated procedure is needed for solid–solid and liquid–liquid phase boundaries because the transition is more difficult to detect. One approach is to use **thermal analysis**, which takes advantage of the heat released during a transition. In a typical thermal analysis experiment, a sample is allowed to cool and its temperature is monitored. When the transition occurs, heat is evolved and the cooling stops until the transition is complete (Fig. 4.4): the transition temperature is obvious from the shape of the graph and is used to mark point D on the phase diagram. This technique is particularly useful for solid–solid transitions, such as that between white and grey tin, where simple visual inspection of the sample may be inadequate.

Any point on a phase boundary of the phase diagram of a substance represents a state of dynamic equilibrium between the two neighbouring phases. We saw in Section 3.4 that a state of dynamic equilibrium is one in which the forward process is taking place at the same rate as the reverse process. For example, any point on the liquid–vapour boundary represents a state of dynamic equilibrium in which vaporization and condensation continue at matching rates. Molecules are leaving the surface of the liquid at a particular rate, and molecules already in the gas phase are returning to the liquid at the same rate so that there is no net change in the number of molecules in the vapour (and hence no net change in its pressure). Similarly, a point on the solid–liquid curve represents a state in which molecules are continuously breaking away from the surface of the solid and contributing to the liquid, but they are doing so at a rate that exactly matches the rate at which molecules already in the liquid are settling on to the surface of the solid and contributing to the solid phase.

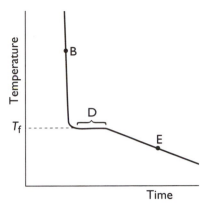

Fig. 4.4 The cooling curve for the BDE section of the horizontal line in Fig. 4.1. The halt at D corresponds to the pause in the fall in temperature while the liquid freezes and releases its enthalpy of transition. The halt enables T_f to be located even if the transition cannot be observed visually.

Critical points and boiling points

When a liquid is heated in an open vessel, the vapour pressure increases, as shown by the rise of the liquid–vapour boundary curve in Fig. 4.1 (and specifically for water in Fig. 4.3). When the vapour pressure of the liquid matches the external pressure, vapour can form throughout the bulk of the liquid and can expand into the surroundings by driving the atmosphere back to make room for its expansion. The formation of bubbles of vapour throughout the body of a liquid is the condition known as **boiling**. Therefore, we can conclude that *a liquid boils when its vapour pressure is equal to the external pressure.*

The temperature at which the vapour pressure of the liquid is equal to the external pressure is called the **boiling temperature**. When the external pressure is 1 atm, the boiling temperature is called the **normal boiling point**, T_b. It follows that we can identify the normal boiling point of a liquid by noting the temperature on the phase diagram at which its vapour pressure is 1 atm.

The vapour pressure of a liquid also rises when a sample is heated in a closed vessel, but under these conditions the liquid does not boil. Instead, the density of the vapour increases as the vapour pressure rises because the vapour is confined to a fixed volume (Fig. 4.5; recall Fig. 1.19, which illustrates the same point from the viewpoint of its bulk appearance). There is a temperature at which the density of the vapour is equal to that of the remaining liquid: at this temperature, the surface between the two phases disappears. The temperature at which the surface disappears is the **critical temperature**, T_c, which we first encountered in Section 1.8. The vapour pressure at the critical

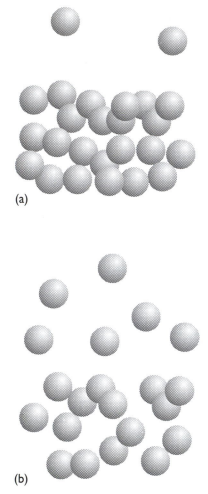

(a)

(b)

Fig. 4.5 (a) Well below the critical temperature, the density of the liquid phase (lower part of diagram (a)) is greater than that of the vapour (shown above the liquid). (b) Just below the critical temperature, so many molecules have escaped into the vapour, and the liquid has expanded, that the density of the vapour is almost the same as that of the liquid. The two phases have the same densities at the critical temperature, and there is no surface between them.

Table 4.1   Critical constants†

	p_c/atm	V_c/cm^3 mol^{-1}	T_c/K
Ammonia, NH_3	111	73	406
Argon, Ar	48	75	151
Benzene, C_6H_6	49	260	563
Bromine, Br_2	102	135	584
Carbon dioxide, CO_2	73	94	304
Chlorine, Cl_2	76	124	417
Ethane, C_2H_6	48	148	305
Ethene, C_2H_4	51	124	283
Hydrogen, H_2	13	65	33
Methane, CH_4	46	99	191
Oxygen, O_2	50	78	155
Water, H_2O	218	55	647

† The critical molar volume V_c is the molar volume at the critical pressure and critical temperature.

temperature is called the **critical pressure**, p_c, and the critical temperature and critical pressure together identify on the phase diagram the **critical point** of the substance (Table 4.1). At and above the critical temperature, a single uniform phase—a gas—fills the container; above this temperature, the liquid does not exist. That is why the liquid–vapour phase boundary terminates at the critical point.

Melting points and triple points

The temperature at which the liquid and solid phases of a substance coexist in equilibrium is called the **melting temperature** of the substance (at a specified pressure). Because a substance melts at the same temperature as it freezes, the melting temperature is the same as the **freezing temperature**. The solid–liquid boundary therefore shows how the melting temperature of a solid varies with pressure (Fig. 4.6). The melting temperature, when the pressure on the sample is 1 atm, is called the **normal melting point** or the **normal freezing point**, T_f. A liquid freezes when the energy of the molecules in the liquid is so low that they cannot escape from the attractive forces of their neighbours, and their mobility is lost.

There is a set of conditions under which three different phases (typically solid, liquid, and vapour) all simultaneously coexist in equilibrium. It is represented by the **triple point**, where the three phase boundaries meet (Fig. 4.6). The triple point of a pure substance is a characteristic physical property of the substance. For water the triple point lies at 273.16 K and 611 Pa (4.6 Torr), and ice, liquid water, and water vapour coexist in equilibrium at no other combination of pressure and temperature. At the triple point, the rate at which molecules break away from the solid to form a liquid or a vapour, the rate at which the molecules of the liquid form a solid, and the rate at which molecules in the vapour condense to a liquid or a solid, are all equal, and there is dynamic equilibrium between all three phases.

As we see from Fig. 4.7(a), if the slope of the solid–liquid phase boundary is as shown in the diagram:

- The triple point marks the lowest temperature at which the liquid can exist.
- The critical temperature marks the highest temperature at which the liquid can exist.

We shall see in the following section that for a few materials (most notably water) the solid–liquid phase boundary slopes as shown in Fig. 4.7(b), and so then only the second of these conclusions is applicable.

The question might arise about the possibility that *four* phases might be in equilibrium (such as the two solid forms of tin, molten tin, and tin vapour). However, it can be shown by very general arguments that, for systems consisting of one substance, no more than three phases can coexist. The general result is a consequence of the **phase rule**. This celebrated consequence of thermodynamics is explained in *Further information 4* (p. 124).

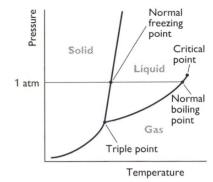

Fig. 4.6 The significant points of a phase diagram. The liquid–vapour phase boundary terminates at the critical point. At the triple point, solid, liquid, and vapour are in dynamic equilibrium. The normal freezing point is the temperature at which the liquid freezes when the pressure is 1 atm; the normal boiling point is the temperature at which the vapour pressure of the liquid is 1 atm.

4.2 Phase diagrams of typical materials

We shall now show how these general features appear in the phase diagrams of a selection of pure substances.

Water

Figure 4.8 is the phase diagram for water. The liquid–vapour phase boundary summarizes how the vapour pressure of liquid water varies with temperature. The curve (which is shown in more detail in Fig. 4.3) also summarizes how the boiling temperature varies with changing external pressure: when the external pressure is 149 Torr (at an altitude of 12 km), for instance, water boils at 60 °C.

Exercise E4.3 What is the minimum pressure at which liquid is the thermodynamically stable phase of water at 25 °C?

[23.8 Torr, 3.22 kPa]

The solid–liquid boundary line shows how the melting temperature depends on the pressure. For example, although ice melts at $0\,°C$ at 1 atm, it melts at $-1\,°C$ when the pressure is 130 atm. The very steep slope of the boundary indicates that enormous pressures are needed to bring about significant changes. Notice that the line slopes down from left to right, which means that the melting temperature of ice falls as the pressure is raised. The reason for this behaviour (which is very uncommon) can be traced to the decrease in volume that occurs when ice melts into water: it is favourable for the solid to transform into the denser liquid as the pressure is raised. The decrease in volume is a result of the very open structure of the ice crystal lattice: the water molecules are held apart (as well as together) by the hydrogen bonds between them, but the structure partially collapses on melting and the liquid is denser than the solid.

FURTHER INFORMATION 4: The phase rule

The **phase rule** is celebrated as one of the most elegant results of chemical thermodynamics. It was derived by Josiah Gibbs, and is applicable to any system at equilibrium. It states that, for a system at equilibrium,

$$F = C - P + 2$$

where F is the number of degrees of freedom, C is the number of components, and P is the number of phases.

A gas, or a gaseous mixture, is a single phase, a crystal is a single phase, and two totally miscible liquids form a single phase. The **number of phases**, P, is the number of such regions in the system. Ice is a single phase ($P = 1$) even though it might be chipped into small fragments. A slurry of ice and water is a two-phase system ($P = 2$) even though it is difficult to map the boundaries between the phases. The **number of components**, C, in a system is the minimum number of independent species necessary to define the composition of all the phases present in the system. The definition is easy to apply when the species present in a system do not

react, for then we simply count their number. For instance, pure water is a one-component system ($C = 1$) and a mixture of ethanol and water is a two-component system ($C = 2$). The **number of degrees of freedom**, F, of a system is the number of intensive variables (such as the pressure, temperature, or mole fractions) that can be changed independently without disturbing the number of phases in equilibrium.

For a one-component system, such as pure water,

$$F = 3 - P$$

When only one phase is present, $F = 2$ and both p and T can be varied independently. In other words, a single phase is represented by an *area* on a phase diagram. When two phases are in equilibrium, $F = 1$, which implies that pressure is not freely variable if we have set the temperature. That is, the equilibrium of two phases is represented by a *line* in the phase diagram. Instead of selecting the temperature, we can select the pressure, but having done so the two phases come into equilibrium at a single definite temperature. Therefore, freezing (or any other phase transition) occurs at a definite temperature at a given pressure. When three phases are in equilibrium, $F = 0$. This special invariant condition can therefore be established

JUSTIFICATION

The explanation of the effect of pressure is based on the following equation for the effect of an increase in pressure Δp on the free energy of each phase:

$$G' = G + V\Delta p$$

where V is the volume of the phase. (This expression was derived in Section 3.4.) The volume of a given mass of ice is larger than the volume occupied by the same mass of liquid water. Therefore, as the pressure is increased, the free energy of the ice rises more rapidly than that of the liquid and it may rise above the free energy of liquid water if the pressure is great enough. Then liquid water becomes the thermodynamically stable phase, the transition ice → liquid water becomes spontaneous, and the ice melts. For a more conventional substance (one for which the volume of the liquid is greater than that of the solid, such as carbon dioxide) application of pressure raises the free energy of the liquid more than that of the solid, the solid becomes the thermodynamically stable phase, and the liquid solidifies.

---

only at a definite temperature and pressure. The equilibrium of three phases is therefore represented by a *point*, the triple point, on the phase diagram. Four phases cannot be in equilibrium in a one-component system because F cannot be negative.

When two components are present in a system, $C = 2$ and

$$F = 4 - P$$

For simplicity we shall keep the pressure constant (at 1 atm, for instance), which uses up one of the degrees of freedom, and write $F' = 3 - P$ for the number of remaining degrees of freedom. One of these remaining degrees of freedom is the temperature, the other is the composition (as expressed by the mole fraction of one component). Hence we should be able to depict the phase equilibria of the system on a temperature–composition diagram of the kind introduced in Chapter 5. As an illustration, consider the phase diagram shown in Fig. 1. In a region where there is only one phase (as at a), $F' = 2$ and both the temperature and the composition can be varied. If two phases are present at equilibrium (as at b), $F' = 1$, and only one of the two variables may be changed at will. For example, if the composition is changed, then to maintain equilibrium between the two phases,

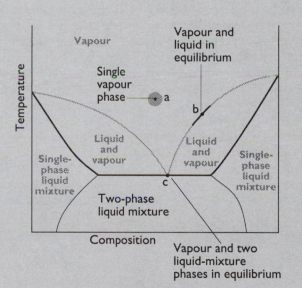

Fig. 1

the temperature must be adjusted. Such two-phase equilibria therefore define a line in the phases diagram. If three phases are present, as at c, $F' = 0$ and there is no degree of freedom for the system. That is, to achieve a three-phase equilibrium, we must adopt a specific temperature and composition. Such a condition is therefore represented by a point on the phase diagram.

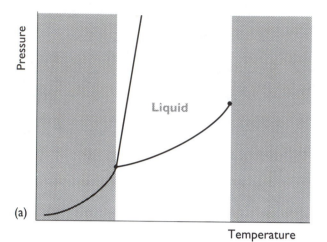

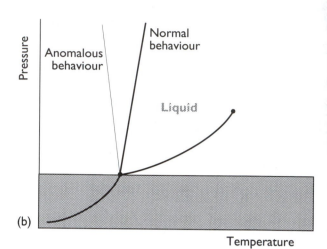

Fig. 4.7 (a) For substances which have phase diagrams that resemble the one shown here (which is common for most substances, with the exception of water), the triple point and the critical point mark the range of temperatures over which the substance may exist as a liquid. The shaded areas show the regions of temperature in which a liquid cannot exist as a stable phase. (b) A liquid cannot exist as a stable phase if the pressure is below that of the triple point for normal or anomalous liquids.

The phase diagram in Fig. 4.8 shows that water has many different solid phases other than the ordinary ice ('ice I') with which we are familiar. The solid phases differ in the arrangement of the water molecules: under the influence of very high pressures, hydrogen bonds buckle and the H_2O molecules adopt different arrangements. Some of these ices have bizarre properties: ice VII (which is not shown in the illustration) melts at 100 °C, but it exists only if the pressure is greater than 25 kbar. These different ices may be responsible for the advance of glaciers, for ice at the bottom of glaciers experiences very high pressures where it rests on jagged rocks. The sudden apparent explosion of Halley's comet in 1991 may have been due to the conversion of one form of ice into another in its interior.

Carbon dioxide

The phase diagram for carbon dioxide is shown in Fig. 4.9. The features to notice include the slope of the solid–liquid boundary (this positive slope is typical of almost all substances), which indicates that the melting temperature of solid carbon dioxide rises as the pressure is increased. As the triple point (217 K, 5.11 bar) lies well above ordinary atmospheric pressure, the liquid does not exist at normal atmospheric pressures whatever the temperature, and the solid sublimes when left in the open (hence the name 'dry ice'). To obtain liquid carbon dioxide, it is necessary to exert a pressure of at least 5.11 bar.

Cylinders of carbon dioxide generally contain the liquid or compressed gas; if both gas and liquid are present inside the cylinder, then at 20 °C the pressure must be about 65 atm. When the gas squirts through the throttle it cools by the Joule–Thomson effect, so when it emerges into a region where the pressure is only 1 atm, it condenses into a finely divided snow-like solid.

Carbon

Figure 4.10 shows a simplified version of the phase diagram for carbon. It is ill-defined and incomplete because the various phases

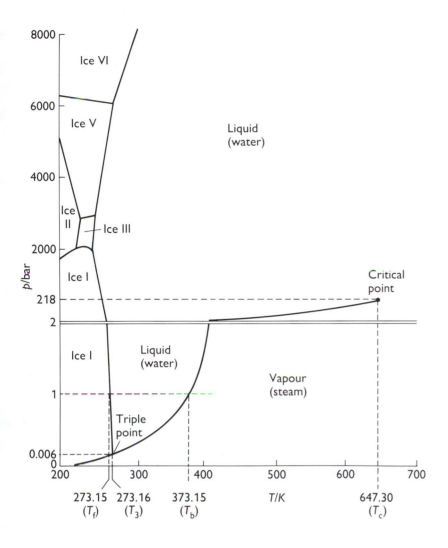

come into stability at extremes of temperature and pressure, and gathering the data is very difficult. For instance, at atmospheric pressure, carbon gas is the stable phase only at temperatures well over 4000 K. To obtain liquid carbon it is necessary to work at about 4000 K and 1 kbar (about 1000 atm), or at 2000 K and 1000 kbar. Making diamonds is a minor problem in comparison, because the diamond phase becomes stable at about 0.1 kbar (about 100 atm) and 1000 K.

Small diamonds are synthesized and are widely used in industry, but the phase diagram does not reveal the full problem. The rate of conversion is an important factor, and pure graphite changes into diamond at a useful rate only when the temperature is about 4000 K and the pressure exceeds 200 kbar; but then the apparatus tends to disappear first. Therefore catalysts are added in commercial syntheses, and then the conversion proceeds at 70 kbar and 2300 K, which are attainable conditions. The contamination by the metal catalysts, such

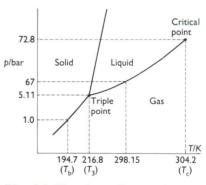

Fig. 4.9 The phase diagram for carbon dioxide. Note that, as the triple point lies well above atmospheric pressure, liquid carbon dioxide does not exist under normal conditions (a pressure of at least 5.11 bar must be applied).

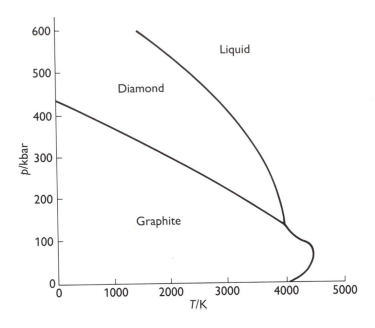

Fig. 4.10 A simplified version of the phase diagram for carbon. Note that the pressure axis is expressed in kilobars (1 kbar ≈ 1000 atm). There are very large uncertainties about the precise form of this phase diagram because the data are so difficult to obtain. The location of the fullerene (C_{60}) phase has not yet been determined.

as molten nickel (which also acts as a solvent for the carbon), enables commercial and natural diamonds to be distinguished.

Although diamond and graphite were long believed to be the only two solid phases of carbon attainable under normal conditions, a third phase, buckminsterfullerene, which consists of C_{60} clusters, was characterized in 1985. It is possible to trap electrons and ions inside the C_{60} cages, and the resulting materials show a variety of unusual properties, among them superconductivity.

Helium

The phase diagram of helium is shown in Fig. 4.11. Helium behaves unusually at low temperatures. For instance, the solid and gas phases of helium are never in equilibrium however low the temperature: the atoms are so light that they vibrate with a large-amplitude motion even at very low temperatures, and the solid simply shakes itself apart. Solid helium can be obtained, but only by holding the atoms together by applying pressure. A second unique feature of helium is that pure helium-4 has two liquid phases. The phase marked He-I in the diagram behaves like a normal liquid; the other phase, He-II, is a **superfluid**; it is so called because it flows without viscosity.

The properties of nonelectrolyte solutions

We now leave pure materials and the limited but important changes they can undergo, and examine solutions. We shall consider mainly

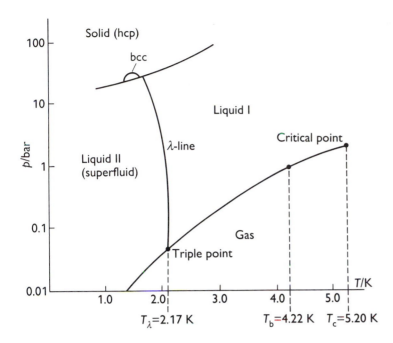

Fig. 4.11 The phase diagram for helium (specifically, ^{4}He). The λ-line marks the conditions under which the two liquid phases are in equilibrium: helium-I is a conventional liquid, but helium-II is superfluid. Note that a pressure of at least 20 bar must be exerted before solid helium can be obtained. The labels hcp and bcc refer to the packing of atoms in the solid (see Chapter 10).

nonelectrolyte solutions, where the solute is not present as ions, such as sucrose in water or sulfur in carbon disulfide. We shall delay until Chapter 6 the special problems of **electrolyte solutions**, in which the solute is ionized and where the ions generally interact strongly with each other.

4.3 The thermodynamic description of mixtures

In this and the following chapters we need a set of concepts that enable us to apply thermodynamics to mixtures of variable composition. We have already seen that the partial pressure, the contribution to the total pressure of one component in a gaseous mixture, is used to discuss the properties of mixtures of gases. For a more general description of the thermodynamics of mixtures we need to introduce other 'partial' properties, each one being the contribution that a particular component makes to the designated property.

Measures of concentration

At the outset we need to establish a way of reporting the composition of mixtures. There are essentially three measures of concentration that we shall employ: one, the molar concentration, emphasizes the amount of solute per unit volume; the other two, the molality and the mole fraction, emphasize the relative numbers of solute and solvent molecules.

The **molar concentration**, [X], of a solute X in a solution is the amount of substance per unit volume:

$$[X] = \frac{n_X}{V_{solution}} \qquad (1a)$$

Molar concentration is typically reported in moles per litre, $mol\,L^{-1}$ (more formally, as $mol\,dm^{-3}$). The concentration $1\,mol\,L^{-1}$ is widely denoted by the symbol $1\,M$. In practice, a solution of given molar concentration is prepared by measuring out the appropriate mass to solute into a volumetric flask, dissolving the solute in a little solvent, and then adding enough solvent to produce the desired volume. Thus, to prepare $1.0\,M\;C_6H_{12}O_6(aq)$, $180\,g$ of glucose ($1.0\,mol\;C_6H_{12}O_6$) would be dissolved in enough water to produce $1.0\,L$ of solution (note that the solute is not dissolved in $1.0\,L$ of water).

Exercise E4.4 What mass of glycine (NH_2CH_2COOH) should be used to make $250\,mL$ of a $0.015\,M$ $NH_2CH_2COOH(aq)$ solution?

[0.282 g]

Once the molar concentration of a solute is known, it is easy to calculate the amount of that substance in a given volume of solution by multiplying the volume of solution by the molar concentration:

$$n_X = [X] \times V_{solution} \qquad (1b)$$

The **molality**, m_X, of a solute X in a solution is the amount of substance per unit mass of solvent:

$$m_X = \frac{n_X}{M_{solvent}} \qquad (2)$$

where M is the mass of solvent. Molality is typically reported in moles per kilogram of solvent, $mol\,kg^{-1}$. There is an important distinction between molar concentration and molality: whereas the former is defined in terms of the volume of the solution, the latter is defined in terms of the mass of pure solvent used to prepare the solution. Thus, a $1.0\,mol\,kg^{-1}$ solution of glucose in water would be prepared by measuring out $180\,g$ of glucose and dissolving it in $1.0\,kg$ of water.

The advantage of the molality over the molar concentration is that it is independent of temperature (the molar concentration varies with temperature because the volume of the solution changes; the mass of solvent, though, remains constant). Moreover, molality is used when it is important to emphasize the relative amounts of solute and solvent molecules. The mass of solvent is proportional to the amount of its molecules present, so from the expression above we see that the molality is proportional to the ratio of the amounts of solute and solvent molecules. For example, a $1.0\,mol\,kg^{-1}\;C_6H_{12}O_6(aq)$ solution contains $1.0\,mol\;C_6H_{12}O_6$ molecules and $55.5\,mol\;H_2O$ molecules. Indeed, any $1.0\,mol\,kg^{-1}$ aqueous nonelectrolyte solution contains

1.0 mol solute particles and 55.5 mol H_2O molecules, so in each case there is 1 solute molecule per 55.5 solvent molecules.

Closely related to the molality of a solute is the mole fraction x, which was introduced in Section 1.3 in connection with mixtures of gases:

$$\text{mole fraction of a species} = \frac{\text{amount of substance of the species}}{\text{total amount of substance in the mixture}}$$

In brief,

$$x_A = \frac{n_A}{n}$$

where n_A is the amount (in moles) of a species A and n is the total amount of species in the sample. As remarked in Section 1.3, the mole fraction of a species A is the fraction of the total number of particles present that are the species A.

Example Relating mole fraction and molality
What is the mole fraction of glycine in an aqueous solution of molality $0.140 \text{ mol kg}^{-1}$?
Answer A solution of the specified molality contains $0.140 \text{ mol } NH_2CH_2COOH$ in 1.00 kg of water. The mass of water corresponds to $55.48 \text{ mol } H_2O$. The total amount of species present in the solution is $55.48 + 0.140 = 55.62 \text{ mol}$. The mole fraction of glycine is therefore

$$x_{NH_2CH_2COOH} = \frac{0.140 \text{ mol}}{55.62 \text{ mol}} = 2.52 \times 10^{-3}$$

That is, about 1 molecule in 400 is a glycine molecule.

Exercise E4.5 Calculate the mole fraction of sucrose in an aqueous sample of molality 1.22 mol kg^{-1}.

$$[2.15 \times 10^{-2}]$$

Partial molar properties

A **partial molar property** is the contribution (per mole) that a substance makes to a property of the system when it is part of a mixture. The easiest partial molar property to visualize is the **partial molar volume**, the contribution to the volume that a component in a sample makes to the total volume of the sample. We have to be alert to the fact that although 1 mol of a substance has a characteristic volume when it is pure, 1 mol of a substance can make different contributions to the total volume of a sample when it is a part of a mixture.

To grasp the meaning of the concept, we first imagine a huge volume of pure water. When a further 1 mol H_2O is added, the volume

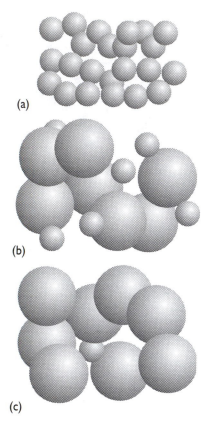

(a)

(b)

(c)

Fig. 4.12 A depiction of the environment of a molecule of one substance (A, small spheres) in a liquid mixture with another substance (B, large spheres). (a) The partial molar volume of A in pure A is the volume per mole of A when A is surrounded by A molecules. (b) The partial molar volume of A in a typical mixture is the molar volume of A when it is surrounded by both A and B molecules. (c) The partial molar volume of A in pure B is the molar volume of A when it is surrounded by only B molecules.

Fig. 4.13 The partial molar volumes of water and ethanol at 25 °C. Note the different scales (water on the left, ethanol on the right).

increases by $18 \, cm^3$. However, when we add $1 \, mol \, H_2O$ to a huge volume of pure ethanol, the volume increases by only $14 \, cm^3$. The quantity $18 \, cm^3 \, mol^{-1}$ is the volume occupied per mole of water molecules in a sample when that sample is pure water; $14 \, cm^3 \, mol^{-1}$ is the volume occupied per mole of water molecules when that sample is pure ethanol. In other words, the partial molar volume of water in pure water is $18 \, cm^3 \, mol^{-1}$ whereas the partial molar volume of water in pure ethanol is only $14 \, cm^3 \, mol^{-1}$. The reason for the different increase in volume is that the volume occupied by a given number of water molecules depends on the molecules that surround them (Fig. 4.12). In the latter case there is so much ethanol present that each H_2O molecule is surrounded by ethanol molecules and the packing of the molecules results in the water molecules occupying only $14 \, cm^3$. The partial molar volume at an intermediate composition of the mixture reflects in a similar way the volume the H_2O molecules occupy when they are surrounded by a mixture of molecules that is representative of the overall composition. The partial molar volume of ethanol also varies as the composition of the mixture is changed, because the environment of an ethanol molecule changes from pure ethanol to pure water as the proportion of water increases, and the volume occupied by the ethanol molecules varies accordingly. The variation of the two partial molar volumes across the full composition range at 25 °C is shown in Fig. 4.13.

Once we know the partial molar volumes V_A and V_B of the two components A and B of a mixture at the composition (and temperature) of interest we can state the total volume V of the mixture using

$$V = n_A V_A + n_B V_B \tag{3}$$

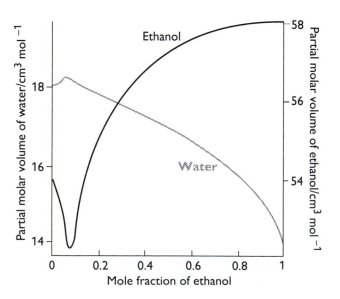

JUSTIFICATION

To show the validity of eqn (3), consider a very large sample of the mixture of the specified composition. Then, when an amount n_A of A is added, the composition remains virtually unchanged but the volume of the sample changes by $n_A V_A$; similarly, when an amount n_B of B is added, the volume changes by $n_B V_B$. The total change of volume is therefore $n_A V_A + n_B V_B$. The mixture now occupies a larger volume, but the proportions of the components are still the same. Next, scoop out of this enlarged volume a sample containing n_A of A and n_B of B. Its volume is $n_A V_A + n_B V_B$. Because volume is a state property, the same sample could have been prepared simply by mixing the appropriate amounts of A and B.

Example Using partial molar volumes

What is the total volume of a mixture of 50.0 g of ethanol and 50.0 g of water at 25 °C?

Answer We need the mole fractions of each substance and the corresponding partial molar volumes. The mole fractions are calculated in the same way as illustrated in Section 1.3 (by using the molar masses of the components to calculate the amounts), and are

$$x_{C_2H_5OH} = 0.282 \qquad x_{H_2O} = 0.718$$

According to Fig. 4.13, the partial molar volumes of the two substances in a mixture of this composition are $56 \text{ cm}^3 \text{ mol}^{-1}$ and $17 \text{ cm}^3 \text{ mol}^{-1}$, respectively, so the total volume of the mixture is

$$V = 1.09 \text{ mol} \times 56 \text{ cm}^3 \text{ mol}^{-1} + 2.77 \text{ mol} \times 17 \text{ cm}^3 \text{ mol}^{-1}$$
$$= 108 \text{ cm}^3$$

Exercise E4.6 Use Fig. 4.13 to calculate the density of a mixture of 20 g of water with 100 g of ethanol.

$$[0.84 \text{ g cm}^{-3}]$$

The chemical potentials of gases

The concept of a partial molar quantity can be extended to other state properties. Among the most important for our purposes is the **partial molar free energy** of a substance, which is the contribution of that substance to the total free energy of a mixture:

$$G = n_A G_A + n_B G_B \qquad (4a)$$

In this expression, G_A and G_B are the partial molar free energies of A and B *in the mixture* of the specified composition. The partial molar free energy has exactly the same significance as the partial molar volume. For instance, ethanol has a particular partial molar free energy when it is pure (and every molecule is surrounded by other ethanol molecules), and it has another partial molar free energy when

133

it is in an aqueous solution of a certain composition (because then each ethanol molecule is surrounded by a mixture of ethanol and water molecules). For example, if the partial molar free energy of a substance is 22 kJ mol^{-1} in a certain mixture, then we know that the addition of 1 mol of the substance to a large sample of the mixture will result in an increase in the total free energy of the mixture by 22 kJ. If 1 mol is removed instead, the free energy of the mixture will fall by 22 kJ.

The partial molar free energy is usually called the **chemical potential** and denoted μ: we shall use that name and notation from now on, and write the total free energy of a mixture as

$$G = n_A \mu_A + n_B \mu_B \tag{4b}$$

In the course of this chapter and the next we shall see that 'chemical potential' is a very suitable name, for we shall see that μ is a measure of the ability of a substance to bring about physical and chemical change: a substance with a high chemical potential has a high ability (in a sense we shall explore) to drive a reaction or some other physical process forward.

To carry out calculations on mixtures of gases and mixtures of liquids, we need an explicit formula for the variation of the chemical potential of a substance with the composition of the mixture. For a mixture of perfect gases, the appropriate expression is

$$\mu = \mu^{\ominus} + RT \ln a, \quad \text{where} \quad a = \frac{p}{p^{\ominus}} \tag{5}$$

In this expression, p is the partial pressure of the gas, $p^{\ominus}$ is the standard pressure (1 bar), and $\mu^{\ominus}$ is the **standard chemical potential** of the gas, its chemical potential when $p = p^{\ominus}$. The expression shows that, because μ is proportional to the logarithm of the partial pressure, *the lower the partial pressure of a gas, the lower its chemical potential.* This conclusion is consistent with the interpretation of the chemical potential as an indication of the potential of a substance to be active chemically: the lower the partial pressure, the less active chemically the species. In this instance the chemical potential represents the tendency of the substance to react when it is in its standard state (the significance of the term $\mu^{\ominus}$) plus an additional tendency that reflects whether it is compressed to a pressure greater than 1 bar or not. Any additional pressure gives a substance more chemical 'punch', just like winding a spring gives a spring more physical punch (that is, enables it to do more work).

Exercise E4.7 Suppose that the partial pressure of a perfect gas falls from 1.00 bar to 0.50 bar as it is consumed in a reaction at 25 °C. What is the change in chemical potential of the substance?

[-1.7 kJ mol^{-1}]

JUSTIFICATION

The starting point for the derivation of eqn (5) is the equation we have already seen several times before, namely $dG = V\,dp$, but written for partial molar free energy (μ) and molar volume V_m:

$$d\mu = V_m\,dp$$

The molar volume of an ideal gas is $V_m = RT/p$, so

the change in chemical potential when the pressure changes from $p^{\ominus}$, when the chemical potential is $\mu^{\ominus}$, to p, when the chemical potential is μ, is

$$\mu - \mu^{\ominus} = \int_{p^{\ominus}}^{p} V_m\,dp = RT\int_{p^{\ominus}}^{p}\frac{1}{p}\,dp = RT\ln\frac{p}{p^{\ominus}}$$

which rearranges into eqn (5).

Ideal solutions

To obtain the expression for the value of the chemical potential of a substance at a certain concentration in a liquid solution we need to introduce a further concept, that of an 'ideal solution'.

The French chemist François Raoult spent most of his life measuring the vapour pressures of solutions. He measured the **partial vapour pressure** of each component in the mixture, the partial pressure of the vapour of each component in dynamic equilibrium with the solution. Raoult found that *the partial vapour pressure of the solvent in the solution was lower than the vapour pressure of the pure solvent*. His observations suggested that the molecules of one substance partially block the escape of the molecules of the other species, thereby reducing the latter's vapour pressure. Raoult was able to establish a quantitative relationship between the partial vapour pressure of the solution and the vapour pressure of the pure solvent:

Raoult's law: the partial vapour pressure of a substance in a mixture is proportional to its mole fraction and its vapour pressure when pure:

$$p_A = x_A\,p_A^* \qquad (6)$$

In this expression p_A^* is the vapour pressure of the pure substance. For example, when the mole fraction of water in an aqueous solution is 0.9, then, provided Raoult's law is obeyed, the partial vapour pressure of the water in the solution is 90 per cent that of pure water. This conclusion is (approximately) true whatever the identity of the solute and the solvent (Fig. 4.14).

Exercise E4.8 A solution is prepared by dissolving 1.5 mol $C_{10}H_8$ (naphthalene) in 1.00 kg of benzene. The vapour pressure of pure benzene is 94.6 Torr at 25 °C. What is the partial vapour pressure of benzene in the solution?

[85 Torr]

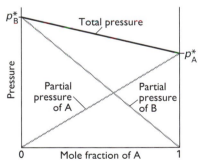

Fig. 4.14 The partial vapour pressures of the two components of an ideal binary mixture are proportional to the mole fractions of the components in the liquid. The total pressure of the vapour is the sum of the two partial vapour pressures.

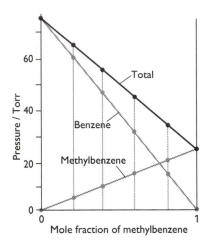

Fig. 4.15 Two similar substances, in this case benzene and toluene (methylbenzene) behave almost ideally and have vapour pressures that closely resemble those for the ideal case depicted in Fig. 4.14.

Raoult's law is only an approximation, but it becomes progressively more accurate as the solution becomes more dilute and the mole fraction of a component and its vapour pressure approach that of the pure substance. Some mixtures obey Raoult's law very well over a wide range of concentrations, especially when the components are structurally similar (Fig. 4.15) in the sense of having similar shapes and being held together in the liquid by similar intermolecular forces, such as two similar hydrocarbons (hexane and heptane, for instance). Hypothetical solutions that obey the law throughout the composition range from pure A to pure B are called **ideal solutions**. A mixture of benzene and methylbenzene (toluene) is a good approximation to an ideal solution, and the vapour pressure of each component satisfies Raoult's law reasonably well throughout the composition range from pure benzene to pure toluene.

No mixture is perfectly ideal, and so all mixtures show deviations from Raoult's law. However, the deviations are small for the component of the mixture that is in large excess (the solvent) and which has a partial vapour pressure that is close to that of the pure liquid. The deviations from Raoult's law for the solvent become smaller as the concentration of solute decreases, and we can usually be reasonably confident that Raoult's law is reliable when the solution is very dilute. Because Raoult's law shows that the vapour pressure of the solvent is proportional to the mole fraction of the liquid, it should not be too surprising that the expression for the chemical potential of the liquid bears a close relation to eqn (5) for the chemical potential of a gas. Indeed, it is quite easy to show that for dilute solutions the chemical potential of the solvent A is given by the expression

$$\mu = \mu^* + RT \ln a_A \qquad a_A = x_A \qquad (7)$$

where x_A is the mole fraction of the solvent (and is close to 1) and μ^* is the chemical potential of the pure liquid. The essential feature of eqn (7) is that the chemical potential of a solvent is *lower* in a solution than when it is pure: x_A is then less than 1, and $\ln x_A$ is negative. A solvent in which a solute is present has less chemical 'punch' (including a lower ability to generate a vapour pressure) than when it is pure.

JUSTIFICATION

When a liquid A in a mixture is in equilibrium with its vapour at a partial pressure p_A, the chemical potentials of the two phases are equal:

$$\mu_A(l) = \mu_A(g)$$

However, we already have an expression for the chemical potential of a vapour, eqn (5), so at equilibrium,

$$\mu_A(l) = \mu_A^\ominus + RT \ln \frac{p_A}{p^\ominus}$$

According to Raoult's law, $p_A = x_A p_A^*$, so

$$\mu_A(l) = \mu_A^\ominus + RT \ln \frac{x_A p_A^*}{p^\ominus}$$

$$= \mu_A^\ominus + RT \ln \frac{p_A^*}{p^\ominus} + RT \ln x_A$$

The first two terms on the right are constants for a given substance at a given temperature, so they may be combined into a constant μ^*. When $x_A = 1$, it follows that $\mu_A(l) = \mu_A^*$, so μ_A^* can be recognized as the chemical potential of the pure liquid. It then follows that

$$\mu_A = \mu_A^* + RT \ln x_A$$

which is eqn (7). We have replaced x by a to give a more uniform appearance to all the expressions for chemical potentials.

Exercise E4.9 By how much is the chemical potential of benzene reduced at 25 °C by a solute that is present at a mole fraction of 0.10?

$$[-0.26 \text{ kJ mol}^{-1}]$$

Insight into the lowering of the chemical potential of a solvent by a solute comes from the recognition that the presence of a solute increases the disorder of the solvent: we cannot be confident that if we blindly select a molecule from the solution, that it will be a solvent molecule. Because the disorder of the solvent is higher, it has a lower tendency to achieve disorder than the pure solvent has. But the hunt for maximum disorder is the motive power of change, so the solvent is chemically more impotent than when it is pure. Its potential to achieve change has been reduced: μ is lower than μ^*.

Ideal-dilute solutions

Raoult's law is a good description of the properties of the almost pure solvent in a real solution, such as the partial vapour pressure of water in a dilute solution of ethanol in water. However, in most cases it is not a good description of the partial vapour pressure of the solute, the component present in relatively small amount, for that substance is far from its pure form: each solute molecule is surrounded by nearly pure solvent, so its environment is quite unlike that in the pure solute and it is very unlikely that the vapour pressure of the solute in solution will be related to that of the pure solute. However, it is found experimentally that the partial vapour pressure of the solute (such as the partial vapour pressure of ethanol in a dilute solution of ethanol in water) is in fact proportional to its mole fraction of the solute. Unlike for the solvent, though, the constant of proportionality is not equal to the vapour pressure of the pure substance (Fig. 4.16). This linear but different dependence was discovered by the English chemist William Henry and is summarized as follows:

Henry's law: the vapour pressure of a volatile solute, B, is proportional to its mole fraction in a solution:

$$p_B = x_B K_B \qquad (8)$$

where K_B is a constant that is characteristic of the solute.

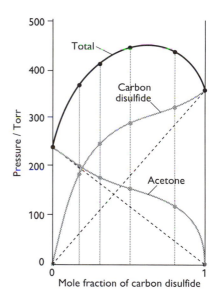

Fig. 4.16 Strong deviations from ideality are shown by dissimilar substances, in this case carbon disulfide and acetone. Note, however, that Raoult's law is obeyed by acetone when only a small amount of carbon disulfide is present (on the left) and by carbon disulfide when only a small amount of acetone is present (on the right).

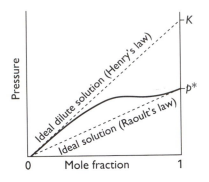

Fig. 4.17 When a component (the solvent) is almost pure, it behaves in accord with Raoult's law and has a vapour pressure that is proportional to the mole fraction in the liquid mixture, and a slope p^*, the vapour pressure of the pure substance. When the same substance is the minor component (the solute), its vapour pressure is still proportional to its mole fraction, but the constant of proportionality is now K.

The constant K_B is chosen so that the plot of the vapour pressure of B against its mole fraction is tangent to the experimental curve at $x_B = 0$ (Fig. 4.17). Henry's law is usually obeyed only at low concentrations of the solute (close to $x_B = 0$), and solutions that are dilute enough to obey Henry's law are called **ideal-dilute solutions**. The essential *molecular* content of Henry's law is that rate at which solute molecules can leave the solution is proportional to their concentration, but that the rate bears little relation to the rate at which they escape from molecules of their own kind in the pure solute.

Example Verifying Raoult's and Henry's laws

The partial vapour pressures of each component in a mixture of propanone (acetone, A) and chloroform (trichloromethane, C) were measured at 35 °C with the following results:

x_C	0	0.20	0.40	0.60	0.80	1
p_C/Torr	0	35	82	142	219	293
p_A/Torr	347	270	185	102	37	0

Confirm that the mixture conforms to Raoult's law for the component in large excess and to Henry's law for the minor component. Find the Henry's law constants.

Answer We need to plot the partial vapour pressures against mole fraction. Raoult's law is tested by comparing the data with the straight line $p = xp^*$ for each component in the region in which it is in excess (and acting as the solvent). Henry's law is tested by finding a straight line $p = xK$ that is tangential to each partial vapour pressure at low x where the component can be treated as the solute. The data are plotted in Fig. 4.18 together with the Raoult's law lines. Henry's law requires $K_A = 175$ Torr and $K_C = 165$ Torr. (Notice how the data deviate from both Raoult's and Henry's laws even for quite small departures from $x = 1$ and $x = 0$, respectively.)

Exercise E4.10 The vapour pressure of methyl chloride at various mole fractions in a mixture at 25 °C was found to be as follows:

x	0.005	0.009	0.0019	0.0024
p/Torr	205	363	756	946

Estimate Henry's law constant.

[4×10^5 Torr]

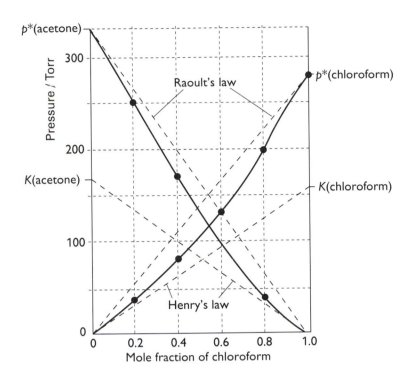

Fig. 4.18 The experimental partial vapour pressures of a mixture of chloroform ($CHCl_3$) and acetone (CH_3COCH_3) based on the data in the example. The values of K, the Henry's law constants for the two components, are obtained by extrapolating the dilute solution vapour pressures as explained in the example.

Some Henry's law constants of gases are listed in Table 4.2. They are often used in calculations relating to gas solubilities, as in the estimation of the concentration of O_2 in river water or the concentration of carbon dioxide in blood plasma. To apply Henry's law to this kind of problem, we treat the gas as the solute, and use its partial pressure above the solvent to calculate the mole fraction in the

Table 4.2 Henry's law constants for gases at 25 °C, K/Torr

	Solvent	
	Water	**Benzene**
Methane, CH_4	3.14×10^5	4.27×10^5
Carbon dioxide, CO_2	1.25×10^6	8.57×10^4
Hydrogen, H_2	5.34×10^7	2.75×10^6
Nitrogen, N_2	6.51×10^7	1.79×10^6
Oxygen, O_2	3.30×10^7	

solution by rearranging $p_B = x_B K_B$ to

$$x_B = \frac{p_B}{K_B}$$

For instance, when the partial pressure of oxygen is 190 Torr and the solvent is water, the mole fraction in solution in equilibrium with the gas at 25 °C is

$$x_{O_2} = \frac{190 \text{ Torr}}{3.30 \times 10^7 \text{ Torr}} = 5.8 \times 10^{-6}$$

That is, about one molecule in 170 000 is an O_2 molecule, the others being water. A knowledge of Henry's law constants for gases in fats and lipids is important for the discussion of respiration, especially when the partial pressure of oxygen is abnormal, as in diving and mountaineering.

Example Determining whether a natural water can support life

The concentration of O_2 in water required to support aquatic life is about 4 mg L^{-1}. What is the minimum partial pressure of oxygen in the atmosphere that can achieve this concentration?

Answer The strategy of the calculation is to determine the partial pressure of oxygen that, according to Henry's law, corresponds to the concentration specified. First, we convert $4 \text{ mg } O_2$ to a mole fraction. The amount of O_2 in 1.00 L of water is

$$n(O_2) = \frac{4 \times 10^{-3} \text{ g}}{32.00 \text{ g mol}^{-1}} = 1 \times 10^{-4} \text{ mol}$$

The mole fraction of O_2, noting that 1.00 L of water corresponds to 55.5 mol H_2O, is therefore

$$x(O_2) = \frac{1 \times 10^{-4} \text{ mol}}{55.5 \text{ mol} + 1 \times 10^{-4} \text{ mol}} = 2 \times 10^{-6}$$

Because Henry's law constant for oxygen in water at 20 °C is 3.3×10^7 Torr, the partial pressure required to achieve a mole fraction of oxygen of 2×10^{-6} is

$$p_{O_2} = (2 \times 10^{-6}) \times (3.3 \times 10^7 \text{ Torr}) = 7 \times 10^1 \text{ Torr}$$

The partial pressure of oxygen in air at sea level is $0.21 \times 760 \text{ Torr} = 160 \text{ Torr}$, so the required concentration can be maintained under normal conditions.

Exercise E4.11 What partial pressure of methane is needed to achieve 21 mg of methane in 100 g of benzene at 25 °C?

[4.3×10^2 Torr]

4.4 Colligative properties

In this section we see how to calculate the effect of a solute on certain properties of a solution. We shall see that a nonvolatile solute

- raises the boiling point of a solution,
- lowers the freezing point of a solution,
- produces an osmotic pressure.

(The meaning of the last will be explained shortly.) In dilute solutions, all the properties we consider depend only on the number of solute particles present, not their identity, and for this reason they are called **colligative properties** (where *colligative* denotes 'depending on the collection'). Thus a $0.001\ mol\ kg^{-1}$ aqueous solution of any non-electrolyte should have the same boiling point, freezing point, and osmotic pressure.

The common features of colligative properties

We shall make two assumptions:

- The solute is not volatile, and so does not contribute to the vapour.
- The solute does not dissolve in the solid solvent.

For example, a solution of sucrose in water consists of a solute (sucrose, $C_{12}H_{22}O_{11}$) that is not volatile, and so never appears in the vapour (which is therefore pure water vapour) and is also left behind in the liquid solvent when ice begins to form (so the ice remains pure). It follows that the solute affects the properties of the liquid solvent alone, and leaves the vapour and the solid completely unaffected.

The effect of the solute on the solvent is to lower the latter's chemical potential: we saw in eqn (7) that a liquid that is slightly contaminated by a foreign substance has a lower chemical potential than the pure liquid because foreign particles introduce disorder. The chemical potentials of the solvent vapour and solid solvent are unchanged by the presence of the solute because it is nonvolatile and insoluble in the solid.

Now consider the consequences for the boiling point of the liquid. Because the liquid has a lower chemical potential when the solute is present, it must be heated to a higher temperature before it rises above the chemical potential of the vapour and the vapour becomes the thermodynamically stable phase (Fig. 4.19(a)). Consequently, the boiling point of the solution is higher than that of the pure solvent. In

Fig. 4.19 (a) The chemical potentials of pure solvent vapour and pure liquid solvent decrease with temperature, and the point of intersection, where the chemical potential of the vapour falls below that of the liquid, marks the boiling point of the pure solvent. A solute lowers the chemical potential of the solvent but leaves that of the vapour unchanged. As a result, the intersection point lies further to the right, and the boiling point is therefore raised. (b) The chemical potentials of pure solid solvent and pure liquid solvent also decrease with temperature, and the point of intersection, where the chemical potential of the liquid rises above that of the solid, marks the freezing point of the pure solvent. A solute lowers the chemical potential of the solvent but leaves that of the solid unchanged. As a result, the intersection point lies further to the left, and the freezing point is therefore lowered.

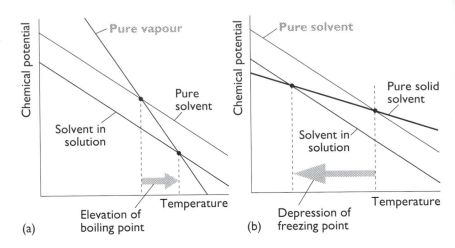

(a) Elevation of boiling point

(b) Depression of freezing point

terms of the entropy, the disorder introduced by the solute has made the solvent relatively chemically lazy (it has a reduced tendency to vaporize, for instance), so a higher temperature must be achieved before boiling occurs.

Next, consider the freezing point of the solvent. Because the chemical potential of the solvent in the solution is lower when the solute is present, the solution must be cooled to a lower temperature before the chemical potential of the solid solvent falls below that of the liquid and the solid becomes the thermodynamically stable phase; consequently the solvent freezes at a *lower* temperature when a solute is present (Fig. 4.19(b)). In terms of the entropy, the disorder introduced by the solute must be overcome by lowering the temperature more than for the pure solvent, so freezing occurs at a lower temperature. One practical consequence of the lowering of freezing point (and hence the lowering of the melting point of the pure solid) is its employment in organic chemistry to judge the purity of a sample, for any impurity lowers the melting point of a substance from its accepted value.

The details of the calculations of the magnitudes of the effects are not important, but the conclusions are that the **elevation of boiling point**, ΔT_b, is proportional to the molality m_B of the solute:

$$\Delta T_b = K_b m_B \tag{9a}$$

In this expression K_b is the **ebulloscopic constant** of the solvent, a characteristic property of the solvent. Similarly the **depression of freezing point**, ΔT_f, is also proportional to the solute molality:

$$\Delta T_f = K_f m_B \tag{9b}$$

Here K_f is the **cryoscopic constant** of the solvent. The two constants can be estimated from other properties of the solvent, but both are best treated as empirical constants, and some experimental values are given in Table 4.3.

Table 4.3 Cryoscopic and ebulloscopic constants

Solvent	$K_f/\text{K kg mol}^{-1}$	$K_b/\text{K kg mol}^{-1}$
Acetic acid	3.90	3.07
Benzene	5.12	2.53
Camphor	40	
Carbon disulfide	3.8	2.37
Carbon tetrachloride	30	4.95
Naphthalene	6.94	5.8
Phenol	7.27	3.04
Water	1.86	0.51

Exercise E4.12 Estimate the lowering of the freezing point of the solution made by dissolving 3.0 g (about one sugar-lump) of sucrose ($C_{12}H_{22}O_{11}$) in 100 g of water.

[−0.16 K]

Osmosis

The phenomenon of **osmosis** (a name derived from the Greek word for 'push') is the passage of a pure solvent into a solution separated from it by a **semipermeable membrane**, which is a membrane (often of cellulose acetate) that is permeable to the solvent but not to the solute (Fig. 4.20); for example, the membrane might have holes that are large enough to allow water molecules to pass through, but not ions or carbohydrate molecules with their bulky coating of hydrating water molecules. The **osmotic pressure**, Π, is the pressure that must be applied to the solution to stop the flow. One of the most important examples of osmosis is transport of fluids through cell membranes, but osmosis is also the basis of the technique called **osmometry**, the determination of molar mass by measurement of osmotic pressure, especially of macromolecules.

In the simple arrangement shown in Fig. 4.21, the pressure opposing the passage of solvent into the solution arises from the head of solution that the osmosis itself produces. This head is formed when the pure solvent flows through the membrane into the solution and pushes the column of solution higher up the tube. Equilibrium is reached when the hydrostatic pressure of the column of solution matches the osmotic pressure. A complication of this arrangement is that the entry of solvent into the solution results in the latter's

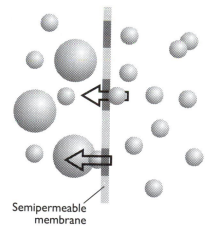

Semipermeable membrane

Fig. 4.20 A semipermeable membrane allows the passage of one type of molecule (such as the solvent, depicted here by small spheres) but not of another substance (the solute, depicted by large spheres). Osmosis is the net flow of solvent into the solution.

dilution, and so it is more difficult to treat than a more sophisticated arrangement in which pressure is applied to oppose any flow of solvent into the solution.

The osmotic pressure is proportional to the amount n_B of solute B present in the solution. In fact, the expression for the pressure bears an uncanny resemblance to the perfect gas equation of state:

$$\Pi V = n_B RT \tag{10a}$$

This equation is called the **van't Hoff equation** for the osmotic pressure. Because $n_B/V = [B]$, the molar concentration of the solute, a simpler form of this equation is

$$\Pi = [B]RT \tag{10b}$$

This equation applies only to solutions that are sufficiently dilute to behave ideally.

JUSTIFICATION

The thermodynamic treatment of osmosis depends on noting that, at equilibrium, the chemical potential of the solvent is the same on each side of the membrane: if it were not, then there would be a tendency for the solvent to flow from the region of high potential into the one of low potential. The chemical potential of the pure solvent, which is at a pressure p, is $\mu_A^*(p)$: this value represents a certain tendency for the solvent molecules to pass through the membrane. The chemical potential of the solvent in the solution is lowered by the solute (which is present at a mole fraction x_B) but is raised on account of the greater pressure $p + \Pi$ that the solution experiences: increasing the pressure on the liquid increases its tendency to pass through the membrane to relieve the stress. At equilibrium the two chemical potentials are equal, because the escaping tendency of the pure liquid is equal to the escaping tendency of the pressurized solvent in the solution, and we can write

$$\mu_A^*(p) = \mu_A(x_A, p + \Pi)$$

The presence of solute is taken into account by using eqn (7):

$$\mu_A(x_A, p + \Pi) = \mu_A^*(p + \Pi) + RT \ln x_A$$

and the effect of pressure on the chemical potential of an (assumed incompressible) liquid is given by the expression first derived in Section 3.4 for the free energy but expressed in terms of the chemical potential (by replacing V by the molar volume V_m of the pure solvent):

$$\mu_A^*(p + \Pi) = \mu_A^*(p) + V_m \Delta p$$

In this expression Δp is the difference in pressure between the two liquids, and is equal to Π. When the last three equations are combined we get

$$-RT \ln x_A = \Pi V_m$$

The mole fraction of the solvent is equal to $1 - x_B$, where x_B is the mole fraction of the solute. In a dilute solution, $\ln(1 - x_B)$ is approximately equal to $-x_B$ (for example, $\ln 0.90 = -0.11$, which is close to -0.10), so

$$RT x_B \approx \Pi V_m$$

When the solution is dilute, $x_B \approx n_B/n_A$. Moreover, because $n_A V_m = V$, the total volume of the solvent, this equation becomes

$$n_B RT \approx \Pi V$$

which is eqn (10a).

One of the most common applications of osmometry is to the measurement of molar masses of proteins and synthetic polymers. As these huge molecules dissolve to produce solutions that are far from ideal, it is assumed that the van't Hoff equation is only the first term of an expansion:

$$\Pi = [B]RT\{1 + B[B] + \cdots\} \qquad (11)$$

Exactly the same extension was used in Section 1.9 to extend the perfect gas equation to real gases, and taking the effects of intermolecular interactions into account by writing the empirical virial equation of state. The empirical parameter B is called the **osmotic virial coefficient**. The osmotic pressure is measured at a series of concentrations, and a plot of $\Pi/[B]$ against $[B]$ is used to find the molar mass of B. In practice, the osmotic pressure is related to the height h of the solution above the level of the solvent through

$$\Pi = \rho gh$$

where ρ is the density of the solution and g is the acceleration of free fall (9.81 m s^{-2}). The molar concentration $[B]$ (in mol L^{-1}) of the solute is also usually converted to the mass concentration c (in g L^{-1}) using

$$c = M[B]$$

where M is the molar mass of the solute. Then

$$\frac{h}{c} = \frac{RT}{\rho gM}\left(1 + \frac{Bc}{M} + \cdots\right)$$

$$= \frac{RT}{\rho gM} + \left(\frac{RTB}{\rho gM^2}\right)c + \cdots$$

It follows that, by plotting h/c against c, the results should fall on a straight line with intercept $RT/\rho gM$ on the vertical axis at $c = 0$. Then, from a determination of this intercept (in practice, by extrapolation of the data to $c = 0$), the molar mass of the solute can be obtained.

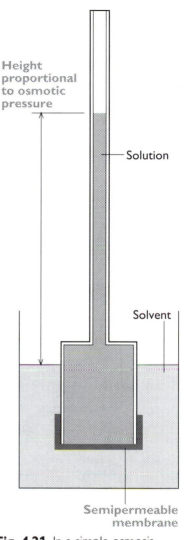

Fig. 4.21 In a simple osmosis experiment, a solution is separated from the pure solvent by a semipermeable membrane. Pure solvent passes through the membrane and the solution rises in the inner tube. The net flow ceases when the pressure exerted by the column of liquid is equal to the osmotic pressure of the solution.

Example Using osmometry to determine molar mass
The osmotic pressures of solutions of poly(vinyl chloride), PVC, in cyclohexanone at 298 K are given below. The pressures are expressed in terms of the heights of solution (of density $\rho = 0.980 \text{ g cm}^{-3}$) in balance with the osmotic pressure. Find the molar mass of the polymer.

$c/\text{g dm}^{-3}$	1.00	2.00	4.00	7.00	9.00
h/cm	0.20	0.71	2.01	5.10	8.00

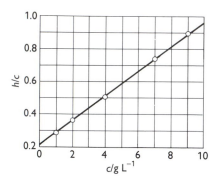

Fig. 4.22 The plot used in the determination of molar mass by osmometry. The molar mass is calculated from the intercept at $c = 0$.

Answer We need the values of h/c:

$c/\text{g dm}^{-3}$	1.00	2.00	4.00	7.00	9.00
$(h/c)/(\text{cm/g dm}^{-3})$	0.28	0.36	0.503	0.739	0.889

The points are plotted in Fig. 4.22 and the intercept of the vertical axis at $c = 0$ is 0.21. Therefore,

$$M = \frac{RT}{\rho g} \times \frac{1}{0.21\ \text{cm g}^{-1}\ \text{dm}^3}$$

Because

$$0.21\ \text{cm g}^{-1}\ \text{dm}^3 = 0.21 \times 10^{-2}\ \text{m}^4\ \text{kg}^{-1}$$

and $\rho = 0.980 \times 10^3\ \text{kg m}^{-3}$, it follows that

$$M = \frac{(8.3145\ \text{J K}^{-1}\ \text{mol}^{-1}) \times (298\ \text{K})}{(980\ \text{kg m}^{-3}) \times (9.81\ \text{m s}^{-2})} \times \frac{1}{0.21 \times 10^{-2}\ \text{m}^4\ \text{kg}^{-1}}$$

$$= 1.2 \times 10^2\ \text{kg mol}^{-1}$$

Note that it is often convenient to express the molar masses of macromolecules in kilograms per mole rather than grams per mole. Biochemists often report molar masses in daltons, Da, with $1\ \text{Da} = 1\ \text{g mol}^{-1}$; this molar mass would be reported as 120 kDa.

Exercise E4.13 The heights of the solution in an osmometry experiment on a solution of an enzyme in water at 25 °C were as follows:

$c/\text{g dm}^{-3}$	0.50	1.00	1.50	2.00	2.50
h/cm	0.18	0.35	0.53	0.71	0.90

The density of the solution is $0.9998\ \text{g cm}^{-3}$. What is the molar mass of the enzyme?

[72 kg mol^{-1}]

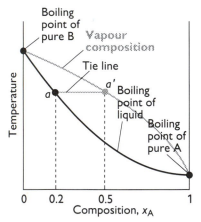

Fig. 4.23 A temperature–composition diagram for a binary mixture of volatile liquids. The tie line connects the points which represent the compositions of liquid and vapour that are in equilibrium at each temperature. The lower curve is a plot of the boiling point of the mixture against composition.

Mixtures of volatile liquids

Now we consider binary mixtures in which both components are volatile, and in particular consider the relation of the boiling point of a liquid mixture to its composition. This kind of information is important when considering the separation of liquids by distillation: indeed, we shall see that some liquids cannot be fully separated in this way.

4.5 Temperature–composition diagrams

Intuitively, we might expect the boiling point of a mixture of two volatile liquids to vary smoothly from the boiling point of one pure component when only that liquid is present to the boiling point of the other pure component when only that liquid is present. This expectation is often borne out in practice, and a typical plot of boiling point against composition is shown in Fig. 4.23 (the lower curve).

The vapour in equilibrium with the boiling mixture will be a mixture of the two components, and we might also expect that it will be richer in the more volatile of the two liquids. This is also often found in practice, and the upper curve in the illustration shows the composition of the vapour in equilibrium with the boiling liquid. To identify the composition of the vapour, we note the boiling point of the liquid mixture (point a, for instance, if the mole fraction of A is 0.2), and draw a horizontal **tie line** across to the upper curve. Its point of intersection (a') gives the composition of the vapour. In this example, we see that the mole fraction of A in the vapour is about 0.5. As expected, the vapour is richer than the liquid in the more volatile component.

The distillation of mixtures

A temperature–composition diagram is useful for discussing the separation of liquids by distillation. Consider what happens when a liquid of composition a is heated (Fig. 4.24). Initially its state is a_1. It boils when the temperature reaches T_2. Then the liquid has composition a_2 (the same as a_1) and the vapour has composition a_2'. The vapour is richer in the more volatile component A, as common sense leads us to expect. In a simple distillation, the vapour is withdrawn and condensed. If the vapour in this example is drawn off and completely condensed, then the first drop gives a liquid of composition a_3 (the same as a_2'), which is richer in the more volatile component (A) than the original liquid.

In the procedure called **fractional distillation**, the boiling and condensation cycle is repeated successively. We can follow the changes that occur by seeing what happens when the condensate of composition a_3 is reheated. The phase diagram shows that this mixture boils at T_3 and yields a vapour of composition a_3' which is even richer in the more volatile component. That vapour is drawn off, and the first drop condenses to a liquid of composition a_4. The cycle can then be repeated until in due course almost pure A is obtained. In practice, the series of vaporizations and condensations are carried out in a single operation: the vapour passes up a vertical **fractionating column** packed with glass rings or beads to give a large surface area, and the successive cycles of vaporization and condensation take place on their surfaces.

Azeotropes

Whereas many binary liquid mixtures have temperature–composition diagrams resembling that shown in Fig. 4.23, in a number of important

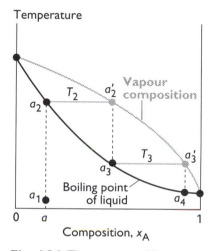

Fig. 4.24 The process of fractional distillation can be represented by a series of steps on a temperature–composition diagram like that in Fig. 4.23. The initial liquid mixture may be at a similar temperature and have a composition like that represented by point a_1. It boils at the temperature T_2, and the vapour in equilibrium with the boiling liquid has composition a_2'. If that vapour is condensed (to a_3 or below), the resulting condensate boils at T_3 and gives rise to a vapour of composition represented by a_3'. As the succession of vaporizations and condensations is continued, the composition of the distillate moves towards pure A (the more volatile component).

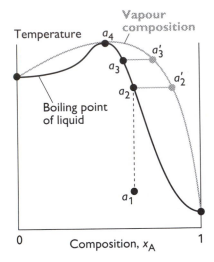

Temperature

Vapour composition

a_4

a'_3

a_3

a'_2

a_2

Boiling point of liquid

a_1

0   Composition, x_A 1

Fig. 4.25 The temperature–composition diagram for a high-boiling azeotrope. As fractional distillation proceeds, the composition of the remaining liquid moves towards a_4; however, once there, the vapour in equilibrium with that liquid has the same composition, so the mixture evaporates with an unchanged composition and no further separation can be achieved.

cases there are marked deviations. When temperature–composition curves are determined experimentally, a maximum in the boiling-point curve (Fig. 4.25) is sometimes found. Such behaviour may occur when the interactions between the components reduce the vapour pressure of the mixture below the ideal value: in effect, the A, B interactions stabilize the liquid and cause it to have a higher boiling point than either component alone. Examples of this behaviour include chloroform/acetone and nitric acid/water mixtures. Temperature–composition curves that pass through a minimum are also found (Fig. 4.26). This behaviour indicates that the mixture is destabilized relative to the ideal solution, the A, B interactions then being unfavourable. Examples include dioxane/water and ethanol/water.

Deviations from ideality are not always so strong as to lead to a maximum or minimum in the phase boundaries, but when they do there are important consequences for distillation. Consider a liquid of composition a_1 on the right of the maximum in Fig. 4.25. It boils at a temperature corresponding to a_2 and its vapour (of composition a'_2) is richer in the more volatile component A. If that vapour is removed (and condensed elsewhere), the remaining liquid will move toward the composition and boiling point represented by the point a_3. The vapour in equilibrium with this boiling liquid has composition a'_3: note that the two compositions are more similar than the original pair (a_3 and a'_3 are closer together than a_2 and a'_2). If that vapour is removed, the composition of the boiling liquid shifts to a_4, and the vapour of that boiling mixture has an identical composition to the liquid. Hence, as evaporation proceeds, the composition of the remaining liquid shifts towards B as A is drawn off. The boiling point of the liquid rises, and the vapour becomes richer in B. When so much A has been evaporated that the liquid has reached the composition a_4, the vapour has the same composition as the liquid. Evaporation then occurs without change of composition. The mixture is said to form an **azeotrope** (which comes from the Greek words for 'boiling without changing').

When the azeotropic composition has been reached, distillation cannot separate the two liquids because the condensate retains the composition of the liquid. One example of azeotrope formation is hydrochloric acid/water, which is azeotropic at 80 per cent water (by mass) and boils unchanged at 108.6 °C.

The system shown in Fig. 4.26 is also azeotropic, but shows its azeotropic character in a different way. Suppose we start with a mixture of composition a_1 and follow the changes in the vapour that rises through a fractionating column. The mixture boils at a_2 to give a vapour of composition a'_2. This vapour condenses in the column to a liquid of the same composition (now marked a_3). That liquid reaches equilibrium with its vapour at a'_3, which condenses higher up the tube to give a liquid of the same composition. The fractionation therefore shifts the vapour towards the azeotropic composition, at a_4, but the composition cannot move beyond a_4 because then the vapour and the liquid have the same composition. Consequently, the azeotropic vapour emerges from the top of the column. An example is ethanol/water, which boils unchanged when the water content is 4 per cent and the temperature is 78 °C.

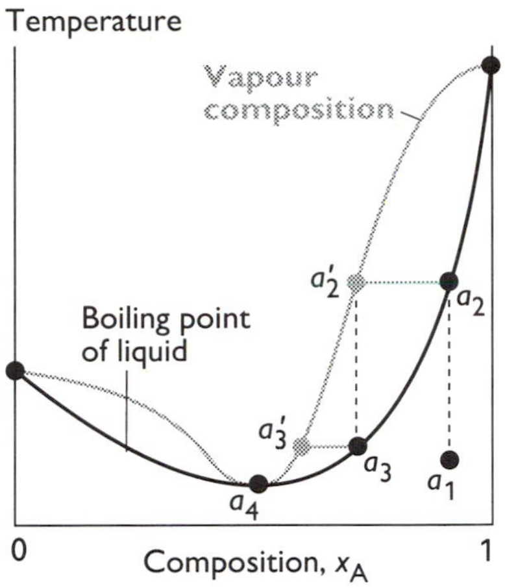

4.6 Immiscible liquids

Finally, we consider the distillation of two immiscible liquids, such as oil and water. As they are immiscible we can regard their 'mixture' as unscrambled with each component in a separate vessel (Fig. 4.27). If the vapour pressures of the two pure components are p_A and p_B, then the total vapour pressure is $p = p_A + p_B$ and the mixture boils when $p = 1$ atm. The presence of the second component means that the system boils at a lower temperature than either would alone because boiling begins when the total pressure reaches 1 atm, not when either vapour pressure reaches 1 atm. This lowering of the boiling temperature is the basis of **steam distillation**, which enables some heat-sensitive organic compounds to be distilled at a lower temperature than their normal boiling point. The only snag is that the composition of the condensate is in proportion to the vapour pressures of the components, and so oils of low volatility distil in low abundance.

4.7 Liquid–liquid phase diagrams

Diagrams analogous to those we have been considering can also be used to discuss the composition of **partially miscible liquids**, which are liquids that do not mix together in all proportions. An example is a mixture of hexane and nitrobenzene: when the two liquids are shaken together, the liquid consists of two phases, one is a saturated solution of hexane in nitrobenzene and the other is a saturated solution of nitrobenzene in hexane. Because the two solubilities vary with temperature, the compositions and proportions of the two phases change as the temperature is changed, and we use a temperature–composition diagram to display the state of the system at each temperature.

Phase separation

We shall introduce liquid–liquid phase diagrams by considering the case of two partially miscible liquids hexane and nitrobenzene. Suppose we add a small amount of nitrobenzene to a sample of hexane

Fig. 4.26 The temperature–composition diagram for a low-boiling azeotrope. As fractional distillation proceeds, the composition of the vapour moves towards a_4; however, once there, the vapour in equilibrium with that liquid has the same composition, so no further separation of the distillate can be achieved.

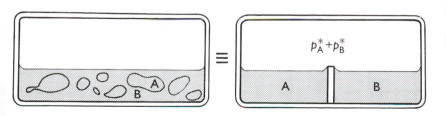

Fig. 4.27 The distillation of two immiscible liquids can be regarded as the simultaneous distillation of the separated components in the same apparatus. Boiling occurs when the sum of the partial pressures reaches 1 atm.

Fig. 4.28 The temperature–composition diagram for hexane and nitrobenzene at 1 atm. The upper critical solution temperature (T_{uc}) is the temperature above which no phase separation occurs. For this system it lies at 293 K (when the pressure is 1 atm).

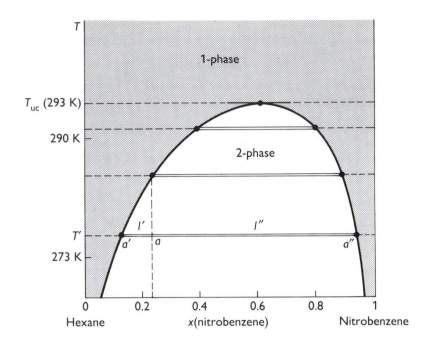

at some temperature T'. It dissolves completely, but, as more nitrobenzene is added, a stage comes when no more dissolves. The sample now consists of two phases in equilibrium with each other, the most abundant one consisting of hexane saturated with nitrobenzene, the minor one a trace of nitrobenzene saturated with hexane. In the temperature–composition diagram drawn in Fig. 4.28, the composition of the former is represented by the point a' and that of the latter by the point a''. The horizontal line joining a' and a'' is another example of a tie line, a line joining two phases that are in equilibrium with each other. The relative abundances of the two phases joined by a tie line are given by the **lever rule**:

$$\frac{\text{Amount of phase of composition } a''}{\text{Amount of phase of composition } a'} = \frac{l'}{l''} \tag{13}$$

JUSTIFICATION

To prove the lever rule, we write $N = n + n'$ and the overall amount of a liquid A as Nz_A where z_A is the mole fraction of A in the entire (liquid + liquid) sample. The overall amount of A is also the sum of its amounts in the two phases, where it has the mole fractions x_A and $_yA$, respectively:

$$Nz_A = n'x_A + n''y_A$$

Because also

$$Nz_A = n'z_A + n''z_A$$

by equating these two expressions it follows that

$$n'(x_A - z_A) = n''(z_A - y_A)$$

or

$$n'l' = n''l''$$

as was to be proved.

When more nitrobenzene is added, hexane dissolves in it slightly. The compositions of the two phases in equilibrium remain a' and a'', but the amount of the second phase increases at the expense of the first. A stage is reached when so much nitrobenzene is present that it can dissolve all the hexane, and the system reverts to a single phase. The addition of more nitrobenzene now simply dilutes the solution, and from then on it remains a single phase.

The temperature affects the compositions of the two phases at equilibrium. For hexane and nitrobenzene, raising the temperature increases their miscibility. The two-phase region is therefore less extensive, because each phase in equilibrium is richer in both components: the hexane-rich phase is richer in nitrobenzene and the nitrobenzene-rich phase is richer in hexane. The entire phase diagram can be constructed by repeating the observations at different temperatures and drawing the envelope of the two-phase region.

Example Interpreting a liquid–liquid phase diagram
A mixture of 50 g (0.59 mol) of hexane and 50 g (0.41 mol) of nitrobenzene was prepared at 290 K. What are the compositions of the phases, and in what proportions do they occur? To what temperature must the sample be heated in order to obtain a single phase?
Answer We denote hexane by H and nitrobenzene by N. Refer to Fig. 4.28 and use the lever rule. The point $x_N = 0.41$, $T = 290$ K occurs in the two-phase region of the phase diagram. The horizontal tie line cuts the phase boundary at $x_N = 0.37$ and $x_N = 0.83$, and so those mole fractions are the compositions of the two phases. The ratio of amounts of each phase is equal to the ratio of the distances l and l':

$$\frac{l'}{l''} = \frac{0.41 - 0.37}{0.83 - 0.41} = \frac{0.04}{0.42} = 0.1$$

Heating the sample to 292 K takes it into the single phase region.

Exercise E4.14 Repeat the problem for 50 g hexane and 100 g nitrobenzene at 273 K.
$[x_N = 0.09$ and 0.95 in ratio $1:1.3$; 290 K]

Critical solution temperatures

The **upper critical solution temperature**, T_{uc}, is the upper limit of temperatures at which phase separation occurs. Above the upper critical solution temperature the two components are fully miscible. This temperature exists because the greater thermal motion leads to greater miscibility of the two components. In thermodynamic terms,

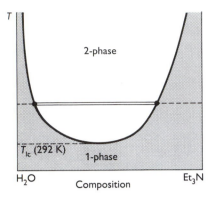

Fig. 4.29 The temperature–composition diagram for water and triethylamine. The lower critical solution temperature (T_{lc}) is the temperature below which no phase separation occurs. For this system it lies at 292 K (when the pressure is 1 atm).

the free energy for mixing becomes negative above a certain temperature, irrespective of the composition. From the expression

$$\Delta G = \Delta H - T\,\Delta S$$

we see that a change of sign is possible if mixing is endothermic ($\Delta H > 0$) and gives rise to an increase in disorder ($\Delta S > 0$) so that, at a sufficiently high temperature, $T\,\Delta S$ exceeds ΔH and ΔG becomes negative.

Some systems show a **lower critical solution temperature**, T_{lc}, below which they mix in all proportions and above which they form two phases. An example is water and triethylamine (Fig. 4.29). In this case, at low temperatures the two components are more miscible because they form a weak complex; at higher temperatures the complexes break up and the two components are less miscible. In thermodynamic terms, the mixing should be exothermic ($\Delta H < 0$) and be accompanied by a *decrease* in disorder ($\Delta S < 0$), perhaps because the two types of molecules stick together by complex formation. Then ΔG will be negative at low temperatures when the entropy contribution is insignificant, but may become positive when $-T\,\Delta S$ is strongly positive.

Some systems have both upper and lower critical solution temperatures. The reason can be traced to the fact that after the weak complexes have been disrupted, leading to partial miscibility, the thermal motion at high temperatures homogenizes the mixture again, just as in the case of ordinary partially miscible liquids. One example is nicotine and water, which are partially miscible between 61 °C and 210 °C (Fig. 4.30).

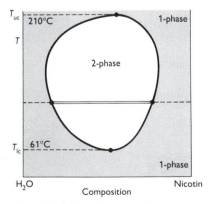

Fig. 4.30 The temperature–composition diagram for water and nicotine, which has both upper and lower critical solution temperatures. Note the high temperatures on the graph: the diagram corresponds to a sample under pressure.

4.8 The distillation of partially miscible liquids

We now consider what happens when the conditions are such that a vapour may be present in equilibrium with a binary mixture of liquids. We shall consider a pair of liquids that are partially miscible and form a low-boiling azeotrope. This combination of properties is quite common because both properties reflect the tendency of the two kinds of molecule to avoid each other. There are two possibilities, one when the liquids become fully miscible before they boil, the other when boiling occurs before mixing is complete.

Miscibility before boiling

Figure 4.31 shows the phase diagram for two components that become fully miscible before they boil. Distillation of a mixture of composition a_1 leads to a vapour of composition b_1 which condenses to the completely miscible single-phase solution at b_2. Phase separation occurs only when this distillate is cooled to b_3. This remark applies only to the first drop of distillate. If distillation continues, the composition of the remaining liquid changes. In the end, when the whole sample has evaporated and condensed, the composition is back to a_1.

Boiling before mixing completely

Figure 4.32 shows the second possibility, in which there is no upper critical solution temperature. The distillate obtained from a liquid initially of composition a_1 has composition b_3 and is a two-phase mixture. One phase has composition b_3' and the other has composition b_3''.

The behaviour of a system of composition e is interesting. A system at e_1 forms two phases, which persist (but with changing proportions) up to the boiling point at e. The vapour of this mixture has the same composition as the liquid (the liquid is an azeotrope). Similarly, condensing a vapour of composition e_2 gives a liquid of the same composition. The mixture vaporizes and condenses like a single substance.

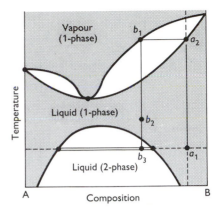

Fig. 4.31 The temperature–composition diagram for binary system in which the upper critical solution temperature is less than the boiling point of the mixture at any composition. The mixture forms a low-boiling azeotrope.

Example Interpreting a phase diagram
State the changes that occur when a mixture of composition $x_B = 0.95$ (a_1 in Fig. 4.32) is boiled and the vapor condensed.
Answer The area occupied by the point gives the number of phases; the compositions of the phases are given by the points at the intersections of the horizontal tie line with the phase boundaries; the relative abundances are given by the lever rule (eqn (13)). The initial point is in the one-phase region. When heated it boils at 370 K (a_2) giving a vapour of composition $x_B = 0.66$ (b_1). The liquid gets richer in B, and the last drop (of pure B) evaporates at 392 K. The boiling range of the liquid is therefore 370 to 392 K. If the initial vapour is drawn off, it has a composition $x_B = 0.66$. This composition would be maintained if the sample were very large, but for a finite sample it shifts to higher values and ultimately to $x_B = 0.95$. Cooling the distillate corresponds to moving down the vertical line at $x_B = 0.66$. At 350 K, for instance, the liquid phase has composition $x_B = 0.87$, the vapour $x_B = 0.49$, in relative proportions 1:1.3. At 340 K the sample is entirely liquid, and consists of three phases, the vapour, and two liquids, one of composition $x_B = 0.44$, the other of composition $x_B = 0.84$ in the ratio 0.85:1. Further cooling moves the system into the two-phase region, and at 298 K the compositions are 0.05 and 0.93 in the ratio 0.46:1. As further distillate boils over, the overall composition of the distillate becomes richer in B. When the last drop has been condensed the phase composition is the same as at the beginning.

Exercise E4.15 Repeat the discussion, beginning at the point $x_B = 0.4$, $T = 298$ K.

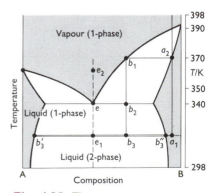

Fig. 4.32 The temperature–composition diagram for a binary system in which boiling occurs below the temperature at which two liquids are fully miscible.

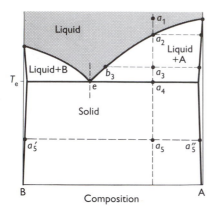

Fig. 4.33 The temperature–composition diagram for two almost immiscible solids and their completely immiscible liquids. Note the similarity to Fig. 4.32. The vertical line through e corresponds to the eutectic composition, the mixture with lowest melting point.

4.9 Liquid–solid phase diagrams

Solid and liquid phases may both be present in a system at temperatures below the boiling point. An example is a pair of metals that are almost completely immiscible right up to their melting points (such as antimony and bismuth). The phase diagram is shown in Fig. 4.33; note how closely it resembles Fig. 4.32.

Consider the two-component liquid of composition a_1. When it is cooled to a_2 it enters the two-phase region labelled 'Liquid + A'. Almost pure solid A begins to come out of solution and the remaining liquid becomes richer in B. On cooling to a_3, more of the solid forms, and the relative amounts of the solid and liquid (which are in equilibrium) are given by the lever rule: at this stage there are roughly equal amounts of each. The liquid phase is richer in B than before (its composition is given by b_3) because A has been deposited. At a_4 there is less liquid than at a_3, and its composition is given by e. This liquid now freezes to give a two-phase system of almost pure A and almost pure B. At a_5, for example, the compositions of the two phases are a_5' and a_5''.

Eutectics

The vertical line through e in Fig. 4.33 corresponds to the **eutectic** composition, the name coming from the Greek words for 'easily melted'. A liquid with the eutectic composition freezes at a single temperature, without previously depositing solid A or B. A solid with the eutectic composition melts, without change of composition, at the lowest temperature of any mixture. Solutions of composition to the right of e deposit A as they cool, and solutions to the left deposit B: only the eutectic mixture (apart from pure A or pure B) solidifies at a single definite temperature without gradually unloading one or other of the components from the liquid.

One technologically important eutectic is solder, which consists of 67 per cent tin and 33 per cent lead by mass and melts at 183 °C. The eutectic formed by 23 per cent NaCl and 77 per cent H_2O melts at −21.1 °C. When salt is added to ice under isothermal conditions (for example, when spread on an icy road) the mixture melts if the temperature is above −21.1 °C (and the eutectic composition has been achieved). When salt is added to ice under adiabatic conditions (for example, in a vacuum flask) the ice melts, but in doing so it absorbs heat from the rest of the mixture. The temperature of the system falls, and if enough salt is added, cooling continues down to the freezing point of the eutectic mixture. Eutectic formation occurs in the great majority of binary alloy systems, and is of great importance for the microstructure of solid materials, for although a eutectic solid is a two-phase system, it crystallizes out in a nearly homogeneous mixture of microcrystals. The two microcrystalline phases can be distinguished by microscopy and structural techniques such as X-ray diffraction.

Thermal analysis is a very useful practical way of detecting eutectics. We can see how it is used by considering the rate of cooling down the vertical line at a_1 in Fig. 4.33. The liquid cools steadily (Fig.

4.34) until it reaches a_2, when A begins to be deposited. Cooling is now slower because the solidification of A is exothermic and retards the cooling. When the remaining liquid reaches the eutectic composition, the temperature remains constant until the whole sample has solidified: this pause in the decrease in temperature is the **eutectic halt**. If the liquid has the eutectic composition e initially, then the liquid cools steadily down to the freezing temperature of the eutectic, when there is a long eutectic halt as the entire sample solidifies (like the freezing of a pure liquid).

Monitoring the cooling curves at different overall compositions gives a clear indication of the structure of the phase diagram. The solid–liquid boundary is given by the points at which the rate of cooling changes. The longest eutectic halt gives the location of the eutectic composition and its melting temperature.

Ultrapurity and controlled impurity

Advances in technology have called for materials of extreme purity. For example, semiconductor devices consist of almost perfectly pure silicon or germanium doped to a precisely controlled extent. For these materials to operate successfully, the impurity level must be kept down to less than 1 in 10^9 (which corresponds to about one grain of salt in 5 tons of sugar).

Consider a liquid of composition a_1 in Fig. 4.35: it is mainly B with some A impurity. On cooling to a_1, a solid of composition b_1 appears. Removing that solid gives a slightly purer material than the original, but not much of it (by the lever rule). That solid could be used as the starting substance for a second stage of this **fractional crystallization** process. In each stage, the composition is shifted towards pure B, in the manner of fractional distillation; but the procedure is slow and wasteful.

We should recognize, however, that Fig. 4.35 applies when the freezing is so slow that the composition of the solid is uniform and has its equilibrium composition. In a real system this is not the case, because A does not have time to disperse throughout the whole solid sample. The technique of **zone refining** makes use of the nonequilibrium properties of the system. It relies on the impurities being more soluble in the molten sample than in the solid, and sweeps them up by passing a molten zone repeatedly from one end to the other along a sample.

Consider a liquid (this represents the molten zone) on the vertical line at a_1, and let it cool without the entire sample coming to overall equilibrium. If the temperature falls to a_2 a solid of composition b_2 is deposited and the remaining liquid (the zone where the heater has moved on) is at a_2'. Cooling that liquid down a vertical line passing through a_2' deposits solid of composition b_3 and leaves liquid at a_3'. The process continues until the last drop of liquid to solidify is heavily contaminated with A. There is plenty of everyday evidence that impure liquids freeze in this way. For example, an ice cube is clear near the surface but misty in the core. This is because the water used

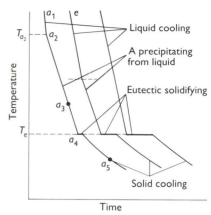

Fig. 4.34 The cooling curves for the system shown in Fig. 4.33. For a sample of composition represented by the vertical line through a_1 to a_5, the rate of cooling decreases at a_2 because solid A comes out of solution. The second cooling curve is for a sample of intermediate composition (between the vertical lines through a and e). If the experiment is repeated using a sample of composition represented by the vertical line through e, then there is a complete halt at e when the eutectic solidifies without change of composition. The halt is longest for the mixture of eutectic composition. The cooling curves can be used to construct the phase diagram.

155

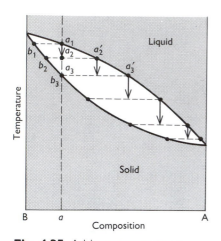

Fig. 4.35 A binary temperature–composition diagram can be used to discuss zone refining, as explained in the text.

to make ice normally contains dissolved air; freezing proceeds from the outside, and air is accumulated in the retreating liquid phase. It cannot escape from the interior of the cube, and so when that freezes it occludes the air in a mist of tiny bubbles.

In zone refining the sample is in the form of a narrow cylinder (Fig. 4.36). The sample is heated in a thin disc-like zone which is swept from one end of the sample to the other. The advancing liquid zone accumulates the impurities as it passes. In practice a hot zone is swept repeatedly from one end to the other. The zone at the end of the sample is the impurity dump: when the heater has gone by, it cools to a dirty solid that can be discarded.

A modification of zone refining is **zone levelling**. It is used to introduce controlled amounts of impurity (for example, of indium into germanium). A sample rich in the required dopant is put at the head of the main sample, and made molten. The zone is then dragged repeatedly in alternate directions through the sample, where it deposits a uniform distribution of the impurity.

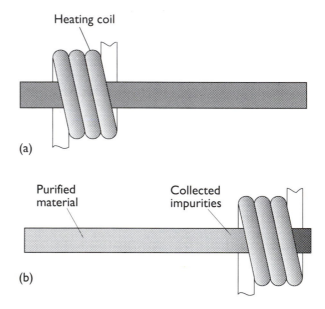

Fig. 4.36 In the zone-refining procedure, a heater is used to melt a small region of a long cylindrical sample of the impure solid, and that zone is swept to the other end of the rod. As it moves, it collects impurities. If a series of passes are made, the impurities accumulate at one end of the rod and can be discarded.

EXERCISES

4.1 An open vessel containing (a) water, (b) benzene, (c) mercury stands in a laboratory measuring 5 m × 5 m × 3 m at 25 °C. What mass of each substance will be found in the air if there is no ventilation? (The vapour pressures are (a) 24 Torr, (b) 98 Torr, (c) 1.7 mTorr.)

4.2 On a cold, dry morning after a frost, the temperature was −5 °C and the partial pressure of water in the atmosphere fell to 2 Torr. Will the frost sublime? What partial pressure of water vapour would ensure that the frost remained?

4.3 Refer to Fig. 4.8 and describe the changes that would be observed when water vapour at 1.0 atm and 400 K is cooled at constant pressure to 260 K. Suggest the appearance of a plot of temperature against time.

4.4 Refer to Fig. 4.7 and describe the changes that would be observed when cooling takes place at the pressure of the triple point.

4.5 Use the phase diagram in Fig. 4.9 to state what would be observed when a sample of carbon dioxide, initially at 1.0 atm and 298 K is subjected to the following cycle: (a) constant-pressure heating to 320 K, (b) isothermal compression to 100 atm, (c) constant-pressure cooling to 210 K, (d) isothermal decompression to 1.0 atm, constant-pressure heating to 298 °C.

4.6 The partial molar volumes of acetone and chloroform in a mixture in which the mole fraction of $CHCl_3$ is 0.4693 are 74.166 cm^3 mol^{-1} and 80.235 cm^3 mol^{-1} respectively. What is the volume of a solution of mass 1.000 kg?

4.7 At 300 K, the vapour pressure of dilute solutions of HCl in liquid $GeCl_4$ are as follows:

x_{HCl}	0.005	0.012	0.019
p/kPa	32.0	76.9	121.8

Show that the solution obeys Henry's law in this range of mole fractions and calculate Henry's law constant at 300 K.

4.8 At 90 °C, the vapour pressure of toluene (methylbenzene) is 400 Torr and that of o-xylene (1,2-dimethylbenzene) is 150 Torr. What is the composition of the liquid mixture that boils at 90 °C when the pressure is 0.50 atm? What is the composition of the vapour produced?

4.9 The vapour pressure of a 500 g sample of benzene was 400 Torr at 60.6 °C, but it fell to 386 Torr when 19.0 g of an nonvolatile organic compound was dissolved in it. Calculate the molar mass of the compound.

4.10 The addition of 100 g of a compound to 750 g of CCl_4 lowered the freezing point of the solvent by 10.5 K. Calculate the molar mass of the compound.

4.11 The osmotic pressure of an aqueous solution at 300 K is 120 kPa. Calculate the freezing point of the same solution.

4.12 Use Henry's law and the data in Table 4.2 to calculate the solubility of CO_2 in water at 25 °C when its partial pressure is (a) 0.10 atm, (b) 1.00 atm.

4.13 the mole fractions of N_2 and O_2 in air at sea level are approximately 0.78 and 0.21. Calculate the molalities of the solution formed in an open flask of water at 25 °C.

4.14 A water-carbonating plant is available for use in the home and operates by providing carbon dioxide at 5.0 atm. Estimate the molar concentration of the soda water it produces. See Table 4.2.

4.15 Calculate the freezing point of a 250 cm^3 glass of water sweetened with 7.5 g of sucrose.

4.16 The osmotic pressure of solutions of polystyrene in toluene were measured at 25 °C and the pressure was expressed in terms of the height of the solvent of density 1.004 g cm^{-3}. The following data were obtained:

$c/(g\,L^{-1})$	2.024	6.613	9.521	12.602
h/cm	0.592	1.910	2.750	3.600

Calculate the molar mass of the polymer.

4.17 The molar mass of an enzyme was determined by dissolving it in water, measuring the osmotic pressure at 20 °C, and extrapolating the data to zero concentration. The following data were obtained:

$c/(mg\,cm^{-3})$	3.221	4.618	5.112	6.722
h/cm	5.746	8.238	9.119	11.990

Calculate the molar mass of the enzyme.

4.18 The following temperature/composition data were obtained for a mixture of octane (O) and toluene (T) at 760 Torr, where x is the mole fraction in the liquid and y the mole fraction in the vapour at equilibrium.

θ/°C	110.9	112.0	114.0	115.8	117.3	119.0	120.0	123.0
x_T	0.908	0.795	0.615	0.527	0.408	0.300	0.203	0.097
y_T	0.923	0.836	0.698	0.624	0.527	0.410	0.297	0.164

The boiling points are 110.6 °C for T and 125.6 °C for O. Plot the temperature/composition diagram of the mixture. What is the composition of the vapour in equilibrium with the liquid of composition (a) $x_T = 0.250$ and (b) $x_O = 0.250$?

4.19 Sketch the phase diagram of the system NH_3/N_2H_4 given that the two substances do not form a compound with each other, that NH_3 freezes at −78 °C and N_2H_4 freezes at +2 °C, and that a eutectic is formed when the mole fraction of N_2H_4 is 0.07 and that the eutectic melts at −80 °C.

157

4.20 Figure 4.31 shows the phase diagram for two partially miscible liquids, which can be taken to be that for water and 2-methyl-1-propanol (B). Describe what will be observed when a mixture of composition b_3 is heated, at each stage giving the number, composition, and relative amounts of the phases present.

4.21 Figure 4.37 is the phase diagram for silver/tin. Label the regions, and describe what will be observed when liquids of compositions a and b are cooled to 200 K.

4.22 Sketch the cooling curves for the compositions a and b in Fig. 4.37.

4.23 Use the phase diagram in Fig. 4.37 to determine (a) the solubility of silver in tin at 800 °C, (b) the solubility of Ag_3Sn in silver at 460 °C, and (c) the solubility of Ag_3Sn in silver at 300 °C.

4.24 Hexane and perfluorohexane show partial miscibility below 22.70 °C. The critical concentration at the upper critical temperature is $x = 0.355$,

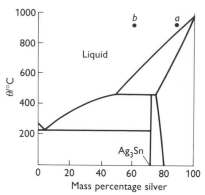

Fig. 4.37 The phase diagram for a binary silver/tin system.

where x is the mole fraction of C_6F_{14}. At 22.0 °C the two solutions in equilibrium have $x = 0.24$ and $x = 0.48$ respectively, and at 21.5 °C the mole fractions are 0.22 and 0.51. Sketch the phase diagram. Describe the phase changes that occur when perfluorohexane is added to a fixed amount of hexane at (a) 23 °C, (b) 22 °C.

5

Chemical equilibrium

CONTENTS

- The interpretation of equilibrium constants

- Acids and bases

- Solubility equilibria

- Coupled reactions

- Exercises

One of the centrally important concepts in chemistry, both for theoretical interpretations and practical applications, is that of the **equilibrium constant**, K, of a reaction. As we saw in Section 3.5, the equilibrium constant (a characteristic of each reaction, and a function of the temperature) specifies the relation between the concentrations or partial pressures of the species taking part in a reaction which guarantees that the reaction is in a state of dynamic equilibrium. As was stressed there, at equilibrium, the reaction free energy is zero ($\Delta G = 0$), so the reaction has no tendency for further change. At equilibrium, also, the rates of the forward and reverse reactions are equal, and the rate at which products are formed is exactly equal to the rate at which those products revert to reactants. In this chapter we examine some of the conclusions that can be drawn from the existence of dynamic equilibria in a variety of important systems in which chemical reactions take place. These systems include living cells, for although overall chemical equilibrium in a biological system occurs only at death, certain reactions (particularly those involving the rapid transfer of hydrogen ions from one species to another) do reach chemical equilibrium and can be discussed by the techniques discussed here. Chemical reactions in industry are only rarely allowed to reach equilibrium (in most cases, products are withdrawn as they are formed, so the plant maximizes its efficiency by chasing an ever-unattainable state of equilibrium). Nevertheless, the techniques we describe here, which are centred on the equilibrium constant, are a very useful guide to the feasibility of a particular reaction and indicate whether it would be worthwhile building a plant to achieve the reaction. We shall see, for instance, that even such large-scale processes as iron making can be discussed in terms of dynamic equilibrium, even though equilibrium is not fully achieved in a blast furnace.

The equilibrium constant of a reaction is defined as follows:

$$aA + bB \rightleftharpoons cC + dD \qquad \frac{a(C)^c a(D)^d}{a(A)^a a(B)^b} = K \text{ at equilibrium} \qquad (1)$$

where

- for a perfect gas, $a = p/p^{\ominus}$, where p is the partial pressure of the gas and $p^{\ominus} = 1$ bar;

- for a pure solid or liquid, $a = 1$ irrespective of the amount present;

- for an ideal solution, $a = [X]/(1 \text{ mol L}^{-1})$.

The equilibrium constant is independent of the overall pressure and it is also independent of the presence or otherwise of a catalyst (including enzymes). For example, for the reaction

$$N_2(g) + 3H_2(g) \rightleftharpoons 2NH_3(g)$$

at equilibrium

$$\frac{a(NH_3)^2}{a(N_2)a(H_2)^3} = K$$

In terms of the partial pressures of the three gases, after substitution of $a = p/p^{\ominus}$, the equilibrium condition is

$$\frac{p(NH_3)^2(p^{\ominus})^2}{p(N_2)p(H_2)^3} = K$$

Exercise E5.1 Write the equilibrium constant in terms of the activities a of the species for the reaction

acetyl CoA$(aq) + H_2O(l)$

$$\rightleftharpoons CH_3CO_2^-(aq) + CoA(aq) + H^+(aq)$$

where acetyl CoA is acetyl coenzyme A and CoA is coenzyme A, one of the proteins that play a key role in metabolism.
$$[K = a(CoA)a(CH_3CO_2^-)a(H^+)/a(\text{acetyl CoA})]$$

If the system is not ideal (if the gases present are not perfect and the solution is not dilute), then the equilibrium constant has the same form, but the activities a are no longer equal to the values quoted here. For example, the value of the activity a for a solute X that forms a nonideal solution is written

$$a = \gamma \times \frac{[X]}{1 \, \text{mol L}^{-1}} \qquad (2)$$

where γ is an empirical parameter called the **activity coefficient**. Values of activity coefficients can be determined experimentally, and some can be estimated theoretically. However, their inclusion in calculations greatly complicates the analysis, and we shall normally confine attention to systems that can be regarded as ideal (and for which $\gamma = 1$). This procedure is reasonably satisfactory for gases and solutions of nonelectrolytes, but it is very risky for ionic solutions (including the environment within biological cells, particularly neurones, where significant accumulations of Na^+, K^+, and Ca^{2+} are present), where deviations from ideality can be large even for very dilute (10^{-2} M) solutions.

In the following sections we shall explore the consequences of the existence of an equilibrium constant, and see how it governs the composition of a variety of different types of reaction system.

The interpretation of equilibrium constants

In broad outline, an equilibrium constant that is much larger than 1 (that is, for K of the order of 10^3 or more) indicates that a reaction has a strong tendency to form products, whereas if K is much smaller than 1 (smaller than about 10^{-3}), then the equilibrium composition will consist of largely unchanged reactants. If K is comparable to 1 (in the range 10^{-3} to 10^3), then significant amounts of both reactants and products will be present at equilibrium.

An equilibrium constant expresses the composition of the equilibrium mixture as a *ratio* of concentrations and partial pressures. It is therefore necessary to do some work to extract the actual concentrations or partial pressures of the reactants and products. This work can be organized into a systematic procedure by constructing an **equilibrium table** in each case. A typical equilibrium table has the format shown at the top of the following page. In most cases, we do not know the explicit change in concentration that must occur for the system to reach equilibrium, and so it is written as x: the reaction stoichiometry is then used to write the changes in concentration of each species given a particular change x in one of them. When the concentrations at equilibrium are substituted into the expression for

> **The balanced reaction**
>
> **The reactant and product species**
>
> 1. The initial concentrations of the species.
> 2. The changes in the concentrations that must take place for the system to reach equilibrium.
> 3. The resulting equilibrium concentrations of reactants and products.

the equilibrium constant, we obtain an equation for x in terms of K. This equation can be solved for x, and hence the concentrations of all the species at equilibrium may be found. In simple cases the equation can be solved exactly, but when the composition at equilibrium differs only slightly from the initial composition, it is possible to solve the equation by making an approximation.

As an illustration of the procedure, suppose that in an industrial process, N_2 at a partial pressure of 1.00 bar is mixed with H_2 at a partial pressure of 3.00 bar and the two gases are allowed to come to equilibrium with the product, ammonia (in the presence of a catalyst). The reaction and the equilibrium constant are given above. At the temperature of the reaction, it has been determined experimentally that $K = 977$. The equilibrium table then has the form

Reaction:	$N_2(g) + 3H_2(g) \rightleftharpoons 2NH_3(g)$		
Species:	N_2	H_2	NH_3
Initial partial pressure/bar	1.00	3.00	0
Change to reach equilibrium/bar	$-x$	$-3x$	$+2x$
Equilibrium partial pressure/bar	$1.00 - x$	$3.00 - 3x$	$2x$

The equilibrium constant for the reaction is expressed in terms of

$$a(N_2) = 1.00 - x$$
$$a(H_2) = 3.00 - 3x$$
$$a(NH_3) = 2x$$

and is

$$K = \frac{(2x)^2}{(1.00 - x) \times (3.00 - 3x)^3} = 977$$

This equation rearranges first to

$$\frac{4}{27}\left(\frac{x}{(1.00-x)^2}\right)^2 = 977$$

and then, after taking the square root of both sides, to

$$\frac{x}{(1.00-x)^2} = 81.2$$

This expression rearranges into the quadratic equation

$$81.2x^2 - 162.4x + 81.2 = 0$$

The solution of a quadratic equation of the form

$$ax^2 + bx + c = 0 \tag{3a}$$

is

$$x = \frac{-b \pm \sqrt{b^2 - 4ac}}{2a} \tag{3b}$$

Therefore, for this example,

$$x = \frac{163.4 \pm \sqrt{(163.4)^2 - 4(81.2)(81.2)}}{2(81.2)} = 1.12 \text{ or } 0.895$$

Because $a(N_2)$ cannot be negative (it is proportional to the partial pressure of nitrogen), and $a(N_2) = 1.00 - x$ (from the equilibrium table), we know that x cannot be greater than 1.00; therefore, we select $x = 0.895$ as the acceptable solution. It then follows from the last line of the equilibrium table that

$$p(N_2) = 0.10 \text{ bar} \qquad p(H_2) = 0.32 \text{ bar} \qquad p(NH_3) = 1.8 \text{ bar}$$

as the composition of the reaction mixture at equilibrium. Note that, because K is large (of the order of 10^3), the product dominates at equilibrium. It is always wise to confirm the accuracy of the calculation by substituting the calculated equilibrium partial pressures into the expression for the equilibrium constant to verify that the value so calculated is equal to the experimental value used in the calculation:

$$\frac{p(NH_3)^2(p^{\ominus})^2}{p(N_2)p(H_2)^3} = \frac{(1.8 \text{ bar})^2(1 \text{ bar})^2}{(0.10 \text{ bar})(0.31 \text{ bar})^3} = 9.9 \times 10^2$$

which is close to the experimental value (the discrepancy stems from rounding errors).

All equilibrium calculations can be treated similarly, but it is sometimes adequate to make approximations, especially when the procedure leads to an equation for x that is more complex than a quadratic equation. For example, when nitrogen and hydrogen are mixed in nonstoichiometric proportions (in a mole ratio of other than $1:3$) the resulting equation is not quadratic and is quite difficult to solve. Although graphical or numerical procedures may be used, it is usually worth checking whether an approximation can be used to arrive at a reasonably reliable conclusion with much less effort. This procedure is illustrated in the following example.

Example Solving an equilibrium calculation by approximation

Suppose that in an industrial process 2.00 mol N_2 is mixed with 2.00 mol H_2 and the two gases are allowed to come to equilibrium at a temperature at which $K = 1.0 \times 10^{-6}$. What is the equilibrium composition of the reaction mixture?

Answer The equilibrium table now has the form

Reaction:	$N_2(g) + 3H_2(g) \rightleftharpoons 2NH_3(g)$		
Species:	N_2	H_2	NH_3
Initial partial pressure/bar	2.00	2.00	0
Change to reach equilibrium/bar	$-x$	$-3x$	$+2x$
Equilibrium partial pressure/bar	$2.00 - x$	$2.00 - 3x$	$2x$

The equilibrium constant for the reaction is expressed in terms of

$$a(N_2) = 2.00 - x \qquad a(H_2) = 2.00 - 3x \qquad a(NH_3) = 2x$$

Now, because K is so small, we can expect only a small amount of NH_3 to be produced at equilibrium; that is, x will be very small. Specifically, we suppose that $x \ll 2.00$ and $3x \ll 2.00$ (if the second is true, then the first certainly is), and write

$$a(N_2) \approx 2.00 \qquad a(H_2) \approx 2.00 \qquad a(NH_3) = 2x$$

Then,

$$\frac{(2x)^2}{(2.00) \times (2.00)^3} \approx 1.0 \times 10^{-6}$$

which solves (by rearranging the expression and taking the square root of both sides) to $x \approx 2.0 \times 10^{-3}$. (This value of x is consistent with the approximations made.) Therefore, at equilibrium, the partial pressures of N_2, H_2, and NH_3 are 2.00 bar, 2.00 bar, and 4.0 mbar, respectively.

Exercise E5.2 Find the equilibrium partial pressures for the same reaction, but starting with partial pressures of 3.00 bar, 1.00 bar, and 0.500 bar of N_2, H_2, and NH_3, respectively, and at a temperature for which $K = 2.50 \times 10^{-4}$.

[2.99 bar, 0.96 bar, 0.528 bar]

The approximation procedure is valid when the equilibrium composition differs only slightly from the initial (given) composition. This is typically the case when *reactants* are mixed initially and K is much smaller than 1 ($K < 10^{-3}$), for then only a small amount of product need be formed to reach equilibrium (so x is very small). It may also be the case when *products* are mixed initially and the equilibrium constant is much larger than 1 ($K > 10^{3}$), because only a small amount of reactants are formed to reach equilibrium. *Whenever the approximation procedure is employed, it should be verified that the answer is consistent with the approximations that have been made.* For example, if it is assumed that the change in concentration, x, is much smaller than the initial concentration or partial pressure of a reactant, then it should be verified that the calculated change is no more than about 5 per cent of the initial concentration. If the calculated change is more than 5 per cent of the initial concentration, then there is no choice but to use a more exact method of solution (for example, a graphical solution or solution by successive approximation).

Acids and bases

One of the most important examples of chemical equilibrium is one that exists when acids and bases are present in solution. As we shall see, the equilibration of acids and bases depends on the transfer of protons between species. In the context of acids and bases, a 'proton' is a hydrogen ion, H^+, and the theory of acids and bases that we shall develop depends on the transfer of hydrogen ions between species. This transfer is normally so facile that we can be confident that any solution of acid or base is at equilibrium and that it is appropriate to calculate the composition of the solution using the equilibrium constant for the proton-transfer reaction.

The principal reason for carrying out a calculation of the equilibrium composition of a solution of an acid or a base is to find the concentration of hydrogen ions in solution. This concentration is of the greatest importance in many applications of chemistry, for hydrogen ions govern the processes of life, modify landscapes by their influence on geochemical processes, and determine the success of many industrial syntheses. Much of our environment, outside us in our physical surroundings and inside us in our biochemical composition, is a manifestation of the role of hydrogen ion concentration.

The concentration of hydrogen ions in a solution is normally reported in terms of the **pH** of the solution where[†]

$$pH = -\log a(H^+) \quad a(H^+) = [H^+]/mol\ L^{-1} \qquad (4)$$

For example, if the molar concentration of H^+ is $2.0 \times 10^{-3}\ mol\ L^{-1}$,

---

[†] The precise relation of $a(H^+)$ to the concentration should make use of the activity coefficient; the equations that we shall derive in the following sections are valid only if *all* the ions in the solution are present at *very* low concentrations.

then

$$pH = -\log(2.0 \times 10^{-3}) = 2.70$$

If the molar concentration were ten times less, at $2.0 \times 10^{-4}\,\text{mol}\,\text{L}^{-1}$, then the pH would be 3.70. Notice that *the higher the pH, the lower the concentration of hydrogen ions in the solution.*

Exercise E5.3 Death is likely if the pH of human blood plasma changes by more than ± 0.4 from its normal value of 7.4. What is the range of molar concentrations of hydrogen ions for which life can be sustained?

$[1.6 \times 10^{-8}\,\text{mol}\,\text{L}^{-1}$ to $1.0 \times 10^{-7}\,\text{mol}\,\text{L}^{-1}]$

5.1 The Brønsted–Lowry theory

Most modern work on the reactions of acids and bases is expressed in terms of a theory proposed by the Danish chemist Johannes Brønsted and the English chemist Thomas Lowry in 1923. According to the **Brønsted–Lowry theory** of acids and bases:

A **Brønsted acid**, HA, is a proton donor, $HA \rightarrow H^+ + A^-$

A **Brønsted base**, B, is a proton acceptor, $B + H^+ \rightarrow BH^+$

In the context of Brønsted–Lowry theory, a proton is a hydrogen ion, H^+. Hydrogen chloride, HCl, is an acid because it can donate a proton to another molecule. Methane, CH_4, is not a Brønsted acid because, despite its hydrogen atoms, it is not a proton donor. Ammonia, NH_3, is a base because it can accept a proton from another molecule and become NH_4^+. The definitions make no mention of the solvent (and apply even if no solvent is present); however, by far the most important medium is aqueous solution, and we shall confine our attention to that.

Exercise E5.4 Identify H_2SO_4, HSO_4^-, and SO_4^{2-} as an acid or a base.

[acid, acid and base, base]

Acid–base equilibria in water

An acid HA (for example, HCl or CH_3COOH) takes part in the following proton-transfer equilibrium in water:

$$HA(aq) + H_2O(l) \rightleftharpoons H_3O^+(aq) + A^-(aq)$$

In the forward reaction, the H_2O molecule accepts a proton from the acid, and therefore acts as a Brønsted base. The H_3O^+ ion, which is formed when H_2O accepts a proton from an acid, is called a **hydronium ion**. In the reverse reaction, the A^- ion accepts a proton from H_3O^+ and is converted back into the acid HA. Therefore, in the reverse reaction, H_3O^+ acts as a Brønsted acid and A^- acts as a Brønsted base.

The proton donor that results from the transfer of a proton to a species is called the **conjugate acid** of the original base. In the forward reaction above, the H_3O^+ ion that results from proton transfer to H_2O is the conjugate acid of the base H_2O. Similarly, the proton acceptor A^- that remains after a Brønsted acid has donated a proton is called the **conjugate base** of the acid HA. For example, in the equilibrium established when HF is present in water,

$$HF(aq) + H_2O(l) \rightleftharpoons H_3O^+(aq) + F^-(aq)$$

the H_3O^+ ion is the conjugate acid of the base H_2O and the F^- ion is the conjugate base of the acid HF. Similarly, in the equilibrium

$$CH_3COOH(aq) + H_2O(l) \rightleftharpoons H_3O^+(aq) + CH_3CO_2^-(aq)$$

the acetate ion, $CH_3CO_2^-$, is the conjugate base of the acid CH_3COOH. The general form of the proton-transfer equilibrium in water is therefore

$$acid_1 + base_2 \rightleftharpoons acid_2 + base_1 \tag{5}$$

with ($acid_1$, $base_1$) one conjugate acid–base pair and ($acid_2$, $base_2$) another conjugate acid–base pair. The conjugate acid–base pairs for the examples above are shown in the margin table.

Acid	Base
H_3O^+	H_2O
HF	F^-
CH_3COOH	$CH_3CO_2^-$

Exercise E5.5 What is the conjugate base of ammonia when it acts as an acid?

[NH_2^-, the amide ion]

For a base B, such as ammonia, NH_3, in water, the characteristic proton-transfer equilibrium is

$$H_2O(l) + B(aq) \rightleftharpoons BH^+(aq) + OH^-(aq)$$

The proton donor formed when the base B accepts a proton and becomes BH^+ is another example of a conjugate acid of a base.

Likewise, OH^-, the ion that results from the loss of a proton from the proton donor, H_2O, is another example of a conjugate base. An example of this equilibrium is

$$H_2O(l) + NH_3(aq) \rightleftharpoons NH_4^+(aq) + OH^-(aq)$$

The conjugate acid of NH_3 is the ammonium ion, NH_4^+. As can be seen, the form of this equilibrium is exactly the same as in the general case, but now H_2O acts as a Brønsted acid:

$$H_2O(l) + NH_3(aq) \rightleftharpoons NH_4^+(aq) + OH^-(aq)$$
$$acid_1 + base_2 \rightleftharpoons acid_2 + base_1.$$

Acid	Base
H_2O	OH^-
NH_4^+	NH_3

The table of conjugate acid–base pairs can therefore be expanded to include the pairs shown in the margin.

The fact that water can act as both a Brønsted acid and a Brønsted base means that even in the absence of any solute, hydronium ions and hydroxide ions exist in water as a result of the **autoprotolysis equilibrium**

$$H_2O(l) + H_2O(l) \rightleftharpoons H_3O^+(aq) + OH^-(aq)$$
$$acid_1 + base_2 \rightleftharpoons acid_2 + base_1$$

At 25 °C, the molar concentration of hydronium ions in water (and the molar concentration of OH^- ions) arising from this autoprotolysis equilibrium is $1.0 \times 10^{-7} \, \text{mol L}^{-1}$, so only 1 in 550 million H_2O molecules has donated a proton to another H_2O molecule. Nevertheless, this tiny concentration of hydronium ions is of crucial importance to the properties of aqueous solutions of acids, bases, and salts, as we shall see.

It follows that, as a result of autoprotolysis, the pH of pure water at 25 °C is

$$pH = -\log(1.0 \times 10^{-7}) = 7.00$$

A **neutral solution** is an aqueous solution with this pH at 25 °C. Because the autoprotolysis reaction is endothermic, the formation of hydronium ions is favoured by increased temperature, and at 37 °C (body temperature), the equilibrium concentration of hydronium ions has risen to $1.5 \times 10^{-7} \, \text{mol L}^{-1}$; the pH of a neutral solution is then 6.84.

Acidity constants

The proton-transfer equilibrium of a Brønsted acid in water can be expressed in terms of an equilibrium constant:

$$HA(aq) + H_2O(l) \rightleftharpoons H_3O^+(aq) + A^-(aq)$$
$$\frac{a(H_3O^+)a(A^-)}{a(HA)a(H_2O)} = K \text{ at equilibrium}$$

where, as usual, $a(X)$ denotes $[X]/\text{mol L}^{-1}$ for a solute (at concentrations so low that deviations from ideality can be ignored) and is equal to 1 for a pure liquid or solid. The same expression applies to the proton-transfer equilibrium established by the conjugate acid of a Brønsted base—there is no fundamental distinction between an acid and a conjugate acid of a base, for both are proton donors. Strictly speaking, there are no pure liquids or solids in the proton-transfer equilibrium, but the solutions we shall consider will always be so dilute that it is almost always the case that the water present can be regarded as being a nearly pure liquid. If we make the approximation that $a(H_2O) = 1$ for all the solutions we consider, then the resulting equilibrium constant is called the **acidity constant**, K_a, of the acid HA:

$$K_a = \left(\frac{a(H_3O^+)a(A^-)}{a(HA)}\right)_{\text{equilibrium}} \qquad (6)$$

The acidity constants of a number of acids (and the conjugate acids of bases) are given in Table 5.1 together with additional information that will be explained below.

Exercise E5.6 Write the expression for the acidity constant of $H_2PO_4^-$.

$$[K_a = a(H_3O^+)a(HPO_4^{2-})/a(H_2PO_4^-)]$$

Proton transfer is so fast in aqueous solution that we can be confident that all solutions that we consider are at equilibrium, and from now on we shall not note 'equilibrium' explicitly when writing acidity constants. An explicit example of an acidity constant is that for hydrofluoric acid:

$$HF(aq) + H_2O(aq) \rightleftharpoons H_3O^+(aq) + F^-(aq) \qquad K_a = \frac{a(H_3O^+)a(F^-)}{a(HF)}$$

It is found experimentally that $K_a = 3.5 \times 10^{-4}$ at 25 °C.

The value of the acidity constant indicates the extent to which proton transfer has occurred: *the smaller the value of K_a, the weaker the proton donating power of the acid.* In the context of the Brønsted–Lowry theory, the term 'ionization' is used to denote proton transfer, so the small value of K_a for HF (3.5×10^{-4}) indicates that hydrogen fluoride is only slightly ionized in solution. The value of K_a for hydrocyanic acid, HCN, is only 4.9×10^{-10}, which indicates that in water it is ionized (that is, has undergone proton transfer) to a far smaller extent even than hydrofluoric acid.

Acidity constants are widely reported as their logarithms, and by analogy with the definition of pH we define

$$pK_a = -\log K_a \qquad (7)$$

For hydrofluoric acid

$$pK_a = -\log(3.5 \times 10^{-4}) = 3.46$$

Table 5.1   Acidity and basicity constants† at 25 °C

Acid/Base		K_b	pK_b	K_a	pK_a
Strongest weak acids	Trichloracetic acid, CCl_3COOH	3.3×10^{-14}	13.48	3.0×10^{-1}	0.52
	Benzenesulfonic acid $C_6H_5SO_3H$	5.0×10^{-14}	13.30	2×10^{-1}	0.70
	Iodic acid, HIO_3	5.9×10^{-14}	13.23	1.7×10^{-1}	0.77
	Sulfurous acid, H_2SO_3	6.3×10^{-13}	12.19	1.6×10^{-2}	1.81
	Chlorous acid, $HClO_2$	1.0×10^{-12}	12.00	1.0×10^{-2}	2.00
	Phosphoric acid, H_3PO_4	1.3×10^{-12}	11.88	7.6×10^{-3}	2.12
	Chloroacetic acid, $CH_2ClCOOH$	7.1×10^{-12}	11.15	1.4×10^{-3}	2.85
	Lactic acid, $CH_3CH(OH)COOH$	1.2×10^{-11}	10.92	8.4×10^{-4}	3.08
	Nitrous acid, HNO_2	2.3×10^{-11}	10.63	4.3×10^{-4}	3.37
	Hydrofluoric acid, HF	2.9×10^{-11}	10.55	3.5×10^{-4}	3.46
	Formic acid, HCOOH	5.6×10^{-11}	10.25	1.8×10^{-4}	3.75
	Benzoic acid, C_6H_5COOH	1.5×10^{-10}	9.81	6.5×10^{-5}	4.19
	Acetic acid, CH_3COOH	5.6×10^{-10}	9.25	1.8×10^{-5}	4.75
	Carbonic acid, H_2CO_3	2.3×10^{-8}	7.63	4.3×10^{-7}	6.37
	Hypochlorous acid, HClO	3.3×10^{-7}	6.47	3.0×10^{-8}	7.53
	Hypobromous acid, HBrO	5.0×10^{-6}	5.31	2.0×10^{-9}	8.69
	Boric acid, $B(OH)_3$‡	1.4×10^{-5}	4.86	7.2×10^{-10}	9.14
	Hydrocyanic acid, HCN	2.0×10^{-5}	4.69	4.9×10^{-10}	9.31
Weakest weak acids	Phenol, C_6H_5OH	7.7×10^{-5}	4.11	1.3×10^{-10}	9.89
	Hypoiodous acid, HIO	4.3×10^{-4}	3.36	2.3×10^{-11}	10.64
Weakest weak bases	Urea, $CO(NH_2)_2$	1.3×10^{-14}	13.90	7.7×10^{-1}	0.10
	Aniline, $C_6H_5NH_2$	4.3×10^{-10}	9.37	2.3×10^{-5}	4.63
	Pyridine, C_5H_5N	1.8×10^{-9}	8.75	5.6×10^{-6}	5.35
	Hydroxylamine, NH_2OH	1.1×10^{-8}	7.79	9.1×10^{-7}	6.03
	Nicotine, $C_{10}H_{11}N_2$	1.0×10^{-6}	5.98	1.0×10^{-8}	8.02
	Morphine, $C_{17}H_{19}O_3N$	1.6×10^{-6}	5.79	6.3×10^{-9}	8.21
	Hydrazine, NH_2NH_2	1.7×10^{-6}	5.77	5.9×10^{-9}	8.23
	Ammonia, NH_3	1.8×10^{-5}	4.75	5.6×10^{-10}	9.25
	Trimethylamine, $(CH_3)_3N$	6.5×10^{-5}	4.19	1.5×10^{-10}	9.81
	Methylamine, CH_3NH_2	3.6×10^{-4}	3.44	2.8×10^{-11}	10.56
	Dimethylamine, $(CH_3)_2NH$	5.4×10^{-4}	3.27	1.9×10^{-11}	10.73
Strongest weak bases	Ethylamine, $C_2H_5NH_2$	6.5×10^{-4}	3.19	1.5×10^{-11}	10.81
	Triethylamine, $(C_2H_5)_3N$	1.0×10^{-3}	2.99	1.0×10^{-11}	11.01

† Values for polyprotic acids—those capable of donating more than one proton—refer to the first ionization.
‡ The proton-transfer equilibrium is

$$B(OH)_3(aq) + 2H_2O(l) \rightleftharpoons H_3O^+(aq) + B(OH)_4^-(aq)$$

Note that the smaller the value of the acidity constant, the larger the value of pK_a. For example, hydrocyanic acid, HCN, has $K_a = 4.9 \times 10^{-10}$, which is less than the value for hydrofluoric acid, and its pK_a is 9.31, which is larger than for hydrofluoric acid. In general, *the larger the value of pK_a, the weaker the proton-donating power of the acid*. The HCN molecule is a far weaker proton donor (to water) than an HF molecule; a CH_4 molecule has a completely negligible proton-donating power to water.

The autoprotolysis constant of water

The equilibrium constant for the water autoprotolysis equilibrium is

$$K = \frac{a(H_3O^+)a(OH^-)}{a(H_2O)^2}$$

However, in the dilute solutions that we shall always consider, the water is almost pure, and to a very good approximation we can replace $a(H_2O)$ by 1. The resulting expression is called the **autoprotolysis constant** of water, and is denoted K_w:

$$K_w = a(H_3O^+)a(OH^-) \qquad (8)$$

(Throughout this section, $a = [X]/\text{mol L}^{-1}$ refers to the *equilibrium* concentration of the species X.) We have already seen that the molar concentrations of H_3O^+ and OH^- ions in pure water at 25 °C are 1.0×10^{-7} mol L^{-1}, so it follows that at 25 °C

$$K_w = (1.0 \times 10^{-7}) \times (1.0 \times 10^{-7}) = 1.0 \times 10^{-14}$$

The corresponding logarithmic expression is

$$pK_w = -\log K_w = 14.00 \text{ at } 25 °C$$

The importance of the autoprotolysis equilibrium is that, *although the individual hydronium ion and hydroxide ion concentrations may change as acid or base is added to a solution, the product of the concentrations must remain equal to K_w* (for otherwise the proton transfer between water molecules would not be at equilibrium).

Exercise E5.7 The molar concentration of OH^- ions in a certain solution is 1.0×10^{-4} mol L^{-1}. What is the pH of the solution?

[10.00]

The relation between the pH and the concentration of OH^- ions in an aqueous solution is most easily expressed by introducing the **pOH** of the solution, which is defined as

$$pOH = -\log a(OH^-), \qquad a(OH^-) = [OH^-]/\text{mol L}^{-1} \qquad (9)$$

if the solution is dilute (and by using $a = \gamma[OH^-]/\text{mol L}^{-1}$ if it is not ideal). Then, taking logarithms of both sides of the expression for the autoprotolysis constant gives

$$-\log K_w = -\log a(H_3O^+) - \log a(OH^-)$$

because $\log xy = \log x + \log y$. It follows that

$$pK_w = pH + pOH \tag{10}$$

For example, in a solution in which the molar concentration of OH^- ions is $1.0 \times 10^{-4} \, mol \, L^{-1}$, so $pOH = 4.00$, the pH is $pH = 14.00 - 4.00 = 10.00$, as stated in the exercise above.

Weak and strong acids

Acidity constants are used to classify acids as weak or strong. A **weak acid** is an acid with K_a smaller than about 1: only a small proportion of its molecules are ionized in solution at normal concentrations. A **strong acid** is an acid with K_a appreciably larger than 1, and at normal concentrations its molecules are fully ionized.

Hydrochloric acid is a strong acid. When hydrogen chloride dissolves in water, every HCl molecule that dissolves donates a proton to a water molecule and the solution consists of only H_3O^+ and Cl^- ions. Therefore, if the concentration of HCl is reported as $1.0 \times 10^{-3} \, mol \, L^{-1}$, then we know that in fact the solution consists of no HCl molecules but $1.0 \times 10^{-3} \, mol \, L^{-1}$ of H_3O^+ and the same concentration of Cl^- ions. The pH of the solution will be 3.00 because the actual concentration of H_3O^+ ions will be the same as the nominal concentration of HCl (that is, the stated concentration of the solution). There are very few strong acids in water: the ones to remember are HCl, HBr, HI, HNO_3, H_2SO_4 (with respect to the donation of one proton), and $HClO_4$.

Most acids are weak (see Table 5.1), and in solution exist largely as the nonionized acid molecules with only a small proportion ionized. The extent of ionization depends on the acidity constant and the concentration of the solution, and it may be calculated by using the equilibrium-table technique presented earlier in the chapter. As an example, consider acetic acid, for which $pK_a = 4.75$ (corresponding to $K_a = 1.8 \times 10^{-5}$, a value that indicates a low degree of ionization into the products H_3O^+ and $CH_3CO_2^-$). To calculate the pH and the proportion of CH_3COOH molecules that are ionized in a solution of molar concentration $A \, mol \, L^{-1}$ we draw up the following equilibrium table:

Reaction:	$CH_3COOH(aq) + H_2O(l) \rightleftharpoons H_3O^+(aq) + CH_3CO_2^-(aq)$		
Species:	CH_3COOH	H_3O^+	$CH_3CO_2^-$
Initial concentration/mol L^{-1}	A	0	0
Change in concentration/mol L^{-1}	$-x$	$+x$	$+x$
Equilibrium concentration/mol L^{-1}	$A - x$	x	x

The value of x can be found by inserting the equilibrium concentrations into the expression for the acidity constant:

$$K_a = \frac{a(H_3O^+)a(CH_3CO_2^-)}{a(CH_3COOH)} = \frac{x \times x}{A - x}$$

This expression rearranges to the quadratic equation

$$x^2 + K_a x - A K_a = 0$$

with the solution

$$x = \frac{-K_a \pm \sqrt{K_a^2 + 4A K_a}}{2}$$

Because x is equal to the concentration of hydronium ions (see the last line of the equilibrium table), it must be positive, and so we must take the solution with the positive square root (for only that solution can give a positive value of x). Suppose that $A = 0.010$ (corresponding to an initial CH_3COOH concentration of $0.010 \ mol \ L^{-1}$), then

$$x = \frac{-(1.8 \times 10^{-5}) + \sqrt{(1.8 \times 10^{-5})^2 + 4(0.010)(1.8 \times 10^{-5})}}{2}$$

$$= 4.2 \times 10^{-4}$$

from which it follows that

$$pH = -\log(4.2 \times 10^{-4}) = 3.38$$

pH calculations of this kind are rarely accurate to more than one decimal place (and even that may be over optimistic) because the effects of ion–ion interactions have been ignored, so this answer would be reported as $pH = 3.4$.

Exercise E5.8 Calculate the pH of a solution of 0.01 M lactic acid. Before carrying out the numerical calculation, do you expect the pH to be higher or lower than that calculated for the same concentration of acetic acid?

[3.0]

The **fraction ionized**, the fraction of acetic acid molecules that have donated a proton in the solution, can now also be calculated:

$$\text{fraction ionized} = \frac{\text{concentration of conjugate base}}{\text{initial concentration of acid}}$$

$$= \frac{x}{A} = \frac{4.2 \times 10^{-4}}{0.010} = 4.2 \times 10^{-2}$$

That is, only 4.2 per cent of the acetic acid molecules have donated a proton. For lactic acid, a somewhat stronger acid, the fraction is 9.1 per cent, which shows that a higher proportion of acid molecules have donated protons.

We saw earlier in the chapter that an equilibrium calculation could be greatly simplified if the change in composition needed to attain equilibrium was very small. Because weak acids have acidity constants that are usually much smaller than 1, that technique can often be applied to them. Specifically, the calculation of the pH of a weak acid is greatly simplified if we assume that the fraction ionized is so small that the denominator $A - x$ in the expression for K_a can be approximated by A itself. Then

$$K_a \approx \frac{x^2}{A}$$

which solves to

$$x \approx \sqrt{AK_a}$$

Hence, by taking logarithms and changing the sign throughout,

$$-\log x \approx -\tfrac{1}{2} \log K_a - \tfrac{1}{2} \log A$$

Because $-\log x$ is the pH of the solution, it follows that

$$pH \approx \tfrac{1}{2} pK_a - \tfrac{1}{2} \log A \tag{11}$$

For example, the pH of $0.010\,\text{M}\,CH_3COOH(aq)$ is predicted to be

$$pH \approx \tfrac{1}{2} \times 4.75 - \tfrac{1}{2} \times \log 0.010 = 2.38 + 1.00 \approx 3.4$$

in good agreement with the value found without making the approximation. This approximation procedure can be used when the fraction of ionization is smaller than about 0.05 (5 per cent).

To verify that the fraction ionized is small, we can substitute the approximate expression for x given above into the definition of fraction ionized, and obtain

$$\text{fraction ionized} = \frac{x}{A} \approx \frac{\sqrt{AK_a}}{A} = \sqrt{\frac{K_a}{A}} \tag{12}$$

It is now easy to verify that the fraction of CH_3COOH molecules ionized in $0.010\,\text{M}\,CH_3COOH(aq)$ is

$$\text{fraction ionized} \approx \sqrt{\frac{1.8 \times 10^{-5}}{0.010}} = 0.042$$

which is within the limit of 5 per cent and in good agreement with the 'precise' value.

The procedure for calculating the pH of the solution of an acid can now be summarized as follows:

1. If the acid is strong, assume that it is fully ionized in solution, and calculate the pH from the stated concentration of the acid by using $[H_3O^+] = [HA]_{\text{added initially}}$.

2. If the acid is weak, use eqn (12) to decide whether, for the concentration specified and the numerical value of K_a of the acid, the fraction ionized is less than 0.05.

3. If the fraction ionized is greater than 0.05, use the 'exact' equilibrium-table procedure to calculate x and thence the pH.

4. If the fraction ionized is not greater than 0.05, use the approximate formula, eqn (11), for the pH.

Even if step 4 is permissible, it is good practice to set up the equilibrium table and make the approximations rather than to rely on memory to recall eqn (11). In any event, though, it should be remembered that the procedures outlined here result in *estimates* of the pH, for in no case have we considered the ion–ion interactions that result in departures from ideality.

Example Calculating the pH of a solution of a weak acid
What is the pH of 0.25 M HCN(aq) at 25°C?
Answer The pK_a of hydrocyanic acid, HCN, is given in Table 5.1 as 9.31, which corresponds to a very small value of K_a (4.9×10^{-10}, in fact). We can expect the fraction of acid ionized to be very small, and confirm that this is so by substituting the data into eqn (12), to obtain

$$\text{fraction ionized} = \sqrt{\frac{4.9 \times 10^{-10}}{0.25}} = 4.4 \times 10^{-5}$$

Because this fraction is very much smaller than 1, it is valid to use eqn (11) for the pH:

$$\text{pH} = \tfrac{1}{2} \times 9.31 - \tfrac{1}{2} \times \log 0.25 = 4.96$$

or about 5.0.

Exercise E5.9 Calculate the pH of 0.20 M HClO(aq).

[4.1]

Polyprotic acids

With **polyprotic acids**, which are species with more than one donatable proton, it is necessary to distinguish the successive acidity constants.†
For a species with two donatable protons (such as H_2SO_4), the

---

† A polyprotic acid is best considered to be a molecular species that can give rise to a series of ionic Brønsted acids as it donates its succession of protons. Thus, H_3PO_4 is a polyprotic acid, which is the parent of two other Brønsted acids, namely $H_2PO_4^-$, and HPO_4^{2-}.

Table 5.2 Successive acidity constants of polyprotic acids

Acid	K_{a1}	pK_{a1}	K_{a2}	pK_{a2}	K_{a3}	pK_{a3}
Carbonic acid, H_2CO_3	4.3×10^{-7}	6.37	5.6×10^{-11}	10.25		
Hydrosulfuric acid, H_2S	1.3×10^{-7}	6.88	7.1×10^{-15}	14.15		
Oxalic acid, $(COOH)_2$	5.9×10^{-2}	1.23	6.5×10^{-5}	4.19		
Phosphoric acid, H_3PO_4	7.6×10^{-3}	2.12	6.2×10^{-8}	7.21	2.1×10^{-13}	12.67
Phosphorous acid, H_2PO_3	1.0×10^{-2}	2.00	2.6×10^{-7}	6.59		
Sulfuric acid, H_2SO_4	Strong		1.2×10^{-2}	1.92		
Sulfurous acid, H_2SO_3	1.5×10^{-2}	1.81	1.2×10^{-7}	6.91		
Tartaric acid, $C_2H_4O_2(COOH)_2$	6.0×10^{-4}	3.22	1.5×10^{-5}	4.82		

equilibria are

$$H_2A(aq) + H_2O(l) \rightleftharpoons H_3O^+(aq) + HA^-(aq)$$

$$K_{a1} = \frac{a(H_3O^+)a(AH^-)}{a(H_2A)}$$

(13a)

in which HA^- is the conjugate base of H_2A, and

$$HA^-(aq) + H_2O(l) \rightleftharpoons H_3O^+(aq) + A^{2-}(aq)$$

$$K_{a2} = \frac{a(H_3O^+)a(A^{2-})}{a(HA^-)}$$

(13b)

in which HA^- now acts as the acid and A^{2-} is the conjugate base of the acid HA^-. Generally K_{a2} is smaller than K_{a1}, typically by three orders of magnitude, because the second proton is more difficult to remove, partly on account of the negative charge on HA^-. Enzymes are polyprotic acids, for they possess many protons that can be donated to a substrate molecule or to the surrounding aqueous medium of the cell. For them, successive acidity constants vary much less because the molecules are so large that the loss of a proton from one part of the molecule has little effect on the ease with which another some distance away may be lost. The values of successive acidity constants of some polyprotic acids are given in Table 5.2.

Example Calculating the concentration of carbonate ion in carbonic acid

Calculate the molar concentration of CO_3^{2-} ions in carbonic acid.

Answer The CO_3^{2-} ion, the conjugate base of the acid HCO_3^-, is produced in the equilibrium

$$HCO_3^-(aq) + H_2O(l) \rightleftharpoons H_3O^+(aq) + CO_3^{2-}(aq)$$

$$K_{a2} = \frac{a(H_3O^+)a(CO_3^{2-})}{a(HCO_3^-)}$$

Hence,

$$a(CO_3^{2-}) = \frac{a(HCO_3^-)K_{a2}}{a(H_3O^+)}$$

Because the second ionization hardly affects the concentration of H_3O^+ and HCO_3^- ions produced in the first ionization, we set $[H_3O^+] \approx [HCO_3^-]$ (we are assuming that the hydronium ions that come from the ionization of H_2CO_3 greatly outnumber those provided by the autoprotolysis of water). As a result, $a(HCO_3^-)$ and $a(H_3O^+)$ cancel, and

$$a(CO_3^{2-}) \approx K_{a2}$$

Because we know from Table 5.2 that $pK_{a2} = 10.25$, it follows that $a(CO_3^{2-}) = 5.6 \times 10^{-11}$, and therefore that $[CO_3^{2-}] \approx 5.6 \times 10^{-11}$ mol L^{-1}.

Exercise E5.10 Calculate the molar concentration of S^{2-} ions in $H_2S(aq)$.

$$[7.1 \times 10^{-15} \text{ mol L}^{-1}]$$

5.2 Weak and strong bases

According to the Brønsted–Lowry theory, when a base (a proton acceptor) such as ammonia is dissolved in water it participates in the proton transfer equilibrium

$$H_2O(l) + NH_3(aq) \rightleftharpoons NH_4^+(aq) + OH^-(aq)$$

$$\text{acid}_1 + \text{base}_2 \rightleftharpoons \text{acid}_2 + \text{base}_1$$

Proton transfer is so fast that a solution of a base is always at equilibrium with its conjugate acid, and the concentrations of the species are described by the equilibrium constant

$$K = \frac{a(NH_4^+)a(OH^-)}{a(NH_3)a(H_2O)}$$

As usual, if we confine our attention to dilute solutions, the water may be treated as pure liquid with $a = 1$, and the resulting equilibrium constant is called the **basicity constant** (or 'base ionization constant') K_b:

$$K_b = \frac{a(NH_4^+)a(OH^-)}{a(NH_3)}$$

177

In general, a basicity constant is defined as follows:

$$H_2O(aq) + B(aq) \rightleftharpoons HB^+(aq) + OH^-(aq)$$

$$K_b = \frac{a(HB^+)a(OH^-)}{a(B)} \tag{14}$$

The basicity constants of a number of species (including the conjugate bases of some acids) are given in Table 5.1. For ammonia in water, for example, $K_b = 1.8 \times 10^{-5}$ at 25 °C. As in the case of acidity constants, it is convenient to report the values of basicity constants in terms of their negative logarithms:

$$pK_b = -\log K_b \tag{15}$$

Thus, for ammonia at 25 °C,

$$pK_b = -\log(1.8 \times 10^{-5}) = 4.75$$

The numerical value of the basicity constant indicates the extent to which a base is protonated in aqueous solution: *the greater the value of K_b, the stronger the proton-accepting power of the base.* Conversely, *the greater the value of pK_b, the weaker the proton-accepting power of the base.* The weak base morphine, for example, has $K_b = 1.6 \times 10^{-6}$ and $pK_b = 5.79$, and is a weaker proton acceptor than ammonia. As a result, when morphine is present in water at the same concentration as ammonia, a smaller proportion of its molecules are protonated.

Exercise E5.11 Which will be more fully protonated in aqueous solutions of the same concentration, methylamine or ethylamine?

[ethylamine]

The value of K_b distinguishes strong bases from weak bases. A **strong base** is a proton acceptor with K_b appreciably larger than 1; strong bases are fully protonated in solutions with concentrations typical of those used in laboratories. A **weak base** is a proton acceptor with K_b less than about 1; weak bases are only partially protonated in water at normal concentrations. The strong bases include the oxide and hydroxide ions, O^{2-} and OH^-, and weak bases include NH_3 and other amines. They also include (as we shall see in more detail shortly) the conjugate bases of strong acids: thus, the Cl^- ion (the conjugate base of the strong acid HCl) is a *very* weak base.

The extent to which a base is protonated is reported in terms of the **fraction protonated**:

$$\text{fraction protonated} = \frac{\text{molar concentration of conjugate acid}}{\text{initial molar concentration of base}}$$

For strong bases, the fraction protonated is close to 1; for weak bases it is typically much less than 1.

The extent of protonation

As an example of the use of basicity constants, we can calculate the fraction of NH_3 protonated in a solution of concentration B mol L^{-1}. This type of calculation is another version of the equilibrium calculations presented earlier in the chapter, and we draw up the following equilibrium table:

Reaction:	$H_2O(l) + NH_3(aq) \rightleftharpoons NH_4^+(aq) + OH^-(aq)$		
Species:	NH_3	NH_4^+	OH^-
Initial concentration/mol L^{-1}	B	0	0
Change in concentration/mol L^{-1}	$-x$	$+x$	$+x$
Equilibrium concentration/mol L^{-1}	$B - x$	x	x

The concentrations at equilibrium are now substituted into the expression for the basicity constant:

$$K_b = \frac{a(NH_4^+)a(OH^-)}{a(NH_3)} = \frac{x \times x}{B - x}$$

This expression rearranges into the quadratic equation

$$x^2 + K_b x - BK_b = 0$$

with the solutions

$$x = \frac{-K_b \pm \sqrt{K_b^2 + 4BK_b}}{2}$$

The positive root must be selected because x is the molar concentration of OH^- ions (see the equilibrium table), which is positive. For a $0.10 \, \text{M} \, NH_3(aq)$ solution,

$$x = \frac{-(1.8 \times 10^{-5}) + \sqrt{(1.8 \times 10^{-5})^2 + 4(0.10)(1.8 \times 10^{-5})}}{2}$$

$$= 1.3 \times 10^{-3}$$

It follows that the fraction of NH_3 protonated is

$$\text{fraction protonated} = \frac{x}{B} = \frac{1.3 \times 10^{-3}}{0.10} = 1.3 \times 10^{-2}$$

That is, 1.3 per cent of the NH_3 molecules (just over 1 in 100) is protonated.

Just as in the calculation of the pH of a weak acid, we can estimate the fraction protonated more quickly. To do so, we anticipate that, because $K_b \ll 1$, the concentration of protonated species is so low that the concentration of unprotonated base is almost the same as the concentration of base added to prepare the solution. If that is so, we can replace $B - x$ in the expression for the basicity constant by B. Then

$$x \approx \sqrt{K_b B} \tag{16a}$$

This approximation is valid so long as x does not exceed about 5 per cent of B. It follows that the fraction protonated is

$$\text{fraction protonated} = \frac{x}{B} \approx \sqrt{\frac{K_b}{B}} \tag{16b}$$

For the example we have just treated exactly, we would estimate

$$x \approx \sqrt{(1.8 \times 10^{-5}) \times 0.10} = 1.3 \times 10^{-3}$$

in excellent agreement with the exact value (and with far less work).

Exercise E5.12 What is the minimum concentration at which it would be valid (to within about 5 per cent accuracy) to estimate the fraction of morphine that is protonated in a solution by using eqn (16b)?

$$[6 \times 10^{-4} \,\text{mol L}^{-1}]$$

The pH of a basic solution

The calculation of the pH of a solution of a base involves one more step than that for the pH of a solution of an acid. The first step is to calculate the concentration of OH^- ions in the solution and to express it as the pOH of the solution. The additional step is to convert that pOH into a pH using the water autoprotolysis equilibrium, eqn (10), in the form

$$\text{pH} = \text{p}K_w - \text{pOH}$$

with $\text{p}K_w = 14.00$ at 25 °C.

The procedure is straightforward in the case of a strong base, such as a hydroxide, because every formula unit (such as NaOH or $Ca(OH)_2$) present in the solution is present as OH^- ions. For example, in a 0.0010 M $NaOH(aq)$ solution the concentration of OH^- ions is 0.0010 mol L^{-1}, and so the pOH of the solution is

$$\text{pOH} = -\log 0.0010 = +3.00$$

The pH of the solution can then be deduced from the autoprotolysis equilibrium:

$$\text{pH} = 14.00 - 3.00 = 11.00$$

Care should be taken to note that if a compound such as $Ca(OH)_2$ is present in solution, each formula unit that dissolves gives rise to *two* OH^- ions.

For a solution of a weak base in water, the pH is given by

$$\text{pH} = \text{p}K_w - \tfrac{1}{2}\text{p}K_b + \tfrac{1}{2}\log B \tag{17}$$

when the fraction of protonation is low. This expression shows that the pH of the solution increases as the concentration of base increases (because $\log B$ increases) and as the pK_b of the base *decreases*. The latter trend is consistent with the strength of a base increasing as pK_b becomes smaller (as from trimethylamine to methylamine).

JUSTIFICATION

When the fraction of protonation of a weak base is low, the molar concentration of OH^- ions is equal to x mol L^{-1}, with x given by eqn (16a). It follows that

$$pOH = -\log x$$
$$\approx -\log \sqrt{(K_b B)} \quad \text{(because } x \approx \sqrt{(K_b B)})$$
$$\approx -\tfrac{1}{2}\log K_b B \quad \text{(because } \log \sqrt{y} = \tfrac{1}{2}\log y)$$

$$\approx -\tfrac{1}{2}\log K_b - \tfrac{1}{2}\log B \quad \text{(because } \log yz = \log y + \log z)$$
$$\approx \tfrac{1}{2}pK_b - \tfrac{1}{2}\log B \quad \text{(because } -\log K_b = pK_b)$$

It then follows that

$$pH = pK_w - (\tfrac{1}{2}pK_b - \tfrac{1}{2}\log B)$$
$$\text{(because } pH = pK_w - pOH)$$

which can be rearranged into the expression given in the text.

Exercise E5.13 Calculate the pH of 0.10 M $NH_3(aq)$, for which $pK_b = 4.75$ at 25 °C.

[11.1]

5.3 Conjugate acids and bases

There is an important relation between the acidity and basicity constants of conjugate acid–base pairs:

$$pK_a + pK_b = pK_w \tag{18a}$$

The proof of the relation is very straightforward: we need to show that

$$K_a K_b = K_w \tag{18b}$$

because taking the logarithm of both sides of this equation and changing the sign throughout leads to the relation above. To confirm this expression, we multiply together the explicit expressions for K_a and K_b:

$$K_a K_b = \frac{a(H_3O^+)a(B)}{a(HB^+)} \times \frac{a(HB^+)a(OH^-)}{a(B)}$$
$$= a(H_3O^+)a(OH^-) = K_w$$

The relation implies that from the basicity constant of NH_3

$$H_2O(l) + NH_3(aq) \rightleftharpoons NH_4^+(aq) + OH^-(aq) \qquad pK_b = 4.75$$

we can conclude that the acidity constant of NH_4^+, the conjugate acid of NH_3,

$$NH_4^+(aq) + H_2O(l) \rightleftharpoons H_3O^+(aq) + NH_3(aq)$$

is given by

$$pK_a = 14.00 - 4.75 = 9.25$$

The great advantage of this relation is that all proton-transfer equilibrium constants can be expressed as acidity constants, even those of bases. For example, instead of treating ammonia as a weak base, we can think of it as the conjugate base of an acid (NH_4^+, for which $pK_a = 9.25$) that is slightly stronger than hydrocyanic acid ($pK_a = 9.31$).

Exercise E5.14 The acidity constant of HPO_4^{2-} is reported as $pK_a = 12.67$. Write the equilibrium of its conjugate base and give the value of pK_b for that base.
[$H_2O(l) + PO_4^{3-}(aq) \rightleftharpoons HPO_4^{2-}(aq) + OH^-(aq)$, $pK_b = 1.33$]

The relation in eqn (18) implies that, because the sum of pK_a and pK_b is a constant for a given conjugate acid–base pair, a large pK_a implies a smaller pK_b, and vice versa. That is,

---

- the stronger an acid, the weaker its conjugate base;

- the stronger a base, the weaker its conjugate acid.

It follows that the conjugate base of a strong acid is a *very* weak base, and the conjugate acid of a strong base is a *very* weak acid. For example, the fact that HCl is a strong acid in water implies that the Cl^- ion is a very weak base and has almost no tendency to acquire a proton. On the other hand, the conjugate of a weak acid has a tendency to accept a proton, and the weaker the acid, the greater this proton-accepting ability. Similar remarks may be made about bases: the greater the tendency of a base to accept a proton, the weaker its conjugate acid.

5.4 Salts in water

The remarks in the last paragraph explain why a solution of ammonium chloride is acidic even though the salt provides both an acid (NH_4^+) and a base (Cl^-) when it dissolves in water. The NH_4^+ ion is the conjugate acid of a weak base, and although it is still a weak acid, it does have an appreciable proton-donating power ($pK_a = 9.25$). The

Cl^- ion is the conjugate base of a strong acid, and hence is a very weak proton acceptor. The solution therefore consists of a weak acid and a very weak base, and the net effect is that the solution is acidic. Similarly, a solution of sodium acetate consists of an essentially neutral ion (the Na^+ ion) and the conjugate base of a weak acid (acetic acid). The net effect is that the solution is basic, and its pH is greater than 7.

Exercise E5.15 Is a solution of potassium tartrate likely to be acidic or basic?

[basic]

When an ammonium salt is dissolved in water, the NH_4^+ ions supplied by the salts participate in the equilibrium

$$NH_4^+(aq) + H_2O(l) \rightleftharpoons H_3O^+(aq) + NH_3(aq)$$

and the equilibrium composition of the solution can be calculated from the acidity constant of the NH_4^+ ion exactly as we have illustrated for other weak acids:

$$K_a = \frac{a(H_3O^+)a(NH_3)}{a(NH_4^+)}$$

There is no distinction between acids and conjugate acids: in the Brønsted–Lowry theory, any proton donor is an acid.

Example Calculating the pH of a solution of a salt
Calculate the pH of 0.010 M $NH_4Cl(aq)$ at 25 °C.
Answer We anticipate that the solution will be acidic, with pH < 7, because NH_4^+ is a weak acid and Cl^- is effectively neutral. Next, we note from Table 5.1 that the pK_a of NH_4^+ is 9.25. This high value (which signifies that NH_4^+ is a very weak acid) suggests that only a small fraction of NH_4^+ will have undergone proton donation. We can therefore use the approximate formula, eqn (11):

$$pH = \tfrac{1}{2} \times 9.25 - \tfrac{1}{2} \log 0.010 = 5.63$$

That is, the pH of the solution will be about 5.6, on the acidic side of neutrality.

Exercise E5.16 Calculate the pH of an aqueous 0.0025 M solution of $[NH(CH_3)_3]Cl(aq)$ at 25 °C.

[6.2]

The addition of ammonium ions to water (by dissolving an ammonium salt) is the addition of an acid, and as a result the pH of the solution is reduced below 7. The same is true of all salts that contain the conjugate acid of a base, but ammonium salts are the most common. In the earlier literature (and still sometimes today), the influence of a dissolved salt on the pH of a solution is ascribed to the 'hydrolysis' of the salt, in which the ions are supposed to react with water and produce an acidic or basic solution. However, it is far simpler, and much more in the spirit of Brønsted–Lowry theory, to regard the ions of the salt as acids or bases in their own right.

The pH of a solution that contains ions that are weak bases (such as solutions of sodium acetate) can be calculated in much the same way as we have demonstrated for solutions of weak acids. The difference is that we use a table that allows for the protonation of a base, as in the reaction

$$H_2O(l) + CH_3CO_2^-(aq) \rightleftharpoons CH_3COOH(aq) + OH^-(aq)$$

For example, to calculate the pH of a solution of acetate ions of molar concentration B mol L^{-1}, we use eqn (17):

$$pH \approx pK_w - \tfrac{1}{2}pK_b + \tfrac{1}{2}\log B$$

When rewritten in terms of the pK_a of the conjugate acid of the base (that is, in terms of the pK_a of acetic acid if the salt is an acetate), this expression becomes

$$pH \approx pK_w - \tfrac{1}{2}(pK_w - pK_a) + \tfrac{1}{2}\log B \quad \text{(because } pK_b = pK_w - pK_a)$$

which simplifies to

$$pH \approx \tfrac{1}{2}pK_w + \tfrac{1}{2}pK_a + \tfrac{1}{2}\log B \tag{19}$$

We see that the pH of the solution increases as the concentration of the salt increases, which is as expected because the salt is providing a base (the acetate ion, for instance). Similarly, the pH increases as the pK_a of the parent acid increases: this is also as expected, because a higher pK_a indicates a stronger conjugate base.

Exercise E5.17 Calculate the pH of 0.010 M NaCH$_3$CO$_2$(aq).
[8.4]

5.5 Acid–base titrations

One application in which acidity constants play an important role is in acid–base titrations, for they can be used to decide the value of the pH that signals the **stoichiometric point**, the stage at which a stoichiometrically equivalent amount of acid has been added to a given amount of base. The plot of the pH of the analyte solution (the solution in the flask that is being analysed) against the volume of titrant (the solution in the burette) added is called the **pH curve**.

First, consider the titration of a strong acid with a strong base, such as that between hydrochloric acid and sodium hydroxide. The reaction is

$$HCl(aq) + NaOH(aq) \rightarrow NaCl(aq) + H_2O(l)$$

Initially, the analyte (hydrochloric acid) has a low pH. The ions present at the stoichiometric point (the Na^+ ions from the strong base and the Cl^- ions from the strong acid) barely affect the pH, so the pH is that of almost pure water, namely pH = 7. After the stoichiometric point, when base is added to a neutral solution, the pH rises sharply to a high value. The pH curve for such titration is shown in Fig. 5.1.

At the stoichiometric point of a titration of a weak acid (such as CH_3COOH) and strong base (NaOH), the solution contains $CH_3CO_2^-$ ions and Na^+ ions together with any ions stemming from autoprotolysis. The presence of the Brønsted base $CH_3CO_2^-$ in the solution means that we can expect pH > 7. In a titration of a weak base (such as NH_3) and a strong acid (HCl), the solution contains NH_4^+ ions and Cl^- ions at the stoichiometric point. Because Cl^- is only a very weak Brønsted base and NH_4^+ is a weak Brønsted acid the solution is acidic and its pH will be less than 7.

The pH curve

Now we consider the pH curve quantitatively. The approximations we shall make are based on the fact that the acid is weak, and therefore that HA is more abundant than any A^- ions in the solution. Furthermore, when HA is present, it provides so many H_3O^+ ions (even though it is a weak acid) that they greatly outnumber any H_3O^+ ions that come from the autoprotolysis of water. Finally, when excess base is present, the OH^- ions it provides dominate any that come from the autoprotolysis of the water.

To be definite, we shall suppose that we are titrating 25.00 mL of 0.10 M HClO(aq) with 0.20 M NaOH(aq) at 25 °C. The pH at the start of a titration of a weak acid (the analyte) with a strong base (the titrant) can be calculated from eqn (11). Because $pK_a = 7.53$ for hypochlorous acid, it follows that initially

$$pH \approx \tfrac{1}{2} \times 7.53 - \tfrac{1}{2} \log 0.10 \approx 4.3$$

(The value is only an estimate of the actual pH because we are treating the solution as ideal, which is not particularly appropriate.)

The addition of titrant converts some of the acid to its conjugate base in the reaction

$$HClO(aq) + OH^-(aq) \rightarrow H_2O(l) + ClO^-(aq)$$

Suppose we add enough titrant to produce a concentration, [base], of the conjugate base and reduce the concentration of acid to [acid]. Then (because the solution remains at equilibrium)

$$K_a = \frac{a(H_3O^+)a(ClO^-)}{a(HClO)} = \frac{a(H_3O^+)[base]}{[acid]}$$

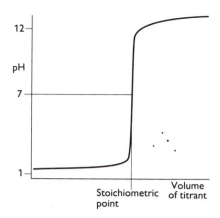

Fig. 5.1 The pH curve for the titration of a strong acid (the analyte) with a strong base (the titrant). There is an abrupt change in pH near the stoichiometric point; the stoichiometric point itself is at pH = 7. The final pH of the medium approaches that of the titrant.

185

which rearranges first to

$$a(H_3O^+) = \frac{K_a[\text{acid}]}{[\text{base}]}$$

and then, by taking logarithms and changing the sign, to

$$pH = pK_a - \log\frac{[\text{acid}]}{[\text{base}]} \qquad (20)$$

This expression is widely called the **Henderson–Hasselbalch equation**.

Example Estimating the pH at an intermediate stage in a titration

Calculate the pH of the solution after the addition of 5.00 mL of the titrant to the analyte in the titration described above.

Answer The addition of 5.00 mL of titrant corresponds to the addition of

$$n(OH^-) = 5.00 \times 10^{-3}\,L \times 0.200\,mol\,L^{-1} = 1.00 \times 10^{-3}\,mol$$

This amount of OH^- converts 1.00×10^{-3} mol of HClO to ClO^-. The initial amount of HClO in the analyte is

$$n(HClO) = 25.00 \times 10^{-3}\,L \times 0.100\,mol\,L^{-1} = 2.50 \times 10^{-3}\,mol$$

so the amount remaining is 1.50×10^{-3} mol. The total volume of the solution is now 30.00 mL (because 5.00 mL of titrant has been added to 25.00 mL of solution), so the concentrations of acid and base are

$$[HClO] = \frac{1.50 \times 10^{-3}\,mol}{30.00 \times 10^{-3}\,L} = 5.00 \times 10^{-2}\,mol\,L^{-1}$$

$$[ClO^-] = \frac{1.00 \times 10^{-3}\,mol}{30.00 \times 10^{-3}\,L} = 3.33 \times 10^{-2}\,mol\,L^{-1}$$

It then follows from the Henderson–Hasselbalch equation that

$$pH = 7.53 - \log\frac{5.00 \times 10^{-2}}{3.33 \times 10^{-2}} \approx 7.4$$

As expected, the addition of base has resulted in an increase in pH from 4.3.

Exercise E5.18 Calculate the pH after the addition of a further 5.00 mL of titrant.

[8.1]

Half way to the stoichiometric point (when enough base has been added to neutralize half the acid), the concentrations of acid and base are equal and the Henderson–Hasselbalch equation gives

$$pH = pK_a \qquad (21)$$

In the present titration, we see that at this stage of the titration, pH ≈ 7.5. Note from the pH curve in Fig. 5.2 how much more slowly the pH is changing compared with initially: this point will prove important shortly. Equation (21) implies that the pK_a of the acid can be measured directly from the pH of the mixture. In practice this is done by recording the pH during a titration and then examining the record for the pH half way to the stoichiometric point.

At the stoichiometric point, enough base has been added to convert all the acid to its base, and so the solution consists only of ClO^- ions. These ions are Brønsted bases, so the solution will be basic. We have already seen how to calculate the pH of a solution of a weak base in terms of its concentration, B (eqn (17); here B is the concentration of ClO^- ions), so all that remains to be done is to calculate the concentration of ClO^- at the stoichiometric point. Because the analyte initially contained 2.50×10^{-3} mol HClO, the volume of titrant needed to neutralize it is the volume that contains the same amount of base:

$$V(\text{base}) = \frac{2.50 \times 10^{-3}\,\text{mol}}{0.200\,\text{mol L}^{-1}} = 1.25 \times 10^{-2}\,\text{L, or 12.5 mL}$$

The total volume of the solution at this stage is therefore 37.5 mL, so the concentration of base is

$$[ClO^-] = \frac{2.50 \times 10^{-3}\,\text{mol}}{37.5 \times 10^{-3}\,\text{L}} = 6.67 \times 10^{-2}\,\text{mol L}^{-1}$$

It then follows from eqn (19) that the pH of the solution at the stoichiometric point is

$$pH = \tfrac{1}{2} \times 14.00 + \tfrac{1}{2} \times 7.53 + \tfrac{1}{2}\log(6.67 \times 10^{-2}) \approx 10.2$$

It is very important to note that *the pH at the stoichiometric point of a weak-acid–strong-base titration is on the basic side of neutrality.* At the stoichiometric point, the solution consists of a weak base (the conjugate base of the weak acid, here the ClO^- ions) and neutral cations (the Na^+ ions from the titrant).

The general form of the pH curve throughout a weak-acid–strong-base titration is illustrated in Fig. 5.2. The pH rises slowly from the value given by eqn (11), passing through the values given by the Henderson–Hasselbalch equation (eqn (20)) when the acid and its conjugate base are both present, until the stoichiometric point is approached. It then changes rapidly to and through the value given by eqn (17), which takes into account the effect on the pH of a solution of a weak base (the conjugate base of the original acid). The pH then climbs less rapidly towards the value corresponding to a solution consisting of excess base, and finally approaches the pH of the original base solution when (a point never reached in practice) so much titrant has been added that the solution is virtually the same as the titrant itself. The stoichiometric point is detected by observing where the pH changes rapidly through the value given by eqn (17).

A similar sequence of changes occurs when the analyte is a weak base (such as ammonia) and the titrant is a strong acid (such as

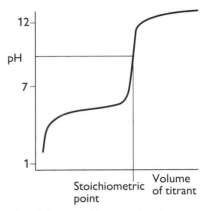

Fig. 5.2 The pH curve for the titration of a weak acid (the analyte) with a strong base (the titrant). Note that the stoichiometric point occurs at pH > 7 and that the change in pH near the stoichiometric point is less abrupt than in Fig. 5.1.

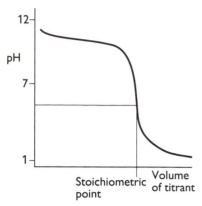

Fig. 5.3 The pH curve for the titration of a weak base (the analyte) with a strong acid (the titrant). The stoichiometric point occurs at pH < 7. The final pH of the solution approaches that of the titrant.

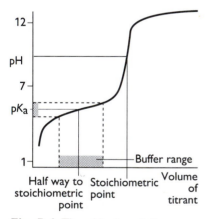

Fig. 5.4 The pH of a solution changes only slowly in the region of half way to the stoichiometric point, and in this region the solution is buffered to a pH that is close to pK_a.

hydrochloric acid). In this case the pH curve is like that shown in Fig. 5.3: the pH falls as acid is added, plunges through the pH corresponding to a solution of a weak acid (the conjugate acid of the original base, such as NH_4^+), and then slowly approaches the pH of the original strong acid. The pH of the stoichiometric point is that of a solution of a weak acid, and is calculated as illustrated in the example on p. 175.

Buffers

The slow variation of the pH when the concentrations of the conjugate acid and base are equal, when $pH = pK_a$, is the basis of **buffer action**, the ability of a solution to oppose changes in pH when small amounts of acids and bases are added (Fig. 5.4). An **acid buffer** solution, one that stabilizes the solution at a pH below 7, is typically prepared by making a solution of a weak acid (such as acetic acid) and a salt that supplies its conjugate base (such as sodium acetate). A **base buffer**, one that stabilizes a solution at a pH above 7, is prepared by making a solution of a weak base (such as ammonia) and a salt that supplies its conjugate acid (such as ammonium chloride).

The *mathematical* basis of buffer action is the logarithmic dependence of the pH as given by the Henderson–Hasselbalch equation (eqn (20)). This logarithmic dependence results in a very flat curve near $pH = pK_a$ because $\log x$ changes much more slowly than x (for example, if x changes from 10 to 1000, $\log x$ changes only from 1 to 3). The *physical* basis of the action of an acid buffer action is that the existence of an abundant supply of A^- ions (supplied by the salt) can remove any H_3O^+ ions brought by additional acid; furthermore, the existence of an abundant supply of HA molecules can supply H_3O^+ ions to react with any base that is added. Similarly, the physical basis of the action of base buffer is the ability of the base to accept protons when an acid is added, and the ability of the conjugate acid (NH_4^+, for instance) to supply protons if a base is added.

Example Estimating the pH of a buffer solution

Estimate the pH of a buffer formed from 0.010 M $KH_2PO_4(aq)$ and 0.0200 M $K_2HPO_4(aq)$.

Answer The buffer consists of an acid ($H_2PO_4^-$) and a salt of that acid (K_2HPO_4); so we can expect the solution to buffer on the acid side of neutrality. The strategy is to calculate the pH of the solution of an acid and its salt using the Henderson–Hasselbalch equation. To do so, we must identify the acid HA and its conjugate base A^-. In this case the acid is the anion $H_2PO_4^-$ and its conjugate base is the anion HPO_4^{2-}:

$$H_2PO_4^-(aq) + H_2O(l) \rightleftharpoons H_3O^+(aq) + HPO_4^{2-}(aq)$$

The acidity constant we require is therefore pK_a for $H_2PO_4^-$, which is pK_{a2} for H_3PO_4; from Table 5.2, $pK_{a2} = 7.21$. Then, eqn (20) gives the pH of the solution as

$$pH = 7.21 - \log \frac{0.0200}{0.0100} \approx 6.9$$

Hence, the solution should buffer close to pH = 7.

Exercise E5.19 Calculate the pH of a buffer solution which is 0.010 M $NH_3(aq)$ and 0.020 M $NH_4Cl(aq)$.

[8.95; more realistically: 9]

Indicators

The rapid change of pH near the stoichiometric point of an acid–base titration is the basis of indicator detection. An **acid–base indicator** is normally some large, water-soluble, organic molecule with acid (HIn) and conjugate base (In^-) forms that differ in colour. The two forms are in equilibrium in solution:

$$HIn(aq) + H_2O(l) \rightleftharpoons H_3O^+(aq) + In^-(aq)$$

$$K_{In} = \frac{a(H_3O^+)a(In^-)}{a(HIn)}$$

and the ratio of the concentrations is

$$\frac{[In^-]}{[HIn]} = \frac{K_{In}}{a(H_3O^+)}$$

which can be rearranged (after taking logarithms) to

$$\log \frac{[In^-]}{[HIn]} = pH - pK_{In} \qquad (22)$$

The pK_{In} of some indicators are listed in Table 5.3.

Exercise E5.20 What is the ratio of the yellow and blue forms of bromocresol green in solutions of pH (a) 3.7, (b) 4.7, and (c) 5.7?

[(a) 10:1, (b) 1:1, (c) 1:10]

At the stoichiometric point, the pH changes sharply through several pH units, so the molar concentration of H_3O^+ changes through several orders of magnitude. The indicator equilibrium changes so as to accommodate the change of pH, with HIn the dominant species on the acid side of the stoichiometric point, when H_3O^+ ions are abundant, and In^- dominant on the basic side, when the base can remove protons from HIn. The accompanying colour change signals the stoichiometric point of the titration. The colour in fact changes over a range of pH (typically from $pH = pK_{In} - 1$ when HIn is ten times as

Table 5.3 Indicator colour changes

Indicator	Acid colour	pH range of colour change	pK_{In}	Base colour
Thymol blue	Red	1.2 to 2.8	1.7	Yellow
Methyl orange	Red	3.2 to 4.4	3.4	Yellow
Bromophenol blue	Yellow	3.0 to 4.6	3.9	Blue
Bromocresol green	Yellow	4.0 to 5.6	4.7	Blue
Methyl red	Red	4.8 to 6.0	5.0	Yellow
Bromothymol blue	Yellow	6.0 to 7.6	7.1	Blue
Litmus	Red	5.0 to 8.0	6.5	Blue
Phenol red	Yellow	6.6 to 8.0	7.9	Red
Thymol blue	Yellow	9.0 to 9.6	8.9	Blue
Phenolphthalein	Colourless	8.2 to 10.0	9.4	Pink
Alizarin yellow	Yellow	10.1 to 12.0	11.2	Red
Alizarin	Red	11.0 to 12.4	11.7	Purple

Fig. 5.5 The range of pH over which an indicator changes colour is depicted by the tinted band. (a) For a strong-acid–strong-base titration, the stoichiometric point is indicated accurately by an indicator that changes colour at pH = 7 (such as bromothymol blue). However, the change in pH is so sharp that accurate results are also obtained even if the indicator changes colour in neighbouring values. Thus, phenolphthalein (which has pK_{In} = 9.4, see Table 5.3) is also often used. (b) In a weak-acid–strong-base titration, an indicator with $pK_{In} \approx 7$ (the lower band, like bromothymol blue) would give a false indication of the stoichiometric point; it is necessary to use an indicator that changes colour close to the pH of the stoichiometric point. If that lies at about pH = 9, phenolphthalein would be appropriate.

abundant as In^-, to pH = pK_{In} + 1, when In^- is ten times as abundant as HIn; see Exercise E5.20). The pH half way through a colour change (when pH $\approx pK_{In}$, and the two forms, HIn and In^-, are in equal abundance) is the **end point** of the indicator. With a well-chosen indicator, the end point coincides with the stoichiometric point of the titration.

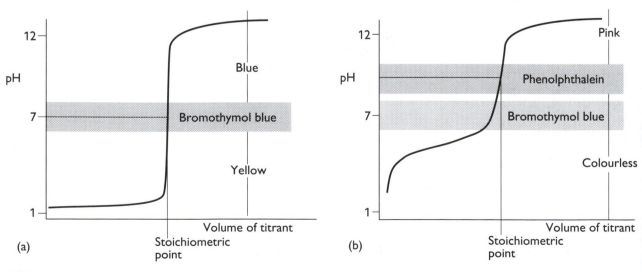

(a) / (b)

Care must be taken to use an indicator that changes colour at the pH appropriate to the type of titration. Specifically, the end point should be matched to the stoichiometric point, and so an indicator should be selected for which pK_{In} is close to the pH at the stoichiometric point. Thus, in a weak-acid–strong-base titration, the stoichiometric point lies at the pH given by eqn (19), and so an indicator that changes at that pH must be selected (Fig. 5.5). Similarly, in a strong-acid–weak-base titration, an indicator changing near the pH given by eqn (17) should be used. Qualitatively, an indicator with $pK_{In} \approx 7$ should be chosen for strong-acid–strong-base titrations, one with $pK_{In} < 7$ for strong-acid–weak-base titrations, and one with $pK_{In} > 7$ for weak-acid–strong-base titrations.

Exercise E5.21 Vitamin C is a weak acid (ascorbic acid), and the amount in a sample may be determined by titration with sodium hydroxide solution. Should you use methyl red or phenophthalein as the indicator?

[phenolphthalein]

Solubility equilibria

A solid dissolves in a solvent until the solution and the solid solute are in equilibrium. At this stage, the solution is said to be **saturated**, and its molar concentration is the **molar solubility**, S, of the solid. That the two phases—the solid and the solution—are in dynamic equilibrium implies that we can use equilibrium concepts to discuss the composition of the saturated solution. The properties of aqueous solutions of electrolytes are widely treated in terms of equilibrium constants, and in this section we shall confine our attention to them. We shall also limit our attention to **sparingly soluble compounds**, which are compounds that dissolve only slightly in water. This restriction is applied because the effects of ion–ion interactions are a complicating feature of more concentrated solutions, and more advanced techniques are needed before the calculations are reliable.

5.6 The solubility constant

The equilibrium between a solid ionic compound, such as calcium hydroxide, $Ca(OH)_2$, and its ions in aqueous solution is

$$Ca(OH)_2(s) \rightleftharpoons Ca^{2+}(aq) + 2OH^-(aq)$$

The equilibrium constant for an ionic equilibrium such as this, bearing in mind that the solid does not appear in the equilibrium expression, is called the **solubility constant** (sometimes the 'solubility product

Table 5.4 Solubility constants at 25 °C

Compound	Formula	K_s	Compound	Formula	K_s
Aluminium hydroxide	$Al(OH)_3$	1.0×10^{-33}	Lead(II) fluoride	PbF_2	3.7×10^{-8}
Antimony sulfide	Sb_2S_3	1.7×10^{-93}	iodate	$Pb(IO_3)_2$	2.6×10^{-13}
Barium carbonate	$BaCO_3$	8.1×10^{-9}	iodide	PbI_2	1.4×10^{-8}
fluoride	BaF_2	1.7×10^{-6}	sulfate	$PbSO_4$	1.6×10^{-8}
sulfate	$BaSO_4$	1.1×10^{-10}	sulfide	PbS	3.4×10^{-28}
Bismuth sulfide	Bi_2S_3	1.0×10^{-97}	Magnesium ammonium phosphate	$MgNH_4PO_4$	2.5×10^{-13}
Calcium carbonate	$CaCO_3$	8.7×10^{-9}	carbonate	$MgCO_3$	1.0×10^{-5}
fluoride	CaF_2	4.0×10^{-11}	fluoride	MgF_2	6.4×10^{-9}
hydroxide	$Ca(OH)_2$	5.5×10^{-6}	hydroxide	$Mg(OH)_2$	1.1×10^{-11}
sulfate	$CaSO_4$	2.4×10^{-5}	Mercury(I) chloride	Hg_2Cl_2	1.3×10^{-18}
Copper(I) bromide	$CuBr$	4.2×10^{-8}	iodide	Hg_2I_2	1.2×10^{-28}
chloride	$CuCl$	1.0×10^{-6}	Mercury(II) sulfide	HgS (black)	1.6×10^{-52}
iodide	CuI	5.1×10^{-12}	(red)		1.4×10^{-53}
sulfide	Cu_2S	2.0×10^{-47}	Nickel(II) hydroxide	$Ni(OH)_2$	6.5×10^{-18}
Copper(II) iodate	$Cu(IO_3)_2$	1.4×10^{-7}	Silver bromide	$AgBr$	7.7×10^{-13}
oxalate	CuC_2O_4	2.9×10^{-8}	carbonate	Ag_2CO_3	6.2×10^{-12}
sulfide	CuS	8.5×10^{-45}	chloride	$AgCl$	1.6×10^{-10}
Iron(II) hydroxide	$Fe(OH)_2$	1.6×10^{-14}	hydroxide	$AgOH$	1.5×10^{-8}
sulfide	FeS	6.3×10^{-18}	iodide	AgI	1.5×10^{-16}
Iron(III) hydroxide	$Fe(OH)_3$	2.0×10^{-39}	sulfide	Ag_2S	6.3×10^{-51}
Lead(II) bromide	$PbBr_2$	7.9×10^{-5}	Zinc hydroxide	$Zn(OH)_2$	2.0×10^{-17}
chloride	$PbCl_2$	1.6×10^{-5}	sulfide	ZnS	1.6×10^{-24}

constant'), and denoted K_s:

$$K_s = a(Ca^{2+})a(OH^-)^2$$

As usual, for very dilute solutions $a(X)$ is equal to the molar concentration of X divided by $1\,mol\,L^{-1}$. Experimental values for solubility constants are given in Table 5.4.

Example Writing the expression for a solubility constant
Write the expression for the solubility constant of aluminium sulphide, Al_2S_3.
Answer We write the equation for the solubility equilibrium, which is

$$Al_2S_3(s) \rightleftharpoons 2Al^{3+}(aq) + 3S^{2-}(aq)$$

and then write the equilibrium constant in the usual way as

$$K_s = a(Al^{3+})^2 a(S^{2-})^3$$

where $a(Al^{3+}) = [Al^{3+}]/\text{mol L}^{-1}$ and $a(S^{2-}) = [S^{2-}]/\text{mol L}^{-1}$. The solid does not appear in the expression because it is pure, and $a = 1$ for all pure solids.

Exercise E5.22 Write the expression for the solubility constant of mercury(I) sulphate, Hg_2SO_4.

$[K_s = a(Hg_2^{2+})a(SO_4^{2-})]$

The numerical value of a solubility constant can be interpreted in terms of the molar solubility of a sparingly soluble substance. For instance, it follows from the stoichiometry of the equilibrium equation written above that the molar concentration of Ca^{2+} ions in solution is equal to that of the $Ca(OH)_2$ present in solution, so $S = [Ca^{2+}]$. Likewise, because the concentration of OH^- ions is twice that of $Ca(OH)_2$ formula units, it follows that $S = \frac{1}{2}[OH^-]$. Therefore,

$$K_s = (S/\text{mol L}^{-1}) \times (2S/\text{mol L}^{-1})^2 = 4(S/\text{mol L}^{-1})^3$$

from which it follows that

$$S = \left(\frac{K_s}{4}\right)^{1/3} \text{mol L}^{-1} \qquad (23)$$

This expression should be regarded as very approximate because ion–ion interactions have been ignored. However, because the solid is sparingly soluble, the concentrations of the ions are low and the inaccuracy is acceptable. Thus, from Table 5.4, $K_s = 5.5 \times 10^{-6}$, so $S \approx 1 \times 10^{-2}\,\text{mol L}^{-1}$. Solubility constants (which are determined by electrochemical measurements of the kind described in Chapter 6) provide a more accurate way of measuring solubilities of very sparingly soluble compounds than the direct measurements of the mass that dissolves.

Exercise E5.23 Copper occurs in many minerals, one of which is chalcocite, Cu_2S. What is the molar solubility of this compound in water at 25 °C?

$[1.7 \times 10^{-16}\,\text{mol L}^{-1}]$

5.7 The common-ion effect

The principle that an equilibrium constant remains unchanged whereas the individual concentrations of species may change is applicable to solubility constants, and may be used to assess the effects of the addition of species to solutions. An example of particular importance is the effect on the solubility of a compound of the presence of another compound that has an ion in common with the sparingly soluble compound already present. For example, we may consider the effect on the solubility of adding sodium chloride to a saturated solution of silver chloride, the common ion in this case being Cl^-.

The molar solubility of silver chloride in pure water is related to its solubility constant by

$$S = \sqrt{K_s}\, mol\, L^{-1}$$

(This equation is obtained from the expression $K_s = a(Ag^+)a(Cl^-)$ and the fact that $S = [AgCl] = [Ag^+] = [Cl^-]$.) To assess the effect of the common ion, we suppose that Cl^- ions are added to a concentration $C\, mol\, L^{-1}$, which greatly exceeds the concentration of the same ion that stems from the presence of the silver chloride. Therefore, we can write

$$K_s = a(Ag^+)a(Cl^-) \approx a(Ag^+) \times C$$

It is very dangerous to neglect deviations from ideal behaviour in ionic solutions, so from now on the calculation will only be indicative of the kinds of changes that occur when a common ion is added to a solution of a sparingly soluble salt: the qualitative trends are re-produced, but the quantitative calculations are unreliable. With these remarks in mind, it follows that the solubility S' of silver chloride in the presence of added chloride ions is

$$S' = \frac{K_s}{C}\, mol\, L^{-1} \tag{24}$$

The solubility is greatly reduced by the presence of the common ion. For example, whereas the solubility of silver chloride in water is $1.3 \times 10^{-5}\, mol\, L^{-1}$, in the presence of $0.10\, M\, NaCl(aq)$ it is only

$$S' \approx \frac{1.6 \times 10^{-10}}{0.10} \approx 2 \times 10^{-9}\, mol\, L^{-1}$$

which is nearly ten thousand times less. The reduction of the solubility of a sparingly soluble salt by the presence of a common ion is called the **common-ion effect**.

Exercise E5.24 What is the solubility of calcium fluoride, CaF_2, in (a) water, (b) $0.010\, M\, NaF(aq)$?

[(a) $2 \times 10^{-4}\, mol\, L^{-1}$, (b) $4 \times 10^{-7}\, mol\, L^{-1}$]

An *indirect* version of the common-ion effect is the removal of Ca^{2+} ions from hard water by the addition of more Ca^{2+} ions. The mechanism is slightly different, because the reduction in solubility occurs as a result of the anions that accompany the Ca^{2+} ions, and not directly as a result of the cations themselves. The Ca^{2+} ions are supplied to the hard water in the form of calcium hydroxide (lime).

The role of the lime is to supply OH^- ions that act as a Brønsted base and remove protons from HCO_3^- ions:

$$HCO_3^-(aq) + OH^-(aq) \rightarrow H_2O(l) + CO_3^{2-}(aq)$$

The formation of CO_3^{2-} ions results in the precipitation of $CaCO_3$ in the reaction

$$Ca^{2+}(aq) + CO_3^{2-}(aq) \rightarrow CaCO_3(s)$$

which removes both the Ca^{2+} ions present initially and those added as lime; overall there is a net reduction of Ca^{2+} in the water.

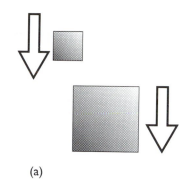

(a)

Coupled reactions

A special kind of approach to equilibrium occurs when a reaction that is not spontaneous is driven forward by a reaction that is more aggressively spontaneous. Thus, although the equilibrium composition of the first reaction may lie strongly in favour of the reactants, that of the overall pair of coupled reactions lies strongly in favour of the products. A simple mechanical analogy is a pair of weights joined by a string (Fig. 5.6): the smaller of the pair of weights will be pulled up as the heavier weight falls, and although the smaller weight has a natural tendency to move downwards, its coupling to the heavier weight results in it being raised. The thermodynamic analogue is an **endergonic reaction**, a reaction with a positive free energy ΔG, (the analogue of the smaller weight) being forced to occur by virtue of its coupling to an **exergonic reaction**, a reaction with a negative free energy $\Delta G'$ (the analogue of the heavier weight falling to the ground), because the sum $\Delta G + \Delta G'$ is negative.

We shall consider two examples of the 'coupled weights' analogy: one is drawn from biology and the other from industrial chemistry.

5.8 Biological activity: the thermodynamics of ATP

The function of adenosine triphosphate, ATP (**1**), is to store the energy made available when food is metabolized and then to supply it on demand to a wide variety of processes, including muscular contraction, reproduction, and vision. The essence of ATP's action is its ability to lose its terminal phosphate group by hydrolysis and to form adenosine diphosphate (ADP):

$$ATP(aq) + H_2O(l) \rightarrow ADP(aq) + P_i^-(aq) + H^+(aq)$$

(P_i^- denotes an inorganic phosphate group, such as $H_2PO_4^-$.) This reaction is exergonic and can drive an endergonic reaction if suitable enzymes are available to couple the reactions.

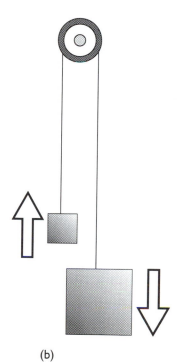

(b)

Fig. 5.6 (a) The two weights shown here both have a tendency to fall downwards in a gravitational field, and will do so if released. (b) However, if the two weights are coupled as shown here, then the larger weight will move the lighter weight in its nonspontaneous direction: overall, the process is still spontaneous. The weights are the analogues of two chemical reactions: a reaction with a large negative ΔG can force another reaction with a smaller ΔG to run in its non-spontaneous direction.

1 Adenosine triphosphate, ATP

Biological standard states

The conventional standard state of hydrogen ions ($a = 1$, corresponding to pH $= 0$, an acidic solution) is not appropriate to normal biological conditions inside cells, where the cell medium has a pH close to 7. Therefore, in biochemistry it is common to adopt the **biological standard state**, in which pH $= 7$, a neutral solution. We shall adopt this convention in this section, and label the corresponding standard thermodynamic function as $G^{\oplus}$, $H^{\oplus}$, and $S^{\oplus}$.

The standard values for the ATP hydrolysis at 37 °C (310 K, body temperature) are $\Delta G^{\oplus} = -30\,\text{kJ mol}^{-1}$, $\Delta H^{\oplus} = -20\,\text{kJ mol}^{-1}$, and $\Delta S^{\oplus} = +34\,\text{J K}^{-1}\,\text{mol}^{-1}$. The hydrolysis is therefore exergonic ($\Delta G < 0$) under these conditions, and $30\,\text{kJ mol}^{-1}$ is available for driving other reactions. Moreover, because the reaction entropy is large, the reaction free energy is sensitive to temperature (recall that $\Delta G = \Delta H - T\,\Delta S$, so ΔS acts as an amplification factor for changes in T). On account of its exergonicity the ADP–phosphate bond has been called a **high-energy phosphate bond**. The name is intended to signify a high tendency to undergo reaction, and should not be confused with 'strong' bond in its normal chemical sense (that of a high bond enthalpy). In fact, even in the biological sense it is not of very 'high energy'. The action of ATP depends on the bond being intermediate in activity. Thus ATP acts as a phosphate donor to a number of acceptors (such as glucose), but is recharged with a new phosphate group by more powerful phosphate donors in the respiration cycle.

Anaerobic and aerobic metabolism

The efficiency of some biological processes can be gauged in terms of the value of $\Delta G^{\oplus}$ given above. The energy source of anaerobic cells is glycolysis, the enzymatic conversion of glucose to lactic acid:

$$C_6H_{12}O_6(aq) \rightarrow 2CH_3CH(OH)COOH(aq)$$

$$\Delta G^{\oplus} = -218\,\text{kJ at 37 °C}$$

The glycolysis (the 'heavy weight') is coupled to a reaction (the 'light weight') in which two ADP molecules are converted into two ATP molecules:

$$C_6H_{12}O_6(aq) + 2P_i^-(aq) + 2ADP(aq) \rightarrow 2CH_3CH(OH)CO_2^-(aq)$$
$$+ 2ATP(aq) + 2H_2O(l)$$

The biological standard reaction free energy is $(-218\,\text{kJ}) - 2(-30\,\text{kJ}) = -158\,\text{kJ}$. The overall reaction is exergonic, and therefore spontaneous: the metabolism of the food has been used to 'recharge' the ATP.

Metabolism by aerobic respiration is much more efficient, as is indicated by the standard free energy of combustion of glucose, which is $-2880\,\text{kJ}$. Terminating its oxidation at lactic acid is therefore, a poor use of resources. In aerobic respiration the oxidation is carried out to completion, and an extremely complex set of reactions preserves as much of the energy released as possible. In the overall reaction, 38 ATP molecules are generated for each glucose molecule consumed. Each mole of ATP extracts 30 kJ from the 2880 kJ supplied by 1 mol $C_6H_{12}O_6$ (180 g of glucose), and so 1140 kJ per mole of glucose has been stored for later use.

Each ATP molecule can be used to derive an endergonic reaction for which $\Delta G^{\oplus}$ does not exceed 30 kJ mol^{-1}. For example, the biosynthesis of sucrose from glucose and fructose can be driven (if a suitable enzyme system is available) because the reaction is endergonic to the extent $\Delta G^{\oplus} = +23$ kJ mol^{-1}. The biosynthesis of proteins is strongly endergonic, not only on account of the enthalpy change but also on account of the large decrease in entropy that occurs when many amino acids are assembled into a precisely determined sequence. For instance, the formation of a peptide link is endergonic, with $\Delta G^{\oplus} = +17$ kJ mol^{-1}, but the biosynthesis occurs indirectly and is equivalent to the consumption of three ATP molecules for each link. In a moderately small protein like myoglobin, with about 150 peptide links, the construction alone requires 450 ATP molecules, and therefore about 12 mol of glucose molecules for 1 mol of protein molecules.

5.9 The extraction of metals from their oxides

Metals can be obtained from their oxides by reduction with carbon if either of the equilibria

$$MO(s) + C(s) \rightleftharpoons M(s) + CO(g)$$

$$MO(s) + \tfrac{1}{2}C(s) \rightleftharpoons M(s) + \tfrac{1}{2}CO_2(g)$$

has $K > 1$ or, equivalently $\Delta G^{\ominus} < 0$. These equilibria can be discussed in terms of the thermodynamic functions of the reactions

(i) $M(s) + \tfrac{1}{2}O_2(g) \rightarrow MO(s)$

(ii) $\tfrac{1}{2}C(s) + \tfrac{1}{2}O_2(g) \rightarrow \tfrac{1}{2}CO_2(g)$

(iii) $C(s) + \tfrac{1}{2}O_2(g) \rightarrow CO(g)$

(iv) $CO(g) + \tfrac{1}{2}O_2(g) \rightarrow CO_2(g)$

All these reactions have been written for the reaction of 1 mol O (in the form of $\tfrac{1}{2}$ mol O_2), for then they can be added and subtracted without further modification.

The temperature dependences of the standard reaction free energies written above depend on the reaction entropy through

$$\text{change in } \Delta G^{\ominus} = -(\text{change in } T) \times \Delta S^{\ominus} \qquad (25)$$

(The equation comes from $\Delta G^{\ominus} = \Delta H^{\ominus} - T\Delta S^{\ominus}$ and the assumption that both $\Delta H^{\ominus}$ and $\Delta S^{\ominus}$ are independent of temperature.) In each case we can decide whether there is an increase or decrease in entropy by noting whether there is a net formation or net consumption of gas.

Because in reaction (iii) there is a net increase in the amount of gas molecules, from $\frac{1}{2}$ mol to 1 mol, the standard reaction entropy is large and positive; therefore, $\Delta G^{\ominus}$ decreases sharply with increasing temperature. In reaction (iv), there is a similar net decrease in the amount of gas molecule, from $\frac{3}{2}$ mol to 1 mol, and so $\Delta G^{\ominus}$ increases sharply with increasing temperature. In reaction (ii), the amount of gas is constant, and so the entropy change is small and $\Delta G^{\ominus}$ changes only slightly with temperature. These remarks are summarized in Fig. 5.7, which is called an **Ellingham diagram** (note that $\Delta G^{\ominus}$ decreases upwards!).

The standard reaction free energy of reaction (i) indicates the metal's affinity for oxygen. At room temperature the contribution of the reaction entropy to $\Delta G^{\ominus}$ is dominated by the reaction enthalpy, and so the order of increasing $\Delta G^{\ominus}$ is the same as the order of increasing $\Delta H^{\ominus}$ (once again, this conclusion follows from the relation from $\Delta G^{\ominus} = \Delta H^{\ominus} - T \Delta S^{\ominus}$). The reaction enthalpy therefore gives the order of values on the left of the diagram (Al_2O_3 is most exothermic, Ag_2O is least exothermic). The standard reaction entropy is similar for all metals because in each case gaseous oxygen is eliminated and a compact, solid oxide is formed. This similarity implies that the temperature dependence of the standard free energy

Fig. 5.7 An Ellingham diagram for a number of common metal oxides.

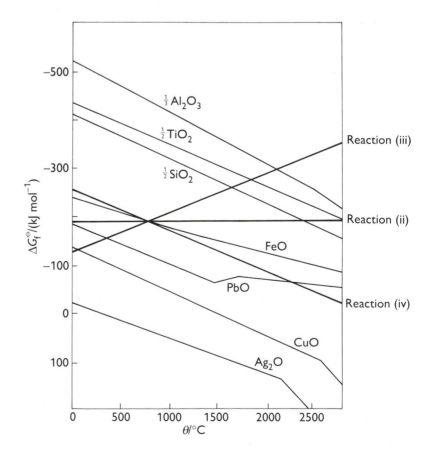

of oxidation should be similar for all metals, as is shown by the similar slopes of the lines in the diagram. The kinks at high temperatures correspond to the evaporation of the metals; less pronounced kinks occur at the melting temperatures of the metals and the oxides.

Reduction of the oxide depends on the competition of the carbon for the oxygen bound to the metal. The standard free energies of the reductions can be expressed in terms of the standard free energies of the reactions above:

$$MO(s) + C(s) \rightarrow M(s) + CO(g) \qquad \Delta G^{\ominus} = \Delta G^{\ominus}(iii) - \Delta G^{\ominus}(i)$$
$$MO(s) + \tfrac{1}{2}C(s) \rightarrow M(s) + \tfrac{1}{2}CO_2(g) \qquad \Delta G^{\ominus} = \Delta G^{\ominus}(ii) - \Delta G^{\ominus}(i)$$
$$MO(s) + CO(g) \rightarrow M(s) + CO_2(g) \qquad \Delta G^{\ominus} = \Delta G^{\ominus}(iv) - \Delta G^{\ominus}(i)$$

The equilibrium lies to the right if $\Delta G^{\ominus} < 0$. This is the case when the line for reaction (i) lies below (is more positive than) the line for one of the carbon reactions (ii) to (iv).

We can now predict the success of a reduction at any temperature simply by looking at the diagram: *a metal oxide is reduced by any carbon reaction lying above it* because the overall reaction then has $\Delta G^{\ominus} < 0$. For example, copper(II) oxide, CuO, can be reduced to copper at any temperature above room temperature. Even in the absence of carbon, Ag_2O decomposes when heated above $200\,^{\circ}C$ because then the standard free energy of reaction (i) becomes positive (so that the reverse reaction is then spontaneous).

Exercise E5.25 What is the minimum temperature at which carbon could be used to reduce alumina, Al_2O_3, to aluminium?

[2000 °C]

EXERCISES

5.1 Write the expressions for the equilibrium constants of the following reactions:
(a) $CO(g) + Cl_2(g) \rightleftharpoons COCl(g) + Cl(g)$
(b) $2SO_2(g) + O_2(g) \rightleftharpoons 2SO_3(g)$
(c) $H_2(g) + Br_2(g) \rightleftharpoons 2HBr(g)$
(d) $2O_3(g) \rightleftharpoons 3O_2(g)$

5.2 When $1.00\,g$ of I_2 is heated to $1000\,K$ in a $1.00\,L$ sealed container, the resulting equilibrium mixture contains $0.830\,g$ of I_2. Calculate K for the dissociation equilibrium $I_2(g) \rightleftharpoons 2I(g)$.

5.3 In a gas-phase equilibrium mixture of $SbCl_5$, $SbCl_3$, and Cl_2 at $500\,K$, $p_{SbCl_5} = 0.15$ bar and

$p_{SbCl_3} = 0.20$ bar. Calculate the equilibrium partial pressure of Cl_2 given that $K = 3.5 \times 10^{-4}$ for the reaction $SbCl_5(g) \rightleftharpoons SbCl_3(g) + Cl_2(g)$.

5.4 The equilibrium constant $K = 0.36$ for the reaction $PCl_5(g) \rightleftharpoons PCl_3(g) + Cl_2(g)$ at $400\,K$. (a) Given that $1.0\,g$ of PCl_5 was initially placed in a $250\,mL$ reaction vessel, determine the molar concentrations in the mixture at equilibrium. (b) What is the percentage of PCl_5 decomposed at $400\,K$?

5.5 The equilibrium pressure of H_2 over solid uranium and uranium hydride at $500\,K$ is 1.04 Torr.

Calculate the standard free energy of formation of $UH_3(s)$ at 500 K.

5.6 The equilibrium constant for the gas-phase isomerization of borneol ($C_{10}H_{17}OH$) to isoborneol at 503 K is 0.106. A mixture consisting of 7.50 g of borneol and 14.0 g of isoborneol in a 5.0 L container is heated to 503 K and allowed to come to equilibrium. Calculate the mole fractions of the two substances at equilibrium.

5.7 The dissociation vapour pressure of NH_4Cl at 427 °C is 608 kPa but at 459 °C it has risen to 1115 kPa. Calculate (a) the equilibrium constant, (b) the standard reaction free energy, (c) the standard enthalpy, (d) the standard entropy of dissociation, all at 427 °C. Assume that the vapour behaves as a perfect gas and that $\Delta H^{\ominus}$ and $\Delta S^{\ominus}$ are independent of temperature in the range given.

5.8 In the Haber process for ammonia synthesis, $K = 0.036$ for the reaction $N_2(g) + 3H_2(g) \rightleftharpoons 2NH_3(g)$ at 500 K. If a reactor is charged with partial pressure of 0.010 atm of N_2 and 0.010 atm of H_2, what will be the equilibrium partial pressure of the components?

5.9 Express the equilibrium $N_2O_4(g) \rightleftharpoons 2NO_2(g)$ in terms of the fraction α of N_2O_4 that has dissociated and the total pressure p of the reaction mixture, and show that when the extent of dissociation is small ($\alpha \ll 1$), α is inversely proportional to the square root of the total pressure ($\alpha \propto 1/\sqrt{p}$).

5.10 Write the proton-transfer equilibria for the following acids in aqueous solution and identify the conjugate acid–base pairs in each one: (a) H_2SO_4, (b) HF (hydrofluoric acid), (c) $C_6H_5NH_3^+$ (anilinium ion), (d) $H_2PO_4^-$ (dihydrogenphosphate ion), (e) HCOOH (formic acid), (f) $NH_2NH_3^+$ (hydrazinium ion).

5.11 The value of K_w for water at body temperature (37 °C) is 2.5×10^{-14}. (a) What is the value of $[H_3O^+]$ and the pH of neutral water at 37 °C? (b) What is the molar concentration of OH^- ions and the pOH of neutral water at 37 °C?

5.12 Heavy water, D_2O, is used in some nuclear reactors. The K_w for heavy water at 25 °C is 1.35×10^{-15}, (a) Write the chemical equation for the autoprotolysis of D_2O. (b) Evaluate pK_w for D_2O at 25°C. (c) Calculate the molar concentra-

tions of D_3O^+ and OD^- in neutral heavy water at 25 °C. (d) Evaluate the pD and pOD of neutral heavy water at 25 °C. (e) Find the relation between pD, pOD, and pK_w.

5.13 The molar concentration of H_3O^+ ions in the following solutions was measured at 25 °C. Calculate the pH and pOH of the solutions: (a) 2.0×10^{-5} mol L^{-1} (sample of rain water), (b) 1.0 mol L^{-1}, (c) 5.0×10^{-14} mol L^{-1}, (d) 5.02×10^{-5} mol L^{-1}.

5.14 Calculate the molar concentration of H_3O^+ ions and the pH of the following solutions: (a) 25.0 mL of 0.30 M HCl was added to 25.0 mL of 0.20 M NaOH, (b) 25.0 mL of 0.15 M HCl was added to 50.0 mL of 0.15 M KOH, (c) 21.7 mL of 0.10 M HNO_3 was added to 10.0 mL of 0.30 M NaOH.

5.15 Determine whether aqueous solutions of the following salts have a pH equal to, greater than, or less than 7; if pH > 7 or pH < 7, write a chemical equation to justify your answer. (a) NH_4Br, (b) Na_2CO_3, (c) KF, (d) KBr, (e) $AlCl_3$, (f) $Co(NO_3)_2$.

5.16 (a) A 10.0 g sample of potassium acetate, KCH_3CO_2, is dissolved in 250 mL of solution. What is the pH of the solution? (b) What is the pH of a solution resulting from the dissolution of 5.75 g of ammonium bromide, NH_4Br, in 100 mL of solution? (c) An aqueous solution filled to 'the mark' in a 1.0 L volumetric flask contains 10.0 g of potassium bromide. What is the percentage of Br^- ions that are protonated?

5.17 A solution of equal concentrations of lactic acid and sodium lactate was found to have pH = 3.08. (a) What are the values of pK_a and K_a of lactic acid? (b) What would the pH be if the acid had twice the concentration of the salt?

5.18 Sketch reasonably accurately the pH curve for the titration of 20.0 mL of 0.10 M $Ba(OH)_2$ with 0.20 M HCl. Mark on the curve (a) the initial pH, (b) the pH at the stoichiometric point.

5.19 Determine the fraction of solute ionized in (a) 0.20 M C_6H_5COOH, (b) 0.20 M NH_2NH_2 (hydrazine), (c) 0.20 M $(CH_3)_3N$ (trimethylamine).

5.20 Calculate the pH, pOH, and fraction of solute ionized in the following aqueous solutions: (a) 0.20 M $CH_3CH(OH)COOH$ (lactic acid), (b) 1.0×10^{-5} M $CH_3CH(OH)COOH$, (c) 0.10 M $C_6H_5SO_3H$ (benzenesulfonic acid).

5.21 Calculate the pH of the following acid solutions at 25 °C; ignore second ionizations only when that approximation is justified. (a) 1.0×10^{-4} M H_3BO_3 (boric acid acts as a monoprotic acid), (b) 0.015 M H_3PO_4, (c) 0.10 M H_2SO_3.

5.22 (a) Calculate the molar concentrations of $(COOH)_2$, $HOOCCO_2^-$, $(CO_2)_2^{2-}$, H_3O^+, and OH^- in a 0.10 M $(COOH)_2$ aqueous solution. (b) Calculate the molar concentrations of H_2S, HS^-, S^{2-}, H_3O^+, and OH^- in a 0.050 M H_2S aqueous solution.

5.23 A 25.0 mL sample of 0.10 M $CH_3COOH(aq)$ is titrated with 0.10 M $NaOH(aq)$. K_a for CH_3COOH is 1.8×10^{-5}. (a) What is the initial pH of the 0.10 M $CH_3COOH(aq)$ solution? (b) What is the pH after the addition of 10.0 mL of 0.10 M $NaOH(aq)$? (c) What volume of 0.10 M $NaOH(aq)$ is required to reach half way to the stoichiometric point? (d) Calculate the pH at that half-way point. (e) What volume of 0.10 M $NaOH(aq)$ is required to reach the stoichiometric point? (f) Calculate the pH at the stoichiometric point.

5.24 A 100 mL buffer solution consists of 0.10 M $CH_3COOH(aq)$ and 0.10 M $NaCH_3CO_2(aq)$. (a) What is the pH of the buffer solution? (b) What is the pH and pH change resulting from the addition of 3.0 mmol NaOH to the buffer solution? (c) What is the pH and pH change resulting from the addition of 6.0 mmol of HNO_3 to the initial buffer solution?

5.25 Predict the pH region in which each of the following buffers will be effective, assuming equal molar concentrations of the acid and its conjugate base: (a) sodium lactate and lactic acid, (b) sodium benzoate and benzoic acid, (c) potassium hydrogenphosphate and potassium phosphate, (d) potassium hydrogenphosphate and potassium dihydrogenphosphate, (e) hydroxylamine and hydroxylammonium chloride.

5.26 At the half-way point in the titration of a weak acid with a strong base the pH was measured as 5.40. What is the acidity constant and the pK_a of the acid? What is the pH of the solution that is 0.015 M in the acid?

5.27 Calculate the pH of (a) 0.10 M $NH_4Cl(aq)$, (b) 0.10 M $NaCH_3CO_2(aq)$, (c) 0.100 M $CH_3COOH(aq)$.

5.28 Calculate the pH at the stoichiometric point of the titration of 25.00 mL of 0.100 M lactic acid with 0.150 M $NaOH(aq)$.

5.29 Sketch the pH curve of a solution containing 0.10 M $NaCH_3CO_2(aq)$ and a variable amount of acetic acid.

5.30 From the information in Tables 5.1 and 5.2, select suitable buffers for (a) pH = 2.2 and (b) pH = 7.0.

5.31 Write the expression for the solubility constant of the following substances: (a) AgBr, (b) Ag_2S, (c) $Ca(OH)_2$, (d) Ag_2CrO_4.

5.32 Use the data in Table 5.4 to determine the molar solubility of (a) $PbSO_4$, (b) Ag_2CrO_4, (c) $Fe(OH)_2$, (d) $SrSO_4$.

5.33 Limestone is composed primarily of calcium carbonate. A 1.0 mm³ pebble limestone was accidentally dropped into a swimming pool measuring 10 m × 7 m × 2 m and filled with water. Assuming that the carbonate ion does not function as a Brønsted base, will the pebble dissolve entirely? The density of calcium carbonate is 2.71 g cm⁻³.

5.34 Use the data in Table 5.3 to calculate the solubility of each sparingly soluble substance in its respective solution: (a) silver bromide in a 1.0×10^{-3} M $NaBr(aq)$ solution, (b) magnesium carbonate in a 4.2×10^{-5} M $Na_2CO_3(aq)$ solution, (c) lead(II) sulfate in a 0.10 M $CaSO_4(aq)$ solution, (d) nickel hydroxide in a 3.7×10^{-5} M $NiSO_4(aq)$ solution.

6

Electrochemistry

Such apparently unrelated processes as combustion, respiration, photosynthesis, and corrosion are actually all closely related, for in each of them an electron (sometimes accompanied by a group of atoms) is transferred from one species to another. Indeed, together with the acid–base reactions, in which a proton is transferred, reactions in which electrons are transferred, the so-called **redox reactions**, account for almost all the reactions encountered in chemistry. Redox reactions are of immense practical significance, not only because they underlie many biochemical and industrial processes, but also because they are the basis of the generation of electricity by chemical reactions and the investigation of reactions by making electrical measurements. Measurements like the ones we describe in this chapter lead to a collection of data that are very useful for discussing the characteristics of electrolyte solutions and of a wide range of different types of equilibria in solution. They are also used throughout inorganic chemistry to assess the thermodynamic feasibility of reactions and the stabilities of compounds, and in physiology to discuss the details of the propagation of signals in neurones.

Electrochemical cells

The device used to study reactions electrically is called an **electrochemical cell**. Such a cell consists of two **electrodes**, or metallic conductors, dipping into an **electrolyte**, an ionic conductor, which may be a solution, a liquid, or a solid. An electrode and its electrolyte comprise an **electrode compartment**. The two electrodes may share the same compartment (Fig. 6.1(a)). If the electrolytes are different, then the two compartments may be joined by a **salt bridge**, which is an electrolyte solution that completes the electrical circuit by permitting ions to move between the compartments, and so enables the cell to function (Fig. 6.1(b)). Alternatively, the two solutions may be in direct physical contact (for example, through a porous membrane), but the presence of such **liquid junctions** introduces complications into

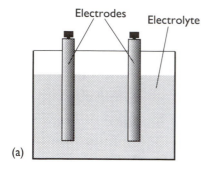

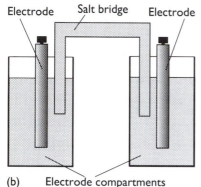

Fig. 6.1 (a) The basic arrangement for an electrochemical cell in which the two electrodes share a common electrolyte. (b) If the electrolytes in the electrode compartments of a cell are different, then they need to be joined so that ions can travel from one compartment to another. One device for joining the two compartments is a salt bridge.

the interpretation of measurements, and we shall not consider them further.

An electrochemical cell that produces electricity as a result of the spontaneous reaction occurring inside it is called a **galvanic cell**. An electrochemical cell in which a nonspontaneous reaction is driven by an external source of current is called an **electrolytic cell**. The commerically available dry cells, mercury cells, and nickel–cadmium cells that are used to power electrical equipment are all galvanic cells, and produce electricity as a result of the spontaneous chemical reaction between the substances built into them at manufacture. A **fuel cell** is a galvanic cell in which the reagents (such as hydrogen and oxygen or methane and oxygen) are supplied from outside. Fuel cells are used on manned spacecraft and one day may be used as a convenient, compact source of electricity in homes. Electric eels and electric catfish are biological versions of fuel cells in which the fuel is food and the cells are adaptations of muscle cells. Electrolytic cells include the arrangement used to electrolyse water into hydrogen and oxygen (a nonspontaneous process) and to obtain aluminium from its oxide in the Hall process. Electrolysis is the only commercially viable means for the production of fluorine.

The link between this chapter and the earlier chapters on thermodynamics lies in the fact that *a galvanic cell is a device for extracting the energy of a spontaneous reaction as nonexpansion work*. We saw in Section 3.3 that the maximum nonexpansion work is given by the value of ΔG for a process; because we shall now identify that nonexpansion work with electrical work, the key equation of this chapter will have the form

$$\text{maximum electrical work available} = \text{change in free energy}$$

which is valid at constant temperature and pressure. (The precise form of this expression, and its signs, will be given later.) Most of this chapter will be based on this relation.

This chapter necessarily makes use of a number of concepts related to electricity; they are reviewed in *Further information 5* (p. 206).

6.1 Half-reactions and electrodes

As we have remarked, a redox reaction is a process in which there is a transfer of electrons from one substance to another. As a result of this transfer of electrons, there is a change in the oxidation numbers of the species. (The concept of oxidation number is reviewed in *Further information 6,* p. 208.)

A redox reaction is the outcome of two contributions, the loss of electrons (and perhaps atoms) from one species, and their gain by another species. The loss of electrons is called **oxidation**, and is identified by noting whether an element has undergone an increase in oxidation number The gain of electrons is called **reduction**, and it is identified by noting whether an element has undergone a decrease in oxidation number. As we also remarked, electron transfer may be accompanied by atom transfer (as in the conversion of PCl_3 to PCl_5 or of NO_2^- to NO_3^-); the requirement to break and form covalent bonds in some redox reactions is one of the reasons why they often achieve equilibrium quite slowly (often much more slowly than acid–base proton-transfer reactions). The **reducing agent** (or 'reductant') is the species that acts as electron donor, and at least one of the elements in the species undergoes an increase in oxidation number when it acts. The **oxidizing agent** (or 'oxidant') is the species that acts as an electron acceptor, and at least one of its elements undergoes a decrease in oxidation number.

Examples of redox reactions include the combustion of magnesium in oxygen

$$2Mg(s) + O_2(g) \rightarrow 2MgO(s)$$

in which magnesium is the reducing agent (and becomes oxidized to Mg^{2+} ions) and oxygen is the oxidizing agent (and becomes reduced to O^{2-} ions). Another example is the reaction of a metal oxide with hydrogen:

$$CuO(s) + H_2(g) \rightarrow Cu(s) + H_2O(g)$$

In this reaction hydrogen is the reducing agent and copper(II) oxide the oxidizing agent. Both these reactions involve the transfer of atoms as well as electrons (but the net effect is a change of oxidation number). A reaction which is purely a transfer of electrons is the displacement of copper from solution by zinc

$$Cu^{2+}(aq) + Zn(s) \rightarrow Cu(s) + Zn^{2+}(aq)$$

in which Cu^{2+} ions are the oxidizing agents and zinc metal is the reducing agent.

Exercise E6.1 Identify the species that have undergone oxidation and reduction in the reaction

$$CuS(s) + O_2(g) \rightarrow Cu(s) + SO_2(g)$$

[Cu(II) reduced, S^{2-} oxidized to S(IV), O reduced]

The breadth of the scope of redox reactions can be judged by noting that because a combustion reaction is a redox reaction, then its reverse is also a redox reaction (with the electrons and atoms transferred in the opposite direction). The reverse of the combustion of glucose,

$$C_6H_{12}O_6(s) + 6O_2(g) \rightarrow 6CO_2(g) + 6H_2O(l)$$

is nothing other than the net outcome of the photosynthesis reaction

$$6CO_2(g) + 6H_2O(l) \rightarrow C_6H_{12}O_6(s) + 6O_2(g)$$

Therefore, the net reaction that we call photosynthesis is a redox reaction. (The actual reaction proceeds by a very complicated mechanism, some steps of which involve proton transfer and others electron and atom transfer; but the *net* outcome is a redox reaction.) Therefore, the generation of carbohydrates on this planet and their consumption in respiration are both aspects of redox reactions.

Half-reactions

An important step in the analysis of redox reactions draws an analogy with acid–base reactions, which may be expressed in terms of proton loss from one species and proton gain by another. Thus, any redox reaction may be expressed as the sum of two **half-reactions**, one involving electron loss by a species and the other electron gain. Two examples are

$$\text{Oxidation of Zn: } \text{Zn}(s) \rightarrow \text{Zn}^{2+}(aq) + 2e^-$$

$$\text{Reduction of Cu}^{2+}\text{: } \text{Cu}^{2+}(aq) + 2e^- \rightarrow \text{Cu}(s)$$

A half-reaction in which atom transfer accompanies electron transfer is

---

FURTHER INFORMATION 5: *Concepts of electrostatics*

The fundamental expression in **electrostatics**, the interaction of stationary electric charges, is the **Coulomb potential energy** of one charge of magnitude q at a distance r from another charge q':

$$V = \frac{1}{4\pi\varepsilon_0} \times \frac{qq'}{r}$$

That is, the potential energy is inversely proportional to the separation of the charges. The fundamental constant ε_0 is the **vacuum permittivity**; its value is

$$\varepsilon_0 = 8.854 \times 10^{-12} \, \text{J}^{-1} \, \text{C}^2 \, \text{m}^{-1}$$

(Note that with r in metres, m, and the charges in coulombs, C, the potential energy is in joules, J.) The potential energy is equal to the work that must be done to bring up a charge q from infinity to a distance r from a charge q'.

The potential energy of a charge q in the presence of another charge q' can be expressed in terms of the **Coulomb potential**, ϕ:

$$V = q \times \phi, \quad \text{where } \phi = \frac{1}{4\pi\varepsilon_0} \times \frac{q'}{r}$$

The units of potential are joules per coulomb, $\text{J} \, \text{C}^{-1}$, so that when ϕ is multiplied by a charge in coulombs, the result is in joules. The combination joules per coulomb occurs widely in electrostatics, and is called a *volt*, V:

$$1 \, \text{V} = 1 \, \text{J} \, \text{C}^{-1}$$

(which implies that $1 \, \text{V} \, \text{C} = 1 \, \text{J}$). If there are several charges $q_1, q_2, \ldots$ present in the system, then the total potential experienced by the charge q is the sum of the potential generated by each charge:

$$\phi = \phi_1 + \phi_2 + \cdots$$

the reduction of MnO_4^- ions:

$$MnO_4^-(aq) + 8H^+(aq) + 5e^- \rightarrow Mn^{2+}(aq) + 4H_2O(l)$$

Half-reactions are *conceptual* reactions showing the loss and gain of electrons: the reactions do not actually occur (except in special cases) by one species releasing an electron and then another species accepting it; electron transfer reactions normally proceed by a much more complex mechanism in which the electron is never free. The electrons in these conceptual reactions are regarded as being 'in transit' and are not ascribed a state.

A redox reaction is the sum of an oxidation and a reduction half-reaction. For the reduction of Cu^{2+} ions by zinc metal, for example, the redox reaction is

$$Cu^{2+}(aq) + Zn(s) \rightarrow Cu(s) + Zn^{2+}(aq)$$

It is common practice, however, to write all half-reactions as reductions; so the zinc-oxidation half-reaction is reversed and expressed as a reduction:

reduction of Zn^{2+}: $Zn^{2+}(aq) + 2e^- \rightarrow Zn(s)$

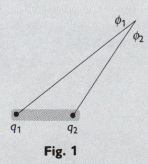

Fig. 1

for example, the potential generated by a dipole is the sum of the potentials of the two equal and opposite charges: these potentials do not in general cancel because the point of interest is at different distances from the two charges (Fig. 1).

The motion of charge is an electric **current**, I. Electric current is measured in amperes, A, where

$$1\,A = 1\,C\,s^{-1}$$

If the electric charge is that of electrons (as it is through metals and semiconductors), then a current of 1 A represents the flow of 6×10^{18} electrons per second. If the current flows from a region of potential ϕ_i to ϕ_f, through a **potential difference** $\Delta\phi = \phi_f - \phi_i$, then the rate of doing work is the current (the rate of transfer of charge) multiplied by the potential difference, $I \times \Delta\phi$. The rate of doing work is called **power**, P, so

$$P = I \times \Delta\phi$$

With current in amperes and the potential difference in volts, the power works out in joules per second, or *watts*, W:

$$1\,W = 1\,J\,s^{-1}$$

The total energy supplied in a time t is the power (the energy per second) multiplied by the time:

$$E = P \times t = I \times \Delta\phi \times t$$

The energy is obtained in joules with the current in amperes, the potential difference in volts, and the time in seconds.

FURTHER INFORMATION 6:
Oxidation numbers

A simple way of judging whether a *monatomic* species has undergone oxidation or reduction is to note if the charge number of the species has changed. For example, an increase in the charge number of a monatomic ion (which corresponds to electron loss), as in the conversion of Fe^{2+} to Fe^{3+}, is an oxidation. A decrease in charge number (to a less positive or more negative value, as a result of electron gain), as in the conversion of Br to Br^-, is a reduction.

It is possible to assign to an atom in a polyatomic species an *effective* charge number, called the **oxidation number**, ω (there is no standard symbol for this quantity). The oxidation number is defined so that an increase in its value ($\Delta\omega > 0$) corresponds to oxidation, and a decrease ($\Delta\omega < 0$) corresponds to reduction.

An oxidation number is assigned to an element in a compound by applying the rules in the table (the rules are not exhaustive, but they cover many compounds and the pattern for other compounds should be clear). These rules are based on arguments of the kind presented in Chapter 9 relating to the sharing of electrons in molecules and the role of electronegativity. The rules must be applied in the order given, and we must stop as soon as the oxidation number has been obtained (because a later rule might contradict an earlier one). The rules imply the following two simple points:

(1) The oxidation number of an elemental substance is zero, $\omega(\text{element}) = 0$.
(2) The oxidation number of a monatomic ion is equal to the charge number of that ion: $\omega(E^{z\pm}) = \pm z$.

Thus, hydrogen, oxygen, iron, and all the elements have $\omega = 0$ in their elemental forms; $\omega(Fe^{3+}) = +3$ and $\omega(Br^-) = -1$. It follows that the conversion of Fe to Fe^{3+} is an oxidation (because $\Delta\omega > 0$) and the conversion of Br to Br^- is a reduction (because $\Delta\omega < 0$). The definition of oxidation number and its relation to oxidation and reduction are consistent with the definitions in terms of electron loss and gain.

Work through the following rules in *the order given*. Stop as soon as the oxidation number has been assigned

---

1. The sum of the oxidation numbers (ω) of all the atoms in the species is equal to its total charge.

2. For atoms in their elemental form, $\omega = 0$.

3. For elements of

	Group 1	$\omega = +1$
	Group 2	$\omega = +2$
	Group 13/III (except B)	$\omega = +3$ for M^{3+} $\omega = +1$ for M^+
	Group 14/IV (except C, Si)	$\omega = +4$ for M^{4+} $\omega = +2$ for M^{2+}

4. For hydrogen $\omega = +1$ in combination with nonmetals
   $\omega = -1$ in combination with metals

5. For fluorine $\omega = -1$ in all its compounds

6. For oxygen $\omega = -2$ unless combined with F
   $\omega = -1$ in peroxides (O_2^{2-})
   $\omega = -\frac{1}{2}$ in superoxides (O_2^-)
   $\omega = -\frac{1}{3}$ in ozonides (O_3^-)

Example Determining oxidation numbers
Determine the oxidation numbers of the elements in (a) SO_2, (b) SO_4^{2-}.

Answer (a) SO_2: By rule 1 in the table, the sum of oxidation numbers of the atoms in the neutral compound must be 0:

$$\omega(S) + 2\omega(O) = 0$$

Rules 2 to 5 are not relevant. According to rule 6, each O atom has $\omega = -2$. Hence,

$$\omega(S) + 2 \times (-2) = 0$$

which solves to $\omega(S) = +4$.
(b) SO_4^{2-}: By rule 1, the sum of oxidation numbers of the atoms in the ion is -2:

$$\omega(S) + 4\omega(O) = -2$$

Rules 2 to 5 are not relevant. According to rule 6, $\omega(O) = -2$. Hence,

$$\omega(S) + 4 \times (-2) = -2,$$
$$\text{which solves to } \omega(S) = +6$$

The sulfur is more highly oxidized in the sulfate ion than in sulfur dioxide.

Exercise Calculate the oxidation numbers of the elements in (a) H_2S, (b) PO_4^{3-}, (c) NO_3^-

$$[(a) \ \omega(H) = +1, \ \omega(S) = -2;$$
$$(b) \ \omega(P) = +5, \ \omega(O) = -2;$$
$$(c) \ \omega(N) = +5, \ \omega(O) = -2]$$

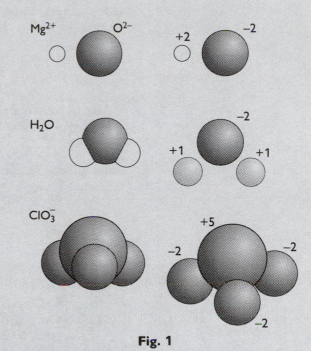

Fig. 1

Some insight into the significance of oxidation numbers for elements in polyatomic species comes from noting that the oxidation number assigned to an element is equal to the charge that the atom would have if every O atom in the compound were present as an O^{2-} ion (Fig. 1). For example, if the oxygen atom in H_2O is treated as an O^{2-} ion, then for overall charge neutrality each H atom must be treated as an H^+ ion; hence, the oxidation number of H in H_2O is +1, the value that would be obtained by applying the rules in the table. It follows that in the formation of water from its elements, hydrogen is oxidized because its oxidation number changes from 0 to +1. Because the oxidation number of oxygen changes from 0 to -2 in the reaction, it is reduced. This conclusion is consistent with our understanding of oxidation and reduction in terms of electron loss and gain, respectively.

Another example is the chlorate ion, ClO_3^-, in which (for understanding the significance of the oxidation number, but not for any other purpose) each O atom is thought of as being present as a O^{2-} ion (see Fig. 1). That adds up to a charge of $3 \times (-2) = -6$. Because the overall charge number of the ion is -1, the effective charge of the Cl atom must be $+5$; therefore, $\omega(Cl) = +5$.

Now the overall reaction is the *difference* of the two reduction half-reactions.

The reduced and oxidized substances in a half-reaction form a **redox couple**, denoted Ox/Red. Thus, the redox couples mentioned so far are Cu^{2+}/Cu, Zn^{2+}/Zn, and $MnO_4^-,H^+/Mn^{2+}$. In general, we adopt the notation

Couple: Ox/Red

Half reaction: $Ox + ne^- \rightarrow Red$

The overall reaction need not be a redox reaction for it to be expressed in terms of half-reactions. For instance, the expansion of a gas

$$H_2(g, p_i) \rightarrow H_2(g, p_f)$$

is not a redox reaction (there is no change of oxidation number) but it can be expressed as the difference of two reductions:

$$2H^+(aq) + 2e^- \rightarrow H_2(g, p_f)$$
$$2H^+(aq) + 2e^- \rightarrow H_2(g, p_i)$$

The two couples in this case are both H^+/H_2.

Example Expressing a reaction in terms of half-reactions
Express the dissolution of $AgCl(s)$, another example of a process that is not a redox reaction (the oxidation numbers of the elements do not change in the *overall* reaction), as the difference of two reduction half-reactions.
Answer The overall process is

$$AgCl(s) \rightarrow Ag^+(aq) + Cl^-(aq)$$

If we select as one half-reaction the reduction of AgCl (more precisely, the reduction of the Ag(I) in AgCl to the metal),

$$AgCl(s) + e^- \rightarrow Ag(s) + Cl^-(aq)$$

then the second half-reaction, when subtracted from this one, must give the overall reaction. It is therefore the reduction of $Ag^+(aq)$ to Ag metal:

$$Ag^+(aq) + e^- \rightarrow Ag(s)$$

Exercise E6.2 Express the formation of H_2O from H_2 and O_2 in acid solution as the difference of two reduction half-reactions.

$$[4H^+(aq) + 4e^- \rightarrow 2H_2(g),$$
$$O_2(g) + 4H^+(aq) + 4e^- \rightarrow 2H_2O(l)]$$

We have already seen that, for thermodynamic considerations, a natural way to express the composition of a system is in terms of the reaction quotient Q (because Q occurs in a number of thermodynamic formulas, particularly the formula for the reaction free energy, eqn (24) in Chapter 3). We shall, in fact, find it useful to express the composition of a single electrode compartment in terms of the reaction quotient Q for the half-reaction that takes place in the compartment. The half-reaction quotient is defined like the quotient for the overall reaction, but with the electron ignored (their contribution always cancels in the overall reaction). Thus, for the half-reaction of the Cu^{2+}/Cu couple we would write

$$Cu^{2+}(aq) + 2e^- \rightarrow Cu(s) \qquad Q = \frac{1}{a(Cu^{2+})}$$

where, as explained in Chapter 3, $a(Cu^{2+}) = [Cu^{2+}]/mol\,L^{-1}$, if the solution is very dilute, and $a = 1$ for a pure solid (the copper metal).

Exercise E6.3 Express the oxidation of NADH (nicotinamide adenine dinucleotide, which participates in the chain of oxidations that constitutes respiration) to NAD^+ by oxygen in aqueous solution as the difference of two reduction half-reactions.

$$[\tfrac{1}{2}O_2(g) + 2H^+(aq) + 2e^- \rightarrow H_2O(l);$$
$$NAD^+(aq) + H^+(aq) + 2e^- \rightarrow NADH(aq)]$$

Reactions at electrodes

In an electrochemical cell, the oxidation half-reaction takes place in one electrode compartment and the reduction half-reaction takes place in the other compartment. As the reaction proceeds, the electrons released in the oxidation half-reaction

$$reduced\ species \rightarrow oxidized\ species + ne^-$$

in one compartment travel through the external circuit, and enter the cell through the other electrode, where they bring about the reduction

$$oxidized\ species + ne^- \rightarrow reduced\ species$$

For example, if one electrode is zinc, then the oxidation reaction is the release of two electrons from each atom in the metal, with the formation of Zn^{2+} ions, which go into solution:

$$Zn(s) \rightarrow Zn^{2+}(aq) + 2e^-$$

These electrons travel through the external circuit and, if the other electrode is copper in aqueous copper(II) sulfate, reduce the Cu^{2+} ions in solution:

$$Cu^{2+}(aq) + 2e^- \rightarrow Cu(s)$$

The electrode at which oxidation occurs is called the **anode**; the electrode at which reduction occurs is called the **cathode** (Fig. 6.2).

211

Fig. 6.2 The flow of electrons in the external circuit is from the anode of a galvanic cell, where they have been lost in the oxidation reaction, and to the cathode, where they are used in the reduction reaction. Electrical neutrality is preserved in the electrolytes by the flow of cations and anions in opposite directions through the salt bridge.

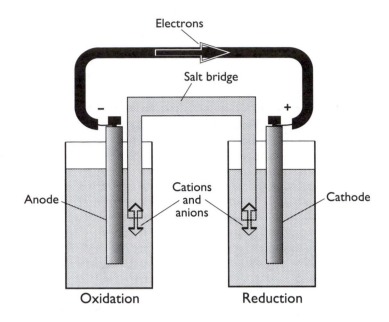

Anode reaction (oxidation):

$$\text{reduced species} \rightarrow \text{oxidized species} + n\text{e}^-$$

Cathode reaction (reduction):

$$\text{oxidized species} + n\text{e}^- \rightarrow \text{reduced species}$$

In the example just described, the zinc electrode would be the anode (the site of oxidation) and the copper electrode would be the cathode (the site of reduction).

In a galvanic cell, the cathode has a higher potential than the anode because the species undergoing reduction withdraws electrons from its

Fig. 6.3 The flow of electrons and ions in an electrolytic cell. An external supply forces electrons into the cathode, where they are used to bring about a reduction, and withdraws them from the anode, which results in the oxidation reaction at that electrode. Cations migrate towards the negatively charged cathode and anions migrate towards the positively charged anode. An electrolytic cell always consists of a single compartment.

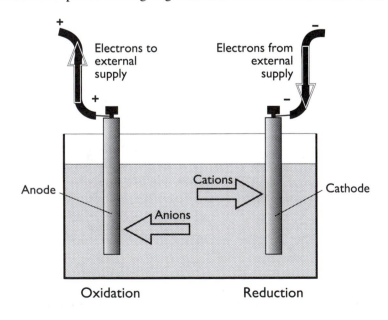

electrode (the cathode), so leaving a relative positive charge on it (corresponding to a high potential). At the anode, oxidation results in the transfer of electrons to the electrode, so giving it a relative negative charge (corresponding to a low potential). In an electrolytic cell, the anode is also the location of oxidation (by definition), but now electrons must be withdrawn from the species in that compartment because oxidation does not occur spontaneously, and at the cathode there must be a supply of electrons to drive the reduction. Therefore, in an electrolytic cell the anode must be made relatively positive to the cathode so that electrons are effectively sucked out of the anode and pushed on to the cathode (Fig. 6.3).

Varieties of electrode

In a **gas electrode** (Fig. 6.4), a gas is in equilibrium with a solution of its ions in the presence of an inert metal. The inert metal, which is often platinum, acts as a source, or 'sink', of electrons, but takes no other part in the reaction (but may act as a catalyst for it). One important example is the **hydrogen electrode**, in which hydrogen is bubbled through an aqueous solution of hydrogen ions and the redox couple is H^+/H_2. This electrode is denoted

$$Pt \mid H_2(g) \mid H^+(aq)$$

where the vertical lines denote junctions between phases (in this electrode, the junction between the metal and the gas, and between the gas and the liquid containing its ions). Note that the electrode description runs in the order $Red \mid Ox$, which is opposite to the order in which the couple is denoted.

The hydrogen electrode may be either a cathode or an anode, depending on the other electrode in the cell and the direction in which the overall reaction is taking place. The half-reaction at the electrode when it is acting as a cathode (undergoing reduction) is

$$2H^+(aq) + 2e^- \rightarrow H_2(g) \qquad Q = \frac{a(H_2)}{a(H^+)^2}$$

where $a(H^+) = [H^+]/mol\,L^{-1}$, if the solution is dilute, and $a(H_2) = p(H_2)/p^{\ominus}$ (Section 3.4). As remarked many times in Chapter 4, the replacement of a by a molar concentration is valid only in solutions that have very low concentrations of *all* ions; in practical electrodes, an activity coefficient should be included to accommodate the effects of ion–ion interactions. In many cases it is advisable to leave the activities as a and not to make the final approximate conversion to molar concentrations; then the equations are exact (and can sometimes be used in that form, as we shall see).

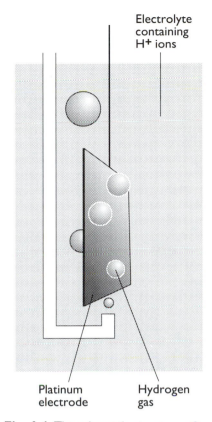

Electrolyte containing H^+ ions

Platinum electrode   Hydrogen gas

Fig. 6.4 The schematic structure of a hydrogen electrode (which is like other gas electrodes). Hydrogen is bubbled over a platinum surface which is in contact with a solution containing hydrogen ions. The platinum, as well as acting as a source (or sink) for electrons, speeds the electrode reaction because hydrogen attaches to (adsorbs on) the surface as atoms.

Example Writing the half-reaction for a gas electrode
Write the half-reaction and the reaction quotient for the reduction of oxygen to water in acidic solution.

Answer The balanced equation for the reduction of O_2 in acidic solution is

$$O_2(g) + 4H^+(aq) + 4e^- \rightarrow 2H_2O(l)$$

The reaction quotient for the half-reaction is therefore

$$Q = \frac{1}{a(O_2)a(H^+)^4}$$

where we have supposed that the solution is so dilute that the water is effectively pure (and set $a = 1$).

Exercise E6.4 Write the half-reaction and the reaction quotient for a chlorine-gas electrode.

$$[Cl_2(g) + 2e^- \rightarrow 2Cl^-(aq), \ Q = a(Cl^-)^2/a(Cl_2)]$$

An **insoluble-salt electrode** consists of a metal M covered by a porous layer of insoluble salt MX, the whole being immersed in solution containing X^- ions (Fig. 6.5). The electrode is denoted $M\,|\,MX\,|\,X^-$. An example is the silver–silver-chloride electrode, $Ag\,|\,AgCl\,|\,Cl^-$, for which the reduction half-reaction is

$$AgCl(s) + e^- \rightarrow Ag(s) + Cl^-(aq) \qquad Q = a(Cl^-)$$

Note that the reaction quotient (and consequently, as we shall see later, the potential of the electrode) depends on the molar concentration of chloride ions in the electrolyte solution.

Example Writing the half-reaction for an insoluble-salt electrode

Write the half-reaction and the reaction quotient for the lead–lead-sulfate electrode of the lead–acid battery, in which Pb(II), as lead(II) sulfate is reduced to metallic lead in the presence of hydrogensulfate ions in the electrolyte.

Answer The electrode is $Pb\,|\,PbSO_4(s)\,|\,HSO_4^-(aq)$, in which Pb(II) is reduced to metallic lead. The reduction half-reaction is

$$PbSO_4(s) + H^+(aq) + 2e^- \rightarrow Pb(s) + HSO_4^-(aq)$$
$$Q = a(HSO_4^-)a(H^+)$$

Exercise E6.5 Write the half-reaction and the reaction quotient for the calomel electrode, $Hg(l)\,|\,Hg_2Cl_2(s)\,|\,Cl^-(aq)$, in which mercury(I) chloride (calomel) is reduced to mercury metal in the presence of chloride ions. This electrode is a component of instruments used to measure pH, as explained later.

$$[Hg_2Cl_2(s) + 2e^- \rightarrow 2Hg(l) + 2Cl^-(aq), \ Q = a(Cl^-)^2]$$

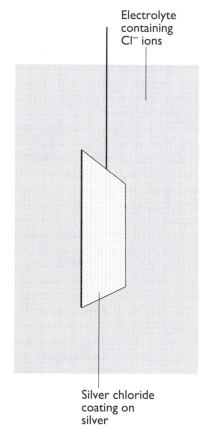

Electrolyte containing Cl^- ions

Silver chloride coating on silver

Fig. 6.5 The schematic structure of a silver–silver-chloride electrode (as an example of a metal–insoluble-salt electrode). The electrode consists of metallic silver coated with a layer of silver chloride in contact with a solution containing Cl^- ions.

The term **redox electrode** is normally reserved for an electrode in which the couple consists of two oxidation states of the same element (Fig. 6.6): an example is an electrode in which the couple is Fe^{3+}/Fe^{2+}. Thus, in general the equilibrium is

$$Ox + ne^- \rightarrow Red \qquad Q = \frac{a(Red)}{a(Ox)}$$

A redox electrode is denoted $M \mid Red, Ox$, where M is an inert metal (typically platinum) making electrical contact with the solution. The electrode corresponding to the Fe^{3+}/Fe^{2+} couple is therefore denoted $Pt \mid Fe^{2+}(aq), Fe^{3+}(aq)$ and the reduction half-reaction is

$$Fe^{3+}(aq) + e^- \rightarrow Fe^{2+}(aq) \qquad Q = \frac{a(Fe^{2+})}{a(Fe^{3+})}$$

6.2 Varieties of cell

The simplest type of galvanic cell has a single electrolyte common to both electrodes (as in Fig. 6.1(a)). In some cases it is necessary to immerse the electrodes in different electrolytes, as in the Daniell cell (Fig. 6.7) in which the redox couple at one electrode is Cu^{2+}/Cu and at the other is Zn^{2+}/Zn. In an **electrolyte concentration cell** (which would be constructed like the cell in Fig. 6.1(b)), the electrode compartments are of identical composition except for the concentrations of the electrolytes. An electrolyte concentration cell is a model of a neurone, which consists of a cell membrane with different concentrations of Na^+ and K^+ ions on either side. In an **electrode concentration cell** the electrodes themselves have different concentrations, either because they are gas electrodes operating at different pressures or because they are amalgams (solutions in mercury) with different concentrations.

Liquid junction potentials

In a cell with two different electrolyte solutions in contact, as in the Daniell cell or in a cell in which the compartments have different concentrations of hydrochloric acid, there is an additional source of potential difference, the **liquid junction potential**, E_{lj}, across the interface of the two electrolytes. Electrolyte concentration cells always have a liquid junction; electrode concentration cells do not.

The contribution of the liquid junction to the potential can be reduced (to about 1 to 2 mV) by joining the electrolyte compartments through a salt bridge consisting of a saturated electrolyte solution (usually KCl) in agar jelly (as in Fig. 6.1(b)). The reason for the success of the salt bridge is that the liquid junction potentials at either end are largely independent of the concentrations of the two more dilute solutions in the electrode compartments, and so nearly cancel.

Notation

In the notation for cells, phase boundaries are denoted by a vertical bar. For example, a cell in which the left-hand electrode is a hydrogen

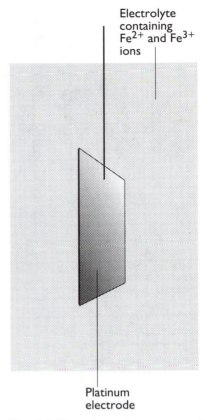

Electrolyte containing Fe^{2+} and Fe^{3+} ions

Platinum electrode

Fig. 6.6 The schematic structure of a redox electrode. The platinum metal acts as a source (or sink) of electrons required for the interconversion of (in this case) Fe^{2+} and Fe^{3+} ions in the surrounding solution.

215

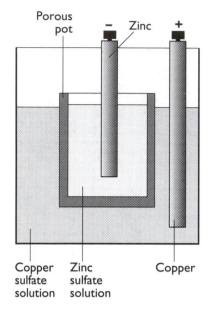

Porous pot — Zinc +

Copper sulfate solution   Zinc sulfate solution   Copper

Fig. 6.7 A Daniell cell consists of copper in contact with copper(II) sulfate solution and zinc in contact with zinc sulfate solution; the two compartments are in contact through the porous pot which contains the zinc sulfate solution. The copper electrode is the cathode and the zinc electrode is the anode.

electrode and the right-hand electrode is a silver–silver-chloride electrode is denoted

$$\text{Pt} \mid \text{H}_2(g) \mid \text{HCl}(aq) \mid \text{AgCl}(s) \mid \text{Ag}$$

A double vertical line ∥ denotes an interface for which it is assumed that the junction potential has been eliminated. Thus a cell in which the left-hand electrode, in an arrangement like that in Fig. 6.1(b), is zinc in contact with aqueous zinc sulfate, and the right-hand electrode is copper in contact with aqueous copper sulfate, is denoted

$$\text{Zn}(s) \mid \text{ZnSO}_4(aq) \parallel \text{CuSO}_4(aq) \mid \text{Cu}(s)$$

Exercise E6.6 Give the notation for a cell in which the oxidation of NADH by oxygen could be studied.
[$\text{Pt} \mid \text{NADH}(aq), \text{NAD}^+(aq), \text{H}^+(aq) \parallel \text{H}^+(aq) \mid \text{O}_2(g) \mid \text{Pt}$]

The cell reaction

The current produced by a galvanic cell arises from the spontaneous reaction taking place inside it. The **cell reaction** is the reaction in the cell written on the assumption that the right-hand electrode is the cathode, and hence that the spontaneous reaction is one in which reduction is taking place in the right-hand compartment. Later we will see how to predict whether the right-hand electrode is in fact the cathode; if it is, then the cell reaction is spontaneous as written. If the left-hand electrode turns out to be the cathode, then the reverse of the cell reaction is spontaneous and we rewrite the cell accordingly.

To write the cell reaction corresponding to the cell diagram, we first write the right-hand half-reaction as a reduction and then subtract from it the left-hand reduction half-reaction. Thus, in the cell

$$\text{Zn}(s) \mid \text{ZnSO}_4(aq) \parallel \text{CuSO}_4(aq) \mid \text{Cu}(s)$$

the two reduction half-reactions are

$$\text{right (R): } \text{Cu}^{2+}(aq) + 2e^- \rightarrow \text{Cu}(s)$$
$$\text{left (L): } \text{Zn}^{2+}(aq) + 2e^- \rightarrow \text{Zn}(s)$$

Hence, the overall cell reaction is the difference:

$$\text{overall (R − L): } \text{Cu}^{2+}(aq) + \text{Zn}(s) \rightarrow \text{Cu}(s) + \text{Zn}^{2+}(aq)$$

Had we written the cell in the opposite order,

$$\text{Cu}(s) \mid \text{CuSO}_4(aq) \parallel \text{ZnSO}_4(aq) \mid \text{Zn}(s)$$

we would have written

$$\text{right (R): } \text{Zn}^{2+}(aq) + 2e^- \rightarrow \text{Zn}(s)$$
$$\text{left (L): } \text{Cu}^{2+}(aq) + 2e^- \rightarrow \text{Cu}(s)$$
$$\text{overall (R − L): } \text{Cu}(s) + \text{Zn}^{2+}(aq) \rightarrow \text{Cu}^{2+}(aq) + \text{Zn}(s)$$

This overall reaction is the reverse of the one above. Because we know (at this stage by experiment) that zinc displaces copper and not vice versa, the first arrangement gives the spontaneous reaction correctly and is the cell diagram we should adopt.

The cell potential

A galvanic cell operates by the oxidation half-reaction depositing electrons in the anode which are then pulled through the external circuit by the reduction half-reaction at the cathode. So long as the overall reaction is not at equilibrium, the oxidation half-reaction effectively pushes the electrons into the external circuit, and the reduction half-reaction effectively pulls them out of it. If the cell reaction is not at equilibrium, the cell can do electrical work as the reaction drives electrons through an external circuit. The work that a given transfer of electrons can do depends on the potential difference between the two electrodes. This potential difference is called the **cell potential** and is measured in volts, V. When the cell potential is large (for instance, 2 V), a given number of electrons travelling between the electrodes can do a large amount of electrical work; when the cell potential is small (such as 2 mV), the same number of electrons can do only a small amount of work. A cell in which the reaction is at equilibrium can do *no* work, and its potential is zero.

According to the discussion in Section 3.3, we know that the maximum electrical work, w', that a system (in this context, the cell) can do is given by the value of ΔG, and in particular that for a spontaneous process (in which both ΔG and w are negative, a negative value of w signifying that energy leaves the cell as work)

$$w'(\text{maximum}) = \Delta G \text{ at constant temperature and pressure} \quad (1)$$

However, maximum work is achieved when a process is occurring *reversibly*. In the present context, reversibility means that the cell should be connected to an external source of potential difference (another cell or a stable power supply) that exactly matches the potential generated by the cell of interest. Then an infinitesimal change of the external potential will allow the reaction to proceed in its spontaneous direction and an opposite infinitesimal change will drive the reaction in its reverse direction. (As explained in Section 3.3, the reversal of a process by an *infinitesimal* change in the external conditions is the criterion of thermodynamic reversibility.) The potential difference measured when a cell is balanced against an external source of potential is called the **zero-current cell potential**, E, (Fig. 6.8). An older name for this quantity (which is still widely used), is the 'electromotive force', or emf, of the cell.

The zero-current cell potential, E, is related to the reaction free energy, ΔG, at a particular composition of the cell compartments by

$$\Delta G = -nFE \quad (2)$$

where n is the amount of electrons (in moles) transferred between the electrodes when the stoichiometric amount of reaction occurs. For

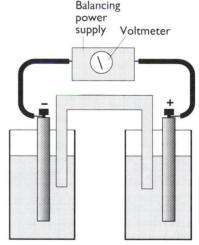

Fig. 6.8 The zero-current cell potential is measured by balancing the cell against an external potential that opposes the reaction in the cell. When there is no current flow, the external potential difference is equal to the cell potential.

example, in the Daniell cell, in which the reaction is

$$Cu^{2+}(aq) + Zn(s) \rightarrow Cu(s) + Zn^{2+}(aq)$$

for the reaction as written, $n = 2$ mol because 2 mol e^- are transferred from zinc metal to Cu^{2+} ions. The constant, F, is **Faraday's constant**, and is the magnitude of electric charge per mole of electrons:

$$F = eN_A = 96.485 \text{ kC mol}^{-1}$$

where e is the charge of one electron. The symbol C denotes the *coulomb*, the unit of electric charge. For future reference it will be useful to note that

$$1 \text{ C} \times 1 \text{ V} = 1 \text{ J}$$

JUSTIFICATION

The relation between the reaction free energy ΔG (the change in free energy at a *fixed* composition of the reaction mixture) and the zero-current cell potential, E, at that composition is derived by considering the change in G when the cell reaction produces an amount n of electrons but in a system that is so large that there is effectively no change in composition. For example, for the Daniell cell in which the reaction is

$$Cu^{2+}(aq) + Zn(s) \rightarrow Cu(s) + Zn^{2+}(aq)$$

the reaction produces 1 mol Zn^{2+} at the expense of 1 mol Cu^{2+} and transfers 2 mol e^- from the anode to the cathode. However, to ensure that there is virtually no change in composition, we have to suppose that the reaction proceeds in such a large volume of solution that the composition remains effectively constant despite the formation of products and consumption of reactants. In this effectively constant composition arrangement the change in the free energy is ΔG. As the reaction takes place, 2 mol e^- travel from the anode to the cathode. The total charge transported between the electrodes when this occurs is $(2 \text{ mol}) \times (-F)$. The electrical work, w', done when this charge travels from the anode to the cathode is equal to the product of the charge and the (constant) potential difference, E:

$$w' = (2 \text{ mol}) \times (-F) \times E$$

Finally, we equate this electrical work to the reaction free energy, for the reaction is occurring reversibly at constant temperature and pressure:

$$-(2 \text{ mol}) \times FE = \Delta G$$

For a general reaction, in which an amount n of electrons is transferred between the electrodes for the reaction as written,

$$-nFE = \Delta G$$

Equation (2) means that, if we know the reaction free energy at a specified composition, then we can state the zero-current cell potential at that composition, and vice versa. It follows that we now have an *electrical* method for measuring a reaction free energy at any composition of the reaction system, because a measurement of E can be converted into a value of ΔG. A point to note is that the negative sign in eqn (2) means that a negative reaction free energy, which corresponds to a spontaneous cell reaction, results in a positive zero-current cell potential.

Equation (2) can be used to estimate the cell potential for a typical reaction for which $\Delta G \approx -100 \text{ kJ}$ and $n = 1 \text{ mol}$:

$$E = -\frac{\Delta G}{nF} \approx -\frac{(-100 \times 10^3 \text{ J})}{(1 \text{ mol}) \times (96 \times 10^3 \text{ C})} \approx 1 \text{ V}$$

Most electrochemical cells bought commercially are in fact rated at between 1 and 2 V.

The Nernst equation

We saw in Chapter 3 that the reaction free energy varies with the composition of the reaction mixture, and that

$$\Delta G_m = \Delta G_m^\ominus + RT \ln Q$$

In this expression (which is eqn (24), in Chapter 3), $\Delta G_m^\ominus$ is the standard molar free energy of the reaction and Q is the reaction quotient for the cell reaction. As explained in Chapter 3, ΔG_m is simply the reaction free energy divided by 1 mol, so the last equation can be written as

$$\Delta G = \Delta G^\ominus + (1 \text{ mol}) \times RT \ln Q$$

if we multiply through by 1 mol. Next, we can express the reaction free energy by dividing through by $-nF$ and using eqn (2), which gives

$$E = E^\ominus - (1 \text{ mol}) \times \frac{RT}{nF} \ln Q$$

$$= E^\ominus - \frac{RT}{(n/\text{mol})F} \ln Q$$

where $E^\ominus$ is the **standard cell potential**:

$$-nFE^\ominus = \Delta G^\ominus \tag{3}$$

The expression n/mol, the numerical value of the amount of electrons transferred (so that if $n = 2 \text{ mol}$, $n/\text{mol} = 2$) is normally simplified to ν, and the final expression is called the **Nernst equation**:

$$E = E^\ominus - \frac{RT}{\nu F} \ln Q \tag{4}$$

Because $RT/F = 25.7 \text{ mV}$ at $25 \,°\text{C}$, a practical form of the Nernst equation at that temperature is

$$E = E^\ominus - \frac{25.7 \text{ mV}}{\nu} \ln Q$$

Hence, for a reaction in which $\nu = 1$, if Q is increased by a factor of 10, then the cell potential decreases by 59.2 mV.

Example Using the Nernst equation to calculate a cell potential

Calculate the zero-current cell potential of a Daniell cell at 25 °C that contains 1.0×10^{-3} M $CuSO_4(aq)$ and 3.0×10^{-3} M $ZnSO_4(aq)$. Use $E^{\ominus} = +1.102$ V.

Answer The cell reaction (the reduction of Cu^{2+} ions by zinc) is

$$Cu^{2+}(aq) + Zn(s) \rightarrow Cu(s) + Zn^{2+}(aq)$$

$$Q = \frac{a(Zn^{2+})}{a(Cu^{2+})} \qquad n = 2 \text{ mol}$$

It follows that, at the stated composition

$$Q = \frac{0.0030}{0.0010} = 3.0$$

Therefore, from the Nernst equation (in its practical form),

$$E = 1.102 \text{ V} - \frac{25.7 \times 10^{-3} \text{ V}}{2} \times \ln 3.00 = +1.09 \text{ V}$$

Exercise E6.7 Calculate the zero-current potential of the cell

$$Pt \mid H_2(g, 2.0 \text{ bar}) \mid HCl(aq, 0.10 \text{ M}) \mid Hg_2Cl_2(s) \mid Hg(l)$$

at 25 °C. Use $E^{\ominus} = +0.27$ V.

[+0.40 V]

Cells at equilibrium

A special case of the Nernst equation has great importance in chemistry. Suppose the reaction has reached equilibrium; then $Q = K$, where K is the equilibrium constant of the cell reaction. However, we have remarked that a chemical reaction at equilibrium cannot do work, and hence a cell in which the reaction is at equilibrium generates zero potential difference between the electrodes. Therefore, setting $E = 0$ and $Q = K$ in the Nernst equation gives

$$\ln K = \frac{vFE^{\ominus}}{RT} \qquad (5)$$

This very important equation lets us predict equilibrium constants from cell potentials. For example, because the standard potential of the zinc/copper cell is

$$Zn(s) \mid ZnSO_4(aq) \parallel CuSO_4(aq) \mid Cu(s) \qquad E^{\ominus} = +1.10 \text{ V}$$

it follows that

$$\ln K = \frac{2 \times (9.6485 \times 10^4 \text{ C mol}^{-1}) \times (1.10 \text{ V})}{(8.3145 \text{ J K}^{-1} \text{ mol}^{-1}) \times (298.15 \text{ K})} = 85.6$$

and therefore that

$$K = 1.5 \times 10^{37}$$

Hence, the displacement of copper by zinc goes virtually to completion in the sense that the ratio of concentrations of Zn^{2+} ions to Cu^{2+} ions at equilibrium is about 10^{37}. Note that if $E^{\ominus}$ is large and positive, then K is also large, and at equilibrium the cell reaction lies strongly in favour of products. The opposite is true if $E^{\ominus}$ is negative, for then K is less than 1 and the reactants are favoured at equilibrium.

Concentration cells

We can use the same procedure to derive an expression for the potential of an electrolyte concentration cell. Consider the cell

$$M \mid M^+(aq, L) \parallel M^+(aq, R) \mid M$$

where the solutions L and R have different concentrations. We shall use this cell as a model of the potential difference across a neurone cell membrane. The cell reaction is

$$M^+(aq, R) \rightarrow M^+(aq, L) \qquad Q = \frac{a_L}{a_R} \qquad n = 1 \text{ mol}$$

The standard potential of an electrolyte concentration cell is zero, because the cell cannot drive a current through a circuit when the two electrode compartments are identical, which is the case when they both have their standard concentrations. Therefore, because $E^{\ominus} = 0$ and $v = 1$, the cell potential is

$$E = -\frac{RT}{F} \ln \frac{a_L}{a_R}$$

If R is the more concentrated solution, $E > 0$. Physically, a positive potential arises because positive ions tend to be reduced, so withdrawing electrons from the electrode, and this process is dominant in the right-hand electrode compartment.

As we have indicated, one important example of a membrane system that resembles a concentration cell is a biological cell wall, which is more permeable to K^+ than to either Na^+ or Cl^-. The concentration of K^+ inside the cell is about 20 to 30 times that on the outside, and is maintained at that level by a specific pumping operation fuelled by ATP and governed by enzymes. If the system is approximately at equilibrium, the potential difference between the two sides is predicted to be

$$E \approx -25.7 \text{ mV} \times \ln \tfrac{1}{20} = 77 \text{ mV}$$

which is broadly correct.

The potential difference across a cell membrane plays a particularly interesting role in the transmission of nerve impulses. When a neurone is inactive, there is a high concentration of K^+ ions inside the cell and a high concentration of Na^+ ions outside. The potential difference across the cell wall is about 70 mV. When the cell is subjected to a pulse of about 20 mV, the structure of the membrane adjusts and it becomes permeable to Na^+ ions. As a result, there is a decrease in membrane potential as the Na^+ ions flood into the interior of the cell. The change in potential difference triggers the adjacent part of the cell membrane, and the pulse of collapsing potential passes along the nerve. Behind the pulse the sodium and potassium pumps restore the concentration difference ready for the next pulse.

6.3 Reduction potentials

Each electrode in a galvanic cell makes a characteristic contribution to the overall cell potential. Although it is not possible to measure the contribution of a single electrode, one electrode can be assigned a value zero and the others assigned values on that basis. The specially selected electrode is the **standard hydrogen electrode** (SHE):

$$Pt \mid H_2(g) \mid H^+(aq) \qquad E^\ominus = 0 \text{ at all temperatures}$$

The **standard reduction potential**, $E^\ominus(Ox/Red)$, of a couple Ox/Red is then measured by constructing a cell in which the couple of interest forms the right-hand electrode and the standard hydrogen electrode is used as the left-hand electrode. For example, the standard reduction potential of the Ag^+/Ag couple is the standard potential of the following cell:

$$Pt \mid H_2(g) \mid H^+(aq) \parallel Ag^+(aq) \mid Ag(s)$$

and is $+0.80$ V. Similarly, the standard reduction potential of the $AgCl/Ag, Cl^-$ couple is the standard potential of the cell

$$Pt \mid H_2(g) \mid H^+(aq) \parallel Cl^-(aq) \mid AgCl(s) \mid Ag(s)$$

and is $+0.22$ V. Table 6.1 lists a selection of standard reduction potentials; a longer list will be found in Appendix 2.

The standard potential of a cell in terms of reduction potentials

We can predict the standard potential of a cell formed from any two electrodes by taking the difference of their standard reduction potentials:

$$E^\ominus = E^\ominus_R - E^\ominus_L \tag{6}$$

where $E^\ominus_R$ is the standard reduction potential of the right-hand electrode and $E^\ominus_L$ is that of the left. This conclusion follows from the fact that a cell such as

$$Ag(s) \mid Ag^+(aq) \parallel Cl^-(aq) \mid AgCl(s) \mid Ag(s)$$

is equivalent to two cells joined back-to-back:

$$Ag(s) \mid Ag^+(aq) \parallel H^+(aq) \mid H_2(g) \mid Pt$$
$$Pt \mid H_2(g) \mid H^+(aq) \parallel Cl^-(aq) \mid AgCl(s) \mid Ag(s)$$

The overall potential of this composite cell, and therefore of the cell of interest, is

$$E^\ominus = E^\ominus(AgCl/Ag, Cl^-) - E^\ominus(Ag^+/Ag) = +0.58 \text{ V}$$

Table 6.1   Standard reduction potentials at 25 ℃

Reduction half-reaction			$E^{\ominus}/V$
Oxidizing agent		**Reducing agent**	
(Strongly oxidizing)			
F_2	$+2e^-$	$\rightarrow 2F^-$	$+2.87$
$S_2O_8^{2-}$	$+2e^-$	$\rightarrow 2SO_4^{2-}$	$+2.05$
Au^+	$+e^-$	$\rightarrow Au$	$+1.69$
Pb^{4+}	$+2e^-$	$\rightarrow Pb^{2+}$	$+1.67$
Ce^{4+}	$+e^-$	$\rightarrow Ce^{3+}$	$+1.61$
$MnO_4^- + 8H^+$	$+5e^-$	$\rightarrow Mn^{2+} + 4H_2O$	$+1.51$
Cl_2	$+2e^-$	$\rightarrow 2Cl^-$	$+1.36$
$Cr_2O_7^{2-} + 14H^+$	$+6e^-$	$\rightarrow 2Cr^{3+} + 7H_2O$	$+1.33$
$O_2 + 4H^+$	$+4e^-$	$\rightarrow 2H_2O$	$+1.23$
			$+0.81$ at pH $= 7$
Br_2	$+2e^-$	$\rightarrow 2Br^-$	$+1.09$
Ag^+	$+e^-$	$\rightarrow Ag$	$+0.80$
Hg_2^{2+}	$+2e^-$	$\rightarrow 2Hg$	$+0.79$
Fe^{3+}	$+e^-$	$\rightarrow Fe^{2+}$	$+0.77$
I_2	$+e^-$	$\rightarrow 2I^-$	$+0.54$
$O_2 + 2H_2O$	$+4e^-$	$\rightarrow 4OH^-$	$+0.40$
			$+0.81$ at pH $= 7$
Cu^{2+}	$+2e^-$	$\rightarrow Cu$	$+0.34$
$AgCl$	$+e^-$	$\rightarrow Ag + Cl^-$	$+0.22$
$2H^+$	$+2e^-$	$\rightarrow H_2$	0, by definition
Fe^{3+}	$+3e^-$	$\rightarrow Fe$	-0.04
$O_2 + H_2O$	$+2e^-$	$\rightarrow HO_2^- + OH^-$	-0.08
Pb^{2+}	$+2e^-$	$\rightarrow Pb$	-0.13
Sn^{2+}	$+2e^-$	$\rightarrow Sn$	-0.14
Fe^{2+}	$+2e^-$	$\rightarrow Fe$	-0.44
Zn^{2+}	$+2e^-$	$\rightarrow Zn$	-0.76
$2H_2O$	$+2e^-$	$\rightarrow H_2 + 2OH^-$	-0.83
			-0.42 at pH $= 7$
Al^{3+}	$+3e^-$	$\rightarrow Al$	-1.66
Mg^{2+}	$+2e^-$	$\rightarrow Mg$	-2.36

Table 6.1   (Continued)

Reduction half-reaction			$E^{\ominus}/V$
Oxidizing agent		**Reducing agent**	
Na^+	$+e^-$	$\rightarrow Na$	-2.71
Ca^{2+}	$+2e^-$	$\rightarrow Ca$	-2.87
K^+	$+e^-$	$\rightarrow K$	-2.93
Li^+	$+e^-$	$\rightarrow Li$	-3.05
		(*Strongly reducing*)	

For a more extensive table, see Appendix 2.

Because $\Delta G^{\ominus} = -nFE^{\ominus}$, it follows that if $E^{\ominus} > 0$ (as it is in this example), then the corresponding cell reaction is spontaneous in the direction written (in the sense that $K > 1$).

Example Identifying the spontaneous direction of a reaction
One of the reactions important in corrosion in an acidic environment is

$$Fe(s) + 2H^+(aq) + \tfrac{1}{2}O_2(g) \rightarrow Fe^{2+}(aq) + H_2O(l)$$

Does the equilibrium constant for this reaction favour the formation of $Fe^{2+}(aq)$?
Answer The two reduction half-reactions are
(a) $Fe^{2+}(aq) + 2e^- \rightarrow Fe(s)$ $\qquad\qquad E^{\ominus} = -0.44$ V
(b) $2H^+(aq) + \tfrac{1}{2}O_2(g) + 2e^- \rightarrow H_2O(l)$ $\quad E^{\ominus} = +1.23$ V
The difference (b) $-$ (a) is

$$Fe(s) + 2H^+(aq) + \tfrac{1}{2}O_2(g) \rightarrow Fe^{2+}(aq) + H_2O(l) \quad E^{\ominus} = +1.67\,V$$

Therefore, because $E^{\ominus} > 0$, it follows that $\Delta G^{\ominus} < 0$ and that the reaction has $K > 1$, favouring products.

Exercise E6.8 Can zinc displace copper from solution when the molar concentration of the ions is 1 mol L^{-1}?
[Yes: $Zn(s) + Cu^{2+}(aq) \rightarrow Zn^{2+}(aq) + Cu(s)$]

Example Calculating an equilibrium constant 1
Calculate the equilibrium constant for the disproportionation $2Cu^+(aq) \rightarrow Cu(s) + Cu^{2+}(aq)$ at 298 K.

Answer The cell is

$$Pt \mid Cu^+(aq), Cu^{2+}(aq) \parallel Cu^+(aq) \mid Cu(s)$$

and so the standard reduction potentials required are

R: $Cu(s) \mid Cu^+(aq)$ $Cu^+(aq) + e^- \rightarrow Cu(aq)$ $E^{\ominus} = +0.52$ V

L: $Pt \mid Cu^{2+}(aq), Cu^+(aq)$ $Cu^{2+}(aq) + e^- \rightarrow Cu^+(aq)$

$$E^{\ominus} = +0.15 \text{ V}$$

The standard cell potential is therefore

$$E^{\ominus} = +0.52 \text{ V} - 0.15 \text{ V} = +0.37 \text{ V}$$

Because $n = 1$ mol (so $\nu = 1$),

$$\ln K = \frac{+0.37 \text{ V}}{25.69 \times 10^{-3} \text{ V}} = 14.4$$

Hence, $K = 1.8 \times 10^6$. The equilibrium lies strongly towards the right of the reaction as written, and so Cu^+ disproportionates almost totally in solution.

Exercise E6.9 Calculate the equilibrium constant for the reaction $Sn^{2+}(aq) + Pb(s) \rightarrow Sn(s) + Pb^{2+}(aq)$ at 298 K.

[0.46]

Example Calculating an equilibrium constant 2

The reduced and oxidized form of riboflavin form a couple with $E^{\ominus} = -0.21$ V in a solution in which pH = 7 and the acetate/acetaldehyde couple has $E^{\ominus} = -0.60$ V under the same conditions. What is the equilibrium constant for the reduction of riboflavin by acetaldehyde in neutral solution at 25 °C? The reaction can be symbolized

$$RibO(aq) + CH_3CHO(aq) \rightleftharpoons Rib(aq) + CH_3COOH(aq)$$

where RibO is the oxidized form of riboflavin and Rib is the reduced form.

Answer The two reduction half-reactions are

$$RibO(aq) + 2H^+(aq) + 2e^- \rightarrow Rib(aq) + H_2O(l)$$

$$E^{\ominus} = -0.21 \text{ V}$$

$$CH_3COOH(aq) + 2H^+(aq) + 2e^- \rightarrow CH_3CHO(aq) + H_2O(l)$$

$$E^{\ominus} = -0.60 \text{ V}$$

and their difference is the redox reaction required. The corresponding cell potential is

$$E = -0.21 \text{ V} - (-0.60 \text{ V}) = +0.39 \text{ V}$$

and $n = 2$ mol (so $\nu = 2$). It follows that

$$\ln K = \frac{2 \times (9.6485 \times 10^4 \text{ C mol}^{-1}) \times (0.39 \text{ V})}{(8.3145 \text{ J K}^{-1} \text{ mol}^{-1}) \times (298.15 \text{ K})} = 30.4$$

225

so

$$K = 1.6 \times 10^{13}$$

We conclude that riboflavin can be reduced by acetaldehyde in neutral solution; however, there may be mechanistic reasons (the energy required to break covalent bonds) that make the reduction too slow to be feasible in practice.

Exercise E6.10 What is the equilibrium constant for the reduction of riboflavin with rubredoxin in the reaction Riboflavin(ox) + rubredoxin(red) ⇌ riboflavin(red) + rubredoxin(ox), given that at pH = 7 the reduction potential for rubredoxin is −0.06 V? Treat the reaction as a two-electron process and comment on your conclusions.
[8.5×10^{-6}; the reduction is not spontaneous]

The variation of potential with pH

The half-reactions of many redox couples involve hydrogen ions. For example, the reduction of fumaric acid, $HOOCCH\!=\!CHCOOH$, to succinic acid, $HOOCCH_2CH_2COOH$, is

$$HOOCCH\!=\!CHCOOH(aq) + 2H^+(aq) + 2e^-$$
$$\rightarrow HOOCCH_2CH_2COOH(aq)$$

Half-reactions of this kind have potentials that depend on the pH of the medium. In this example, where the hydrogen ions occur as reactants, an increase in pH (corresponding to a decrease in hydrogen ion concentration) favours the formation of reactants, so the fumaric acid has a lower thermodynamic tendency to become reduced: we expect, therefore, that the reduction potential of the fumaric/succinic acid couple should *decrease* as the pH is increased.

The quantitative variation of reduction potential with pH for a reaction can be established quite readily by using the Nernst equation for the half-reaction and

$$\ln a(H^+) = 2.303 \log a(H^+) = -2.303\, pH$$

If we suppose that the fumaric acid and succinic acid have their standard concentrations, with only the hydrogen ion concentration variable, then the reduction potential for the fumaric/succinic redox couple is

$$E^{\ominus\prime} = E^{\ominus} - \frac{RT}{2F} \ln Q \qquad Q = \frac{1}{a(H^+)^2}$$

which is easily rearranged into

$$E^{\ominus\prime} = E^{\ominus} - \frac{2.303RT}{F} \times pH$$

226

At 25 °C,

$$E^{\ominus\prime} = E^{\ominus} - (59.2\,\text{mV}) \times \text{pH}$$

We see that each increase in 1 unit of pH *decreases* the reduction potential by 59.2 mV, which is in agreement with the remark above, that the reduction of fumaric acid is discouraged by an increase in pH.

Arguments of this type can be used to convert standard potentials to **biological standard potentials**, $E^{\oplus}$, which correspond to neutral solution (pH = 7). If the hydrogen ions appear as reactants in the reduction half-reaction, then the reduction potential is decreased below its standard value (for the fumaric/succinic couple, by $7 \times 59.2\,\text{mV} = 414\,\text{mV}$, or about 0.4 V). If the hydrogen ions appear as products, then the biological standard potential is higher than the thermodynamic standard potential. The precise change depends on the number of electrons and protons in the half-reaction.

Example Converting a standard reduction potential to a biological standard value

What is the biological standard reduction potential of the NAD^+/NADH couple at 25 °C, where NAD^+ is the oxidized form of nicotinamide adenine dinucleotide and NADH is the reduced form of this species? The reduction half-reaction is

$$\text{NAD}^+(aq) + \text{H}^+(aq) + 2\text{e}^- \rightarrow \text{NADH}(aq) \qquad E^{\ominus} = -0.11\,\text{V}$$

Answer The Nernst equation for the half-reaction, taking the NAD^+ and NADH to have their standard concentrations, and noting that $n = 2\,\text{mol}$, is

$$E^{\ominus\prime} = E^{\ominus} - \frac{25.7\,\text{mV}}{2} \ln Q \qquad Q = \frac{1}{a(\text{H}^+)}$$

This expression rearranges readily to

$$E^{\ominus\prime} = E^{\ominus} - \frac{2.303 \times (25.7\,\text{mV})}{2} \times \text{pH}$$
$$= E^{\ominus} - 29.59\,\text{mV} \times \text{pH}$$

On substitution of the data, the biological standard potential (at pH = 7.0) is

$$E^{\oplus} = -0.11\,\text{V} - (29.59 \times 10^{-3}\,\text{V}) \times 7.0 = -0.32\,\text{V}$$

Exercise E6.11 Calculate the biological standard potential of the half-reaction $\text{O}_2(g) + 4\text{H}^+(aq) + 4\text{e}^- \rightarrow 2\text{H}_2\text{O}(l)$ at 25 °C given its value +1.23 V under thermodynamic standard conditions.

[+0.81 V]

The hydrogen electrode and pH

The potential of a hydrogen electrode is directly proportional to the pH of the solution. For example, for the cell

$$Hg(l) \mid Hg_2Cl_2(s) \mid Cl^-(aq) \parallel H^+(aq) \mid H_2(g) \mid Pt$$

in which the cell reaction is

$$Hg_2Cl_2(s) + 2H^+(aq) \rightarrow 2Hg(l) + 2Cl^-(aq) + H_2(g)$$

the Nernst equation gives

$$E = E^\ominus - \frac{RT}{2F} \ln Q \qquad Q = \frac{a(H_2)a(Cl^-)^2}{a(H^+)^2}$$

We shall suppose that the hydrogen pressure has its standard value of $p^\ominus$ (small variations in pressure make very little change to the cell potential). We also note that the concentration of Cl^- ions is constant (it depends on the composition of the calomel electrode, which is independent of that of the hydrogen electrode), and we find that

$$E = E^\ominus - \frac{RT}{2F} \ln a(Cl^-) - \frac{RT}{2F} \ln \frac{1}{a(H^+)^2}$$

$$= E' + \frac{RT}{F} \ln a(H^+)$$

where E' is a constant. It follows that

$$E = E' - \frac{2.303RT}{F} \times pH \tag{7}$$

and the pH of a solution can be measured by determining the potential of a cell in which a hydrogen electrode is one component.

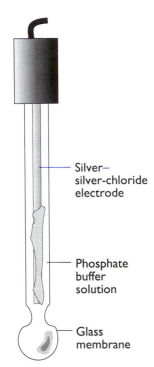

Fig. 6.9 A glass electrode has a potential that varies with the hydrogen-ion concentration in the medium in which it is immersed. It consists of a thin glass membrane containing an electrolyte and a silver–silver-chloride electrode. The electrode is used in conjunction with a calomel (Hg_2Cl_2) electrode that makes contact with the test solution through a salt bridge.

Silver–silver-chloride electrode

Phosphate buffer solution

Glass membrane

> **Exercise E6.12** What range should a voltmeter have (in volts) to display changes of pH from 1 to 14 at 25 °C if it is arranged to give a reading of 0 when pH = 7?
> [from −0.41 V to +0.36 V, a range of 0.77 V]

In practice, indirect methods are much more convenient to use than one based on the standard hydrogen electrode, and the hydrogen electrode is replaced by a **glass electrode**. This electrode (Fig. 6.9) is sensitive to hydrogen ion concentration, and has a potential proportional to pH. It is filled with a phosphate buffer containing Cl^- ions, and conveniently has $E = 0$ when the external medium is at pH = 7. The glass electrode is much more convenient to handle than the gas electrode itself, and can be calibrated using solutions of known pH (for example, one of the buffer solutions described in Section 5.3).

Finally, it should be noted that we now have a method for measuring the pK_a of an acid electrically. As explained in Section 5.5, the pH of a solution containing equal amounts of the acid and its conjugate base is $pH = pK_a$; but we now know how to determine pH electrochemically, and hence can determine pK_a in the same way.

Applications of reduction potentials

The measurement of zero-current cell potential is a convenient source of data on the free energies, enthalpies, and entropies of reactions. In practice the standard values (and the biological standards values) of these quantities are the ones normally determined.

6.4 The electrochemical series

We have seen that a cell reaction is spontaneous if it has a positive cell potential, $E^{\ominus} > 0$. We have also seen that $E^{\ominus}$ may be written as the difference of the reduction potentials of the redox couples in the right and left electrodes:

$$E^{\ominus} = E_R^{\ominus} - E_L^{\ominus}$$

The cell reaction is therefore spontaneous as written if $E_R^{\ominus} > E_L^{\ominus}$. Because the reduced species in the left-hand electrode compartment (the anode, the site of oxidation) reduces the oxidized species in the right-hand electrode compartment (the cathode, the site of reduction) in the cell reaction, we can conclude that

A species with a *low* reduction potential has a thermodynamic tendency to reduce a species with a *high* reduction potential.

More briefly: *low reduces high* and, equivalently, *high oxidizes low*. For example,

$$E^{\ominus}(Zn^{2+}, Zn) = -0.76\,V < E^{\ominus}(Cu^{2+}, Cu) = +0.34\,V$$

and $Zn(s)$ has a thermodynamic tendency to reduce $Cu^{2+}(aq)$ under standard conditions. Hence the reaction

$$Zn(s) + CuSO_4(aq) \rightarrow ZnSO_4(aq) + Cu(s)$$

can be expected to have $K > 1$ (in fact, as we have seen, $K = 1.5 \times 10^{37}$ at 298 K).

Exercise E6.13 Does acidified dichromate ($Cr_2O_7^{2-}$) have a thermodynamic tendency to oxidize mercury to mercury(I)?

[Yes]

Table 6.2 shows a part of the **activity series** of the metals, the redox couples of the metals arranged in the order of their reducing power. In the activity series,

The reduced member of a couple higher in the series (with lower standard reduction potential) can reduce the oxidized member of couples lower in the series.

This is a qualitative conclusion. The quantitative value of K is then obtained by doing the calculations we have introduced previously.

Table 6.2   The metal activity series

	Element	Reduced form	Oxidized form	
Most	Potassium	K	K^+	*Least*
strongly	Calcium	Ca	Ca^{2+}	*strongly*
reducing	Sodium	Na	Na^+	*oxidizing*
	Magnesium	Mg	Mg^{2+}	
	Aluminium	Al	Al^{3+}	
	Zinc	Zn	Zn^{2+}	
	Chromium	Cr	Cr^{2+}	
	Iron	Fe	Fe^{2+}	
	Nickel	Ni	Ni^{2+}	
	Tin	Sn	Sn^{2+}	
	Lead	Pb	Pb^{2+}	
	(Hydrogen)	H_2	H^+	
	Copper	Cu	Cu^{2+}	
	Mercury	Hg	Hg^{2+}	
Least	Silver	Ag	Ag^+	*Most*
strongly	Platinum	Pt	Pt^{2+}	*strongly*
reducing	Gold	Au	Au^+	*oxidizing*

Appendix 2 lists the standard reduction potentials of a wide variety of redox couples.

Example Using the activity series

Can zinc displace magnesium from aqueous solutions at 298 K?

Answer Displacement corresponds to a reduction of $Mg^{2+}(aq)$ by $Zn(s)$. This reduction is spontaneous if $E^{\ominus}(Zn^{2+}/Zn)$ is lower than $E^{\ominus}(Mg^{2+}/Mg)$ and hence if the couple Zn^{2+}/Zn is higher in the activity series than Mg^{2+}/Mg (as displayed in Table 6.2). From Table 6.1 we see that the Zn^{2+}/Zn couple lies below Mg^{2+}/Mg. Therefore, zinc cannot displace magnesium.

Exercise E6.14 Can lead displace (a) iron(II) ions, (b) copper(II) ions, from solution at 298 K?

[(a) No, (b) Yes]

6.5 Thermodynamic functions from cell-potential measurements

The standard cell potential is related to the standard reaction free energy by eqn (3),

$$\Delta G^{\ominus} = -nFE^{\ominus} \tag{8a}$$

Therefore, by measuring the standard potential of a cell driven by the reaction of interest we can obtain the standard reaction free energy. If we were interested in the biological standard state (for which pH = 7), then we would use the same expression, but with the standard potential at pH = 7:

$$\Delta G^{\oplus} = -nFE^{\oplus} \tag{8b}$$

Exercise E6.15 What is the standard free energy of the reaction $Ag^{+}(aq) + \frac{1}{2}H_2(g) \rightarrow H^{+}(aq) + Ag(s)$ given that the standard potential of the cell $H_2 \mid H^{+}(aq) \parallel Ag^{+}(aq) \mid Ag$ is $E^{\ominus} = +0.7996$ V?

[+77.15 kJ]

Example Calculating a standard reaction free energy

Calculate the biological standard free energy of the reaction

$$NADH(aq) + \frac{1}{2}O_2(g) + H^{+}(aq) \rightarrow NAD^{+}(aq) + H_2O(l)$$

which is involved in respiration.

231

Answer We have already calculated the biological standard potentials of the following two half-reactions:

$$NAD^+(aq) + H^+(aq) + 2e^- \rightarrow NADH(aq) \qquad E^\oplus = -0.32 \text{ V}$$

$$O_2(g) + 4H^+(aq) + 4e^- \rightarrow 2H_2O(l) \qquad E^\oplus = +0.81 \text{ V}$$

The latter can be written

$$\tfrac{1}{2}O_2(g) + 2H^+(aq) + 2e^- \rightarrow H_2O(l) \qquad E^\oplus = +0.81 \text{ V}$$

The standard potential is unchanged when the stoichiometric coefficients are multiplied by a factor because E is an intensive property (more explicitly, both ΔG and n in $\Delta G = -nFE$ change by the same amount, which implies that E remains unchanged). The difference between the third and first of these half-reactions gives the overall reaction we require, and shows that $n = 2$ mol. The overall cell potential is therefore

$$E^\oplus = 0.81 \text{ V} - (-0.32 \text{ V}) = +1.13 \text{ V}$$

It follows that the biological standard free energy of the reaction is

$$\Delta G^\oplus = -(2 \text{ mol}) \times (9.6485 \times 10^4 \text{ C mol}^{-1}) \times (1.13 \text{ V})$$

$$= -218 \text{ kJ}$$

(We have used $1 \text{ C} \times 1 \text{ V} = 1 \text{ J}$ and $1 \text{ kJ} = 10^3 \text{ J}$.) This energy can be used in metabolic processes to drive nonspontaneous reactions.

Exercise E6.16 The biological standard potential for the reduction of cytochrome c is

$$Cyt^{3+}(aq) + e^- \rightarrow Cyt^{2+}(aq) \qquad E^\oplus = +0.22 \text{ V}$$

Calculate the biological standard reaction free energy of the oxidation of cytochrome c in the reaction

$$4Cyt^{2+}(aq) + O_2(aq) + 4H^+(aq) \rightarrow 4Cyt^{3+}(aq) + 2H_2O(l)$$
$$[-228 \text{ kJ}]$$

The relation between the standard potential of a cell and the standard reaction free energy is a convenient route for the evaluation of a reduction potential from two others. For example, given that the standard reduction potentials of the Cu^{2+}/Cu and Cu^+/Cu couples are $E^\ominus(Cu^{2+}, Cu) = +0.340 \text{ V}$ and $E^\ominus(Cu^+, Cu) = +0.522 \text{ V}$, we can evaluate $E^\ominus(Cu^{2+}, Cu^+)$ by converting the $E^\ominus$ values to $\Delta G^\ominus$ values using eqn (8a), adding them appropriately, and converting the overall $\Delta G^\ominus$ so obtained to the required $E^\ominus$ using eqn (8a) again. The electrode reactions are as follows:

(a) $Cu^{2+}(aq) + 2E^- \rightarrow Cu(s)$ $E^\ominus = +0.340$ V

$\Delta G^\ominus(a) = -(2 \text{ mol}) \times 0.340 \text{ V} \times F = -0.680 \text{ V mol} \times F$

(b) $Cu^+(aq) + e^- \rightarrow Cu(s)$ $E^\ominus = +0.522$ V

$\Delta G^\ominus(b) = -(1 \text{ mol}) \times 0.522 \text{ V} \times F = -0.522 \text{ V mol} \times F$

The required reaction is

(c) $Cu^{2+}(aq) + e^- \rightarrow Cu^+(aq)$ $\Delta G^\ominus = -(1 \text{ mol}) \times F E^\ominus$

Because (c) = (a) − (b), it follows that

$$\Delta G^\ominus = \Delta G^\ominus(a) - \Delta G^\ominus(b)$$

Therefore, from eqn (8a)

$$E^\ominus = \frac{(-0.680 \text{ V mol}) \times F - (-0.522 \text{ V mol}) \times F}{(-1 \text{ mol}) \times F} = +0.16 \text{ V}$$

Note that we cannot combine the $E^\ominus$ values directly because they are not extensive properties, and we must always work via $\Delta G^\ominus$.

The entropy of the cell reaction can be obtained from the change in the cell potential with temperature using the relation

$$\Delta S^\ominus = nF \times \left(\frac{E^{\ominus\prime} - E^\ominus}{T' - T} \right) \tag{9}$$

where $E^{\ominus\prime}$ is the standard cell potential at a temperature T' and $E^\ominus$ is its value at a temperature T. Hence, we now have an electrochemical technique for obtaining standard reaction entropies of reactions that can be studied in a galvanic cell.

JUSTIFICATION

The thermodynamic relation on which eqn (9) is based is

change in $G = -S \times$ change in T

(at constant pressure)

or, more formally,

$dG = -S \, dT$ (at constant pressure)

This equation is derived from the definition $G = H - TS$ by changing the temperature infinitesimally. Because this equation applies to both the reactants and the products, it follows that

$$d(\Delta G^\ominus) = -\Delta S^\ominus \, dT$$

Substitution of $\Delta G^\ominus = -nFE^\ominus$ then gives

$$nF \, dE^\ominus = \Delta S^\ominus \, dT$$

This equation is exact, but it applies only to infinitesimal changes in the temperature. The equation may be integrated if we suppose that the reaction entropy is constant in the temperature range of interest:

$$nF \int_{E^\ominus}^{E^{\ominus\prime}} dE^\ominus = \Delta S \int_T^{T'} dT$$

Therefore,

$$nF \times (E^{\ominus\prime} - E^\ominus) = \Delta S^\ominus \times (T' - T)$$

which is easily rearranged into eqn (9).

Finally, we can combine the results obtained so far and use them to obtain the standard reaction enthalpy:

$$\Delta H^{\ominus} = \Delta G^{\ominus} + T\,\Delta S^{\ominus} \tag{10}$$

with $\Delta G^{\ominus}$ determined from the cell potential and $\Delta S^{\ominus}$ from its temperature variation. Thus, we now have a *noncalorimetric* method of measuring a reaction enthalpy.

Example Using the temperature coefficient of the cell potential

The standard potential of the cell

$$\text{Pt} \mid \text{H}_2(g) \mid \text{HCl}(aq) \mid \text{Hg}_2\text{Cl}_2(s) \mid \text{Hg}(l)$$

was found to be $+0.2699$ V at 293 K and $+0.2669$ V at 303 K. Evaluate the standard free energy, enthalpy, and entropy at 298 K of the reaction $\text{Hg}_2\text{Cl}_2(s) + \text{H}_2(g) \rightarrow 2\text{Hg}(l) + 2\text{HCl}(aq)$.

Answer We find $\Delta G^{\ominus}$ from eqn (8a) at 298 K with $n = 2$ mol by linear interpolation between the two temperatures (in this case, we take the mean $E^{\ominus}$ because 298 K lies midway between 293 K and 303 K). Because the mean standard cell potential is $+0.2684$ V,

$$\begin{aligned}
\Delta G^{\ominus} &= -nFE^{\ominus} \\
&= -(2\,\text{mol}) \times (9.6485 \times 10^4\,\text{C mol}^{-1}) \times (+0.2684\,\text{V}) \\
&= -51.79\,\text{kJ}
\end{aligned}$$

From eqn (9), the standard entropy of the reaction is

$$\begin{aligned}
\Delta S^{\ominus} &= (2\,\text{mol}) \times (9.6485 \times 10^4\,\text{C mol}^{-1}) \\
&\quad \times \left(\frac{0.2669\,\text{V} - 0.2699\,\text{V}}{293\,\text{K} - 303\,\text{K}}\right) \\
&= +58\,\text{J K}^{-1}, \text{ or } +5.8 \times 10^{-2}\,\text{kJ K}^{-1}
\end{aligned}$$

We now use eqn (10) to form

$$\begin{aligned}
\Delta H^{\ominus} &= \Delta G^{\ominus} + T\,\Delta S^{\ominus} \\
&= -51.79\,\text{kJ} + (298.15\,\text{K}) \times (5.8 \times 10^{-2}\,\text{kJ K}^{-1}) = -35\,\text{kJ}
\end{aligned}$$

One difficulty with this procedure lies in the accurate measurement of small temperature coefficients of cell potential. Nevertheless, it is another example of the striking ability of thermodynamics to relate the apparently unrelated, in this case to relate electrical measurements to thermal properties.

Exercise E6.17 Predict the standard potential of the *Harned cell* at 303 K from tables of thermodynamic data. The Harned cell is $\text{Pt} \mid \text{H}_2(g) \mid \text{HCl}(aq) \mid \text{AgCl}(s) \mid \text{Ag}(s)$.

[0.2191 V]

EXERCISES

6.1 Calculate the solubility of mercury(II) chloride at 25 °C from standard free energies of formation.

6.2 Consider a hydrogen electrode in aqueous HBr solution at 25 °C operating at 1.15 atm. Calculate the change in the electrode potential when the solution is changed from 5.0 mmol L^{-1} to 20.0 mmol L^{-1}.

6.3 Devise a cell in which the cell reaction is $Mn(s) + Cl_2(g) \rightarrow MnCl_2(aq)$. Give the half-reactions for the electrodes and from the standard cell potential of 2.54 V deduce the standard potential of the Mn^{2+}/Mn couple.

6.4 Write the cell reactions and electrode half-reactions for the following cells:
(a) $Zn \mid ZnSO_4(aq) \parallel AgNO_3(aq) \mid Ag$
(b) $Cd \mid CdCl_2(aq) \parallel HNO_3(aq) \mid H_2(g) \mid Pt$
(c) $Pt \mid K_3[Fe(CN)_6](aq), K_4[Fe(CN)_6](aq) \parallel$
   $CrCl_3(aq) \mid Cr$
(d) $Pt \mid Cl_2(g) \mid HCl(aq) \parallel$
   $K_2CrO_4(aq) \mid Ag_2CrO_4(s) \mid Ag$
(e) $Pt \mid Fe^{3+}(aq), Fe^{2+}(aq) \parallel$
   $Sn^{4+}(aq), Sn^{2+}(aq) \mid Pt$
(f) $Cu \mid Cu^{2+}(aq) \parallel$
   $Mn^{2+}(aq), H^+(aq) \mid MnO_2(s) \mid Pt$

6.5 Devise cells in which the following are the reactions;
(a) $Zn(s) + CuSO_4(aq) \rightarrow ZnSO_4(aq) + Cu(s)$
(b) $2AgCl(s) + H_2(g) \rightarrow 2HCl(aq) + 2Ag(s)$
(c) $2H_2(g) + O_2(g) \rightarrow 2H_2O(l)$
(d) $H_2(g) + I_2(s) \rightarrow 2HI(aq)$

6.6 Use standard reduction potentials to calculate the standard potentials of the cells in Exercises 6.4 and 6.5.

6.7 (a) Calculate the standard cell potential of $Hg(l) \mid HgCl_2(aq) \parallel TlNO_3(aq) \mid Tl$ at 25 °C. (b) Calculate the cell potential when the molar concentration of the Hg^{2+} ion is 0.150 mol L^{-1} and that of the Tl^+ ion is 0.93 mol L^{-1}.

6.8 Calculate the standard free energies at 25 °C of the following reactions from the reduction potential data in Appendix 2.
(a) $2Na(s) + 2H_2O(l) \rightarrow 2NaOH(aq) + H_2(g)$
(b) $2K(s) + 2H_2O(l) \rightarrow 2KOH(aq) + H_2(g)$
(c) $K_2S_2O_8(aq) + 2KI(aq) \rightarrow I_2(s) + 2K_2SO_4(aq)$
(d) $Pb(s) + ZnCO_3(aq) \rightarrow PbCO_3(aq) + Zn(s)$

6.9 The standard free energy function of the reaction

$K_2CrO_4(aq) + 2Ag(s) + FeCl_3(aq)$
$\qquad \rightarrow Ag_2CrO_4(s) + 2FeCl_2(aq) + 2KCl(aq)$

is -62.5 kJ mol^{-1} at 298 K. (a) Calculate the standard potential of the corresponding galvanic cell and (b) the standard reduction potential of the $Ag_2CrO_4/Ag, CrO_4^{2-}$ couple.

6.10 Estimate the potential of the cell

$Ag \mid AgBr(s) \mid KBr(aq, 0.050 \text{ mol kg}^{-1}) \parallel$
$\qquad\qquad Cd(NO_3)_2(aq, 0.010 \text{ mol kg}^{-1}) \mid Cd$

at 25 °C.

6.11 Use the information in Appendix 2 to calculate the standard potential of the cell $Ag \mid AgNO_3(aq) \parallel Fe(NO_3)_2(aq) \mid Fe$ and the standard free energy and enthalpy of the cell reaction at 25 °C. Estimate the value of $\Delta G^{\ominus}$ at 35 °C.

6.12 The solubility constant of $Cu_3(PO_4)_2$ is 1.3×10^{-37}. Calculate (a) the solubility of $Cu_3(PO_4)_2$, (b) the potential of the cell

$Pt \mid H_2(g) \mid HCl(aq, pH = 0) \parallel$
$\qquad\qquad Cu_3(PO_4)_2(aq, \text{ sat}) \mid Cu$

at 25 °C.

6.13 Calculate the equilibrium constants of the following reactions at 25 °C from the reduction potential data:
(a) $Sn(s) + Sn^{4+}(aq) \rightleftharpoons 2Sn^{2+}(aq)$
(b) $Sn(s) + 2AgCl(s) \rightleftharpoons SnCl_2(aq) + 2Ag(s)$
(c) $2Ag(s) + Cu(NO_3)_2(aq) \rightleftharpoons Cu(s) + 2AgNO_3(aq)$
(d) $Sn(s) + CuSO_4(aq) \rightleftharpoons Cu(s) + SnSO_4(aq)$
(e) $Cu^{2+}(aq) + Cu(s) \rightleftharpoons 2Cu^{2+}(aq)$

6.14 The solubilities of AgCl and BaSO$_4$ in water are 1.34×10^{-5} mol L^{-1} and 9.51×10^{-4} mol L^{-1} respectively at 25 °C. Calculate their solubility constants from the appropriate standard reduction potentials.

6.15 Derive an expression for the potential of an electrode for which the half-reaction is the reduction of $Cr_2O_7^{2-}$ ions to Cr^{3+} ions in acidic solution.

6.16 The zero-current potential of the cell $Pt \mid H_2(g) \mid HCl(aq) \mid AgCl(s) \mid Ag$ was 0.322 V at 25 °C. What is the pH of the electrolyte solution?

6.17 The solubility of AgBr is $2.6\ \mu mol\ L^{-1}$ at 25 °C. What is the zero-current potential of the cell $Ag \mid AgBr(aq) \mid AgBr(s) \mid Ag$ at that temperature?

6.18 The standard potential of the cell $Ag \mid AgI(s) \mid AgI(aq) \mid Ag$ is 0.9509 V at 25 °C. Calculate (a) the solubility of AgI and (b) its solubility constant.

6.19 Devise a cell in which the overall reaction is

$$Pb(s) + Hg_2SO_4(s) \rightarrow PbSO_4(s) + 2Hg(l)$$

What is its potential when the electrolyte is saturated with both salts at 25 °C?

6.20 A fuel cell develops an electric potential from the chemical reaction between reagents supplied from an outside source. What is the zero-current potential of a cell fuelled by (a) hydrogen and oxygen, (b) the combustion of butane at 1.0 atm and 298 K?

6.21 What is the standard reduction potential of the MnO_4^-, H^+/Mn^{2+} couple at pH = 7.00?

6.22 The biological standard reduction potential of the redox couple pyruvic acid/lactic acid is −0.19 V and that of fumaric acid/succinic acid is +0.03 V at 25 °C. What is the equilibrium constant for the reaction $P + S \rightleftharpoons L + F$ in pH = 7 (where P = pyruvic acid, S = succinic acid, L = lactic acid, and F = fumaric acid)?

6.23 The biological standard reduction potential of the couple pyruvic acid/lactic acid is −0.19 V. What is the thermodynamic standard reduction potential of the couple? Pyruvic acid is $CH_3COCOOH$ and lactic acid is $CH_3CH(OH)COOH$.

7

Chemical kinetics

The branch of physical chemistry called **chemical kinetics** is concerned with the rates of chemical reactions: how rapidly reactants are consumed and products formed, how the rate responds to changes in the conditions or the presence of a catalyst (including enzymes), and the steps by which the reaction takes place. One reason for studying the rates of reactions is the practical importance of being able to predict how quickly a reaction mixture approaches equilibrium. The rate might depend on variables under our control, such as the pressure, the temperature, and the presence of a catalyst, and we might be able to optimize it by the appropriate choice of conditions. Another reason is that the study of reaction rates leads to an understanding of the **mechanism** of a reaction, its analysis into a sequence of elementary steps. For example, we might discover that the reaction of hydrogen and bromine to form hydrogen bromide proceeds by the dissociation of a Br_2 molecule, the attack of a Br atom on an H_2 molecule, and several subsequent steps, and not by a single event in which an H_2 molecule encounters a Br_2 molecule and the atoms exchange partners to form two HBr molecules. The analysis of the effect of an enzyme on the rate of the reaction it catalyses is one of the principal means by which its mode of action may be determined. **Enzyme kinetics**, the study of the effect of enzymes on the rates of biochemically significant reactions, is also an important window on the manner in which the action of enzymes is inhibited, such as by poisons or (from a human perspective) pharmaceuticals.

Empirical chemical kinetics

The first stage in the kinetic analysis of reactions is the determination of the stoichiometry of the reaction and the identification of any side reactions. Once those features have been established, the basic data of chemical kinetics are the concentrations of the reactants and products at different times after a reaction has been initiated. Because the rates of chemical reactions are generally sensitive to the temperature (for reasons we explore later), the temperature of the reaction mixture must be held constant throughout the course of the reaction, for otherwise the observed rate would be a meaningless average of rates at different temperatures (indeed, one of the reasons why we fight infection with a fever is to upset the balance of reaction rates in the infecting organism by the increase in temperature). The requirement to maintain isothermal conditions in laboratory studies of reaction rates puts severe demands on the design of experiments. Gas-phase reactions, for instance, are often carried out in a vessel held in contact with a substantial block of metal. Liquid-phase reactions, including those taking place in flow conditions, must be carried out in an efficient thermostat.

7.1 Experimental techniques

The method selected to monitor the concentrations of reactants and products and their variation with time depends on the substances involved and the rapidity with which they change. Many reactions reach thermodynamic equilibrium over periods of minutes or hours but some reactions reach equilibrium in fractions of a second. Under special conditions, modern techniques are capable of studying reactions that are complete within a few femtoseconds ($1 \, \text{fs} = 10^{-15} \, \text{s}$).

Monitoring the progress of a reaction

A reaction in which at least one component is a gas might result in an overall change in pressure in a container that has a constant volume, and so its progress may be followed by recording the variation of pressure with time. An example is the decomposition of nitrogen(V) oxide,

$$2N_2O_5(g) \rightarrow 4NO_2(g) + O_2(g)$$

For each mole of N_2O_5 molecules destroyed, $\frac{5}{2}$ mol of gas molecules is formed, and so the total pressure increases as the reaction proceeds (if the volume and temperature are constant). A disadvantage of this method is that it is not specific: all the gas-phase particles contribute to the pressure.

Example Monitoring the variation in pressure
Predict how the total pressure varies during the gas-phase decomposition of N_2O_5.
Answer Let the initial pressure be p_0 and the initial amount of N_2O_5 molecules present be n. When a fraction α of the N_2O_5 molecules has decomposed, the amount remaining is $(1 - \alpha)n$. Similarly, if αn of N_2O_5 decomposes, it follows from the reaction stoichiometry that the amount of NO_2 produced will be twice that amount, or $2\alpha n$, and that the amount of O_2 present will be $\frac{1}{2}\alpha n$ (because $\frac{1}{2}$ mol O_2 is formed for 1 mol

N_2O_5 consumed). That is, at an arbitrary stage of the reaction, the amounts of the components in the reaction mixture are:

	N_2O_5	NO_2	O_2	Total
Amount	$n(1-\alpha)$	$2\alpha n$	$\frac{1}{2}\alpha n$	$n(1+\frac{3}{2}\alpha)$

The total pressure (at constant volume and temperature, and assuming perfect gas behaviour) is proportional to the number of gas-phase molecules. When $\alpha = 0$ the pressure is p_0, and so at any later stage the total pressure is

$$p = (1 + \tfrac{3}{2}\alpha)p_0$$

(For example, when the reaction is complete, $\alpha = 1$ and the pressure will have risen to $\frac{5}{2}$ times its initial value.) If the initial amount of N_2O_5 is known, then the amounts of N_2O_5, NO_2, and O_2 can be calculated from

$$\alpha = \frac{2}{3}\left(\frac{p}{p_0} - 1\right)$$

and the expressions in the table above.

Exercise E7.1 Repeat the calculation for the decomposition $2NOBr(g) \rightarrow 2NO(g) + Br_2(g)$.

$$[p = (1 + \tfrac{1}{2}\alpha)p_0]$$

Spectrophotometry, the measurement of the intensity of absorption in a particular spectral region, is widely applicable, and is especially useful when one substance (and only one) in the reaction mixture has a strong characteristic absorption in a conveniently accessible region of the spectrum. For example, the reaction

$$H_2(g) + Br_2(g) \rightarrow 2HBr(g)$$

can be followed by measuring the absorption of visible light by bromine.

If a reaction changes the number or type of ions present in a solution, then it may be followed by monitoring the conductivity of the solution. Reactions that change the concentration of hydrogen ions may be studied by monitoring the pH of the solution with a glass electrode. Other methods of determining composition include titration, mass spectrometry, gas chromatography, and magnetic resonance (Chapter 11). Polarimetry, the observation of the optical activity of a reaction mixture, is occasionally applicable.

Application of the techniques

In a **real-time analysis** the composition of a system is analysed while the reaction is in progress, either by direct spectroscopic observation of the reaction mixture or by withdrawing a small sample and analysing it. In the **quenching method**, the reaction is stopped after it

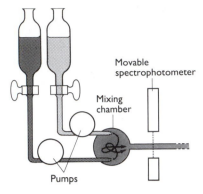

Fig. 7.1 The arrangement used in the flow technique for studying reaction rates. The reactants are pumped into the mixing chamber at a steady rate by the peristaltic pumps (pumps that squeeze the fluid through flexible tubes). The location of the spectrometer corresponds to different times after initiation.

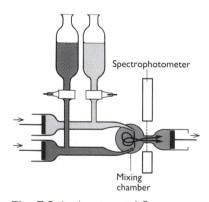

Fig. 7.2 In the stopped-flow technique the reagents are driven quickly into the mixing chamber and then the time dependence of the concentrations is monitored.

has been allowed to proceed for a certain time, and the composition is analysed at leisure. The quenching (of the entire mixture or of a sample drawn from it) can be achieved either by cooling suddenly, by adding the mixture to a large amount of solvent, or by rapid neutralization of an acid reagent. This method is suitable only for reactions that are slow enough for there to be little reaction during the time it takes to quench the mixture.

In the **flow method**, the reactants are mixed as they flow together in a chamber (Fig. 7.1). The reaction continues as the thoroughly mixed solutions flow through the outlet tube, and different points along the tube correspond to different times after the start of the reaction. Therefore, spectroscopic observation of the composition at different positions along the tube is equivalent to the observation of the composition of the reaction mixture at different times after mixing. The disadvantage of conventional flow techniques is that a large volume of reactant solution is necessary because the mixture must flow continuously through the apparatus. This disadvantage is particularly important for reactions that take place very rapidly, because to spread the reaction over an appreciable length of tube the flow must be rapid. The **stopped-flow technique** (Fig. 7.2) avoids this disadvantage. The two solutions are mixed very rapidly by injecting them into a mixing chamber designed to ensure that the flow is turbulent and that complete mixing occurs very rapidly. Behind the reaction chamber there is an observation cell fitted with a plunger that moves back as the liquids flood in, but which comes up against a stop after a certain volume has been admitted. The filling of that chamber corresponds to the sudden creation of an initial sample of the reaction mixture. The reaction then continues in the thoroughly mixed solution and is monitored spectrophotometrically. Because only a small, single charge of the reaction chamber is prepared, the technique is much more economical than the flow method. The suitability of the stopped-flow technique to the study of small samples means that it is appropriate for biochemical reactions, and it has been widely used to study the kinetics of enzyme action.

In **flash photolysis**, the gaseous or liquid sample is exposed to a brief photolytic flash of light, and then the contents of the reaction chamber are monitored spectrophotometrically. Although discharge lamps can be used for flashes of about 10^{-5} s duration, most work is now done with lasers, which can be used to generate nanosecond flashes routinely, picosecond flashes quite readily, and flashes as brief as a few femtoseconds in special arrangements. Both emission and absorption spectroscopy may be used to monitor the reaction, and the spectra are observed electronically or photographically at a series of times following the flash.

7.2 The rates of reactions

The raw data obtained from experiments used to measure reaction rates are information on the variation of the concentrations of reactants, intermediates, and products and how the reaction rates vary with the temperature of the reaction mixture. The next few sections look at these observations in more detail.

The definition of rate

The rate of a reaction is defined in terms of the rate of change of the concentration of a designated species. However, because the rate at which reactants are consumed and products are formed changes in the course of a reaction, it is necessary to consider the **instantaneous rate** of the reaction, its rate at a specific instant. The instantaneous rate of consumption of a reactant is the slope of a graph of its molar concentration plotted against the time, with the slope evaluated at the instant of interest (Fig. 7.3). The steeper the slope, the greater the rate of consumption of the reactant. Similarly, the rate of formation of a product is the slope of the graph of its concentration plotted against time. With the concentration measured in moles per litre and the time in seconds, the reaction rate is reported in moles per litre per second $(\text{mol L}^{-1}\text{s}^{-1})$.

CALCULUS

The formal definition of the instantaneous rate is expressed in terms of the slope, dy/dx, of a graph of concentration versus time at a specified time after the start of the reaction:

$$\text{rate of formation of a product P} = \frac{d[P]}{dt}$$

$$\text{rate of consumption of a reactant R} = -\frac{d[R]}{dt}$$

The minus sign is included in the definition of rate in terms of the reactant because the concentration of R decreases. Then the slope of its graph is negative, and the negative sign then ensures that the rate of consumption is a positive quantity.

In general, the various reactants in a reaction are consumed at different rates, and the various products are also formed at different rates. For example, in a reaction of the form

$$N_2(g) + 3H_2(g) \rightarrow 2NH_3(g)$$

the rate of formation of NH_3 is twice the rate of disappearance of N_2 (because for every mole of N_2 consumed, *two* moles of NH_3 are formed), and the H_2 disappears three times as rapidly as the N_2 (because 3 mol H_2 are consumed for 1 mol N_2). It follows that, to report the reaction rate unambiguously, we must specify the species to which the rate refers. Once the rate of formation or consumption of one substance is known, the reaction stoichiometry can be used to deduce the rates of formation or consumption of the other participants in the reaction.

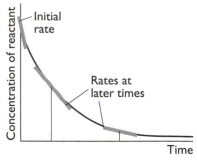

Fig. 7.3 The rate of a chemical reaction is the slope of the tangent to the curve showing the variation of concentration of a species with time. This graph is a plot of the concentration of a reactant, which is consumed as the reaction progresses. The rate of consumption decreases in the course of the reaction as the concentration of reactant decreases.

Example Reporting rates of reaction

The rate of formation of $NO(g)$ in the reaction

$$2NOBr(g) \rightarrow 2NO(g) + Br_2(g)$$

was reported as $1.6 \times 10^{-4} \, \text{mol L}^{-1} \text{s}^{-1}$. What is the rate of consumption of NOBr and the rate of formation of Br_2?

Answer Because 1 mol NOBr is consumed by 1 mol NO, the rate of consumption of NOBr is the same as the rate of formation of NO. Because $\frac{1}{2}$ mol Br_2 is formed when 1 mol NOBr is consumed, the rate of formation of Br_2 is only half that of the rate of consumption of NO, and is therefore $8.0 \times 10^{-5} \, \text{mol L}^{-1} \text{s}^{-1}$.

Exercise E7.2 The rate of consumption of CH_3 radicals in the reaction $2CH_3(g) \rightarrow CH_3CH_3(g)$ was reported as $1.2 \, \text{mol L}^{-1} \text{s}^{-1}$ under particular conditions. What is the rate of formation of CH_3CH_3?

$$[0.60 \, \text{mol L}^{-1} \text{s}^{-1}]$$

Rate laws and rate constants

An empirical observation of the greatest importance is that *the measured rate of reaction is often found to be proportional to the molar concentrations of the reactants raised to a power*. For example, it may be found that the rate is directly proportional to the concentration of the reactant A and to the concentration of another reactant B, so

$$\text{rate of reaction} = k[A][B] \qquad (1)$$

The coefficient k, which is characteristic of the reaction being studied, is called the **rate constant**. It is independent of the concentrations but depends on the temperature. An *experimentally determined* equation of this kind is called the 'rate law' of the reaction. More formally:

A **rate law** is an equation that expresses the rate of reaction as a function of the concentrations of the species in the overall reaction (including the products).

The units of k are always such as to convert the product of concentrations into a rate expressed as a change in concentration per unit time. For example, if the rate law is the one shown in eqn (1), with molar concentrations expressed in moles per litre, then the units of k are $mol^{-1} L s^{-1}$, because

$$(k, \text{ in } mol^{-1} L s^{-1}) \times ([A], \text{ in } mol L^{-1}) \times ([B], \text{ in } mol L^{-1})$$
$$= \text{rate, in } mol L^{-1} s^{-1}$$

Rate laws have three main applications. Once we know the rate law and the rate constant we can predict the rate of reaction from the composition of the mixture. We shall also see that we can use a rate law to predict the concentrations of the reactants and products. Moreover, a rate law is a guide to the mechanism of the reaction, because any proposed mechanism must be consistent with the observed rate law.

Reaction order

A rate law provides a basis for the classification of reactions according to their kinetics. The advantage of having such a classification is that reactions belonging to the same class will have similar kinetic behaviour—their rates will vary with composition in a similar way. The classification of reactions is based on their 'order':

> The **order** of a reaction with respect to each species is the power to which the concentration of that species is raised in the rate law.

For example, a reaction with the rate law

$$\text{rate} = k[A][B] \tag{2}$$

is *first order* in A and first order in B. A reaction with the rate law

$$\text{rate} = k[A]^2 \tag{3}$$

is *second order* in A.

The **overall order** of a reaction is the sum of the orders of all the components. These two rate laws both correspond to reactions that are *second order* overall. Examples of the two different types of reaction are the reaction between persulfate and iodide ions,

$$S_2O_8^{2-}(aq) + 3I^-(aq) \rightarrow 2SO_4^{2-}(aq) + I_3^-(aq) \qquad \text{rate} = k[S_2O_8^{2-}][I^-]$$

which is first order in $S_2O_8^{2-}$ ions, first order in I^- ions, and second order overall; and the reduction of nitrogen dioxide by carbon monoxide,

$$NO_2(g) + CO(g) \rightarrow NO(g) + CO_2(g) \qquad \text{rate} = k[NO_2]^2$$

which is second order in NO_2 and, because no other species occurs in the rate law, second order overall. In the latter reaction, the rate is independent of the concentration of CO in the sense that, so long as

243

some CO is present, then the rate is independent of the precise concentration. This independence of concentration is expressed by saying that the reaction is *zero order* in CO, because a concentration raised to the power zero is 1 ($[CO]^0 = 1$, just as $x^0 = 1$ in algebra).

Exercise E7.3 The reaction between the amino acid tyrosine and iodine obeys the rate law: rate $= k[Tyr][I_2]$. Classify it by order.
[First order in Tyr, first order in I_2, and second order overall]

A reaction need not have an integral order, and many gas-phase reactions do not. For example, if a reaction is found to have the rate law

$$\text{rate} = k[A]^{1/2}[B]$$

then it is *half order* in A, first order in B, and *three-halves* order overall. If a rate law is not of the form $[A]^x[B]^y[C]^z \cdots$ then the reaction does not have an order. Thus, the experimentally determined rate law for the gas-phase reaction $H_2 + Br_2 \rightarrow 2HBr$ is

$$\text{rate of formation of HBr} = \frac{k[H_2][Br_2]^{3/2}}{[Br_2] + k'[HBr]} \quad (4)$$

Although the reaction is first order in H_2, it has an indefinite order with respect to both Br_2 and HBr, and overall. Similarly, a typical rate law for the action of an enzyme E on a substrate S is

$$\text{rate of formation of product} = \frac{k[E][S]}{[S] + K_M} \quad (5)$$

where K_M is a constant.

Under certain circumstances a complex rate law without an overall order may simplify into a law with a definite order. For example, if the concentration of Br_2 is so high that $[Br_2] \gg k'[HBr]$, then the denominator in eqn (4) is equal to $[Br_2]$ to a good approximation, and the rate law simplifies to

$$\text{rate of formation of HBr} = k[H_2][Br_2]^{1/2}$$

which is first order in H_2, half order in Br_2, and three-halves order overall. Likewise, if the substrate concentration in the enzyme-catalysed reaction is so low that $[S] \ll K$, then eqn (5) simplifies to

$$\text{rate of formation of product} = k[S][E]$$

which is first order in S, first order in E, and second order overall.

Rate laws may also be expressed in terms of quantities that are proportional to molar concentration. Among the common of these alternatives is the use of the partial pressure of a gas-phase species. For example, a gas-phase reaction might have a rate law expressed in the form

$$\text{rate of formation of } X = kp_A p_B^2$$

where the rate of formation of X would be interpreted as the rate of change of the partial pressure of X. Most of the time, though, we shall express rate laws in terms of molar concentrations.

It is very important to note that *a rate law is established experimentally, and cannot in general be inferred from the reaction equation*. The reaction of hydrogen and bromine, for example, has a very simple stoichiometry, but its rate law (eqn (4)) is very complicated. Similarly, the thermal decomposition of nitrogen(V) oxide

$$2N_2O_5(g) \rightarrow 4NO_2(g) + O_2(g)$$

has the rate law

$$\text{rate of consumption of } N_2O_5 = k[N_2O_5]$$

and the reaction is first order. In some cases, however, the rate law does happen to reflect the reaction stoichiometry. This is the case with the oxidation of nitrogen oxide, NO, which, under certain conditions, is found to have a third-order rate law:

$$2NO(g) + O_2(g) \rightarrow 2NO_2(g)$$

$$\text{rate of formation of } NO_2 = k[NO]^2[O_2]$$

Some reactions obey a zero-order rate law, with a rate that is independent of the concentration of the reactant (so long as some is present). Thus, the catalytic decomposition of phosphane, PH_3, on hot tungsten at high pressures, is found to obey the rate law

$$\text{rate of decomposition of } PH_3 = k, \text{ independent of concentration}$$

The PH_3 decomposes at a constant rate until it has entirely disappeared, when the reaction stops abruptly.

The determination of the rate law

The determination of a rate law is simplified by the **isolation method**, in which it is ensured that the concentrations of all the reactants except one are present in large excess. If a reactant B is in large excess, for example, it is a good approximation to take its concentration as constant throughout the reaction. Then, although the true rate law might be

$$\text{rate} = k[A][B]^2$$

we can approximate [B] by its initial value $[B]_0$ (which barely changes in the course of the reaction) and write

$$\text{rate} = k'[A] \quad \text{with} \quad k' = k[B]_0^2$$

which has the form of a first-order rate law. Because the true rate law has been forced into first-order form by assuming a constant B concentration, the effective rate law is classified as **pseudo-first order**. If, instead, the concentration of A were in large excess, and virtually constant, then the effective rate law would be

$$\text{rate} = k''[\text{B}]^2 \quad \text{with} \quad k'' = k[\text{A}]_0$$

This **pseudo-second-order** rate law is also much easier to analyse and identify than the complete law. Many reactions in aqueous solution that are reported as first or second order are actually pseudo-first or pseudo-second order, because the water participates in the reaction but is in such large excess that its concentration remains constant. In general, the dependence of the rate on all the reactants may be found by isolating them in turn (by having all the other substances present in large excess), and piecing together a picture of the overall rate law.

In the method of **initial rates**, which is often used in conjunction with the isolation method, the instantaneous rate is measured at the beginning of the reaction for several different initial concentrations of reactants. For example, suppose the rate law for a reaction with A isolated is

$$r = k[\text{A}]^a$$

where r denotes rate. Then the initial rate of the reaction, r_0, is given by the initial concentration of A:

$$r_0 = k[\text{A}]_0^a$$

Taking logarithms gives

$$\log r_0 = \log k + a \log [\text{A}]_0 \qquad (6)$$

It follows that, for a series of initial concentrations, a plot of the logarithms of the initial rates against the logarithms of the initial concentrations of A should be a straight line, and that the slope of the graph will be the order a.

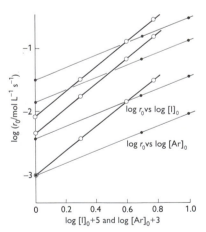

Fig. 7.4 The plot of log r_0 against log $[\text{I}]_0$ for a given $[\text{Ar}]_0$ and against log $[\text{Ar}]_0$ for a given $[\text{I}]_0$.

Example Using the method of initial rates
The recombination of I atoms in the gas phase in the presence of argon (which removes the energy released by the formation of an I–I bond, and so prevents the immediate dissociation of a newly formed I_2 molecule) was investigated and the order of the reaction was determined by the method of initial rates. The initial rates of the reaction $2I(g) + Ar(g) \rightarrow I_2(g) + Ar(g)$ were as follows:

$[\text{I}]_0/10^{-5}\,\text{mol L}^{-1}$	1.0	2.0	4.0	6.0
$r_0/\text{mol L}^{-1}\,\text{s}^{-1}$ (a)	8.70×10^{-4}	3.48×10^{-3}	1.39×10^{-2}	3.13×10^{-2}
(b)	4.35×10^{-3}	1.74×10^{-2}	6.96×10^{-2}	1.57×10^{-1}
(c)	8.69×10^{-3}	3.47×10^{-2}	1.38×10^{-1}	3.13×10^{-1}

The Ar concentrations are (a) $1.0 \times 10^{-3}\,\text{mol L}^{-1}$, (b) $5.0 \times 10^{-3}\,\text{mol L}^{-1}$, and (c) $1.0 \times 10^{-2}\,\text{mol L}^{-1}$. Find the orders of reaction with respect to I and Ar and the rate constant.
Answer Figure 7.4 is a plot of log r_0 against log $[\text{I}]_0$ for a given $[\text{Ar}]_0$ and against log $[\text{Ar}]_0$ for a given $[\text{I}]_0$. The intercepts give

$\log k$ and the slopes give the orders. The slopes are 2 and 1 respectively, so the (initial) rate law is

$$r_0 = k[I]_0^2[Ar]_0$$

This rate law signifies that the reaction is second order in I, first order in Ar, and third order overall. The intercept corresponds to $\log(k/mol^{-2}\,L^2\,s^{-1}) = 9.9$, so $k = 8 \times 10^9\,mol^{-2}\,L^2\,s^{-1}$.

Exercise E7.4 The initial rate of a certain reaction depended on concentration of a substance J as follows:

$[J]_0/10^{-3}\,mol\,L^{-1}$	5.0	8.2	17	30
$r_0/10^{-7}\,mol\,L^{-1}\,s^{-1}$	3.6	9.6	41	130

Find the order of the reaction with respect to J and the rate constant.

$$[2, 1.4 \times 10^{-2}\,mol^{-1}\,L\,s^{-1}]$$

7.3 Integrated rate laws

The method of initial rates might not reveal the full rate law, for in a complex reaction the products themselves might affect the rate. For example, products participate in the synthesis of HBr, for eqn (4) shows that the full rate law depends on the concentration of HBr, none of which is present initially. To avoid this difficulty, the rate law should be fitted to the data throughout the reaction. The fitting may be done, in simple cases at least, by using a proposed rate law to predict the concentration of any component at any time, and comparing it with the data.

Because rate laws are differential equations (equations for the rate $d[A]/dt$ in terms of $[A]$), they must be integrated to find the concentration as a function of time. Now that computers are so widely available, even the most complex rate laws may be integrated numerically. However, in a number of simple cases analytical solutions are easily obtained, and prove to be very useful. Indeed, rates are rarely measured directly (because slopes are difficult to determine accurately), and almost all experimental work in chemical kinetics deals with integrated rate laws: the great advantage of integrated rate laws is that they are directly related to the experimental observables of concentration and time.

First-order reactions

The concentration of a reactant A at a time t for a first-order reaction in which the rate law is

$$\text{rate of consumption of A} = k[A] \tag{7a}$$

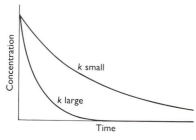

Fig. 7.5 The exponential decay of the reactant in a first-order reaction. The greater the rate constant, the more rapid the decay.

and for which the initial concentration of A, at $t = 0$, is $[A]_0$, is

$$\ln \frac{[A]_0}{[A]} = kt \qquad (7b)$$

Alternative forms of this expression are

$$\ln [A] = \ln [A]_0 - kt \qquad (7c)$$

$$[A] = [A]_0 e^{-kt} \qquad (7d)$$

The last has the form of an **exponential decay** (Fig. 7.5). A common feature of all first-order reactions, therefore, is that *the concentration of the reactant decays exponentially with time.*

JUSTIFICATION

The rate of consumption of a reactant A is $-d[A]/dt$, so a first-order rate equation has the form

$$\frac{d[A]}{dt} = -k[A]$$

This expression can be rearranged to

$$\frac{d[A]}{[A]} = -k \, dt$$

Integration from $t = 0$ when the concentration of A is $[A]_0$ to the time of interest, t, when the concentration of A is $[A]$, gives

$$\int_{[A]_0}^{[A]} \frac{d[A]}{[A]} = -\int_0^t k \, dt$$

and therefore, because the integral of $1/x$ is $\ln x$,

$$(\ln [A])|_{[A]_0}^{[A]} = -kt$$

Equation (7b) follows by rearranging this expression.

Equation (7b) shows that if $\ln ([A]_0/[A])$ is plotted against t, then a first-order reaction will give a straight line. If the plot is straight, then the reaction is first order, and k may be obtained from the slope

Table 7.1   Kinetic data for first-order reactions

Reaction	Phase	$\theta/°C$	k/s^{-1}	$t_{1/2}$
$2N_2O_5 \rightarrow 4NO_2 + O_2$	g	25	3.38×10^{-5}	2.85 h
$2N_2O_5 \rightarrow 4NO_2 + O_2$	$Br_2(l)$	25	4.27×10^{-5}	2.25 h
$C_2H_6 \rightarrow 2CH_3$	g	700	5.46×10^{-4}	21.2 min
Cyclopropane $\rightarrow$ propene	g	500	6.17×10^{-4}	17.2 min

The rate constant is for the rate of formation or consumption of the species in bold type. The rate laws for the other species may be obtained from the reaction stoichiometry.

(which is equal to k). Some rate constants determined in this way are given in Table 7.1.

Example Analysing a first-order reaction
The variation in the partial pressure p of azomethane with time was followed at 4600 K, with the results given below. Confirm that the decomposition

$$CH_3N_2CH_3(g) \rightarrow CH_3CH_3(g) + N_2(g)$$

is first order in $CH_3N_2CH_3$, and find the rate constant at this temperature.

t/s	0	1000	2000	3000	4000
$p/(10^{-2}$ Torr)	8.20	5.72	3.99	2.78	1.94

Answer We plot $\ln(p_0/p)$ against t in Fig. 7.6. The plot is straight, confirming a first-order reaction, and its slope is 3.6×10^{-4}. Therefore, $k = 3.6 \times 10^{-4}\,s^{-1}$.

Exercise E7.5 The concentration of N_2O_5 in liquid bromine varied with time as follows:

t/s	0	200	400	600	1000
$[N_2O_5]/mol\,L^{-1}$	0.110	0.073	0.048	0.032	0.014

Determine the order and rate constant for the reaction $2N_2O_5 \rightarrow 4NO_2 + O_2$.

$$[1, 2.1 \times 10^{-3}\,s^{-1}]$$

Second-order reactions

If the rate law is
$$\text{rate of consumption of A} = k[A]^2 \qquad (8a)$$

and the concentration of A at $t = 0$ is $[A]_0$, then at time t the concentration $[A]$ can be obtained from the expression

$$\frac{1}{[A]} = \frac{1}{[A]_0} + kt \qquad (8b)$$

This expression can be rearranged into

$$[A] = \frac{[A]_0}{1 + kt[A]_0} \qquad (8c)$$

Equation (8b) shows that to test for a second-order reaction we should plot $1/[A]$ against t and expect a straight line. If it is straight,

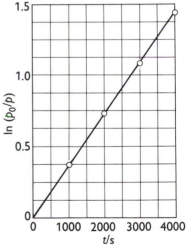

Fig. 7.6 The determination of the rate constant of a first-order reaction: a straight line is obtained when $\ln[A]$ (or $\ln p$, where p is the partial pressure of the species of interest) is plotted against t: the slope gives the rate constant (slope = k).

249

JUSTIFICATION

The differential equation for the rate law

$$\frac{d[A]}{dt} = -k[A]^2$$

can be rearranged to

$$\frac{d[A]}{[A]^2} = -k \, dt$$

This expression is now integrated between $t = 0$, when the concentration of A is $[A]_0$, and the time of interest, t, when the concentration of A is $[A]$:

$$-\int_{[A]_0}^{[A]} \frac{d[A]}{[A]^2} = \int_0^t k \, dt$$

The integral on the left is evaluated using $\int (1/x^2) \, dx = -1/x$, which gives

$$\left. \frac{1}{[A]} \right|_{[A]_0}^{[A]} = kt$$

and a straightforward rearrangement leads to eqn (8b).

then the reaction is second order in A and the slope of the line is equal to the rate constant. Some rate constants determined in this way are given in Table 7.2. Equation (8c) lets us predict the concentration of A at any time after the start of the reaction. When $[A]$ is plotted against t, we see that the concentration of A approaches zero more slowly than in a first-order reaction with the same initial rate (Fig. 7.7).

Similar calculations may be carried out to find the integrated rate laws for other orders, and some are listed in Table 7.3.

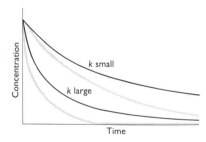

Fig. 7.7 The variation with time of the concentration of a reactant in a second-order reaction. The pale line is the corresponding decay in a first-order reaction with the same initial rate.

Table 7.2 Kinetic data for second-order reactions

Reaction	Phase	$\theta/°C$	$k/\text{L mol}^{-1}\text{s}^{-1}$
$2NOBr \rightarrow 2NO + \mathbf{Br_2}$	g	10	0.80
$2NO_2 \rightarrow 2NO + \mathbf{O_2}$	g	300	0.54
$\mathbf{H_2} + I_2 \rightarrow 2HI$	g	400	2.42×10^{-2}
$\mathbf{D_2} + HCl \rightarrow DH + DCl$	g	600	0.141
$2I \rightarrow \mathbf{I_2}$	g	23	7×10^9
$2I \rightarrow \mathbf{I_2}$	hexane	50	1.8×10^{10}
$\mathbf{CH_3Cl} + CH_3O^-$	$CH_3OH(l)$	20	2.29×10^{-6}
$\mathbf{CH_3Br} + CH_3O^-$	$CH_3OH(l)$	20	9.23×10^{-6}
$\mathbf{H^+} + OH^- \rightarrow H_2O$	water	25	1.5×10^{11}

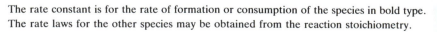

The rate constant is for the rate of formation or consumption of the species in bold type. The rate laws for the other species may be obtained from the reaction stoichiometry.

Table 7.3   Integrated rate laws

Order	Reaction type	Rate law	Integrated rate law
0	$A \rightarrow P$	$r = k$	$[P] = kt$ for $kt \leq [A]_0$
1	$A \rightarrow P$	$r = k[A]$	$[P] = [A]_0(1 - e^{-kt})$
2	$A \rightarrow P$	$r = k[A]^2$	$[P] = \dfrac{kt[A]_0^2}{1 + kt[A]_0}$
	$A + B \rightarrow P$	$r = k[A][B]$	$[P] = \dfrac{[A]_0[B]_0(1 - e^{([B]_0-[A]_0)kt})}{[A]_0 - [B]_0 e^{([B]_0-[A]_0)kt}}$

7.4 Half-lives

A useful indication of the rate of a chemical reaction is the 'half-life' of a species:

> The **half-life**, $t_{1/2}$, of a species is the time it takes for the concentration of the species to fall to half its initial value.

The half-life of a species A that decays in a first-order reaction can be found by substituting $[A] = \frac{1}{2}[A]_0$ and $t = t_{1/2}$ into eqn (7b):

$$kt_{1/2} = -\ln \frac{\frac{1}{2}[A]_0}{[A]_0} = -\ln \frac{1}{2} = \ln 2$$

from which it follows that

$$t_{1/2} = \frac{\ln 2}{k}, \qquad \text{with } \ln 2 = 0.693 \tag{9}$$

For example, because the rate constant for the first-order reaction

$$2N_2O_5(g) \rightarrow 4NO_2(g) + O_2(g)$$

rate of consumption of $N_2O_5 = k[N_2O_5]$

is equal to $6.76 \times 10^{-5}\,\text{s}^{-1}$ at $25\,^\circ\text{C}$, the half-life of N_2O_5 is 2.85 h. Hence, the concentration of N_2O_5 falls to half its initial value in 2.85 h, and then to half that again in a further 2.85 h, and so on.

The main point to note about eqn (9) is that *for a first-order reaction, the half-life of a reactant is independent of its initial concentration*. It follows that if the concentration of A at some arbitrary stage of the reaction is [A], then the concentration will fall to $\frac{1}{2}[A]$ after an interval of $0.693/k$ (Fig. 7.8). Some half-lives are given in Table 7.1.

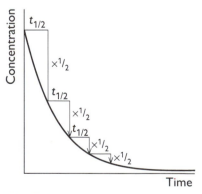

Fig. 7.8 In each successive period of duration $t_{1/2}$, the concentration of a reactant in a first-order reaction decays to half its value at the start of that period. After n such periods, the concentration is $(\frac{1}{2})^n$ of its initial concentration.

251

Example Using the half-life of a species
In acidic solution, the disaccharide sucrose (cane sugar) is converted to a mixture of the monosaccharides glucose and fructose in a pseudo-first-order reaction. Under certain conditions of pH, the half-life of sucrose is 28.4 min. How long will it take for the concentration of a sample to fall from $8.0 \, \text{mmol L}^{-1}$ to $1.0 \, \text{mmol L}^{-1}$?
Answer In the course of successive half-lives, the concentration falls to half its initial value, then to half that value, and so on. Under the conditions specified, the sequence is

concentration/mmol L^{-1}:

$$8.0 \xrightarrow{28.4 \, \text{min}} 4.0 \xrightarrow{28.4 \, \text{min}} 2.0 \xrightarrow{28.4 \, \text{min}} 1.0$$

The total time required is $3 \times 28.4 \, \text{min} = 85.2 \, \text{min}$.

Exercise E7.6 The half-life of a substrate in a certain enzyme-catalysed first-order reaction is 138 s. How long is required for the initial concentration of substrate, which was $1.28 \, \text{mmol L}^{-1}$, to fall to $0.040 \, \text{mmol L}^{-1}$?

[650 s]

Exercise E7.7 Derive an expression for the half-life of a second-order reaction in terms of the rate constant, k.
[see below, eqn (10)]

In contrast to first-order reactions, the half-life of a second-order reaction does depend on the initial concentration of the reactant (see the preceding exercise):

$$t_{1/2} = \frac{1}{k[A]_0} \tag{10}$$

We see that the half-life lengthens as the concentration of A decreases. Whereas the half-life of a first-order reaction is a characteristic of the reaction and not the particular stage that it has reached, that is not true of the half-life of a second- (or higher) order reaction, and so the half-life of a second-order reaction is rarely used.

One application of the concentration-independence of the half-life of a first-order reaction is to the identification of such reactions. Thus, if in a set of data of composition against time it is seen that the initial concentration falls to half its value in a certain time, and that another concentration falls to half its value in the same time, then we can infer that the reaction is first-order; the order can be confirmed by plotting ln [A] against t and obtaining a straight line, as indicated earlier.

7.5 The temperature dependence of reaction rates

It is found that the rates of most reactions increase as the temperature is raised. Many reactions in solution fall somewhere in the range spanned by the hydrolysis of methyl ethanoate (where the rate constant at 35 °C is 1.8 times that at 25 °C) and the hydrolysis of sucrose (where the factor is 4.1). Reactions in the gas phase typically have rates that are only weakly sensitive to the temperature.

The Arrhenius parameters

As data on reaction rates were accumulated towards the end of the nineteenth century, the Swedish chemist Svante Arrhenius noted empirically that almost all of them followed a similar dependence on the temperature. In particular, he noted that $\ln k$, where k is the rate constant for the reaction, varies linearly with $1/T$, with a constant of proportionality that was a characteristic of the reaction. That is, he found that a plot of $\ln k$ against $1/T$ gave a straight line with a slope that was characteristic of the reaction (Fig. 7.9). The mathematical expression of this conclusion is that the rate constant varies with temperature in accord with the expression

$$\ln k = \text{intercept} + \text{slope} \times \frac{1}{T}$$

This expression is normally written as the **Arrhenius equation**:

$$\ln k = \ln A - \frac{E_a}{RT} \tag{11a}$$

The Arrhenius equation is often written as

$$k = A e^{-E_a/RT} \tag{11b}$$

The parameter, A (which has the same units as k), is called the **pre-exponential factor** and E_a (which is a molar energy and has the units of kilojoules per mole) is called the **activation energy**. Collectively the two parameters are called the **Arrhenius parameters** of the reaction, and some experimental values are given in Table 7.4.

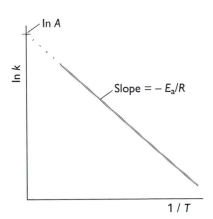

Fig. 7.9 The general form of an Arrhenius plot of $\ln k$ against $1/T$. The slope is equal to $-E_a/R$ and the intercept at $1/T = 0$ is equal to $\ln A$.

Example Determining the Arrhenius parameters
The rate of the second-order decomposition of acetaldehyde (ethanal, CH_3CHO) was measured over the temperature range 700 to 1000 K, and the rate constants are reported below. Find the activation energy and the pre-exponential factor.

T/K	700	730	760	790	810	840	910	1000
$k/mol^{-1}\,L\,s^{-1}$	0.011	0.035	0.105	0.343	0.789	2.17	20.0	145

Table 7.4 Arrhenius parameters

Reactions	A/s^{-1}	$E_a/\text{kJ mol}^{-1}$
First order		
Cyclopropane $\rightarrow$ propene	1.58×10^{15}	272
$CH_3NC \rightarrow CH_3CN$	3.98×10^{13}	160
cis-CHD$=$CHD $\rightarrow$ *trans*-CHD$=$CHD	3.16×10^{12}	256
cyclobutane $\rightarrow 2C_2H_4$	3.98×10^{15}	261
$2N_2O_5 \rightarrow 4NO_2 + O_2$	4.94×10^{13}	103
$N_2O \rightarrow N_2 + O$	7.94×10^{11}	250

	$A/\text{L mol}^{-1}\text{s}^{-1}$	$E_a/\text{kJ mol}^{-1}$
Second order, gas phase		
$O + N_2 \rightarrow NO + N$	1×10^{11}	315
$OH + H_2 \rightarrow H_2O + H$	8×10^{10}	42
$Cl + H_2 \rightarrow HCl + H$	8×10^{10}	23
$CH_3 + CH_3 \rightarrow C_2H_6$	2×10^{10}	0
$NO + Cl_2 \rightarrow NOCl + Cl$	4×10^9	85
Second order, solution		
$NaC_2H_5O + CH_3I$ in ethanol	2.42×10^{11}	81.6
$C_2H_5Br + OH^-$ in water	4.30×10^{11}	89.5
$CH_3I + S_2O_3^{2-}$ in water	2.19×10^{12}	78.7
Sucrose + H_2O in acidic water	1.50×10^{15}	107.9

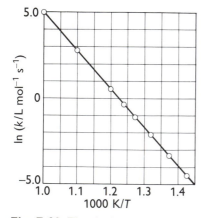

Fig. 7.10 The Arrhenius plot for the decomposition of CH_3CHO, and the best (least squares) straight line fitted to the data points.

Answer We plot $\ln k$ against $1/T$ (Fig. 7.10). The least-squares best fit of the line has slope -2.21×10^4 and intercept 27.0. Therefore, since the slope is $-E_a/R$ and the intercept at $1/T = 0$ is $\ln A$,

$$E_a = (2.21 \times 10^4 \text{ K}) \times (8.3145 \text{ J K}^{-1} \text{ mol}^{-1}) = 184 \text{ kJ mol}^{-1}$$
$$A = e^{27.0} \text{ mol}^{-1} \text{ L s}^{-1} = 5.3 \times 10^{11} \text{ mol}^{-1} \text{ L s}^{-1}$$

Exercise E7.8 Find A and E_a from the following data:

T/K	300	350	400	450	500
$k/\text{mol}^{-1}\text{L s}^{-1}$	7.9×10^6	3.0×10^7	7.9×10^7	1.7×10^8	3.2×10^8

$$[8 \times 10^{10} \text{ mol}^{-1} \text{ L s}^{-1}, 23 \text{ kJ mol}^{-1}]$$

Once the activation energy of a reaction is known, it is a simple matter to predict the value of a rate constant k' at a temperature T' from its value k at another temperature T. To do so, we write

$$\ln k' = \ln A - \frac{E_a}{RT'}$$

and then subtract eqn (11a), obtaining

$$\ln k' - \ln k = -\frac{E_a}{RT'} + \frac{E_a}{RT}$$

This expression can be rearranged to

$$\ln \frac{k'}{k} = \frac{E_a}{R}\left(\frac{1}{T} - \frac{1}{T'}\right) \tag{12}$$

As an illustration, for a reaction with an activation energy of $50\ kJ\ mol^{-1}$, an increase in the temperature from $25\,°C$ to $37\,°C$ (body temperature) corresponds to

$$\ln \frac{k'}{k} = \frac{50 \times 10^3\ J\ mol^{-1}}{8.3145\ J\ K^{-1}\ mol^{-1}} \times \left(\frac{1}{298\ K} - \frac{1}{310\ K}\right) = 0.78$$

Then, by taking antilogarithms,

$$k' = e^{0.78} \times k = 2.18k$$

which corresponds to a doubling of the rate.

Exercise E7.9 The activation energy of one of the reactions in the Krebs citric acid cycle is $87\ kJ\ mol^{-1}$. What is the change in rate constant when the temperature falls from $37\,°C$ to $15\,°C$?

$$[k' = 0.076k]$$

The origin of the Arrhenius parameters: collision theory

The origin of the Arrhenius parameters can be understood most simply in terms of a certain class of gas-phase reactions in which reaction occurs when two molecules encounter one another.† In this **collision theory** it is supposed that such a reaction occurs only if two molecules collide with a certain minimum kinetic energy (Fig. 7.11). In collision theory, a reaction is supposed to resemble the collision of two billiard balls: the balls bounce apart if they collide with only a small

† In the terminology to be introduced in Section 7.7, we are considering *bimolecular* gas-phase reactions.

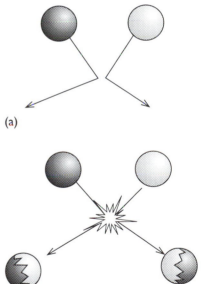

(a)

(b)

Fig. 7.11 In the collision theory of gas-phase chemical reactions, reaction occurs when two molecules collide, but only if the collision is sufficiently vigorous. (a) An insufficiently vigorous collision: the reactant molecules collide but bounce apart unchanged. (b) A sufficiently vigorous collision results in a reaction.

energy, but might smash each other into fragments (products) if they collide with more than a certain minimum energy.

First, we consider the form of the **reaction profile** for the encounter of two molecules. A reaction profile in collision theory is a graph showing the potential energy of two molecules as they approach, react, and then separate as products. A typical example is shown in Fig. 7.12. On the left the horizontal line represents the energy of the two stationary reactant molecules that are far apart from one another. The potential energy rises from this value only when the separation of the molecules is so small that the two molecules are in contact, when it rises as bonds bend and start to break. The potential energy reaches a peak when the two molecules are highly distorted. Then it starts to decrease as new bonds are formed. At separations to the right of the maximum, the potential energy rapidly falls to a low value as the product molecules separate. For the reaction to be successful, therefore, the molecules must approach with sufficient kinetic energy along their line of approach to carry them over the **activation barrier**, the peak in the reaction profile. As we shall see, we can identify the height of the activation barrier with the activation energy of the reaction.

With the reaction profile in mind, it is quite easy to establish that collision theory accounts for Arrhenius behaviour. Thus, the rate of collisions between species A and B is proportional to both of their concentrations: if the concentration of B is doubled, then the rate at which A molecules collide with B molecules is doubled, and if the concentration of A is doubled, the rate at which B molecules collide with A molecules is also doubled. It follows that the rate of collision of A and B molecules is directly proportional to the concentrations of the

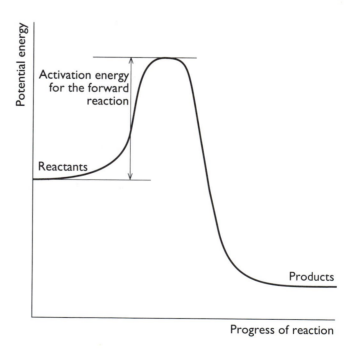

Fig. 7.12 A reaction profile. The graph depicts schematically the changing potential energy of two species that approach, collide, and then go on to form products. The activation energy is the height of the barrier in the potential energy curve above the potential energy of the reactants.

256

two types of molecule, and we can write

$$\text{rate of collision} \propto [\text{A}][\text{B}]$$

Next, we need to multiply the collision rate by a factor f that represents the fraction of collisions that occur with at least a kinetic energy E_a along the line of centres (Fig. 7.13), for only these collisions will lead to the formation of products. Molecules that approach with a kinetic energy less than E_a will behave like a ball that rolls towards the activation barrier, fails to surmount it, and rolls back to form reactants. It follows from very general arguments (see the discussion of the Boltzmann distribution in *Further information 7*, p. 258) concerning the probability that a molecule has a specified energy that the fraction of collisions that occur with at least a kinetic energy E_a is

$$f = e^{-E_a/RT}$$

Fig. 7.13 The criterion for a successful collision is that the two reactant species should collide with a kinetic energy along their line of approach that exceeds a certain minimum value E_a characteristic of the reaction. The two molecules might also have components of velocity (and an associated kinetic energy) in other directions (for example, the two molecules depicted here might be moving up the page as well as towards each other); but only the energy associated with their mutual approach can be used to overcome the activation energy.

Exercise E7.10 What is the fraction of collisions that have sufficient energy for reaction if the activation energy is 50 kJ mol^{-1} and the temperature is (a) 25 °C, (b) 500 °C?
$$[(a)\ 1.7 \times 10^{-9},\ (b)\ 4.2 \times 10^{-4}]$$

At this stage we can conclude that the rate of reaction, which is proportional to the rate of collision multiplied by the fraction of successful collisions, is

$$\text{reaction rate} \propto [\text{A}][\text{B}]e^{-E_a/RT}$$

If we compare this expression with a second-order rate law,

$$\text{reaction rate} = k[\text{A}][\text{B}]$$

it follows that

$$k \propto e^{-E_a/RT}$$

This expression has exactly the Arrhenius form if we identify the constant of proportionality with A. This analysis of collision theory therefore results in the following identifications:

- The pre-exponential factor A is the constant of proportionality between the concentrations of the reactants and the rate at which they collide.

- The activation energy E_a is the minimum kinetic energy required for a collision to result in reaction.

257

The value of A can be calculated from the kinetic theory of gases (Chapter 1). However, it is often found that the experimental value of A is smaller than that calculated. One possible explanation is that not only must the molecules collide with sufficient energy, but they must also come together in a specific relative orientation (Fig. 7.14). It follows that the reaction rate is proportional to the probability that the encounter occurs in the correct relative orientation. The pre-exponential factor A should therefore include a **steric factor**, P, which

FURTHER INFORMATION 7:
The Boltzmann distribution

The **Boltzmann distribution** expresses the probability, p, that a molecule will be found in a state with energy E:

$$p = \frac{e^{-E/kT}}{q}, \quad \text{where} \quad q = \sum_i e^{E_i/kT}$$

where the E_i are the energies of the states of the system. The constant, k, is **Boltzmann's constant**, with the value

$$k = 1.381 \times 10^{-23} \, \text{J K}^{-1}$$

This constant is a fundamental constant of nature, and when multiplied by Avogadro's constant yields the gas constant:

$$R = N_A \times k$$

The Boltzmann distribution shows that the population decreases exponentially with increasing energy (Fig. 1). Specifically, it follows from the distribution that the ratio of the probabilities

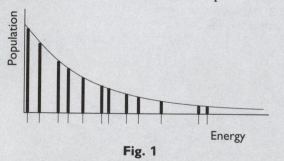

Fig. 1

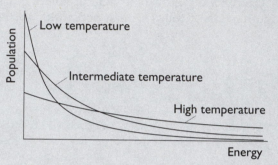

Fig. 2

p' and p of finding a system in states with energies E' and E at a temperature T is

$$\frac{p'}{p} = e^{-(E'-E)/kT}$$

which shows the exponential dependence explicitly. The spread of populations to high energies increases with temperature (Fig. 2).

The quantity q is called the **partition function** for the molecule. Broadly speaking, q is a measure of the number of states of the system that are significantly populated at the temperature of interest. Thus, q varies from 1 at $T = 0$ (when only the state of lowest energy can be populated) to infinity at infinite temperature (when all states of the system are accessible). The basis of this interpretation of q can be appreciated by noting that the explicit form of the partition function (taking $E_0 = 0$ for the state of lowest energy) is

$$q = 1 + e^{-E_1/kT} + e^{-E_2/kT} + \cdots$$

usually lies between 0 (no orientations lead to reaction) and 1 (all orientations lead to reaction). As an example, for the reactive collision

$$NOCl + NOCl \rightarrow NO + NO + Cl_2$$

in which two NOCl molecules collide and break apart into two NO molecules and a Cl_2 molecule, $P \approx 0.16$. For the hydrogen addition reaction

$$H_2 + H_2C{=}CH_2 \rightarrow H_3C{-}CH_3$$

Then, noting that $e^{-1} = 0.37$, we can make the approximation that $e^{-x} \approx 1$ if $x < 1$ and $e^{-x} \approx 0$ if $x > 1$. If the energies of the system up to E_4 are smaller than kT (so $E/kT < 1$) and all other energies are greater than kT (so that $E/kT > 1$), the expression has the form

$$q \approx 1 + 1 + 1 + 1 + 1 + 0 + 0 + 0 + \cdots = 5$$

and about five states are significantly populated. We see that kT (and in molar terms, RT) is the dividing line between states that are significantly populated and states that are not. At 25 °C, $RT = 2.5\,kJ\,mol^{-1}$, and kT corresponds to 207 cm^{-1} or 0.026 eV, so these three values are the frontier between occupation and emptiness at about room temperature.

Under certain circumstances, the Boltzmann distribution can be used to calculate the fraction of molecules that have an energy of at least a certain minimum (such as the activation energy E_a in the collision theory of reactions). Thus, suppose that the system can have any energy E between zero and infinity. The partition function is then given by the integral

$$q = C \int_0^\infty e^{-E/kT}\,dE = C \times kT$$

and the constant of proportionality (which will cancel shortly) is C. The total probability that the molecules have an energy of at least E_a is then the sum (integral, in the case of continuous energies) of the individual probabilities given by the Boltzmann distribution:

fraction with energy greater than E_a

$$= \frac{1}{q} \times C \int_{E_a}^\infty e^{-E/kT}\,dE = \frac{C}{q} \times kT\,e^{-E_a/kT}$$

The constant C cancels the C in the expression for q, and we are left with

fraction with energy greater than $E_a = e^{-E_a/kT}$

That is, the fraction decreases exponentially with the energy E_a and increases sharply as the temperature is raised.

The Boltzmann distribution is of very wide applicability: it applies to *any* system that is at thermal equilibrium at a temperature T. The reason for its generality can be found in its derivation. We shall not present that derivation here, but it is based on the view that any molecule can occupy any of the available states of the system without favour. Thus, if a molecule happens to have a particular energy because it is vibrating energetically but rotating slowly, and can have the same energy because it is rotating very energetically but not vibrating, then both states of motion are held to be equally probable. Then the calculation considers the chances of each state being obtained by a purely random distribution of molecules over the available states, subject to the requirement that the total energy of all the molecules in the system is a constant. On the basis of this ultimate in equal opportunities, it turns out that the most likely distribution of molecules over their available states is the Boltzmann distribution.

Fig. 7.14 Energy is not the only criterion of a successful reactive encounter, for relative orientation may also play a role. (a) In this schematic collision, the reactants approach in an inappropriate relative orientation, and no reaction occurs, though their energy is sufficient. (b) In this encounter, both the energy and the orientation are suitable for reaction.

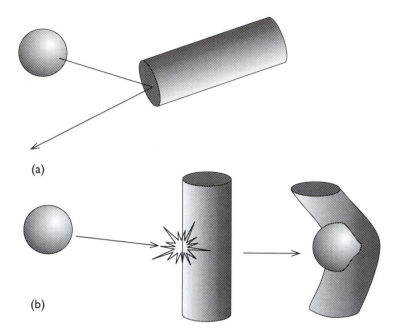

in which a hydrogen molecule attaches directly to an ethene molecule, to form an ethane molecule, P is only 1.7×10^{-6}, which suggests that the reaction has very stringent orientational requirements.

Some reactions have $P > 1$. Such a value may seem absurd, because it appears to suggest that the reaction occurs more often than the molecules meet! An example of a reaction of this kind is

$$K + Br_2 \rightarrow KBr + Br$$

in which a K atom plucks a Br atom out of a Br_2 molecule; for this reaction the experimental value of P is 4.8. In this reaction the distance of approach at which reaction can occur appears to be considerably larger than the distance needed for deflection of the path of the approaching molecules in a nonreactive collision. It has been proposed that the reaction proceeds by a **harpoon mechanism**. This brilliant name is based on a model of the reaction which pictures the K atom as approaching the Br_2 molecules, and when the two are close enough, an electron (the harpoon) flips across to the Br_2 molecule. In place of two neutral particles there are now two ions, and so there is a Coulombic attraction between them: this attraction is the line on the harpoon. Under its influence the ions move together (the line is wound in), the reaction takes place, and $KBr + Br$ emerge. The harpoon extends the cross-section for the reactive encounter, and we greatly underestimate the reaction rate by taking, for the collision cross-section, the value for simple mechanical contact between K and Br_2.

Activated complex theory

A more sophisticated theory of reaction rates can be applied to reactions taking place in solution as well as in the gas phase, and is therefore applicable to a wide range of solution chemistry and

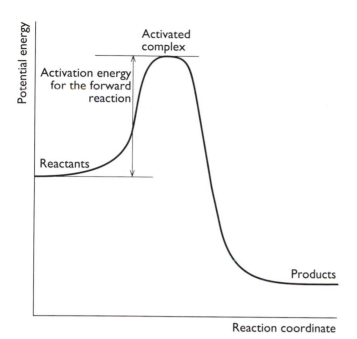

biochemical processes. In the **activated complex theory** of reactions, it is supposed that as two reactants approach, their potential energy rises and reaches a maximum, as illustrated by the reaction profile in Fig. 7.15. This maximum corresponds to the formation of an **activated complex**. Unlike in the collision theory, the activated complex is supposed to have a definite composition and a loose structure. It can be pictured as a cluster of atoms which is poised to pass on to products or to collapse back into the reactants from which it was formed (Fig. 7.16); an activated complex is not a reaction intermediate that can be isolated and studied like ordinary molecules. The concept of an activated complex is applicable to reactions in solutions as well as to the gas phase, because we can think of the activated complex as involving any solvent molecules that may be present.

Initially only the reactants A and B are present. As the reaction

Fig. 7.16 In the activated complex theory of chemical reactions, two reactants encounter each other (either in a gas-phase collision or as a result of diffusing together through a solvent), and if they have sufficient energy, form an activated complex. The activated complex is depicted here by a relatively loose cluster of atoms that may undergo rearrangement into products. In an actual reaction, only some atoms—those at the actual reaction site—might be significantly loosened in the complex, the bonding of the others remaining almost unchanged. This would be the case for CH_3 groups attached to a carbon atom that was undergoing substitution.

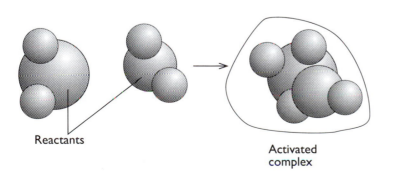

event proceeds, A and B come into contact, distort, and begin to exchange or discard atoms. The potential energy rises to a maximum, and the cluster of atoms which corresponds to the region close to the maximum is the activated complex. The potential energy falls as the atoms rearrange in the cluster, and reaches a value characteristic of the products. The climax of the reaction is at the peak of the potential energy. Here two reactant molecules have come to such a degree of closeness and distortion that a small further distortion will send them in the direction of products. This crucial configuration is called the **transition state** of the reaction. Although some molecules entering the transition state might revert to reactants, if they pass through this configuration it is probable that products will emerge from the encounter.

As an example, consider the approach of an H atom to an F_2 molecule. For simplicity, we imagine the approach as occurring along the F—F bond direction. At great distances the potential energy is the sum of the potential energies of H and F_2. When H and F_2 are so close that their orbitals start to overlap, the F—F bond begins to stretch and a bond begins to form between H and the nearer F. The H atom comes closer, the F—F bond lengthens, the H—F bond shortens and strengthens, and the atoms enter the range of locations characteristic of the activated complex. There comes a stage when the cluster of three atoms which constitutes the activated complex has maximum potential energy and is poised at the transition state. An infinitesimal compression of the H—F bond and a stretch of F—F takes the complex through the transition state. Distances further along the reaction coordinate represent stages at which the H—F bond forms more fully and the F—F bond breaks. Motion along the reaction coordinate from left to right therefore represents the progress of H and F_2 through these configurations. Whether or not a colliding H atom and F_2 molecule actually crosses the potential barrier depends on the kinetic energy the molecules have initially, because they must be able to climb the barrier and attain the transition state.

In an actual reaction, H atoms approach F_2 molecules from all angles and the specification of the reaction coordinate is a subtle problem; it is even more subtle for a reaction taking place in solution, for then the surrounding solvent molecule may be involved in the formation of the activated complex. We shall therefore regard the reaction coordinate simply as an indication of the distortions in the reactant molecules (and the surrounding medium, if that is relevant) as the activated complex is formed, the critical transition state is reached, and the product molecules emerge. At the transition state, motion along the reaction coordinate corresponds to some complicated collective vibration-like motion of all the atoms in the complex (and the motion of the solvent molecules).

In a simple form of activated complex theory, it is supposed that the activated complex is in equilibrium with the reactants, and that its abundance in the reaction mixture can be expressed in terms of an equilibrium constant, which is normally denoted $K^{\ddagger}$:

$$\text{reactants} \rightleftharpoons \text{activated complex} \qquad K^{\ddagger} = \frac{[\text{activated complex}]}{[\text{reactants}]}$$

Then, if we suppose that the rate at which products are formed is proportional to the concentration of the activated complex, we can write

$$\text{rate of formation of products} \propto [\text{activated complex}]$$

$$\propto K^{\ddagger}[\text{reactants}]$$

Therefore, by comparing this expression with the form of the rate law,

$$\text{rate of formation of products} = k[\text{reactants}]$$

we see that the rate constant k is proportional to the equilibrium constant $K^{\ddagger}$ for the formation of the activated complex. We have already seen that an equilibrium constant may be expressed in terms of the standard reaction free energy, which in this case is the free energy, denoted $\Delta G^{\ddagger}$, for the formation of the activated complex from the reactants. It follows from eqn (25) in Chapter 3 that

$$K^{\ddagger} = e^{-\Delta G^{\ddagger}/RT}$$

and therefore, by using

$$\Delta G^{\ddagger} = \Delta H^{\ddagger} - T \, \Delta S^{\ddagger}$$

that

$$k \propto e^{-(\Delta H^{\ddagger} - T \, \Delta S^{\ddagger})/RT} \propto e^{\Delta S^{\ddagger}/R} \, e^{-\Delta H^{\ddagger}/RT} \qquad (13)$$

This expression has the form of the Arrhenius expression, eqn (11b), if we identify the **enthalpy of activation**, $\Delta H^{\ddagger}$, with the activation energy, and the **entropy of activation**, $\Delta S^{\ddagger}$, with the pre-exponential factor (more precisely, with $R \ln A$).

The advantage of activated complex theory over collision theory is that it is applicable to reactions in solution as well as in the gas phase. It also gives some clue to the calculation of the steric factor, P, since the orientation requirements are carried in the entropy of activation. Thus, if there are strict orientation requirements (for example, in the approach of a substrate molecule to an enzyme), then the entropy of activation will be strongly *negative* (representing a decrease in disorder when the activated complex forms), and the pre-exponential factor will be small. In practice, it is occasionally possible to estimate the sign and magnitude of the entropy of activation and hence to estimate the rate constant. The general importance of activated complex theory is that it shows that even a complex series of events—not only a collisional encounter in the gas phase—displays Arrhenius-like behaviour and that the concept of activation energy (and its consequences, such as eqn (12) for the effect of temperature on the rate constant) is applicable.

Exercise E7.11 In a particular reaction in water, it is proposed that two ions of opposite charge come together to form an electrically neutral activated complex. Is the contribution of the solvent to the entropy of activation likely to be positive or negative?

[positive, as H_2O less organized around the neutral species]

Accounting for the rate laws

We now move on to the second stage of the analysis of kinetic data, their explanation in terms of a postulated reaction mechanism—the sequence of elementary molecular events that lead from the reactants to the products.

7.6 Elementary reactions

Many reactions occur in a series of steps called **elementary reactions**, each of which involves only one or two molecules. We shall denote an elementary reaction by writing the chemical equation without displaying the phase of the species, as in

$$H + Br_2 \rightarrow HBr + Br$$

and some of the reactions that we have already discussed in this chapter. This equation signifies that a specific H atom attacks a specific Br_2 molecule to produce a molecule of HBr and a Br atom.

The **molecularity** of an elementary reaction is the number of molecules coming together to react. In a **unimolecular reaction** a single molecule shakes itself apart, or shakes its atoms into a new arrangement (Fig. 7.17). An example is the isomerization of cyclopropane to propene:

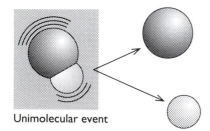

The radioactive decay of nuclei (for example, the emission of a β particle from the nucleus of a tritium atom, which is used in mechanistic studies to follow the course of particular groups of atoms) is 'unimolecular' in the sense that a single nucleus shakes itself apart. In a **bimolecular reaction**, two molecules collide and exchange energy, atoms, or groups of atoms, or undergo some other kind of change, as in the reaction between H and F_2 or between H and Br_2 (Fig. 7.18). It is important to distinguish molecularity from order:

Unimolecular event

Fig. 7.17 In a unimolecular elementary reaction, an energetically excited species decomposes without further interaction with other species present in the system.

- The order is an empirical quantity, and is obtained from the experimental rate law.

- The molecularity refers to an elementary reaction that has been postulated to be an individual step in a proposed mechanism.

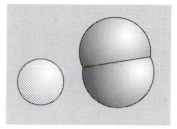

Bimolecular event

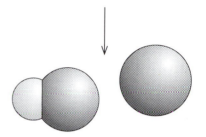

Fig. 7.18 In a bimolecular elementary reaction, two species are involved in the process.

Many substitution reactions in organic chemistry (for instance, S_N2 nucleophilic substitutions) are bimolecular and involve an activated complex that is formed from two reactant species.

The rate law of an elementary reaction (but not of an overall reaction in general) can be written down from its reaction equation. Thus, the rate law of a unimolecular elementary reaction is first order in the reactant:

$$A \rightarrow \text{products} \qquad \text{rate} = k[A] \qquad (14)$$

A unimolecular reaction is first order because the number of A molecules that decay in a short interval is proportional to the number available to decay. For example, ten times as many decay in the same interval when there are initially 1000 A molecules than when there are only 100 present. Therefore, the rate of decomposition of A is proportional to its concentration.

Similarly, the rate law of an elementary bimolecular reaction is second order:

$$A + B \rightarrow \text{products} \qquad \text{rate} = k[A][B] \qquad (15)$$

A bimolecular reaction is second order because its rate is proportional to the rate at which the reactants meet, which is proportional in turn to both their concentrations. Therefore, if we believe (or simply postulate) that a reaction is a single step, bimolecular process, we can write down the rate law (and then go on to test it).

The interpretation of an experimentally determined rate law is full of pitfalls, partly because a simple rate law can also result from a complex reaction scheme. We shall see below how to string simple steps together into a mechanism and how to arrive at the corresponding rate law. For the present we emphasize that if the reaction is an elementary bimolecular process, then it has second-order kinetics, but if the kinetics are second order, then the reaction might be complex. The postulated mechanism can be explored only by detailed detective work on the system, and by investigating whether side products or intermediates appear during the course of the reaction. Detailed analysis of this kind was one of the ways, for example, in which the reaction $H_2(g) + I_2(g) \rightarrow 2HI(g)$ was shown to proceed by a complex mechanism after many years during which people had accepted on good, but insufficiently meticulous evidence, that it was a fine example of a simple bimolecular reaction in which atoms exchanged partners during a collision.

265

7.7 The formulation of rate laws

A rate law is an experimentally determined aspect of a reaction. Once it has been determined, the next step is often to propose a reaction mechanism that accounts for the observed rate law. In this section, we describe the second of these two steps.

A technique has been developed for formulating the rate law of a reaction that is implied by the mechanism that has been proposed. We shall introduce it by example, and consider the rate law for the gas-phase oxidation of nitrogen oxide, NO, which is found experimentally to be third order overall.

$$2NO(g) + O_2(g) \rightarrow 2NO_2(g)$$

$$\text{rate of formation of } NO_2 = k[NO]^2[O_2] \qquad (16)$$

That the reaction is third order accounts for the relatively slow oxidation of nitrogen oxide in the atmosphere (as in the formation of nitrogen oxide pollutants, NO_x) in the absence of other processes, because the rate depends on the square of the concentration of NO, which is very small if the concentration itself is small.

One explanation of the observed reaction order might be that the reaction is a single termolecular elementary step, but a termolecular process requires the simultaneous collision of three particles, which occurs very infrequently. Therefore, although termolecular collisions may occur, the rate of reaction by this mechanism is so slow that another mechanism usually dominates. Indeed, the fact that the reaction rate decreases as the temperature is raised points to a complex reaction mechanism, because simple reactions always go faster at higher temperatures.

The following mechanism has been proposed.

1. Two NO molecules combine to form a dimer

$$NO + NO \rightarrow N_2O_2$$

$$\text{rate of formation of } N_2O_2 = k_a[NO]^2$$

This step is plausible, because NO is an odd-electron species, and two molecules can form a covalent bond when they meet. That the N_2O_2 dimer is also known in the solid makes the suggestion plausible: it is often a good strategy to decide whether a proposed intermediate is the analogue of a known compound.

2. The N_2O_2 dimer decomposes into NO molecules:

$$N_2O_2 \rightarrow NO + NO$$

$$\text{rate of decomposition of } N_2O_2 = k_a'[N_2O_2]$$

This step, the reverse of step 1, is a unimolecular decay: the dimer shakes itself apart. It is a common convention to mark the rate constant of a reverse reaction with a prime (k_a for the forward reaction, k_a' for its reverse).

3. Alternatively, an O_2 molecule collides with the dimer and results in

the formation of NO_2:

$$N_2O_2 + O_2 \rightarrow NO_2 + NO_2$$

$$\text{rate of consumption of } N_2O_2 = k_b[N_2O_2][O_2]$$

The rate at which NO_2 is formed in this step is

$$\text{rate of formation of } NO_2 = 2k_b[N_2O_2][O_2]$$

The 2 appears in the rate law because two NO_2 molecules are formed in each reaction event, and so the concentration of NO_2 increases at twice the rate that the concentration of N_2O_2 decays.

The steady-state assumption

Now we proceed to derive the rate law on the basis of this proposed mechanism. The rate of formation of product comes directly from step 3:

$$\text{rate of formation of } NO_2 = 2k_b[N_2O_2][O_2]$$

However, this expression is not an acceptable overall rate law because it is expressed in terms of the intermediate: *an acceptable rate law for an overall reaction is expressed solely in terms of the species that appear in the overall reaction.* Therefore, we need to find an expression for the concentration of N_2O_2. To do so, we consider the **net rate** of formation of the intermediate, the difference between its rates of formation and decay. Because N_2O_2 is formed by step 1 but decays by steps 2 and 3, its net rate of formation is

$$\text{net rate of formation of } N_2O_2 = k_a[NO]^2 - k_a'[N_2O_2] - k_b[N_2O_2][O_2]$$

At this stage we introduce the steady-state assumption:

In the **steady-state assumption**, it is supposed that the concentrations of all intermediates remain constant throughout the reaction (except right at the beginning and right at the end).

The justification of this assumption is that the intermediate is present in low concentration throughout the reaction, and to a good approximation it is possible to neglect the small variation in its concentration once some has been formed. For our purposes, we identify the intermediate (which, in general, is any species that does not appear in the overall reaction but which has been invoked in the mechanism) as N_2O_2, and write

$$\text{steady-state assumption: net rate of formation of } N_2O_2 = 0$$

It follows from the expression above that

$$k_a[NO]^2 - k_a'[N_2O_2] - k_b[N_2O_2][O_2] = 0$$

This equation can be rearranged to give an equation for the concentration of N_2O_2:

$$[N_2O_2] = \frac{k_a[NO]^2}{k_a' + k_b[O_2]}$$

It follows that the rate of formation of NO_2 is

$$\text{rate of formation of } NO_2 = 2k_b[N_2O_2][O_2] = \frac{2k_a k_b[NO]^2[O_2]}{k_a' + k_b[O_2]} \quad (17a)$$

At this stage, the rate law is more complex than the observed law, but resembles it. The two become identical if we suppose that the rate of decomposition of the dimer is much greater than its rate of reaction with oxygen, for then

$$k_a'[N_2O_2] \gg k_b[N_2O_2][O_2]$$

or, after cancelling the $[N_2O_2]$,

$$k_a' \gg k_b[O_2]$$

When this condition is satisfied, the denominator in the overall rate law can be approximated by k_a' alone, and we conclude that

$$\text{rate of formation of } NO_2 = \left(\frac{2k_a k_b}{k_a'}\right)[NO]^2[O_2] \quad (17b)$$

which has the observed overall third-order form, eqn (16). Moreover, we can identify the observed rate constant as the following combination of rate constants for the elementary reactions:

$$k = 2\frac{k_a k_b}{k_a'} \quad (17c)$$

The proposed mechanism is consistent with the anomalous temperature dependence because although each one of the elementary rate constants increases with temperature, if k_a' increases more than the product $k_a k_b$ increases, then the overall rate constant k will decrease with increasing temperature and the reaction will go more slowly. That k_a' has a strong temperature dependence is consistent with the mechanism, because the decomposition step, which relies on the dissociation of the dimer into NO molecules, can be expected to have a high activation energy (and, as we have seen, a high activation energy implies that a reaction has a rate that depends strongly on temperature). The general conclusion is that, although the rate constants of elementary reactions almost invariably increase with increasing temperature, the observed rate constant, being a composite of several elementary rate constants, may either increase or decrease with temperature. In practice, most composite rate constants increase with temperature, so the rates of most reactions do increase with temperature.

Exercise E7.12 An alternative mechanism that may apply when the concentration of O_2 is high and that of NO is low is one in which the first step is $NO + O_2 \rightarrow NO \cdot O_2$ and its reverse, followed by $NO \cdot O_2 + NO \rightarrow NO_2 + NO_2$. Confirm that this mechanism also leads to the observed rate law when the concentration of NO is low.

$$[\text{rate} = 2k_a k_b [NO]^2 [O_2]/\{k_a' + k_b [NO]\} \approx (2k_a k_b / k_a')[NO]^2 [O_2]]$$

The rate-determining step

The mechanism of oxidation of nitrogen monoxide can be used to introduce another important concept. Suppose that the rate of step 3 (p. 266) is very fast, so that k_a' may be neglected relative to $k_b [O_2]$. (One way to achieve this condition is to increase the concentration of O_2 in the reaction mixture.) Then the eqn (17a) simplifies to

$$\text{rate of formation of } NO_2 = \frac{2k_a k_b [NO]^2 [O_2]}{k_b [O_2]} = 2k_a [NO]^2 \quad (18)$$

Now the reaction is second order in NO, and the concentration of O_2 does not appear in the rate law. The explanation is that the rate of reaction of N_2O_2 is so great (on account of the high concentration of O_2 in the system), that as soon as any N_2O_2 is formed, it reacts. Therefore, the rate of formation of NO_2 is determined by the rate at which N_2O_2 is formed, which is a bimolecular, second-order-elementary process. The formation of N_2O_2 in this mechanism, and in the presence of a high concentration of O_2, is an example of a rate-determining step:

The **rate-determining step** in a reaction mechanism, the slowest step, is the step that controls the rate of the overall reaction.

The rate-determining step is like a slow ferry crossing between two fast highways: the overall rate at which traffic can reach its destination is determined by the rate at which it can make the ferry crossing.

When the concentration of O_2 is reduced to the point that eqn (17b) is applicable, the rate-determining step in the reaction becomes the slow reaction of N_2O_2 with the scarce O_2 molecules. Now the rate of the overall reaction is determined by the value of k_b, the rate constant of the slowest step. It also depends on the ratio of the rate constants for the fast forward and reverse steps, k_a and k_a'. The latter dependence can be explained by considering the extreme case in which these two reactions are so fast compared with the rate-determining step that they reach a state of dynamic equilibrium. The two rates are then equal, and by setting

$$k_a[NO]^2 = k_a'[N_2O_2]$$

we can deduce that the ratio of rate constants is equal to the equilibrium constant for the formation of the intermediate:

$$K = \frac{[N_2O_2]}{[NO]^2} = \frac{k_a}{k_a'}$$

Therefore, the rate constant in eqn (17c) can be written

$$k = 2K \times k_b$$

The first factor K effectively determines the concentration of the reaction intermediate and the second, the rate constant for the slow step, determines the rate at which that intermediate forms products.

That the rate determining step is the N_2O_2 formation reaction when the O_2 concentration is high, but is the rate of the reaction of N_2O_2 when the O_2 concentration is low, emphasizes that the rate-determining step is not necessarily a fixed quantity, but may switch from one elementary step to another as the conditions are changed.

The Michaelis–Menten mechanism of enzyme action

Another example of a reaction in which an intermediate is formed is the **Michaelis–Menten mechanism** of enzyme action. The rate of an enzyme-catalyzed reaction in which a substrate S is converted into products P is found to depend on the concentration of the enzyme E even though the enzyme undergoes no net change, and a typical rate law was given at the start of the chapter:

$$\text{rate of formation of product} = \frac{k[E][S]}{[S] + K} \tag{19}$$

The proposed mechanism (with all species in an aqueous environment) is

1. The bimolecular formation of a combination, ES, of the enzyme and the substrate:

$$E + S \rightarrow ES \qquad \text{rate of formation of ES} = k_a[E][S]$$

2. The unimolecular decomposition of the complex:

$$ES \rightarrow E + S \qquad \text{rate of decomposition of ES} = k_a'[ES]$$

3. The unimolecular formation of products and the release of the enzyme from its combination with the substrate:

$$ES \rightarrow P + E \qquad \text{rate of formation of P} = k_b[ES]$$

$$\text{rate of consumption of ES} = k_b[ES]$$

We seek the rate law for the rate of formation of product, which, according to step 3, is

$$\text{rate of formation of P} = k_b[ES]$$

We see that we need to know the concentration of the intermediate ES (which cannot occur in the overall rate law). Therefore, to employ the steady-state assumption, we set up an expression for the net rate of formation of ES (allowing for its formation in step 1 and its removal in steps 2 and 3), and then set that net rate equal to zero:

net rate of formation of ES $= k_a[E][S] - k_a'[ES] - k_b[ES] = 0$

It follows that

$$[ES] = \frac{k_a[E][S]}{k_a' + k_b}$$

Exercise 7.13 What would be the concentration of ES if it was assumed that the formation and decomposition of ES were a rapid dynamic equilibrium?

$$[[ES] = (k_a/k_a')[E][S]]$$

However, there is now a small complication: [E] and [S] are the concentrations of the *free* enzyme and *free* substrate, and if $[E]_0$ is the total concentration of enzyme,

$$[E] + [ES] = [E]_0$$

Because only a little enzyme is added, the free substrate concentration is almost the same as the total substrate concentration, and we can ignore the fact that [S] differs slightly from [S] + [ES]. Therefore,

$$[ES] = \frac{k_a([E]_0 - [ES])[S]}{k_a' + k_b}$$

which rearranges to

$$[ES] = \frac{k_a[E]_0[S]}{k_a' + k_b + k_a[S]}$$

It follows that the rate of formation of product is

$$\text{rate of formation of P} = k[E]_0 \quad \text{with} \quad k = \frac{k_b[S]}{K_M + [S]} \qquad (20a)$$

where the **Michaelis constant**, K_M, is

$$K_M = \frac{k_a' + k_b}{k_a} \qquad (20b)$$

and the concentration of the complex ES is

$$[ES] = \frac{[E][S]}{K_M} \qquad (20c)$$

According to eqn (20a), the rate of enzymolysis is first order in the enzyme concentration, but the effective rate constant k depends on the concentration of substrate. Thus, when $[S] \gg K_M$, the effective rate constant is equal to k_b, and the rate law in eqn (20a) reduces to

$$\text{rate of formation of P} = k_b[E]_0 \quad (21a)$$

The rate is independent of the concentration of S because there is so much substrate present that it remains at effectively the same concentration even though products are being formed. Moreover, the rate is a maximum, and $k_b[E]_0$ is called the **maximum velocity**, v_{max}, of the enzymolysis:

$$v_{max} = k_b[E]_0 \quad (21b)$$

The constant k_b is called the **maximum turnover number**. The rate-determining step is step 3, because there is ample ES present (because S is so abundant), and the rate is determined by the rate at which ES reacts to form the product. It follows from eqns (20a) and (21b) that the reaction rate, v, at a general substrate composition, is related to the maximum velocity by

$$v = \frac{[S]v_{max}}{K_M + [S]} \quad (22)$$

This expression is the basis of the analysis of enzyme-kinetic data by using a **Lineweaver–Burk plot**, as explained in the following example.

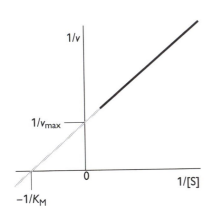

Fig. 7.19 A Lineweaver–Burk plot is used to analyse kinetic data on enzyme-catalysed reactions. The reciprocal of the rate of formation of products ($1/v$) is plotted against the reciprocal of the substrate concentration ($1/[S]$). All the data points (which typically lie in the black region of the line) correspond to the same overall enzyme concentration, $[E]_0$. The intercept of the extrapolated (grey) straight line with the horizontal axis is used to obtain the Michaelis constant K_M. The intercept with the vertical axis is used to determine $v_{max} = k_b[E]_0$, and hence k_b. The slope may also be used, for it is equal to K_M/v_{max}.

Example Analysing enzyme-kinetic data
A *Lineweaver–Burk plot* of kinetic data is a graph of $1/v$ (the reciprocal of the reaction rate) against $1/[S]$, the reciprocal of the substrate concentration. Show how this plot may be used to determine the maximum velocity and the Michaelis constant for the reaction.

Answer The rate of an enzyme-catalysed reaction is given by eqn (22). If we take the reciprocal of both sides the equation becomes

$$\frac{1}{v} = \frac{K_M + [S]}{v_{max}[S]} = \frac{K_M}{v_{max}} \times \frac{1}{[S]} + \frac{1}{v_{max}}$$

Therefore, if $1/v$ is plotted against $1/[S]$, a straight line should be obtained. The intercept of the straight line is K_M/v_{max} and the extrapolated slope at $1/[S] = 0$ is equal to $1/v_{max}$ (Fig. 7.19). Therefore, the slope can be used to find v_{max}, and then that value combined with the intercept to find the value of K_M. Alternatively, note that the extrapolated intercept with the horizontal axis (when $1/v = 0$) occurs at $1/[S] = -1/K_M$.

Exercise E7.14 Show that a plot of v against $v/[S]$ is an alternative route to the value of K_M.

$$[v = v_{max} - K_M \times (v/[S])]$$

When so little S is present that $[S] \ll K_M$, the rate of formation of products is

$$\text{rate of formation of P} = \frac{k_b}{K_M}[E]_0[S] \qquad (23)$$

Now the rate is proportional to $[S]$ as well as to $[E]_0$ because the rate at which ES is formed has become rate determining.

Enzyme inhibition

The action of an enzyme may be partially suppressed by the presence of a foreign substance, which is called an **inhibitor**. An inhibitor, which we denote I, may be a poison that has been administered (perhaps accidently) to the organism, or it may be a substance that is naturally present in a cell, and which is a component of the regulatory mechanism of the cell. The use of the techniques we have been developing for investigating the mode of action of enzymes can be illustrated by the problem of deciding whether the inhibition of an enzyme is competitive or noncompetitive. In **competitive inhibition** the inhibitor competes for the active site and so reduces the ability of the enzyme to bind the substrate (Fig. 7.20(b)). In **noncompetitive inhibition** the inhibitor does not compete for the active site; instead, it attaches to another part of the enzyme molecule, thereby distorting it and reducing its ability to bind the substrate (Fig. 7.20(c)). We shall now show that the two kinds of inhibition can be distinguished by making use of kinetic data on the rate of enzyme action.

First, consider competitive inhibition. We suppose that the inhibitor molecule, I, is in equilibrium with the complex, EI, it forms when it is bound to the active site:

$$EI \rightleftharpoons E + I \qquad K_I = \frac{[E][I]}{[EI]}$$

The rate of formation of product turns out to be

$$\text{rate of formation of P} = \frac{k_b[S][E]_0}{[S] + K_M\left(1 + \dfrac{[I]}{K_I}\right)} \qquad (24)$$

If we compare this expression with eqn (22), we see that the role of the inhibitor is to modify K_M; therefore, in a Lineweaver–Burk plot (see the example above), the slope and the intercept with the horizontal axis change as $[I]$ is changed, but the intercept with the vertical axis (the value of $v_{max} = k_b[E]_0$) remains unchanged (Fig. 7.21(a)).

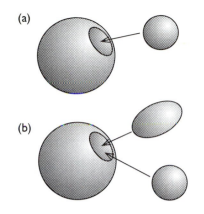

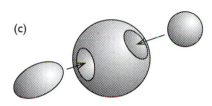

Fig. 7.20 (a) In an enzyme-catalysed reaction in which there is no inhibition, the reaction can be considered to occur by the attachment of the substrate (small sphere) to the active site (the cavity) in the enzyme molecule (the large sphere). (b) In competitive inhibition, both the substrate and the inhibitor (the egg shape) compete for the active site, and reaction ensues only if the substrate is successful in attaching there. (c) In one version of noncompetitive inhibition, the substrate and the inhibitor attach to distant sites of the enzyme molecule, and a complex in which they are both attached does not lead to the formation of product.

JUSTIFICATION

The only change between this case and the case of no inhibition is that some of the enzyme is uselessly linked to the inhibitor, so the total concentration of enzyme is

$$[E]_0 = [E] + [ES] + [EI]$$

in place of $[E] + [ES]$ in the former calculation. It remains the case that

$$[ES] = \frac{[E][S]}{K_M} \quad \text{rate of formation of P} = k_b[ES]$$

and so

$$[E]_0 = \frac{K_M[ES]}{[S]} + [ES] + \frac{K_M[I][ES]}{K_I[S]}$$

$$= \frac{[ES]}{[S]}\left(K_M + [S] + \frac{K_M[I]}{K_I}\right)$$

This expression can be rearranged into an equation for $[ES]$:

$$[ES] = \frac{[S][E]_0}{[S] + K_M\left(1 + \dfrac{[I]}{K_I}\right)}$$

When substituted into the rate equation for the formation of product, we obtain eqn (24).

Now consider noncompetitive inhibition. We suppose that the inhibitor is in equilibrium with a bound state IE, but that the site occupied by I is not the active site for the attachment of S (which is why we write it IE and not EI, to suggest the use of a site that is distant from the one used to form ES). Moreover, because I and S are not in competition for the same site, I may also bind to the complex ES to give a complex that we shall denote IES:

$$IES \rightleftharpoons I + ES \qquad K_I' = \frac{[I][ES]}{[IES]}$$

We suppose that although I and S may both bind to E, the enzyme can bring about a change in S only if I is not present. (The presence of I in

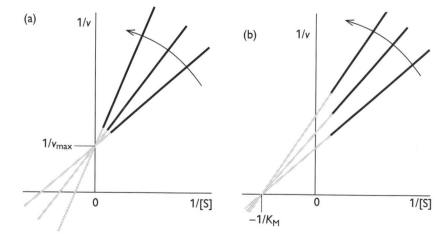

Fig. 7.21 Lineweaver–Burk plots may be used to distinguish competitive and noncompetitive inhibition kinetically. (a) Competitive inhibition: the intercept with the *vertical* axis does not move as the concentration of inhibitor is increased. (b) Noncompetitive inhibition: the intercept with the *horizontal* axis does not move as the concentration of inhibitor is increased.

IE allows S to bind, but so affects the structure of the enzyme that it cannot carry out its function.) Therefore, only ES can give rise to products; IES cannot. In this scenario, the rate of formation of product turns out to follow the rate law

$$\text{rate of formation of P} = \frac{k_b[S][E]_0}{([S] + K_M)\left(1 + \frac{[I]}{K_I}\right)} \tag{25}$$

In a Lineweaver–Burk plot with different values of [I], the straight lines now pass through a common intercept with the horizontal axis (because $1/v = 0$ at $1/[S] = -1/K_M$ independent of the inhibitor properties), but the slope and the intercept with the vertical axis both increase as the concentration of inhibitor is increased (Fig. 7.21(b)).

JUSTIFICATION

The total concentration of enzyme is now given by the expression

$$[E]_0 = [E] + [ES] + [IE] + [IES]$$

The concentrations of the complexes are given by the same equations as before, but with the addition of the equilibrium for the double complex IES:

$$[ES] = \frac{[E][S]}{K_M}$$

$$[I] = \frac{[I][E]}{K_I}$$

$$[IES] = \frac{[I][ES]}{K_I'}$$

From now on we shall suppose that the equilibrium between I and its bound state with the complex is independent of whether S is attached to the enzyme (because the sites are so distant), and hence set K_I' equal to K_I. With this simplification in mind, we can express the total concentration of enzyme as

$$[E]_0 = \frac{K_M[ES]}{[S]} + [ES] + \frac{K_I[I][ES]}{K_I[S]} + \frac{[I][ES]}{K_I}$$

$$= \frac{[ES]}{[S]}\left(K_M + [S] + \frac{K_M[I]}{K_I} + \frac{[I][S]}{K_I}\right)$$

$$= \frac{[ES]}{[S]}\left(1 + \frac{[I]}{K_I}\right)(K_M + [S])$$

This expression can be rearranged into an equation for [ES]:

$$[ES] = \frac{[S][E]_0}{([S] + K_M)\left(1 + \frac{[I]}{K_I}\right)}$$

The rate of formation of product is proportional to [ES] (because we are assuming that IES is inactive), and so at this stage we can use the same expression for the rate as in the noninhibited case:

$$\text{rate of formation of P} = k_b[ES]$$

When the expression for [ES] that we have just calculated is substituted into that rate equation, we obtain eqn (25).

Now we can see how the distinction between competitive and noncompetitive inhibition can be made. A series of Lineweaver–Burk plots are made for different inhibitor concentrations; if the plots resemble those in Fig. 7.21(a), then the inhibition is competitive. On the other hand, if the plots resemble those in Fig. 7.21(b), then the inhibition is noncompetitive.

7.8 Unimolecular reactions

A number of gas-phase reactions follow first-order kinetics, as in the isomerization of cyclopropane mentioned earlier, when the strained triangular molecule bursts apart into an acyclic alkene:

$$cyclo\text{-}C_3H_6 \rightarrow CH_3CH{=}CH_2 \qquad \text{rate} = k[cyclo\text{-}C_3H_6]$$

The problem with first-order rate laws is that, presumably, a molecule acquires enough energy to react as a result of its collisions with other molecules. The cyclopropane molecule, for example, needs energy to overcome the activation barrier to breaking a carbon–carbon bond and reorganizing the arrangement of hydrogen atoms. Collisions, though, are simple bimolecular events, and so how can they result in a first-order rate law? First-order gas-phase reactions are widely called **unimolecular reactions** since the rate-determining step is an elementary unimolecular reaction in which the reactant molecule changes into the product. This term must be used with caution, because the composite mechanism has bimolecular as well as unimolecular steps.

The first successful explanation of unimolecular reactions is ascribed to Frederick Lindemann.† In the **Lindemann mechanism** it is supposed that a reactant molecule, A, becomes energetically excited by collision with another molecule, M:

$$A + M \rightarrow A^* + M \quad \text{rate of formation of } A^* = k_a[A][M]$$

The gas M is an inert gas that is usually present in much greater abundance than the gas actually undergoing reaction. The energized molecule might lose its excess energy by collision with another:

$$A^* + M \rightarrow A + M \quad \text{rate of deactivation of } A^* = k_a'[A^*][M]$$

Alternatively, the excited molecule might shake itself apart (as may happen with vibrationally excited cyclopropane) and form products, P. That is, it might undergo the unimolecular decay

$$A^* \rightarrow P \qquad \text{rate of formation of } P = k_b[A^*]$$

$$\text{rate of consumption of } A^* = k_b[A^*]$$

If the unimolecular step is slow enough to be the rate-determining step, the overall reaction will have first-order kinetics, as observed. We can demonstrate this explicitly by applying the steady-state approximation to the net rate of formation of A^*:

$$\text{net rate of formation of } A^* = k_a[A][M] - k_a'[A^*][M] - k_b[A^*] = 0$$

† Laidler, in his *Chemical kinetics* (Harper and Row, 1987), gives an interesting historical summary of the origin of the mechanism. Apparently, Lindemann sketched the mechanism at a meeting in 1921 and published a brief note; almost simultaneously a Danish doctoral student, J. A. Christiansen, published his Ph.D. thesis in which the same mechanism was proposed and developed in much greater detail. The Lindemann–Christiansen mechanism would appear to be a fairer name.

This equation solves to

$$[A^*] = \frac{k_a[A][M]}{k_b + k_a'[M]}$$

It follows that the rate law for the formation of products is

$$\text{rate of formation of P} = k_b[A^*] = \frac{k_a k_b[M][A]}{k_b + k_a'[M]} \qquad (26a)$$

At this stage the rate law is not first order. However, if the rate of deactivation by (A^*, M) collisions is much greater than the rate of unimolecular decay, in the sense that

$$k_a'[A^*][M] \gg k_b[A^*], \quad \text{which corresponds to} \quad k_a'[M] \gg k_b$$

then we can neglect k_b in the denominator of the rate law and obtain

$$\text{rate of formation of P} = k[A] \quad \text{with} \quad k = \frac{k_a k_b}{k_a'} \qquad (26b)$$

Equation (26b) is a first-order rate law, as we set out to show.

Exercise E7.15 Suppose that the gas M is absent and that excitation of A and de-excitation of A^* occur by collisions with A molecules. Devise the rate law for the formation of products.

$$[\text{rate} = k_a k_b[A]^2/(k_b + k_a'[A])]$$

Chain reactions

Many gas-phase reactions and liquid-phase polymerization reactions are **chain reactions**, reactions in which an intermediate produced in one step generates a reactive intermediate in a subsequent step, then that intermediate generates another reactive intermediate, and so on.

7.9 The structure of chain reactions

The intermediates responsible for the propagation of a chain reaction are called **chain carriers**. In a **radical chain reaction** the chain carriers are radicals. Ions may also propagate chains, and in nuclear fission the chain carriers are neutrons.

The classification of reaction steps

The first chain carriers are formed in the **initiation step** of the reaction. For example, Cl atoms are formed by the dissociation of Cl_2 molecules either as a result of vigorous intermolecular collisions in a thermolysis reaction or as a result of absorption of a photon in a photolysis reaction. The chain carriers produced in the initiation step attack other reactant molecules in the **propagation steps**, and each attack gives rise to a new chain carrier. An example is the attack of a methyl radical on ethane:

$$\cdot CH_3 + CH_3CH_3 \rightarrow CH_4 + \cdot CH_2CH_3$$

(The dot signifies the unpaired electron and marks the radical.) In some cases the attack results in the production of more than one chain carrier. An example of such a **branching step** is

$$\cdot O\cdot + H_2O \rightarrow HO\cdot + HO\cdot$$

where the attack of one O atom on an H_2O molecule forms two $\cdot OH$ radicals (an O atom has the configuration $[He]2s^2 2p^4$, with two unpaired electrons).

The chain carrier might attack a product molecule formed earlier in the reaction. Because this attack reduces the net rate of formation of product, it is called a **retardation step**. For example, in a photochemical reaction in which HBr is formed from H_2 and Br_2, an H atom might attack an HBr molecule, leading to H_2 and Br:

$$\cdot H + HBr \rightarrow H_2 + \cdot Br$$

Retardation does not end the chain, because one radical $(\cdot H)$ gives rise to another $(\cdot Br)$, but it does deplete the concentration of the product. Elementary reactions in which radicals combine and end the chain are called **termination steps**, as in

$$CH_3CH_2\cdot + \cdot CH_2CH_3 \rightarrow CH_3CH_2CH_2CH_3$$

In an **inhibition step**, radicals are removed other than by chain termination, such as by reaction with the walls of the vessel or with foreign radicals:

$$CH_3CH_2\cdot + \cdot R \rightarrow CH_3CH_2R$$

The NO molecule has an unpaired electron and is a very efficient chain inhibitor. The observation that a reaction is quenched when NO is introduced is a good indication that a radical chain mechanism is in operation.

The rate laws of chain reactions

A chain reaction often leads to a complicated rate law (but not always). As a first example, consider the thermal reaction between H_2 and Br_2. The overall reaction and the observed rate law are

$$H_2(g) + Br_2(g) \rightarrow 2HBr(g)$$

$$\text{rate of formation of HBr} = \frac{k[H_2][Br_2]^{3/2}}{[Br_2] + k'[HBr]} \qquad (27)$$

The complexity of the rate law suggests that a complicated mechanism is involved, and the following radical chain mechanism has been proposed:

1. Initiation:

$$Br_2 \rightarrow 2Br\cdot \qquad \text{rate of consumption of } Br_2 = k_a[Br_2]$$

(At low pressures this elementary reaction is bimolecular and second order in Br_2.)

2. Propagation:

$$Br\cdot + H_2 \rightarrow HBr + H\cdot \qquad \text{rate} = k_b[Br][H_2]$$
$$H\cdot + Br_2 \rightarrow HBr + Br\cdot \qquad \text{rate} = k_b'[H][Br_2]$$

In this and the following steps, 'rate' means either the rate of formation of one of the products or the rate of consumption of one of the reactants. We shall specify the species only if the rates differ.

3. Retardation:

$$H\cdot + HBr \rightarrow H_2 + Br\cdot \qquad \text{rate} = k_c[H][HBr]$$

4. Termination:

$$Br\cdot + \cdot Br + M \rightarrow Br_2 + M \quad \text{rate of formation of } Br_2 = k_d[Br]^2$$

(The third body, M, a molecule of an inert gas, removes the energy of recombination; the constant concentration of M has been absorbed into the rate constant, k_d.) Other possible termination steps include the recombination of H atoms to form H_2 and the combination of H and Br atoms; however, it turns out that only Br-atom recombination is important.

Now we establish the rate law for the reaction. The experimental rate law is expressed in terms of the rate of formation of product, HBr, so we start by writing an expression for its net rate of formation. Because HBr is formed in step 2 (by both reactions) and consumed in step 3,

$$\text{net rate of formation of HBr} = k_b[Br][H_2] + k_b'[H][Br_2] - k_c[H][HBr]$$

To make progress, we need the concentrations of the intermediates Br and H. Therefore, we set up the expressions for their net rate of formation and apply the steady-state assumption to both:

$$\text{net rate of formation of H} = k_b[Br][H_2] - k_b'[H][Br_2]$$
$$- k_c[H][HBr] = 0$$
$$\text{net rate of formation of Br} = 2k_a[Br_2] - k_b[Br][H_2]$$
$$+ k_b'[H][Br_2] + k_c[H][HBr]$$
$$- 2k_d[Br]^2 = 0$$

The steady-state concentrations of the intermediates are found by solving these two equations and are

$$[Br] = \sqrt{\frac{k_a[Br_2]}{k_d}}$$

$$[H] = \frac{k_b \left(\dfrac{k_a}{k_d}\right)^{1/2} [Br_2]^{1/2}}{k_b'[Br_2] + k_c[HBr]}$$

When we substitute these concentrations into the equation for the net rate of formation of HBr we obtain

$$\text{rate of formation of HBr} = \frac{2k_b \left(\dfrac{k_a}{k_d}\right)^{1/2} [H_2][Br_2]^{3/2}}{[Br_2] + \dfrac{k_c}{k_b'}[HBr]} \tag{28a}$$

This equation has the same form as the empirical rate law, and we can identify the two empirical rate coefficients as

$$k = 2k_b \left(\frac{k_a}{k_d}\right)^{1/2} \qquad k' = \frac{k_c}{k_b'} \tag{28b}$$

We can conclude that the proposed mechanism is at least consistent with the observed rate law. Additional support for the mechanism would come from the detection of the proposed intermediates (by spectroscopy), and the measurement of individual rate constants for the elementary steps and confirming that they correctly reproduced the observed composite rate constants.

7.10 Explosions

A **thermal explosion** is due to the rapid increase of reaction rate with temperature. If the energy of an exothermic reaction cannot escape, the temperature of the reaction system rises, and the reaction goes faster. The acceleration of the rate results in a faster rise of temperature, and so the reaction goes even faster $\cdots$ catastrophically fast. A **chain-branching explosion** may occur when there are chain branching steps in a reaction, for then the number of chain carriers grows exponentially and the rate of reaction may cascade into an explosion.

An example of both types of explosion is provided by the reaction between hydrogen and oxygen:

$$2H_2(g) + O_2(g) \rightarrow 2H_2O(g)$$

Although the net reaction is very simple, the mechanism is very complex and has not yet been fully elucidated. It is known that a chain reaction is involved, and that the chain carriers include $\cdot H$, $\cdot O \cdot$, $\cdot OH$, and $\cdot O_2H$. Some steps are shown below.

Initiation: $H_2 + O_2 \rightarrow \cdot O_2H + \cdot H$

Propagation: $O_2 + \cdot H \rightarrow \cdot O \cdot + \cdot OH$ (branching)

$\cdot O \cdot + H_2 \rightarrow \cdot OH + \cdot H$ (branching)

$H_2 + \cdot OH \rightarrow \cdot H + H_2O$

The two branching steps can lead to chain-branching explosion.

The occurrence of an explosion depends on the temperature and pressure of the system, and the explosion regions for the reaction are shown in Fig. 7.22. At very low pressures, the system is outside the explosion region and the mixture reacts smoothly. At these pressures the chain carriers produced in the branching steps can reach the walls of the container where they combine (with an efficiency that depends on the composition of the walls). Increasing the pressure of the mixture (along the broken line in the illustration) takes the system through the first explosion limit (if the temperature is greater than about 730 K). The mixture then explodes because the chain carriers react before reaching the walls and branching reactions are explosively efficient. The reaction is smooth when the pressure is above the second explosion limit. The concentration of molecules in the gas is then so great that the radicals produced in the branching reaction combine in the body of the gas, and gas-phase reactions such as $O_2 + \cdot H \rightarrow \cdot O_2H$ can occur. Recombination reactions like this are facilitated by three-body collisions because the third body (M) can remove the excess energy:

$$O_2 + \cdot H + M \rightarrow \cdot O_2H + M^*$$

The radical $\cdot O_2H$ is relatively unreactive and can reach the walls, where it is removed. At low pressures three-particle collisions are unimportant and recombination is much slower. At higher pressures, when three-particle collisions are important, the explosive propagation of the chain by the radicals produced in the branching step is partially quenched because $\cdot O_2H$ is formed in place of $\cdot O \cdot$ and $\cdot OH$. If the pressure is increased to above the third explosion limit the reaction rate increases so much that a thermal explosion occurs.

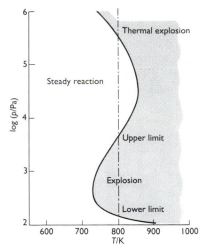

Fig. 7.22 The explosion limits of the $H_2 + O_2$ reaction. In the explosive regions, the reactions proceeds explosively when heated homogenously.

7.11 Photochemical reactions

Many reactions can be initiated by the absorption of light. The most important of all are the photochemical processes that capture the sun's radiant energy. Some of these reactions lead to the heating of the atmosphere during the day by absorption in the ultraviolet region as a result of reactions like those depicted in Fig. 7.23. Others include the absorption of red and blue light by chlorophyll and the subsequent use of the energy to bring about the synthesis of carbohydrates from carbon dioxide and water. Without photochemical processes the world would be simply a warm, sterile, rock.

Quantum yield

A molecule acquires enough energy to react by absorbing photons. The **Stark–Einstein law** states that one photon is absorbed by each molecule responsible for the primary photochemical process. The Stark–Einstein law is valid under normal conditions (when the radiant intensity is not very high), but it leaves open the possibility that, even though a reactant molecule absorbs a photon, it does not lead to products: there are many ways in which the excitation may be lost other than by dissociation or ionization. We therefore speak of the

Chemical kinetics

Fig. 7.23 The temperature profile through the atmosphere and some of the reactions that occur. The temperature peak at about 50 km is due to the absorption of solar radiation by the O_2 and N_2 reactions.

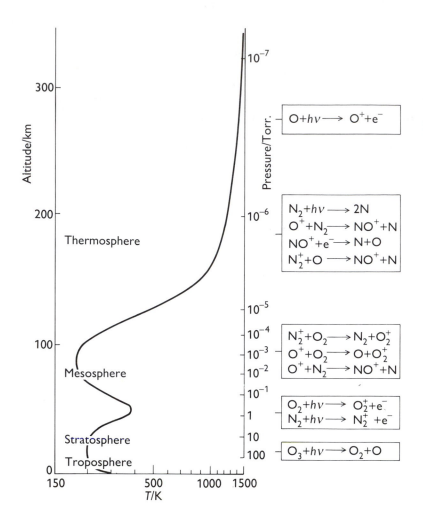

primary quantum yield, ϕ, the number of reactant molecules producing specified primary products (atoms or ions, for instance) for each photon absorbed.

The primary product of photon absorption—a radical, a photoexcited molecule, or an ion—may be successful in initiating a process that leads to products. As a result of one successful initiation, many reactant molecules might be consumed. The **overall quantum yield**, Φ, is the number of reactant molecules that react for each photon absorbed. In the photolysis of HI, for example, the processes are

$$HI + photon \rightarrow H + I$$

$$H + HI \rightarrow H_2 + I$$

$$2I \rightarrow I_2$$

The overall quantum yield is 2 because the absorption of one photon leads to the destruction of two HI molecules. In a chain reaction, Φ may be very large, and values of about 10^4 are common. In such cases the chain reaction acts as a chemical amplifier of the initial absorption step.

Example Using the quantum yield

The overall quantum yield for the formation of ethene from 4-heptanone with 313 nm light is 0.21. How many molecules of the ketone per second, and what amount per second, are destroyed when the sample is irradiated with a 50 W, 313 nm source under conditions of total absorption?

Answer Calculate the number of photons emitted by the lamp per second; all are absorbed (by assertion); the number of molecules destroyed per second is the number of photons absorbed multiplied by the quantum yield, Φ. The energy of a 313 nm photon is

$$\frac{hc}{\lambda} = 6.35 \times 10^{-19}\,\text{J}$$

A 50 W source therefore generates photons at a rate

$$\frac{50\,\text{J s}^{-1}}{6.35 \times 10^{-19}\,\text{J}} = 7.9 \times 10^{19}\,\text{s}^{-1}$$

The number of 4-heptanone molecules destroyed per second is therefore 0.21 times this quantity, or $1.7 \times 10^{19}\,\text{s}^{-1}$, which corresponds to $2.8 \times 10^{-5}\,\text{mol s}^{-1}$.

Exercise E7.16 The overall quantum yield for another reaction at 290 nm is 0.30. For what length of time must irradiation with a 100 W source continue in order to destroy 1.0 mol of molecules?

[3.8 h]

Photochemical rate laws

As an example of how to incorporate the photochemical activation step into a mechanism, consider the photochemical activation of the reaction

$$H_2(g) + Br_2(g) \rightarrow 2HBr(g)$$

In place of the first step in the thermal reaction we have

$$Br_2 \xrightarrow{h\nu} 2Br \quad \text{rate of formation of Br} = I_{abs}$$

where I_{abs} is the number of photons of the appropriate frequency absorbed per unit time per unit volume. It follows that I_{abs} should take the place of $k_a[Br_2]$ in the thermal reaction scheme, and so from eqn

283

(28a) we can write

$$\text{rate of formation of HBr} = \frac{2k_b(1/k_d)^{1/2}[H_2][Br_2]\sqrt{I_{abs}}}{[Br_2] + (k_c/k_b')[HBr]} \quad (29)$$

Although the details of this expression are complex, the essential prediction is clear: the reaction rate should depend on the square root of the absorbed light intensity. The prediction is confirmed experimentally.

EXERCISES

7.1 The rate of formation of C in the reaction $2A + B \rightarrow 2C + 3D$ is $1.0 \, \text{mol L}^{-1}\text{s}^{-1}$. State the rates of formation and consumption of A, B, and D.

7.2 The rate law for the reaction in Exercise 7.1 was reported as rate $= k[A][B][C]$. What are the units of k?

7.3 The rate constant for the first-order decomposition of N_2O_5 in the reaction $2N_2O_5 \rightarrow 4NO_2 + O_2$ is $k = 3.38 \times 10^{-5} \, \text{s}^{-1}$ at $25\,°C$. What is the half-life of N_2O_5? What will be the total pressure, initially 500 Torr for the pure N_2O_5 vapour, after (a) 10 s, (b) 10 min after initiation of the reaction?

7.4 If the rate laws are expressed with (a) concentrations in mol L^{-1}, (b) pressures in atmospheres, what are the units of the second-order and third-order rate constants?

7.5 The half-life for the (first-order) radioactive decay of ^{14}C is $5730 \, \text{y}$ (it emits β rays with an energy of $0.16 \, \text{MeV}$). An archaeological sample contained wood that had only 72 per cent of the ^{14}C found in living trees. What is its age?

7.6 One of the hazards of nuclear explosions is the generation of ^{90}Sr and its subsequent incorporation in place of calcium in bones. This nuclide emits β rays of energy $0.55 \, \text{MeV}$, and has a half-life of $28.1 \, \text{y}$. Suppose $1.00 \, \mu g$ was absorbed by a newly born child. How much will remain after (a) 18 y, (b) 70 y if none is lost metabolically?

7.7 The second-order rate constant for the reaction

$$CH_3COOC_2H_5(aq) + OH^-(aq)$$
$$\rightarrow CH_3CO_2^-(aq) + CH_3CH_2OH(aq)$$

is $0.11 \, \text{L mol}^{-1}\text{s}^{-1}$. What is the concentration of ester after (a) 10 s, (b) 10 min when ethyl acetate is added to sodium hydroxide so that the initial concentrations are $[NaOH] = 0.050 \, \text{mol L}^{-1}$ and $[CH_3COOC_2H_5] = 0.100 \, \text{mol L}^{-1}$?

7.8 A reaction $2A \rightarrow P$ has a second-order rate law with $k = 3.50 \times 10^{-4} \, \text{L mol}^{-1}\text{s}^{-1}$. Calculate the time required for the concentration of A to change from $0.260 \, \text{mol L}^{-1}$ to $0.011 \, \text{mol L}^{-1}$.

7.9 The composition of a liquid phase reaction $2A \rightarrow B$ was followed by a spectrophotometric method with the following results:

t/min	0	10	20	30	40	∞
$[B]/\text{mol L}^{-1}$	0	0.089	0.153	0.200	0.230	0.312

Determine the order of the reaction and its rate constant.

7.10 The rate constant for the decomposition of a certain substance is $2.80 \times 10^{-3} \, \text{L mol}^{-1}\text{s}^{-1}$ at $30\,°C$ and $1.38 \times 10^{-2} \, \text{L mol}^{-1}\text{s}^{-1}$ at $50\,°C$. Evaluate the Arrhenius parameters of the reaction.

7.11 The reaction $2H_2O_2(aq) \rightarrow 2H_2O(l) + O_2(g)$ is catalysed by Br^- ions. If the mechanism is

$$H_2O_2(aq) + Br^-(aq) \rightarrow H_2O(l) + BrO^-(aq)$$
$$\text{(slow)}$$
$$BrO^-(aq) + H_2O_2(aq) \rightarrow H_2O(l) + O_2(g) + Br^-(aq)$$
$$\text{(fast)}$$

give the order of the reaction with respect to the various participants.

7.12 The reaction mechanism

$$A_2 \rightleftharpoons 2A \qquad \text{(fast)}$$
$$A + B \rightarrow P \qquad \text{(slow)}$$

involves an intermediate, A. Deduce the rate law for the formation of P.

7.13 Consider the following mechanism for renaturation of a double helix from its strands A and B:

$$A + B \rightleftharpoons \text{unstable helix} \qquad \text{(fast)}$$
$$\text{unstable helix} \rightarrow \text{stable double helix} \quad \text{(slow)}$$

Derive the rate equation for the formation of the double helix and express the rate constant of the renaturation reaction in terms of the rate constants of the individual steps.

7.14 The enzyme-catalysed conversion of a substrate at 25 °C has a Michaelis constant of 0.035 mol L^{-1}. The rate of the reaction is 1.15×10^{-3} mol L^{-1} s^{-1} when the substrate concentration is 0.110 mol L^{-1}. What is the maximum velocity of this enzymolysis?

7.15 Find the condition for which the reaction rate of an enzymolysis that follows Michaelis–Menten kinetics is half its maximum value.

7.16 Consider the following mechanism for the thermal decomposition of R$_2$:

$$R_2 \rightarrow 2R \qquad (1)$$
$$R + R_2 \rightarrow P_B + R' \qquad (2)$$
$$R' \rightarrow P_A + R \qquad (3)$$
$$2R \rightarrow P_A + P_B \qquad (4)$$

where R$_2$, P$_A$, P$_B$ are stable hydrocarbons and R and R' are radicals. Find the dependence of the rate of decomposition of R$_2$ on the concentration of R$_2$.

7.17 Refer to Fig. 7.22 and determine the pressure range for branching chain explosion in the hydrogen–oxygen reaction at 700 K, 800 K, and 900 K.

7.18 In a photochemical reaction $A \rightarrow 2B + C$, the overall quantum yield with 500 nm light is 2.1×10^2 mol einstein^{-1}, where 1 einstein = 1 mol photons. After exposure of 300 mmol of A to the light, 2.28 mmol of B are formed. How many photons were absorbed by A?

7.19 In an experiment to measure the quantum efficiency of a photochemical reaction, the absorbing substance was exposed to 490 nm light from a 100 W source for 45 minutes. The intensity of the transmitted light was 40 per cent of the intensity of the incident light. As a result of irradiation, 0.344 mol of the absorbing substance decomposed. Find the quantum efficiency.

7.20 The condensation reaction of acetone, $(CH_3)_2CO$ (propanone), in aqueous solution is catalysed by bases, B, which react reversibly with acetone to form the carbanion $C_3H_5O^-$. The carbanion then reacts with a molecule of acetone to give the product. A simplified version of the mechanism is

$$AH + B \rightarrow BH^+ + A^- \qquad (1)$$
$$A^- + BH^+ \rightarrow AH + B \qquad (2)$$
$$A^- + HA \rightarrow \text{product} \qquad (3)$$

where AH stands for acetone and A$^-$ its carbanion. Use the steady-state approximation to find the concentration of the carbanion and derive the rate equation for the formation of the product.

7.21 Consider the acid-catalysed reaction

$$HA + H^+ \underset{(2)}{\overset{(1)}{\rightleftharpoons}} HAH^+ \qquad \text{(fast)}$$

$$HAH^+ + B \overset{(3)}{\longrightarrow} BH^+ + AH \qquad \text{(slow)}$$

Deduce the rate law and show that it can be made independent of the specific term [H$^+$].

7.22 Consider the following chain mechanism:

  (a) $AH \rightarrow A\cdot + H\cdot$
  (b) $A\cdot \rightarrow B\cdot + C$
  (c) $AH + B\cdot \rightarrow A\cdot + D$
  (d) $A\cdot + B\cdot \rightarrow P$

Identify the initiation, propagation, and termination steps, and use the steady-state approximation to deduce that the decomposition of AH is first order in AH.

7.23 The rate v of an enzyme-catalysed reaction was measured when various amounts of substrate S were present and the concentration of enzyme was 15 μmol L^{-1}. The following results were obtained:

[S]/mmol L^{-1}	1.0	2.0	3.0	4.0	5.0
v/μmol L^{-1} s^{-1}	1.1	1.8	2.3	2.6	2.9

285

Determine the Michaelis–Menten constant K_M, the maximum velocity of the reaction, and the maximum turnover number of the enzyme.

7.24 The following results were obtained when the rate of an enzymolysis was monitored (a) without inhibitor, (b) with inhibitor at a concentration of 18 μmol L^{-1}:

$[S]/10^{-4}$ mol L^{-1}		1.0	3.0	7.0	12.0	18.0
v/μmol L^{-1} s^{-1}	(a)	0.49	0.95	1.3	1.5	1.6
	(b)	0.27	0.52	0.71	0.81	0.86

Is the inhibition competitive or noncompetitive?

8

Atomic structure

CONTENTS

Central to the explanation of the physical and chemical properties of elements and compounds is the behaviour of electrons in atoms and molecules. It was once thought that the motion of atomic and subatomic particles could be expressed using the laws of **classical mechanics** introduced in the seventeenth century by Isaac Newton (see *Further information 2*, p. 24), for these laws were very successful at explaining the motion of planets and everyday objects such as pendulums and projectiles. However, towards the end of the nineteenth-century, experimental evidence accumulated showing that classical mechanics failed when it was applied to very small particles, such as individual atoms, nuclei, and electrons, and it took until about 1926 to discover the appropriate concepts and equations for describing them. The new concepts that have been found to be necessary make up the subject known as **quantum mechanics**. We introduce the concepts in this chapter and apply them throughout the remainder of the text.

The failures of classical physics

Classical physics is based on two presumptions:

1. A particle travels in a trajectory with a precise position and a precise velocity at each instant.

2. Any type of motion can be excited to a state of arbitrary energy.

These two presumptions agree with everyday experience. For example, a pendulum swings with a precise oscillating motion and can be made to oscillate with any energy simply by pulling it back to an arbitrary angle and then letting it swing freely. We can predict its position and the speed at which it is swinging at every instant.

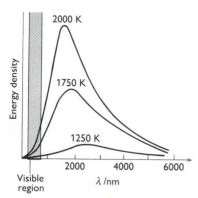

Fig. 8.1 The energy density per unit wavelength range in a black-body cavity at three temperatures. Note how the energy density increases in the visible region (the tinted range of wavelengths) as the temperature is raised, and how the peak maximum moves to shorter wavelengths. The total energy density (the area under the curve) increases as the temperature is increased (as T^4).

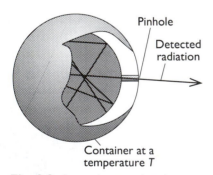

Fig. 8.2 A good approximation to a black-body radiator is a pinhole in a heated cavity. Any radiation inside is reflected many times and comes to thermal equilibrium with the walls at a temperature T. The pinhole allows some radiation to escape and be detected.

Everyday experience, however, does not extend to a familiarity with the behaviour of individual atoms and subatomic particles, and careful experiments of the type described below have shown that the laws of classical mechanics fail to account for the observed behaviour of very small particles. Classical mechanics is in fact only an *approximate* description of the motion of particles, and the approximation is invalid when it is applied to microscopically small particles, such as molecules, atoms, and electrons.

In the following sections, we review the observations that first showed that classical mechanics is not just slightly wrong but *fundamentally* incorrect.

8.1 Black-body radiation

The properties of electromagnetic radiation that are needed to understand this section are summarized in *Further information 8* (p. 290). A hot object emits electromagnetic radiation (this emission is the basis of the operation of incandescent lamps and of sunlight). At high temperatures, an appreciable proportion of the radiation is in the visible region of the spectrum, and a higher proportion of short-wavelength blue light is generated as the temperature is raised. This behaviour is observed when an iron bar glowing red hot become white hot when heated further, because then more blue light mixes into the red light, and changes the perceived colour to white (white is a mixture of all visible colours, from red through yellow and green to blue and violet). The precise dependence is illustrated in Fig. 8.1, which shows how the energy output depends on the wavelength at a series of temperatures. The curves are those of an ideal emitter called a **black body**, which is an object capable of emitting and absorbing all frequencies of radiation uniformly. A good approximation to a black body is a pinhole in a container, because the radiation leaking out of the hole has been absorbed and re-emitted inside so many times that it has come to thermal equilibrium with the walls (Fig. 8.2).

The experimental observations

Figure 8.1 shows two main features. The first is that the peak of energy output shifts to shorter wavelengths as the temperature is raised. As a result, the perceived colour shifts towards the blue, as already mentioned. An analysis of the data led Wilhelm Wien (in 1893) to a conclusion now known as **Wien's displacement law** and which summarizes this behaviour:

$$T\lambda_{max} = \text{constant} \quad (\text{constant} = 0.29 \text{ cm K}) \tag{1}$$

In this expression, λ_{max} is the wavelength of the maximum in the energy when the temperature is T. The displacement law implies that as T increases, λ_{max} decreases enough to preserve the same value of $T\lambda_{max}$. This formula can be used to estimate the temperatures of stars from the wavelength of the light they emit. For example, the maximum emission of the sun occurs at $\lambda_{max} \approx 490$ nm, so its surface temperature must be close to

$$T = \frac{0.29 \times 10^{-2} \text{ m K}}{490 \times 10^{-9} \text{ m}} = 5.9 \times 10^3 \text{ K}$$

or about 6000 K.

Exercise E8.1 Estimate the wavelength at the maximum energy output of an incandescent lamp if the filament is at 3000 °C.

[890 nm]

The second feature of black-body radiation had been noticed in 1879 by Josef Stefan, who considered the sharp rise in the power—the rate of energy output—emitted by a black body per unit area of its surface as the temperature was raised. He established **Stefan's law**:

$$\text{power emitted per unit area} = aT^4 \qquad (2)$$

Stefan's law implies that each square centimetre of the surface of a black body at 1000 K radiates about 5.7 W when all wavelengths are taken into account, but it radiates $3^4 = 81$ times that power (460 W) when the temperature is increased by a factor of 3, to 3000 K. The law is the basis of seeking as high a temperature as possible for an incandescent lamp, for then the emission is as strong as possible.

Exercise E8.2 Suppose technological advances made it possible to produce a ceramic material that could be used as a filament at 3800 °C instead of 3000 °C. By what factor would the power output of a lamp that used the new material increase?

[2.4]

The rate of energy emission from a black body is proportional to the density of the energy inside the cavity depicted in Fig. 8.2, because the higher the energy density (the total energy in the cavity divided by its volume), the greater the rate of emission of energy into the surroundings. Therefore, Stefan's law enables us to monitor the energy density inside the cavity.

The attempted classical interpretation

The physicist Lord Rayleigh studied black-body radiation from a classical viewpoint. In his day (at the end of the nineteenth century), electromagnetic radiation was regarded as waves in a jellylike 'ether'.

If the ether could oscillate at a certain frequency, ν, then radiation of that frequency and corresponding wavelength would be present in it. Rayleigh took the view that the ether could oscillate with any frequency and so waves could exist in it of any wavelength, and calculated the contribution to the energy density from each wavelength. With minor help from James Jeans, Rayleigh arrived at the **Rayleigh–Jeans law**:

$$\text{energy density in the wavelength range } \lambda \text{ to } \lambda + \delta\lambda = \frac{8\pi k T}{\lambda^4}\,\delta\lambda \quad (3)$$

The power emitted in the wavelength range $\delta\lambda$ is proportional to this energy density.

Unfortunately (for Rayleigh, Jeans, and classical physics), although

FURTHER INFORMATION 8:
Electromagnetic radiation and photons

Electromagnetic radiation, which includes γ radiation, ultraviolet radiation, visible light, infrared radiation, microwave radiation, and radio waves, is a wavelike, oscillating electric and magnetic field that propagates through space with a constant speed, c, the 'speed of light'. The radiation is depicted in Fig. 1: the electric field and magnetic fields are perpendicular to each other and vary sinusoidally with a wavelength λ and frequency ν that are related by

$$\lambda\nu = c$$

Therefore, the shorter the wavelength, the

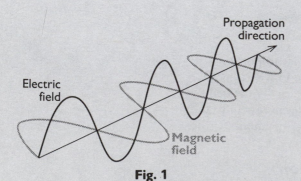

Fig. 1

higher the frequency of the radiation. According to classical physics, the intensity of a ray is proportional to the square of the amplitude of the wave, so an intense ray would be a wave of electromagnetic field that oscillated with a large amplitude. The classification of the electromagnetic spectrum into different regions according to the frequency and wavelength of the radiation is summarized in Table 1.

The wave shown in the illustration is **plane polarized**: it is so called because the electric field oscillates in a single plane. The plane may be orientated in any direction around the direction of propagation (with the electric field perpendicular to that direction). An alternative mode of polarization is **circular polarization**, in which the electric field rotates around the direction of propagation in either a clockwise or a counterclockwise sense.

According to quantum theory, a ray of frequency ν consists of a stream of photons, each one of which has energy

$$E = h\nu$$

Thus, a photon of high-frequency radiation has more energy than a photon of low-frequency radiation. The greater the intensity of the ray, the greater the number of photons in it. In a vacuum, each photon travels with the speed of light. The frequency of the radiation determines the colour of visible light because different visual receptors in the eye respond to photons of the Rayleigh–Jeans formula is quite successful at long wavelengths (low frequencies), it fails badly at short wavelengths. Thus, as λ decreases, the energy density in the cavity, and hence the power emitted per unit area, increases without going through a maximum (Fig. 8.3). The equation therefore predicts that oscillations of extremely small wavelength (high frequency, corresponding to ultraviolet light, X-rays, and even γ-rays) are strongly excited even at room temperature. This absurd result is called the **ultraviolet catastrophe**.

The Planck distribution

The German physicist Max Planck studied black-body radiation from the viewpoint of thermodynamics, in which he was an expert. In 1900

Table 1 The regions of the electromagnetic spectrum†

Regions	Wavelength	Frequency/ Hz
Radio-frequency	>30 cm	<10^9
Microwave	3 mm to 30 cm	10^9 to 10^{11}
Infrared	1000 nm to 3 mm	10^{11} to 3×10^{14}
Visible	400 nm to 800 nm	4×10^{14} to 8×10^{14}
Ultraviolet	300 nm to 3 nm	10^{15} to 10^{17}
X-rays, γ-rays	<3 nm	>10^{17}

† The regions are not sharply defined.

Table 2 Colour, frequency, and wavelength of light†

	Frequency/ 10^{14} Hz	Wavelength/ nm	Energy per photon/ 10^{-19} J photon^{-1}
X-rays and γ rays	10^3 and above	3 and below	660 and above
Ultraviolet	10	300	6.6
Visible light			
Violet	7.1	420	4.7
Blue	6.4	470	4.2
Green	5.7	530	3.7
Yellow	5.2	580	3.4
Orange	4.8	620	3.2
Red	4.3	700	2.8
Infrared	3.0	1000	1.9
Microwaves and radiowaves	3×10^{11} Hz and below	3×10^6 and above	2.0×10^{-22} J and below

† The values given are approximate but typical.

different energy. The relation between colour and frequency is shown in Table 2, which also gives the energy carried by each type of photon.

Photons may also be polarized. A plane-polarized ray of light consists of plane-polarized photons and a circularly polarized ray consists of circularly polarized photons. The latter can be regarded as spinning either clockwise or counterclockwise about their direction of propagation.

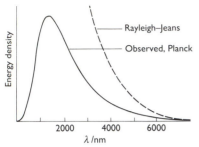

Fig. 8.3 Theoretical attempts to account for black-body radiation. The Rayleigh–Jeans law leads to an infinite energy density at short wavelengths and gives rise to the ultraviolet catastrophe. The Planck distribution is in excellent agreement with experiment.

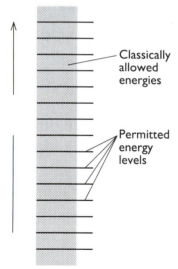

Fig. 8.4 According to classical physics, an oscillator (including the oscillators that correspond to vibrations of the electromagnetic field and correspond to radiation of a particular frequency) can have any energy (as depicted by the tinted range of energies). Planck's proposal implied that an oscillator could be excited only in discrete steps, for it can possess only certain energies (those depicted by the horizontal lines in the illustration).

he found that he could account for the characteristics of black-body radiation by proposing that *the energy of each electromagnetic oscillator is limited to discrete values and cannot be varied arbitrarily.* Thus, the oscillation of the electromagnetic field that corresponds to yellow light, for instance, could be stimulated only if a certain quantity of energy was provided. This limitation of the energy of an object to discrete values is called the **quantization of energy** (from the Latin word *quantum*, meaning amount). In particular, Planck proposed that the energy of an oscillator of frequency v is restricted to an integral multiple of the quantity hv, where h is a fundamental constant now known as **Planck's constant**:

$$E = nhv \quad (h = 6.626 \times 10^{-34} \, \text{J Hz}^{-1}) \tag{4}$$

with $n = 0, 1, 2, \ldots$ (Fig. 8.4).

Exercise E8.3 What is the minimum energy that can be used to excite an oscillator corresponding to yellow light (of frequency 5.2×10^{14} Hz)?

[3.4×10^{-19} J]

When Planck calculated the energy density in the wavelength range λ to $\lambda + \delta\lambda$ on the basis of his quantization postulate, instead of the Rayleigh–Jeans law, he derived the expression

energy density in the

$$\text{wavelength range } \lambda \text{ to } \lambda + \delta\lambda = \frac{8\pi hc}{\lambda^5} \left(\frac{1}{e^{hc/\lambda kT} - 1} \right) \delta\lambda \tag{5}$$

The crucial difference is the exponential function in the denominator, which has the effect of making the energy density approach zero as λ approaches zero,† and hence eliminates the ultraviolet catastrophe. The graph predicted by this **Planck distribution** is shown in Fig. 8.3: it compares very favourably with the experimental curve shown in Fig. 8.1.

Exercise E8.4 Calculate the ratio of the energy densities predicted by the Planck and Rayleigh–Jeans formulas for yellow (580 nm) light at 1000 K.

[Rayleigh : Planck $= 2.4 \times 10^9$]

† As λ approaches zero, $hc/\lambda kT$ approaches infinity and $e^{hc/\lambda kT}$ approaches infinity too; therefore the denominator in the Planck formula becomes infinite and so the expression itself approaches zero.

The reason why Planck's quantization hypothesis is successful is as follows. The atoms in the walls of the black body undergo thermal motion, and this motion excites the oscillators of the electromagnetic field. According to classical mechanics, all the electromagnetic oscillators are excited, even those of very high frequency, and the corresponding short-wavelength radiation is emitted. According to quantum mechanics, however, the oscillators are excited only if they can acquire an energy of at least $h\nu$. This minimum energy is too large for the walls to supply in the case of the high-frequency oscillators, and so the latter remain unexcited. *The effect of quantization is to eliminate the contribution from the high-frequency oscillators, for they cannot be excited with the energy available, and hence the very-short-wavelength radiation is not emitted.*

The Planck distribution accounts for the Stefan and Wien laws. Thus, when the area under the graph in Fig. 8.3 is calculated (to obtain the *total* energy density over the entire wavelength range), the resulting expression is proportional to T^4, which is consistent with Stefan's law. Similarly, when the wavelength that corresponds to the maximum point of the curve, λ_{max}, is calculated, its position is found to be inversely proportional to the temperature, which is in agreement with Wien's law.

Planck's view that an electromagnetic oscillator of frequency ν can possess only the energies $0, h\nu, 2h\nu, \ldots$ has led to an alternative view about the character of electromagnetic radiation. Instead of thinking of radiation as the excitement of an oscillator to one of its permitted energy levels, we can think of radiation as consisting of $0, 1, 2, \ldots$ *particles*, each particle having an energy $h\nu$. If there is only one such particle present, then the energy of the radiation is $h\nu$, if there are two particles of that frequency, then their total energy is $2h\nu$, and so on. These particles are now called **photons** (the name was coined by the chemist G. N. Lewis, whom we meet again in Chapter 9). According to the photon picture of radiation, a ray of light of a certain frequency consists of photons of a definite, fixed energy (the value of $h\nu$), and that is true whatever the intensity of the ray. However, as the intensity of the ray is increased (for example, by increasing the temperature of the source), the *number* of those fixed-energy photons increases. An intense beam of monochromatic (single-frequency) radiation consists of a dense stream of photons; a weak beam of radiation of the same frequency consists of a relatively small number of the same type of photons.

Example Calculating the number of photons
Calculate the number of photons emitted by a 100 W yellow lamp in 1.0 s. Take the wavelength of yellow light as 560 nm and assume 100 per cent efficiency.
Answer A 100 W lamp emits 100 J of energy in 1.0 s (if all the energy it consumes is converted into radiation). Because

560 nm photons have frequency 5.4×10^{14} Hz, each one has an energy

$$E = h\nu = (6.626 \times 10^{-34}\,\text{J Hz}^{-1}) \times (5.4 \times 10^{14}\,\text{Hz})$$
$$= 3.7 \times 10^{-19}\,\text{J}$$

The number of photons required to carry away 100 J of energy is therefore

$$N = \frac{100\,\text{J}}{3.7 \times 10^{-19}\,\text{J}} = 2.7 \times 10^{20}$$

Exercise E8.5 How many 1000 nm photons does a 1 mW infrared rangefinder emit in 0.1 s?

$$[5 \times 10^{14}]$$

8.2 Heat capacities

Heat capacities were first introduced in Section 2.1, where we saw that they are the constant of proportionality C between the rise in temperature (ΔT) of a sample and the heat q that needs to be supplied to bring that increase about:

$$q = C\,\Delta T$$

We should suspect that there will be similarities between black-body radiation and heat capacities, because the former involves the examination of how energy is taken up by the oscillations of the electromagnetic field, whereas the latter involves examining how energy is taken up by the oscillation of atoms (the only way that solids can store energy).

On the basis of some somewhat slender experimental evidence, the French scientists Pierre-Louis Dulong and Alexis-Thérèse Petit had proposed in 1819 that the molar heat capacities of all monatomic solids—such as metals—were equal (in modern units) to about $24\,\text{J K}^{-1}\,\text{mol}^{-1}$. Classical physics was able to account for this value quite readily, for if it is assumed that the atomic oscillators can be excited to any energy, then the predicted molar heat capacity is $3R$, where R is the gas constant, and $3R = 24\,\text{J K}^{-1}\,\text{mol}^{-1}$.

JUSTIFICATION

The theoretical justification of Dulong and Petit's law involves three essential steps. The first is to note that if a solid consists of N atoms, then since each atom can oscillate in any of three perpendicular directions (Fig. 8.5), the solid is equivalent to a collection of $3N$ oscillators. The second step makes use of a conclusion from classical physics known as the **equipartition theorem**, which implies that, at a temperature T, the average energy of an oscillator is kT, where k is **Boltzmann's constant**. Boltzmann's constant is equal to the gas constant R divided by Avogadro's constant N_A, so it follows that $R = N_A \times k$. It now follows from these two steps that the total energy of the N vibrating atoms is $3N \times kT$. The energy per mole of atoms is therefore $3N_A kT$, or $3RT$, because the number of atoms per mole is N_A. In the final step, we note that when the temperature of the sample increases by ΔT, the molar energy increases by $3R\,\Delta T$. This increase in energy must be supplied as heat from the surroundings, so the heat required to raise the temperature by ΔT is $3R\,\Delta T$. It follows by comparison with the definition of heat capacity ($q = C\,\Delta T$) that $C = 3R$.

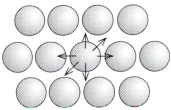

Fig. 8.5 An atom in a solid can oscillate about its position in three perpendicular directions, and when a solid is heated, the vigour of the motion increases. This illustration depicts a view of one layer of atoms in a solid and shows the directions of oscillation of one of the atoms in the plane.

The apparent success of classical mechanics in accounting for observed heat capacities was short-lived, for when technological advances made it possible to measure heat capacities at low temperatures, all substances were found to have values significantly smaller than $24\ \text{J K}^{-1}\,\text{mol}^{-1}$, and at very low temperatures the heat capacity was found to approach zero (Fig. 8.6).

To account for these new observations, Einstein proposed that each atom oscillates about its equilibrium position with a single frequency, v. He then invoked Planck's hypothesis to assert that the only permitted energy of any oscillating atom is an integral multiple of hv (exactly as in Fig. 8.4 for electromagnetic oscillators). On the basis of this model, Einstein was able to deduce the expression

$$C = 3R \times f, \quad \text{with} \quad f = \left(\frac{hv}{kT}\right)^2 \left(\frac{e^{hv/2kT}}{e^{hv/kT} - 1}\right)^2 \qquad (6)$$

The graph of this expression as a function of temperature is shown in Fig. 8.6: we see that it does indeed predict that the heat capacity of a solid should decrease as the temperature is lowered, and that it approaches zero as the temperature approaches zero.

The physical reason for the success of the Einstein model—and the form of eqn (6)—is that, as in Planck's treatment of the electromagnetic field, at low temperatures only a few atoms possess enough energy to be able to oscillate. As a result, the solid is incapable of absorbing heat readily and consequently its heat capacity is low. At higher temperatures there is enough energy available for all the oscillators to become active: all $3N$ oscillators contribute, and the heat capacity approaches its classical value of $3R$.

A more refined approach to the calculation was carried out by the Dutch physicist Peter Debye. He allowed for the atoms to oscillate with a range of frequencies rather than the single frequency supposed by Einstein. The graph of his more complicated expression is similar to Einstein's, but the numerical agreement with the experimental data is better (see Fig. 8.6). An important practical conclusion from Debye's calculation is that *at low temperatures, the heat capacity of a solid is*

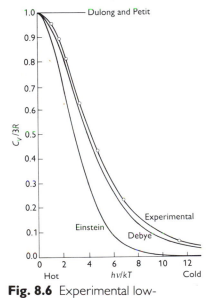

Fig. 8.6 Experimental low-temperature heat capacities and theoretical predictions. Note that the temperture *decreases* from left to right because the horizontal axis refers to $1/T$. The Dulong and Petit law predicts no falling off at low temperatures. The Einstein formula predicts the general temperature dependence quite well, but is everywhere too low. Debye's modification of Einstein's calculation gives much better agreement with experiment.

proportional to T^3. This dependence is used to extrapolate measurements of heat capacities at low temperatures down to values close to $T = 0$ (for the experimental determination of entropies, for instance; see Section 3.2).

8.3 Waves as particles and particles as waves

So far, we have seen that two observations—on the electromagnetic field and the heat capacities of solids—have led to the overthrow of the classical view that oscillators could have any energy. Now we move on to see that classical physics is even more rotten than these conclusions suggest. We shall now see how two other experimental observations upset another central concept of classical physics, that of the distinction between waves and particles. According to quantum mechanics, the concepts of particle and wave blend together: particles have the characteristics of waves and waves have the characteristics of particles.

The photoelectric effect

We have already seen that Planck's work led to the view that electromagnetic radiation consists of streams of particles, or photons. Evidence that confirmed the view that radiation can be interpreted as a stream of particles comes from the **photoelectric effect**, the ejection of electrons from metals when they are exposed to ultraviolet radiation (Fig. 8.7). The three characteristic observations of the photoelectric effect are as follows:

1. No electrons are ejected, *regardless of the intensity of the radiation*, unless the frequency of that radiation exceeds a threshold value characteristic of the metal.

2. The kinetic energy of the ejected electrons *varies linearly with the frequency* of the incident radiation but independent of its intensity.

3. Even at low light intensities, *electrons are ejected immediately* if the frequency is above the threshold value.

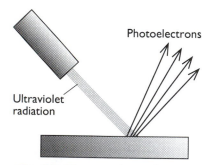

Photoelectrons

Ultraviolet radiation

Fig. 8.7 The experimental arrangement used to demonstrate the photoelectric effect. A beam of ultraviolet radiation is used to irradiate a patch of the surface of a metal, and electrons are ejected from the surface if the frequency of the radiation is above a threshold value that depends on the metal.

These observations strongly suggest an interpretation of the photoelectric effect in which an electron is ejected in a collision with a particle-like projectile so long as the projectile carries enough energy to expel the electron from the metal. If we suppose that the projectile is a photon of energy hv, where v is the frequency of the radiation, then the law of conservation of energy requires that the kinetic energy of the electron (which is equal to $\frac{1}{2}m_e v^2$, when the speed of the electron is v) should be equal to the energy supplied by the photon less the energy, Φ, required to remove the electron from the metal (1):

$$\tfrac{1}{2}m_e v^2 = h\nu - \Phi \qquad (7)$$

The quantity Φ is called the **work function** of the metal (it is the analogue of the ionization energy for atoms).

> **Exercise E8.6** The work function of rubidium is 2.09 eV (1 eV = 1.60×10^{-19} J). Can blue (470 nm) light eject electrons from the metal?
>
> [yes]

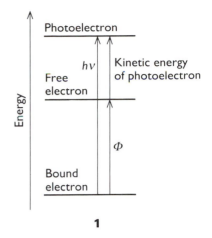

1

If $h\nu < \Phi$, photoejection cannot occur because the photon supplies insufficient energy to expel the electron: this conclusion accounts for observation 1 above. Equation (7) predicts that the kinetic energy of an ejected electron should vary linearly with the frequency, in agreement with the second observation. When a photon collides with an electron, it gives up all its energy, and so we should expect electrons to appear as soon as the collisions begin, provided they have sufficient energy: this conclusion agrees with observation 3. Thus, the photoelectric effect is strong evidence for the existence of photons.

The diffraction of electrons

The photoelectric effect shows that light has certain properties of particles. Although contrary to the long-established wave theory of light, a similar view had been held before, but discarded. No significant scientist, however, had taken the view that matter is wave-like. Nevertheless, experiments carried out in 1925 forced people to even that conclusion. The crucial experiment was performed by the American physicists Clinton Davisson and Lester Germer, who observed the diffraction of electrons by a crystal (Fig. 8.8): diffraction—the interference between waves, caused by an object in their path, and resulting in a series of bright and dark lines where the waves are detected—is a typical characteristic of waves.

The Davisson–Germer experiment, which has since been repeated with other particles (including molecular hydrogen), shows clearly that particles have wave-like properties. We have also seen that waves have particle-like properties. Thus we are brought to the heart of modern physics. *When examined on an atomic scale, the concepts of particle and wave melt together, particles taking on the characteristics of waves, and waves the characteristics of particles.* This joint wave–particle character of matter and radiation is called **wave–particle duality**.

Some progress towards coordinating the properties of waves and particles was made by Louis de Broglie when, in 1924, he suggested that any particle travelling with a momentum p (the product of its

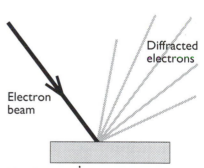

Fig. 8.8 In the Davisson–Germer experiment, a beam of electrons was directed on a single crystal of nickel, and the scattered electrons showed a variation in density with angle which corresponded to the pattern that would be expected if the electrons had a wave character and were diffracted by the layers of atoms in the solid.

Fig. 8.9 According to the de Broglie relation, a particle with low momentum has a long wavelength whereas a particle with high momentum has a short wavelength. A high momentum can result either from a high mass or from a high velocity (because $p = mv$). Macroscopic objects have such large masses that, even if they are travelling very slowly, their wavelengths are undetectably short.

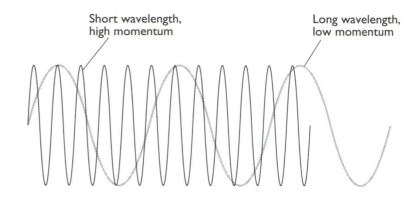

Short wavelength, high momentum

Long wavelength, low momentum

mass and velocity, $p = mv$) should have (in some sense) a wavelength given by the **de Broglie relation**:

$$\lambda = \frac{h}{p} \qquad (8)$$

The de Broglie relation implies that the wavelength of a particle should decrease as its speed increases (Fig. 8.9). It also implies that, for a given speed, heavy particles should have shorter wavelengths than lighter particles. Equation (8) was confirmed by the Davisson–Germer experiment, since the wavelength it predicts for the electrons they used in their experiment agrees with the details of the diffraction pattern they observed. We shall build on the relation, and understand it more, in the next section.

Example Estimating the de Broglie wavelength
Estimate the wavelength of electrons that have been accelerated from rest through a potential difference of 1.00 kV.
Answer The energy acquired by an electron is eV, where e is its charge and V is the potential difference it falls through (see *Further information 5*, p. 206). At the end of the period of acceleration all the acquired energy is in the form of kinetic energy, and we can write

$$\tfrac{1}{2}m_e v^2 = eV$$

Hence, the momentum of the electron is

$$p = m_e v = \sqrt{2m_e eV}$$

and the de Broglie wavelength is

$$\lambda = \frac{h}{\sqrt{2m_e eV}}$$

Substituting the data and the fundamental constants (from

inside the front cover) gives

$$\lambda =$$

$$\frac{6.626 \times 10^{-34}\,\text{J s}}{\{2 \times (9.110 \times 10^{-31}\,\text{kg}) \times (1.602 \times 10^{-19}\,\text{C}) \times (1.00 \times 10^{3}\,\text{V})\}^{1/2}}$$

$$= 3.88 \times 10^{-11}\,\text{m}$$

(We have used $1\,\text{C V} = 1\,\text{J}$ and $1\,\text{J} = 1\,\text{kg m}^2\,\text{s}^{-2}$.) The wavelength of 38.8 pm is comparable to typical bond lengths in molecules (about 100 pm). Electrons accelerated in this way are used in the technique of electron diffraction for the determination of molecular structure.

Exercise E8.7 Calculate the wavelength of an electron in a 10 GeV particle accelerator ($1\,\text{GeV} = 10^9\,\text{eV}$).

[0.12 pm]

Fig. 8.10 The spectrum of radiation emitted by energetically excited mercury atoms consists of radiation at a series of discrete frequencies, called *spectral lines*. The lines are images of the slit that the radiation passes through before being separated into its components by a prism or diffraction grating.

8.4 Atomic and molecular spectra

The most directly compelling evidence for the quantization of energy comes from the frequencies of radiation absorbed or emitted by atoms and molecules. A typical atomic emission spectrum is shown in Fig. 8.10 and a typical molecular absorption spectrum is shown in Fig. 8.11. The obvious feature of both is that *radiation is emitted at a series of discrete frequencies*. This observation can be understood if the energy of the atoms or molecules is also confined to discrete values, for then energy can be discarded only in discrete amounts (Fig. 8.12). For example, if the energy of an atom decreases by ΔE, then the energy is carried away as a photon of frequency $\nu = \Delta E/h$, and a line of that frequency appears in the spectrum.

Classical mechanics utterly failed in its attempts to account for the appearance of spectra, just as it failed to account for the other experiments described above. Such total failure showed that the basic concepts of classical mechanics were false. A new mechanics, quantum mechanics, had to be devised to take its place.

Fig. 8.11 When a molecule changes its state it does so by absorbing radiation at definite frequencies. The discrete character of the transitions suggests that the molecule can possess only discrete energies, not an arbitrary energy. The illustration shows a part of the ultraviolet absorption spectrum of ScF: the lines stem from excitations of the electrons of the molecule and its vibrational and rotational states.

299

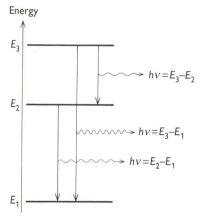

Fig. 8.12 Spectral lines can be accounted for if we assume that a molecule emits a photon as it 'jumps' between discrete energy levels. High-frequency radiation is emitted when the two states involved in the transition are widely separated in energy; low-frequency radiation is emitted when the two states are close in energy.

Fig. 8.13 According to classical mechanics, a particle may have a well-defined trajectory, with a precisely specified position and momentum at each instant (as represented by the precise path in the upper diagram). According to quantum mechanics, a particle cannot have a precise trajectory; instead, there is only a probability that it may be found at a specific location at any instant. The wavefunction that determines its probability distribution is a kind of blurred version of the trajectory. In the lower diagram the wavefunction is represented by areas of shading: the darker the area, the greater the probability of finding the particle there.

The dynamics of microscopic systems

We shall take the de Broglie relation as our starting point, and abandon the classical concept of particles moving on trajectories. From now on, we shall adopt the quantum mechanical view that *the position of a particle is distributed through space like the amplitude of a wave.* As for a wave in water, where the water accumulates in some places but is low in others, there are regions where the particle is more likely to be found than others. To describe this distribution, we introduce the concept of **wavefunction**, ψ, in place of the trajectory, and then set up a scheme for calculating and interpreting ψ. To a very crude first approximation, a wavefunction is a blurred version of a trajectory (Fig. 8.13); however, we shall refine this picture in the following sections.

8.5 The Schrödinger equation

In 1926, the Austrian physicist Erwin Schrödinger proposed an equation for finding the wavefunction of any system. The Schrödinger equation for a particle of mass m moving in one dimension with energy

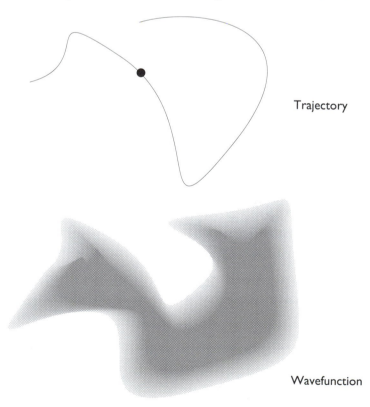

Trajectory

Wavefunction

E is

$$-\frac{\hbar^2}{2m}\frac{\mathrm{d}^2\psi}{\mathrm{d}x^2} + V\psi = E\psi \qquad (9)$$

In this expression, V, which may depend on the position, x, of the particle, is the potential energy; $\hbar$ (which is read 'h-bar') is a convenient modification of Planck's constant:

$$\hbar = \frac{h}{2\pi} = 1.054\ 59 \times 10^{-34}\ \mathrm{J\ s}$$

The fact that the Schrödinger equation is a differential equation should not cause too much consternation. We shall not need to solve it explicitly, and the rare cases where we shall need to show the explicit forms of its solution will involve mathematical functions no more complicated than $\sin x$ and e^{-x}.

JUSTIFICATION

The form of the Schrödinger equation can be justified to a certain extent by the following remarks. Consider first the case of motion in a region where the potential energy is zero. Then the equation simplifies to

$$-\frac{\hbar^2}{2m}\frac{\mathrm{d}^2\psi}{\mathrm{d}x^2} = E\psi$$

and a solution is

$$\psi = \sin kx, \text{ with } k = \sqrt{\frac{2mE}{\hbar^2}}$$

as may be verified by substitution of the solution into both sides of the equation. The function $\sin kx$ is a wave of wavelength $\lambda = 2\pi/k$, as can be seen by comparing $\sin kx$ with the standard form of a harmonic wave of wavelength λ, which is $\sin(2\pi x/\lambda)$. Next, we note that the energy of the particle is entirely kinetic (because $V = 0$ everywhere), and so

$$E = \tfrac{1}{2}mv^2 = \frac{(mv)^2}{2m} = \frac{p^2}{2m}$$

Because the energy is related to k by

$$E = \frac{k^2\hbar^2}{2m}$$

it follows from a comparision of the two equations that

$$p = k\hbar$$

Therefore, the linear momentum is related to the wavelength of the wavefunction by

$$p = \frac{2\pi}{\lambda} \times \frac{h}{2\pi} = \frac{h}{\lambda}$$

which is de Broglie's relation. We see, in the case of freely moving particles, that the Schrödinger equation has led to an experimentally verified conclusion.

One feature of the solution of the Schrödinger equation (and which is common to all differential equations) is that there is an infinite number of possible solutions allowed *mathematically*. For instance, if $\sin x$ is a solution of a particular Schrödinger equation, then so too is $a \sin bx$, where a and b are arbitrary constants. (This property is easily verified by substituting $\psi = a \sin bx$ into the equation in the Justification box above, and checking that it is a solution for all values of a and

Fig. 8.14 Although an infinite number of solutions of the Schrödinger equation exist, not all of them are physically acceptable. Acceptable wavefunctions have to satisfy certain boundary conditions that vary from system to system. In the example shown here, where the particle is confined between two impenetrable walls, the only acceptable wavefunctions are those that fit between the walls (like the vibrations of a stretched string). Because each wavefunction corresponds to a characteristic energy, and the boundary conditions rule out many solutions, only certain energies are permissible.

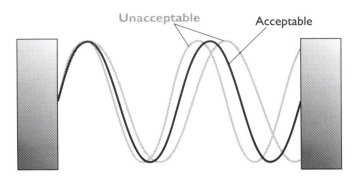

b.) However, it turns out that only some of these solutions are acceptable *physically*. The reason is that the solutions must satisfy certain constraints called **boundary conditions** (Fig. 8.14). Suddenly, we are at the heart of quantum mechanics, because the fact that only *some* solutions are acceptable, together with the fact that each solution corresponds to a characteristic value of *E*, implies that only certain values of the energy are acceptable. That is, *when the Schrödinger equation is solved subject to the boundary conditions that the solutions must satisfy, we find that the energy of the system is quantized.* Planck and his immediate successors had to *postulate* the quantization of energy for each system they considered: now we see that quantization is an automatic feature of a single equation, the Schrödinger equation, that is universally applicable to all systems. Later in this chapter we shall see exactly which energies are allowed in a variety of systems, the most important of which (for chemistry) are atoms.

The Born interpretation

Before we demonstrate the role of boundary conditions, it will be helpful to understand the physical significance of a wavefunction. The interpretation of ψ that is widely used is based on a suggestion made by the German physicist Max Born. He made use of an analogy with the wave theory of light, in which the square of the amplitude of an electromagnetic wave is interpreted as its intensity and therefore (in quantum terms) as the number of photons present. The **Born interpretation** asserts the following:

> The probability of finding a particle in a small (strictly, infinitesimal) region of volume δV is equal to $\psi^2 \times \delta V$.

That is, if the *square* of ψ is large at a particular location, then there is a high probability of finding the particle at that point; if the square of ψ is small at a particular location, then there is only a small chance of finding the particle at that point. The density of shading in Fig. 8.13

represents this probabilistic interpretation. It is very important to note at this point how the precise location of a particle, in classical physics, has been replaced by a *probability* that a particle is at each location.

Example Interpreting a wavefunction

The wavefunction of an electron in the lowest energy state of a hydrogen atom is proportional to e^{-r/a_0}, with $a_0 = 52.9$ pm and r the distance from the nucleus. Calculate the relative probabilities of finding the electron inside a small volume of magnitude 1.0 pm^3 located at (a) the nucleus, and (b) a distance a_0 from the nucleus.

Answer The probability is equal to $\psi^2 \times \delta V$ evaluated at the location in question. The volume of interest is so small (even on the scale of the atom) that we can ignore the variation of ψ within it and write

$$\text{probability} = \psi^2 \times 1.0\,\text{pm}^3$$

with ψ^2 evaluated at the point in question. (a) At the nucleus, $r = 0$, and so there $\psi^2 \propto 1.0$ and

$$\text{probability} \propto 1.0\,\text{pm}^3$$

(b) At a distance $r = a_0$ in an arbitrary direction, $\psi^2 \propto e^{-2} = 0.14$ and

$$\text{probability} \propto 0.14 \times 1.0\,\text{pm}^3$$

Therefore, the ratio of probabilities is $1.0/0.14 = 7.1$. It is more probable (by a factor of 7.1) that the electron will be found at the nucleus than in same volume element located at a distance a_0 from the nucleus.

Exercise E8.8 The wavefunction for the lowest energy state in the ion He$^+$ is proportional to e^{-2r/a_0}. Repeat the calculation for this ion. Any comment?

[55; a more compact wavefunction]

The uncertainty principle

We have seen that, according to the de Broglie relation, a wave of constant wavelength (that is, the wavefunction $\sin 2\pi x/\lambda$) corresponds to a particle with a definite linear momentum $p = h/\lambda$. However, a wave does not have a definite location at a single point in space, so we cannot speak of the precise position of the particle if it has a definite momentum. Indeed, because a sine wave spreads throughout the whole of space we cannot say *anything* about the location of the particle: because the wave spreads everywhere, the particle may be

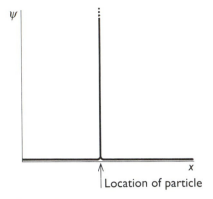

Fig. 8.15 The wavefunction for a particle with a well-defined position is a sharply spiked function that has zero amplitude everywhere except at the particle's position.

found anywhere in the whole of space. This statement is one half of the **uncertainty principle** proposed by Werner Heisenberg in 1927, in one of the most celebrated results of quantum mechanics:

> It is impossible to specify simultaneously, with arbitrary precision, both the momentum and the position of a particle.

Before discussing the principle further, we must establish the other half: that if we know the position of a particle exactly, then we can say nothing about its momentum. If we know that the particle is at a definite location, then its wavefunction must be nonzero there and zero everywhere else (Fig. 8.15). Such a wavefunction can be simulated by forming a **superposition** of many wavefunctions; that is, it may be reproduced by adding together the amplitudes of a large number of sine functions. This procedure is successful because the amplitudes of the waves add together at one location to give a nonzero amplitude, but cancel everywhere else to give a net amplitude of zero. In other words, we can create a sharply localized wavefunction by adding together wavefunctions corresponding to many different wavelengths, and therefore, by the de Broglie relation, of many different linear momenta.

The superposition of a few sine functions gives a broad, ill-defined wavefunction (Fig. 8.16(a)), but as the number used increases, the wavefunction becomes sharper because of the more complete interference between the positive and negative regions of the components (Fig. 8.16(b)). When an infinite number of components is used, the wavefunction is a sharp, infinitely narrow spike (Fig. 8.16(c)), which corresponds to perfect localization of the particle. Now the particle is perfectly localized, but at the expense of discarding all information about its momentum.

The quantitative version of the position–momentum uncertainty relation is

$$\Delta p \, \Delta x \geq \tfrac{1}{2}\hbar \tag{10}$$

The quantity Δp is the 'uncertainty' in the linear momentum† and Δx is the uncertainty in position (which is proportional to the width of the peak in Fig. 8.16). Equation (10) expresses quantitatively the feature of nature that the more closely the location of a particle is specified (the smaller the value of Δx), then the greater the uncertainty in the momentum (the larger the value of Δp) parallel to that coordinate, and vice versa (Fig. 8.17). Although position on the x axis and

† Strictly, the uncertainty in momentum is the root-mean-square (r.m.s.) deviation of the momentum from its mean value, $\Delta p = \sqrt{\langle p^2 \rangle - \langle p \rangle^2}$, where the angle brackets denote mean values. Likewise, the uncertainty in position is the r.m.s. deviation in the mean value of position, $\Delta x = \sqrt{\langle x^2 \rangle - \langle x \rangle^2}$.

(a)

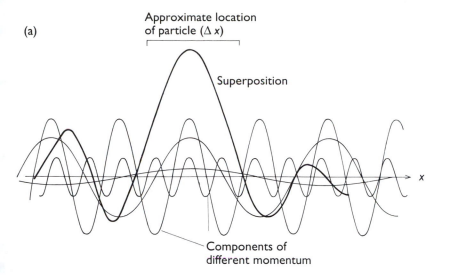

(b)

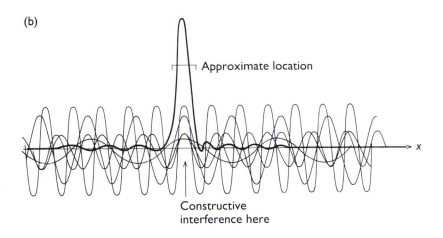

(c)

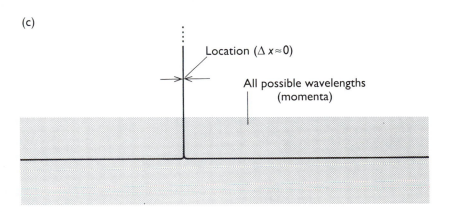

Fig. 8.16 (a) The wavefunction for a particle with an ill-defined location can be regarded as the sum (superposition) of several wavefunctions of different wavelengths that interfere constructively in one place but destructively elsewhere. (b) As more waves are used in the superposition, the location becomes more precise at the expense of uncertainty in the particle's momentum. (c) An infinite number of waves are needed to construct the wavefunction of a perfectly localized particle.

Fig. 8.17 A representation of the content of the uncertainty principle. The range of locations of a particle is shown by the circles, and the range of momenta by the arrows. In the upper diagram, the position is quite uncertain, and the range of momenta is small. In the lower diagram, the location is much better defined, but now the momentum of the particle is quite uncertain.

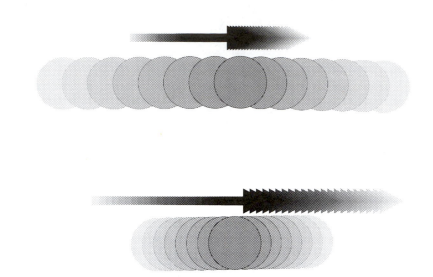

Properties that can be specified simultaneously

Position along		x	y	z
Momentum along	x	No	Yes	Yes
	y	Yes	No	Yes
	z	Yes	Yes	No

momentum parallel to that axis are restricted by the uncertainty relation, simultaneous location of position on x and motion along y or z are not restricted (see the summary in the margin).

Example Using the uncertainty principle
The speed of a 1.0 g projectile is known to within $1 \times 10^{-6}\,\mathrm{m\,s^{-1}}$. Calculate the minimum uncertainty in its position. **Answer** Estimate Δp from $m\,\Delta v$, where Δv is the uncertainty in the speed; then use eqn (10) to estimate the minimum uncertainty in position, Δx. The uncertainty in position is

$$\Delta x \geq \frac{\hbar}{2m\,\Delta v} = 5 \times 10^{-26}\,\mathrm{m}$$

This degree of uncertainty is completely negligible for all practical purposes. However, when the mass is that of an electron, then the same uncertainty in speed implies an uncertainty in position far larger than the diameter of an atom, and so the concept of a trajectory, the simultaneous possession of a precise position and momentum, is untenable on an atomic scale.

Exercise E8.9 Estimate the minimum uncertainty in the speed of an electron in a hydrogen atom (taking its diameter as 100 pm).

$$[500\,\mathrm{km\,s^{-1}}]$$

The uncertainty principle summarizes the difference between classical and quantum mechanics. Classical mechanics supposed, falsely as we now know, that the position and momentum of a particle could be specified simultaneously with arbitrary precision. However, quantum mechanics shows that position and momentum are **complementary**, that is, not simultaneously specifiable, and that we have to make a choice: we can specify position at the expense of momentum, or momentum at the expense of position.

8.6 Applications of quantum mechanics

We shall now illustrate some of the concepts that have been introduced. We shall describe two concrete examples. One is a treatment of translational motion (motion in a straight line), and the other is a treatment of rotation (motion in a circle).

Translational motion: the particle in a box

First, we consider the translational motion of a particle that is confined to a region between two walls and can travel in a straight line in one dimension (along the x axis). We shall suppose that this **particle in a box** has mass m and can travel freely from $x = 0$ to $x = L$. The potential energy of the particle is zero inside the box but rises abruptly to infinity at the walls (Fig. 8.18).

The boundary conditions for this system are that each acceptable wavefunction of the particle must fit inside the box exactly (like the vibrations of a violin string, as illustrated in Fig. 8.14). Therefore, the wavelength, λ, of the wavefunctions must be one of the values

$$\lambda = 2L, L, \tfrac{2}{3}L, \tfrac{1}{2}L, \ldots = \frac{2L}{1}, \frac{2L}{2}, \frac{2L}{3}, \frac{2L}{4}, \ldots$$

In general, the wavelength must be one of the values given by the expression

$$\lambda = \frac{2L}{n}, \quad \text{with} \quad n = 1, 2, \ldots$$

Each wavefunction is a sine wave with one of these wavelengths; therefore, since a sine wave of wavelength λ is proportional to

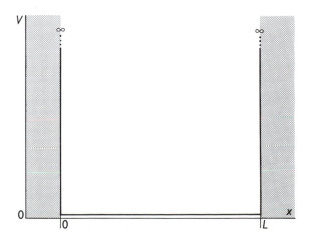

Fig. 8.18 A particle in a one-dimensional region with impenetrable walls at either end. Its potential energy is zero between $x = 0$ and $x = L$ and rises abruptly to infinity as soon as the particle touches either wall.

$\sin 2\pi x/\lambda$, the permitted vavefunctions are

$$\psi_n = N \sin \frac{n\pi x}{L}, \quad \text{with} \quad n = 1, 2, \ldots \qquad (11)$$

The constant N is called the **normalization constant**. It is chosen so that the *total* probability of finding the particle inside the box is 1. For a particle in a box, $N = (2/L)^{1/2}$.

JUSTIFICATION

According to the Born interpretation, the probability of finding the particle in an infinitesimal region of length dx is equal to $\psi^2 \, dx$. Therefore, the total probability of finding the particle in a region between $x = 0$ and $x = L$ is the sum of all the contributions from each infinitesimal region in that range:

$$\int_0^L \psi^2 \, dx = 1$$

Substitution of the form of the wavefunction gives

$$N^2 \int_0^L \sin^2\left(\frac{n\pi x}{L}\right) dx = 1$$

Because

$$\int \sin^2 ax \, dx = \frac{x}{2} - \frac{\sin 2ax}{4a}$$

it follows that

$$N^2 \times \tfrac{1}{2}L = 1$$

and therefore that

$$N = \sqrt{\frac{2}{L}}$$

It is now a simple matter to find the permitted energy levels because the only contribution to the energy is the kinetic energy of the particle (the potential energy is zero everywhere inside the box). First, we note that it follows from the de Broglie relation, eqn (8), that the only values of the linear momentum that are acceptable are

$$p = \frac{h}{\lambda} = \frac{nh}{2L}, \quad \text{with} \quad n = 1, 2, \ldots$$

Then, because the kinetic energy of a particle of momentum p and mass m is

$$E = \frac{p^2}{2m}$$

it follows that the permitted energies of the particle are

$$E_n = \frac{n^2 h^2}{8mL^2}, \quad \text{with} \quad n = 1, 2, \ldots \qquad (12)$$

The energies and wavefunctions are labelled with the quantum number, n. A **quantum number** is an integer (in certain cases, as we shall see, a half-integer) which labels the state of the system. As well as acting as a label, a quantum number specifies certain physical properties of the system: in the present example, n specifies the energy of the particle through eqn (12). The permitted energies of the particle

The dynamics of microscopic systems

are shown in Fig. 8.19 together with the shapes of the wavefunctions for $n = 1$ to 5. An important feature that can be seen in the illustration (and which will be encountered again when we consider atoms and molecules) is that as the number of **nodes** increases, so the energy increases. A node in a wavefunction is a point at which the wavefunction passes through zero, so the number of nodes in the wavefunctions shown in the illustration increases from 0 (for $n = 1$) to 4 (for $n = 5$), and is $n - 1$ in for a particle in a box in general. (The points at the edges of the box where $\psi = 0$ are not nodes, because the wavefunction does not pass *through* zero there.)

Because n cannot be zero for a particle in a box, the lowest energy that the particle may possess is not zero (as would be allowed by classical mechanics) but $E_1 = h^2/8mL^2$ (the energy when $n = 1$). This lowest, irremovable energy is called the **zero-point energy**. The existence of a zero-point energy is consistent with the uncertainty principle because, if a particle is confined to a finite region, its location is not completely indefinite; consequently its momentum cannot be specified precisely as zero, and therefore its kinetic energy cannot be precisely zero either.

The energy difference between adjacent levels is

$$\Delta E = E_{n+1} - E_n = (2n + 1)\frac{h^2}{8mL^2} \qquad (13)$$

This difference decreases as the length L of the container increases and it is extremely small when the container is large: the separation becomes zero when the walls are infinitely far apart. Atoms and molecules free to move in laboratory-sized vessels may therefore be treated as though their translational energy is not quantized. The separation also decreases as the mass of the particle increases, and particles of macroscopic mass (like balls and planets, and even minute specks of dust) behave as though their translational motion is unquantized. In general:

1. The larger the size of the system, the less important are the effects of quantization.

2. The greater the mass of the particle, the less important are the effects of quantization.

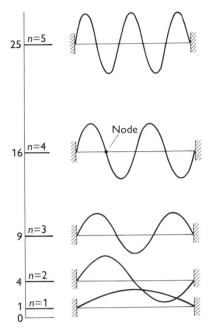

Fig. 8.19 The allowed energy levels and the corresponding (sine wave) wavefunctions for a particle in a box. Note that the energy levels increase as n^2, and so their spacing increases as n increases. Each wavefunction is a standing wave, and successive functions possess one more half-wave and a correspondingly shorter wavelength.

Exercise E8.10 Consider an electron that is a part of a conjugated polyene (such as a carotene molecule) of length 2.0 nm. What is the energy required (in electronvolts, $1\,eV = 1.602 \times 10^{-19}$ J) to excite it from the level with $n = 5$ to the next higher level?

[1.0 eV]

Rotational motion: the particle on a ring

One of the principal characteristics of rotational motion is the **angular momentum**. The magnitude of the angular momentum of a particle that is travelling on a circular path of radius r is defined as

$$\text{angular momentum} = p \times r \qquad (14)$$

where p is its linear momentum (the product of its mass and velocity, $p = mv$) at any instant. A particle that is travelling at high speed in a circle has a higher angular momentum than a particle of the same mass travelling more slowly. An object with a high angular momentum (like a flywheel) requires a strong braking force (more precisely, a strong *torque*) to bring it to a standstill. Angular momentum is the circular analogue of the linear momentum of translational motion.

Consider a particle of mass m moving in a circular path of radius r. The energy of the particle is entirely kinetic because the potential energy is zero everywhere, and we can therefore write $E = p^2/2m$. By using eqn (14), this energy can be expressed in terms of the angular momentum as

$$E = \frac{(\text{angular momentum})^2}{2mr^2}$$

The quantity mr^2 is called the **moment of inertia** of the particle, and denoted I: a heavy particle in a path of large radius has a large moment of inertia (Fig. 8.20). It follows that the energy of the particle is

$$E = \frac{(\text{angular momentum})^2}{2I}$$

Now we use the de Broglie relation to see that the energy of rotation is quantized. To do so, we express the angular momentum in

Fig. 8.20 A particle travelling on a circular path has a moment of inertia I that is given by mr^2. (a) This heavy particle has a large moment of inertia about the central point. (b) This light particle is travelling on a path of the same radius, but it has a smaller moment of inertia. The moment of inertia plays a role in circular motion that is the analogue of the mass for linear motion: a particle with a high moment of inertia is difficult to accelerate into a given state of rotation, and requires a strong braking force to stop its rotation.

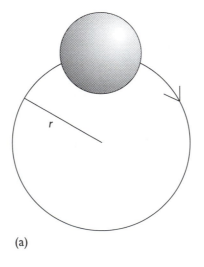

(a)

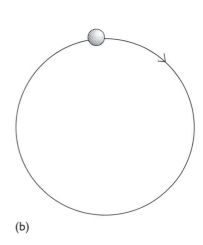

(b)

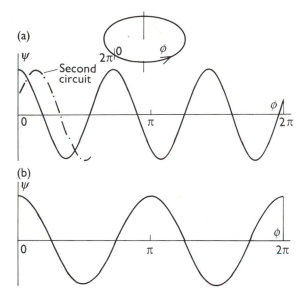

terms of the wavelength of the particle:

$$\text{angular momentum} = p \times r = \frac{h}{\lambda} \times r$$

Suppose for the moment that λ can takes an arbitrary value. In that case, the amplitude of the wavefunction depends on the angle, as shown in Fig. 8.21(a). When the angle increases beyond 360°, the wavefunction continues to change, but for an arbitrary wavelength it gives rise to an amplitude at each point that is different from the amplitude on the first circuit, and the interference between the waves on successive circuits cancels the amplitude of the wave. Thus, this particular arbitrary wave does not exist in the system. *An acceptable solution is obtained only if the wavefunction reproduces itself on successive circuits*, as in Fig. 8.21(b). The acceptable wavefunctions have wavelengths that match after each circuit, and therefore have wavelengths that are given by the expression

$$\lambda = \frac{2\pi r}{n}, \quad \text{with} \quad n = 0, \pm 1, \pm 2, \ldots$$

Both positive and negative values of n are permitted to allow for the fact that the particle may rotate clockwise or counterclockwise around the ring. The value $n = 0$, which gives an infinite wavelength, corresponds to a uniform amplitude. It follows that the permitted energies are

$$E = n^2 \left(\frac{h}{2\pi}\right)^2 \frac{1}{2I} = \frac{n^2\hbar^2}{2I}$$

with $n = 0, \pm 1, \pm 2, \ldots$. It is conventional in the discussion of

311

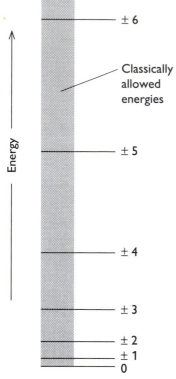

Fig. 8.22 The energy levels of a particle that can move on a circular path. Classical physics allowed the particle to travel with any energy (as represented by the continuous grey band); quantum mechanics, however, allows only discrete energies. Each energy level, other than the one with $m_l = 0$, is doubly degenerate, because the particle may rotate either clockwise or counterclockwise with the same energy.

rotational motion (for reasons that will become clear later) to denote the quantum number by m_l; therefore, the final expression for the energy levels is

$$E = \frac{m_l^2 \hbar^2}{2I}, \quad \text{with} \quad m_l = 0, \pm 1, \pm 2, \ldots \quad (15)$$

These energy levels are drawn in Fig. 8.22.

Exercise E8.11 The moment of inertia of an HCl molecule is 2.6×10^{-47} kg m^2. What is the minimum energy needed to start it rotating in a plane?

$$[2.1 \times 10^{-22} \, \text{J}]$$

The occurrence of m_l as its square in the expression for the enegy means that *two* states of motion, such as that with $m_l = +1$ and $m_l = -1$, both correspond to the same energy. Such a condition, where more than one state has the same energy, is called **degeneracy**. All the states with $|m_l| > 0$ are **doubly degenerate**, because two states correspond to the same energy for each value of $|m_l|$. The state with $m_l = 0$, the lowest energy state of the particle, is **nondegenerate**. The origin of the degeneracy is easy to identify: a particle may travel either clockwise or counterclockwise on the circle, and the different signs of m_l correspond to the two opposite directions of motion. The state with $m_l = 0$ is nondegenerate because the particle is stationary, so the question of its direction of travel does not arise.

An important additional conclusion is that *the angular momentum of the particle is quantized*. This conclusion is an example of an observable other than the energy being confined to discrete values. In the present case, we can use the relation between angular momentum and linear momentum (angular momentum $= p \times r$), and between linear momentum and the allowed wavelengths of the particle ($\lambda = 2\pi r/m_l$), to conclude that the angular momentum of a particle around an axis is confined to the values

$$\text{angular momentum} = pr = \frac{hr}{\lambda} = \frac{hr}{(2\pi r/m_l)} = \frac{m_l h}{2\pi}$$

That is, the angular momentum of the particle around the axis is confined to the values

$$\text{angular momentum around axis} = m_l \hbar, \text{ with } m_l = 0, \pm 1, \pm 2, \ldots \quad (16)$$

(We have used the constant $\hbar = h/2\pi$, which was introduced earlier.) Positive values of m_l correspond to counterclockwise rotation (as seen from above) and negative values correspond to clockwise rotation (Fig. 8.23). The quantized motion can be thought of in terms of the motion of a bicycle wheel that can rotate only with a discrete series of angular momenta, so that as the wheel is accelerated, the angular momentum jerks from the values 0 (when the wheel is stationary) to $\hbar, 2\hbar, \ldots$ but can have no intermediate value.

A final point concerning the rotational motion of a particle is that *it does not have a zero-point energy*: m_l may take the value 0, so $E = 0$. This conclusion is also consistent with the uncertainty principle, because although the particle is certainly between the angles 0 and 360° on the ring, that range is equivalent to not knowing where it is on the ring. Consequently, the angular momentum may be specified exactly and a value of zero is possible. Because the angular momentum is zero precisely, the energy of the particle is also zero precisely.

The structures of atoms

We now have enough background to move on to the major purpose of this chapter: the discussion of the electronic structures of atoms. When considering the structures of atoms we need to distinguish between hydrogenic atoms and many-electron atoms. **Hydrogenic atoms** are one-electron atoms and ions of general atomic number Z; they include H, He^+, Li^{2+}, C^{5+}, and U^{91+} (such very highly ionized atoms may be found in the outer regions of stars). **Many-electron atoms** are atoms and ions with more than one electron, and include all neutral atoms other than H (for instance, helium, with its two electrons, is a many-electron atom in this sense). Hydrogenic atoms, and H in particular, are important because their structures can be discussed exactly. They also provide a set of concepts that are used to describe the structures of many-electron atoms and (as we shall see in the next chapter) the structures of molecules too.

One of the principal experimental techniques for determining the structures of atoms is spectroscopy, and in this chapter we shall meet some of the principles involved. Some of the essential properties of electromagnetic radiation are summarized in *Further information 8* (p. 290). We shall often refer to the **wavenumber**, $\bar{\nu}$, of the radiation. The wavenumber is related to the wavelength, λ, and frequency, ν, by

$$\bar{\nu} = \frac{1}{\lambda} = \frac{\nu}{c} \tag{17}$$

where c is the speed of light. A wavenumber has dimensions of 1/length, and it is almost always expressed in the units of reciprocal centimetres, cm^{-1}; a wavenumber of $1000\ cm^{-1}$ then signifies that there are 1000 complete wavelengths per centimetre (Fig. 8.24). Visible light is electromagnetic radiation with wavenumbers in the range 14 000 to 24 000 cm^{-1} (corresponding to red and violet light, respectively). The record of wavenumbers (or frequencies and wavelengths) of the radiation emitted or absorbed by an atom is called its **spectrum** (from the Greek word for 'appearance').

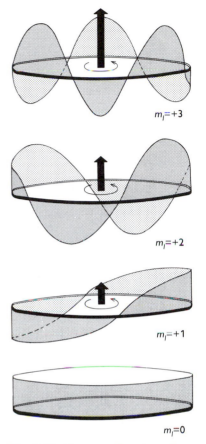

Fig. 8.23 The wavefunction of a particle on a ring. As the wavelength shortens, the angular momentum grows in steps of $\hbar$. This momentum can be represented by an arrow of length m_l units along the z axis. Note the sign convention (the right-hand screw rule) for the direction of motion.

$m_l = +3$

$m_l = +2$

$m_l = +1$

$m_l = 0$

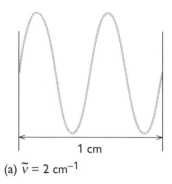

(a) $\tilde{v} = 2$ cm^{-1}

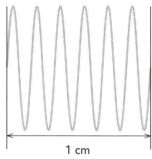

(b) $\tilde{v} = 6$ cm^{-1}

Fig. 8.24 The *wavenumber* of radiation is the number of complete oscillations per unit length (usually, per centimetre). The illustrations depict radiation of wavenumber (a) 2 cm^{-1} and (b) 6 cm^{-1}. Note that as the wavenumber increases so the wavelength decreases.

8.7 The experimental basis: the spectra of hydrogenic atoms

When an electric discharge is passed through gaseous hydrogen, the H_2 molecules are dissociated and the energetically excited H atoms that are produced emit light of discrete frequencies as they discard energy and return to their **ground state**, their state of lowest energy (Fig. 8.25). The first important contribution to understanding this spectrum was made by the Swiss schoolteacher Johann Balmer, who pointed out in 1885 that (in modern terms) the wavenumbers of the light in the visible region of the electromagnetic spectrum fit the expression

$$\tilde{v} = R_H\left(\frac{1}{2^2} - \frac{1}{n^2}\right), \quad \text{with} \quad n = 3, 4, \ldots$$

where R_H is a constant. The set of lines described by this formula is now called the **Balmer series**. Later, another set of lines was discovered in the ultraviolet region of the spectrum, and is called the **Lyman series**; yet another set in the infrared region was also discovered when detectors became available for that region, and is called the **Paschen series**. With this additional information available, the Swedish spectroscopist Johannes Rydberg noted (in 1890) that all the lines could be fitted to the expression

$$\tilde{v} = R_H\left(\frac{1}{n_1^2} - \frac{1}{n_2^2}\right), \quad \text{with } R_H = 109\,677\,\text{cm}^{-1} \quad \begin{cases} n_1 = 1, 2, \ldots, \text{ and} \\ n_2 = n_1 + 1, n_1 + 2, \ldots \end{cases}$$

(18)

The constant R_H is now called the **Rydberg constant** for the hydrogen atom. The first five series of lines then correspond to

$$n_1 = 1 \qquad 2 \qquad 3 \qquad 4 \qquad 5$$

Lyman   Balmer   Paschen   Brackett   Pfund

Fig. 8.25 The spectrum of atomic hydrogen. The spectrum is shown at the top, and is analysed into overlapping series below. The Balmer series lies largely in the visible region.

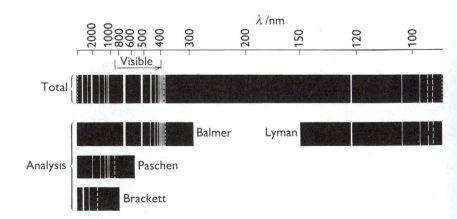

Example Calculating the wavelengths of spectral lines
Calculate the longest wavelength transition in the Lyman series of atomic hydrogen.

Answer The Lyman series ($n_1 = 1$) has wavenumbers

$$\tilde{v} = R_H\left(1 - \frac{1}{n_2^2}\right), \text{ with } n_2 = 2, 3, 4, \ldots$$

The transition with the longest wavelength is the one with the smallest wavenumber, which is the transition with $n_2 = 2$. Hence the wavenumber of this transition is

$$\tilde{v} = \tfrac{3}{4} \times R_H = 82\,258 \text{ cm}^{-1}$$

The wavelength is the reciprocal of the wavenumber:

$$\lambda = \frac{1}{8.2258 \times 10^6 \text{ m}^{-1}} = 1.2157 \times 10^{-7} \text{ m, or } 121.57 \text{ nm}$$

The transition occurs in the ultraviolet region of the spectrum.

Exercise E8.12 Calculate the *shortest* wavelength transition in the Paschen series.

[821 nm]

The existence of spectral lines of specific frequencies strongly suggests that the energy of the electron in the hydrogen atom is quantized. Thus, because energy is conserved, when an atom changes its energy by ΔE the difference is carried away as a photon of frequency v (as illustrated in Fig. 8.12), where

$$\Delta E = hv \qquad (19)$$

This relation is called the **Bohr frequency condition**. We can then expect to observe only certain wavenumbers in the spectrum if only certain energy states of atoms are permitted. Our tasks in this section are to determine the origin of this quantization, to find the permitted energy levels, and to account for the value of R_H.

8.8 The interpretation: the structures of hydrogenic atoms

The quantum mechanical description of the structure of a hydrogenic atom is based on Rutherford's **nuclear model**, in which the atom is pictured as a central nucleus of charge Ze and one extranuclear electron. To derive the details of the structure of the atom, and its energy levels, we should set up and solve the Schrödinger equation for the atom, using for the potential energy the **Coulomb potential energy** of the interaction between the nucleus of charge $+Ze$ and the electron of charge $-e$:

$$V = -\frac{Ze^2}{4\pi\varepsilon_0 r} \qquad (20)$$

315

where ε_0 is the vacuum permittivity (a fundamental constant, see *Further information 5* on p. 206) and e is the fundamental electric charge.

8.9 Atomic orbitals and their energies

The wavefunctions of hydrogenic atoms, which are obtained as solutions of the Schrödinger equation with the Coulomb potential energy, are called **orbitals**. The name expresses something less definite than the 'orbits' of classical mechanics. It turns out that each orbital is defined by *three* quantum numbers that are restricted to certain discrete values.

1. The **principal quantum number**, $n = 1, 2, \ldots$, determines the energy of the electron in the atom.

2. The **azimuthal quantum number**, $l = 0, 1, 2, \ldots, n-1$, determines the orbital angular momentum of the electron (and is also called the orbital angular momentum quantum number).

3. The **magnetic quantum number**, $m_l = 0, \pm 1, \ldots, \pm l$, determines the amount of angular momentum of the electron around a particular axis.

We shall describe the energy levels first, and then turn to a consideration of the shapes of the orbitals.

The solution of the Schrödinger equation leads to the conclusion that the allowed energy levels of a hydrogenic atom of atomic number Z are given by the expression

$$E_n = -\frac{hcR}{n^2}, \quad \text{with} \quad hcR = \frac{Z^2 m_e e^4}{32\pi^2 \varepsilon_0^2 \hbar^2} \qquad (21)$$

(with, as stated above, $n = 1, 2, \ldots$; R is the Rydberg constant and is equal to R_H for $Z = 1$). These energy levels are depicted in Fig. 8.26: note how the energy levels are widely separated at low values of n, but then converge as n increases. Close to the state of zero energy (the widely separated nucleus and electron), the separation of energy levels is almost zero: now the atom is very large, and energy levels are close together.

Exercise E8.13 The shortest wavelength transition in the Paschen series occurs at 821 nm in hydrogen; at what wavelength does the same transition occur in Li^{2+}?

$$[\tfrac{1}{9} \times 821 \text{ nm} = 91.2 \text{ nm}]$$

The zero of energy (which occurs at $n = \infty$) corresponds to the infinitely widely separated and stationary electron and nucleus. The infinite separation implies that, from the Coulomb formula, the electron has zero potential energy, and the fact that it is motionless implies that it has zero kinetic energy too. All the energies given by eqn (21) are negative: the states it describes are the **bound states** of the atom, in which the energy of the electron is lower than when it is infinitely far from the nucleus and at rest. The state of lowest energy is that with $n = 1$ (the lowest permitted value of n and hence the largest negative value of the energy), and the energy of this state is

$$E_1 = -hcR$$

The state with $n = 1$ is the ground state of the atom and it lies at an energy hcR below that of the infinitely separated electron and nucleus. The first excited state of the atom, the state with $n = 2$, lies at

$$E_2 = -\tfrac{1}{4}hcR$$

This energy level is $\tfrac{3}{4}hcR$ above the ground state.

Although each orbital require *three* labels (the values of n, l, and m_l), the energy is determined by the principal quantum number, n, alone and is independent of the values of l and m_l. That is, *in a hydrogenic atom, all orbitals with the same value of n are degenerate.* It is for this reason that all orbitals of the same principal quantum number are said to belong to the same **shell** of the atom. Thus, all orbitals with $n = 1$ (there is in fact only one) belongs to one shell with a particular energy, all orbitals with $n = 2$ (there are four) belong to another shell with another energy, and so on. It is common to refer to successive shells by the following letters:

$$n = 1 \quad 2 \quad 3 \quad 4 \cdots$$
$$K \quad L \quad M \quad N \cdots$$

Thus, all the orbitals of the shell with $n = 2$ form the L shell of the atom. The energy levels with the shell structure shown explicitly are displayed in Fig. 8.27.

Ionization energies

The **ionization energy**, I, of an atom is the minimum energy required to remove an electron from the ground state of the atom. The ground state of hydrogen is the state with $n = 1$, which has energy $E_1 = -hcR_H$. The atom is ionized when the electron has been excited to the level with zero energy, corresponding to $n = \infty$ (see Fig. 8.26). Therefore, the energy that must be supplied is

$$I = hcR_H = 2.179 \times 10^{-18} \, \text{J}$$

which corresponds to $1312 \, \text{kJ mol}^{-1}$ or $13.59 \, \text{eV}$.

Example Measuring an ionization energy spectroscopically
The spectrum of atomic hydrogen has lines at the wavenumbers 82 259, 97 492, 102 824, 105 292, 106 632, and

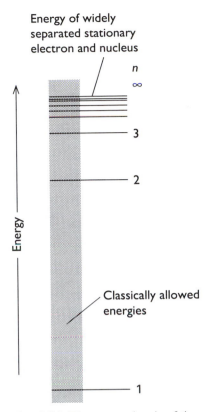

Fig. 8.26 The energy levels of the hydrogen atom. The energies are relative to a proton and an infinitely distant, stationary electron.

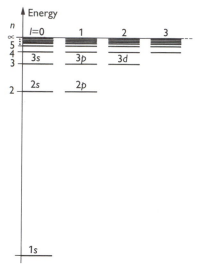

Energy

n
$l=0$ 1   2   3

5
4
3   3s   3p   3d

2   2s   2p

1s

Fig. 8.27 A more detailed analysis of the energy levels of a hydrogen atom, in which the subshells of each shell are shown individually. The degeneracy of the orbitals with the same value of n but different values of l (that is, the degeneracy of orbitals in different subshells but the same shell of the atom) is a special characteristic of one-electron, hydrogenic atoms.

$107\,440\,\text{cm}^{-1}$. Find (a) the ionization energy of the lower state, and (b) the value of the Rydberg constant.

Answer For a given lower state of energy E_1, the lines of a spectral series are given by the expression

$$\tilde{v} = -\frac{R_H}{n^2} - \frac{E_1}{hc}$$

with n a sequence of integers (such as $n = 2, 3, \ldots$ if E_1 corresponds to the state with $n = 1$). It follows that if we plot wavenumber against $1/n^2$ with $n = 2, 3, \ldots$ (this step assumes that the lower state is $n = 1$; if a straight line is not obtained, try $n = 3, 4, \ldots$, etc.), then the intercept at $1/n^2 = 0$ is $-E_1/hc$, and the slope is $-R_H$. The ionization energy of the lower state is E_1. It is best to make a least-squares fit of the data to obtain a result that reflects the precision of the data. The wavenumbers are plotted against $1/n^2$ in Fig. 8.28. The (least-squares) intercept lies at $-109\,679\,\text{cm}^{-1}$, and so the ionization energy is $-hc$ times this quantity, or $2.1788 \times 10^{-18}\,\text{J}$ ($1312.1\,\text{kJ mol}^{-1}$). The slope is, in this instance, numerically the same, and so $R_H = 109\,679\,\text{cm}^{-1}$.

Exercise E8.14 The spectrum of atomic deuterium shows lines at the following wavenumbers: $15\,238$, $20\,571$, $23\,039$, $24\,380\,\text{cm}^{-1}$. Determine (a) the ionization energy of the lower state, and (b) the ionization energy of the ground state.
[(a) $328.1\,\text{kJ mol}^{-1}$, (b) $1312.4\,\text{kJ mol}^{-1}$]

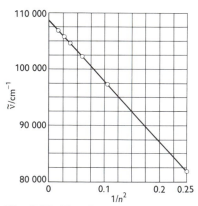

Fig. 8.28 The determination of the ionization energy of an atom by plotting spectral wavenumbers against $1/n^2$ and extrapolating the straight line to $n = \infty$ and noting its intercept with the vertical axis.

Exercise E8.15 Predict the ionization energy of He^+ given that the ionization energy of H is $13.59\,\text{eV}$.
Hint: Decide how the energy of the $1s$ orbital varies with Z.
[$54.36\,\text{eV}$]

Subshells and orbitals

The orbitals with the same value of n but different values of l form the **subshells** of a given shell. These subshells are generally denoted by the letters $s, p, \ldots$ using the correspondence

$$l = 0 \quad 1 \quad 2 \quad 3 \quad 4 \cdots$$
$$ s \quad p \quad d \quad f \quad g \cdots$$

(the letters then run alphabetically, with the omission of i). Thus, the subshell with $l = 1$ of the shell with $n = 2$ is called the $2p$ subshell, and the orbitals that belong to this subshell are called the $2p$ orbitals. For

$n = 1$, there is only one subshell, the one with $l = 0$. When $n = 2$, there are two subshells, namely the $2s$ subshell (with $l = 0$) and the $2p$ subshell (with $l = 1$). The general pattern of the first four shells and their subshells is shown in the margin.

In general, each subshell contains $2l + 1$ orbitals (corresponding to the $2l + 1$ values of m_l for each value of l). Thus, in any given subshell, the number of orbitals is given by

n	Shell	Subshells
1	K	s
2	L	s, p
3	M	s, p, d
4	N	s, p. d, f

Subshell: s p d f

Number of orbitals: 1   3   5   7

It follows that when $n = 1$, there is only one orbital, that with $l = 0$ and $m_l = 0$ (the only value of m_l permitted). When $n = 2$, there are four orbitals, one in the s subshell with $l = 0$ and $m_l = 0$, and three in the $l = 1$ subshell with $m_l = 0, +1$, and -1. When $n = 3$ there are nine orbitals (one with $l = 0$, three with $l = 1$, and five with $l = 2$). We remark again that, *for hydrogenic atoms*, all the orbitals of a given shell have the same energy. The relation between shells, subshells, and orbitals is summarized in Fig. 8.29.

Exercise E8.16 How many orbitals are there in a shell with $n = 5$?

[25]

We shall now consider the shapes of individual types of orbital.

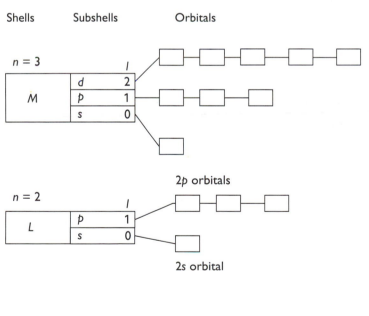

Shells   Subshells   Orbitals

$n = 3$

M   d   2
      p   1
      s   0

2p orbitals

$n = 2$

L   p   1
      s   0

2s orbital

$n = 1$

K   s   0

1s orbital

Fig. 8.29 The structures of atoms are described in terms of *shells* of electrons that are labelled by the principal quantum number *n*, and a series of *n* subshells of these shells, with each subshell of a shell being labelled by the quantum number *l*. Each subshell consists of $2l + 1$ orbitals.

s orbitals

The orbital corresponding to the ground state of a hydrogenic atom is the one with $n = 1$ (and therefore necessarily with $l = 0$ and $m_l = 0$, the only possible values of these quantum numbers when $n = 1$). The mathematical form of the 1s orbital for a hydrogen atom is

$$\psi = \left(\frac{1}{\pi a_0^3}\right)^{1/2} e^{-r/a_0} \tag{22}$$

This expression is independent of angle, so the orbital has the same amplitude at all points of constant radius; that is, the 1s orbital is spherically symmetrical. However, ψ does depend on distance from the nucleus (Fig. 8.30), and its amplitude decays exponentially from a maximum value of $(1/\pi a_0^3)^{1/2}$ at the nucleus (at $r = 0$). The structure of

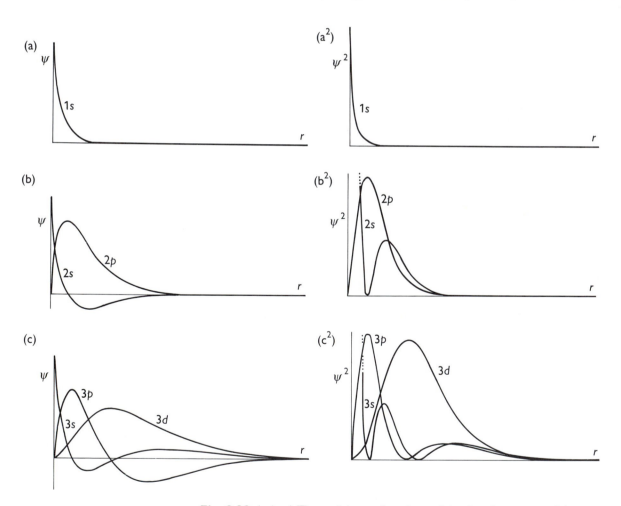

Fig. 8.30 (a, b, c) The radial wavefunctions of the first few states of the hydrogen atom and (a², b², c²) the probability densities. Note that the s orbitals have a nonzero and finite value at the nucleus. The vertical scales are different in each case.

Electron densities

1s   2s   3s

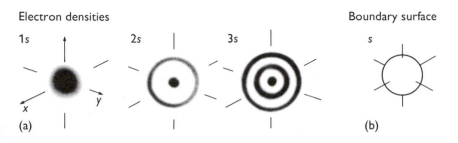

Boundary surface

s

(a)   (b)

Fig. 8.31 Representations of the first three s hydrogenic atomic orbitals in terms of (a) the electron densities (as represented by the density of shading) and (b) the boundary surfaces, within which there is a 90 per cent probability of finding the electron.

the ground state of the hydrogen atom can therefore be pictured as one in which an electron clusters closely to the nucleus.

It follows from the exponentially decaying form of the 1s orbital that, in the ground state of a hydrogenic atom, *the most probable point at which the electron will be found is at the nucleus itself*. A method of depicting the probability of finding the electron at each point in space is to represent ψ^2 by the density of shading in a diagram (Fig. 8.31(a)). A simpler procedure is to show only the **boundary surface**, the shape that captures about 90 per cent of the probability. For the 1s orbital, the boundary surface is a sphere (Fig. 8.31(b)).

Exercise E8.17 What is the probability of finding the electron in a volume of $1\,\text{pm}^3$ centred on the nucleus in a hydrogen atom?

$$[2.2 \times 10^{-6}, \text{ one observation in } 455\,000]$$

We often need to know the probability that an electron will be found at a given distance from a nucleus regardless of its angular position. The probability that an electron in an s orbital will be found between a radius r and a slightly greater radius $r + \delta r$ is calculated by evaluating $4\pi r^2 \psi^2 \times \delta r$. The factor $4\pi r^2 \psi^2$ multiplying the width of the region is called the **radial distribution function**. Because r^2 increases as r increases, but ψ^2 decreases exponentially, the radial distribution function goes through a maximum (Fig. 8.32): this maximum marks the most probable radius (not point) at which the electron will be

Probe, volume $4\pi r^2 \delta r$

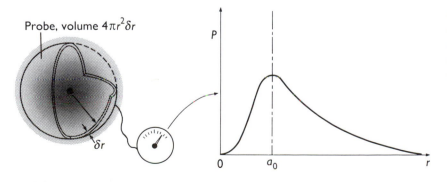

Fig. 8.32 The radial distribution function gives the probability that the electron will be found anywhere in a shell of radius r, regardless of angle.

found. For a $1s$ orbital of hydrogen, the maximum occurs at the **Bohr radius**, a_0, which is at 53 pm from the nucleus.†

JUSTIFICATION

The mathematical expression for the radial distribution function is justified by considering the probability of finding the electron in a spherical shell of radius r and thickness δr. The volume of such a shell is its surface area, $4\pi r^2$, multiplied by its thickness, δr, and is therefore $4\pi r^2 \, \delta r$.

According to the Born interpretation, the probability of finding an electron inside a small volume of magnitude δV is given by the value of $\psi^2 \times \delta V$. Therefore, interpreting δV as the volume of the shell gives $\psi^2 \times 4\pi r^2 \, \delta r$ for the probability of finding the particle anywhere in its volume.

A $2s$ orbital is also spherical, and its boundary surface is a sphere. It differs from a $1s$ orbital in its radial dependence (Fig. 8.30(b)), for although the wavefunction has a nonzero value at the nucleus, it passes through zero before commencing its exponential decay towards zero at large distances. Because a $2s$ electron spreads further out from the nucleus than a $1s$ electron, its boundary surface is a sphere of larger radius. The fact that the wavefunction passes through zero at a particular radius is summarized by saying that the orbital has a **radial node**: there is zero probability of finding the electron at the location of a radial node. A $3s$ orbital has two radial nodes (Fig. 8.30(c)), a $4s$ orbital has three radial nodes, and so on.

p orbitals

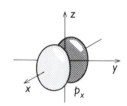

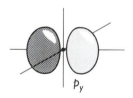

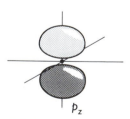

Fig. 8.33 The boundary surfaces of p orbitals. The nodal plane passes through the nucleus and separates the two lobes of each orbital. The dark and pale areas denote regions of opposite sign of the wavefunction.

All **p orbitals** (orbitals with $l = 1$) have a double-lobed appearance like that shown in Fig. 8.33. The two lobes are separated by a **nodal plane** that cuts through the nucleus. There is zero probability of finding an electron on this plane, and therefore (because the nucleus lies in the plane) zero probability of finding the electron at the nucleus. The exclusion of the electron from the nucleus is a common feature of all p orbitals. A p orbital corresponds to a state of angular momentum of the electron around the nucleus. In general, the orbital angular momentum of an electron in an orbital with quantum number l is given by the expression

$$\text{angular momentum} = \sqrt{l(l+1)} \times \hbar \tag{23}$$

† An analogy that might help to fix the significance of the radial distribution function for an electron is the corresponding distribution for the population of the Earth regarded as a perfect sphere. The distribution is zero at the centre of the Earth and for the next 6400 km (to the surface of the planet), when it peaks sharply and then rapidly decays again to zero. It remains virtually zero for all radii more than about 10 km above the surface. Almost all the population will be found very close to $r = 6400$ km, and it is not relevant that people will be dispersed over a very wide range of latitudes and longitudes. The small probabilities of finding people above and below $r = 6400$ km corresponds to the population that happens to be down mines or living in places as high as Denver or Tibet at the time.

Therefore, because $l = 1$, the angular momentum of an electron in a p orbital is $\sqrt{2} \times \hbar$.

The reason for the difference of p orbitals from s orbitals (the former's possession of nonzero orbital angular momentum and its exclusion from the nucleus) can be related to the angular shape of the orbital. An s orbital, with no angular nodes, is a three-dimensional version of a wavefunction for a particle on a ring with zero angular momentum; eqn (23) shows that an electron in an s orbital has zero orbital angular momentum (see $l = 0$). A p orbital, with one angular node, is a three-dimensional version of a wavefunction with angular momentum (compare the shape of a p orbital with the wavefunction in Fig. 8.23). The fact that an electron in a p orbital possesses orbital angular momentum means that it is flung away from the nucleus by the centrifugal force arising from its motion. The same centrifugal effect appears in all orbitals with angular momentum (those for which $l > 0$), such as d orbitals and f orbitals, and all such orbitals have a node at the nucleus.

Exercise E8.18 By what factor is the orbital angular momentum of a d electron increased when it is promoted to an f orbital?

[1.4]

There are three p orbitals in each shell of the atom (for $n \geq 2$). They are distinguished by the three different values that m_l can take when $l = 1$. The three orbitals are normally represented by their boundary surfaces, as depicted in Fig. 8.33. The p_x orbital has a dumb-bell shape directed along the x-axis, and similarly the p_y and p_z orbitals are directed along the y and z axes, respectively.

d orbitals

Orbitals with $l = 2$ are called **d orbitals**. There are five d orbitals in each shell (for $n \geq 3$), each one corresponding to one of the values $m_l = 0, \pm 1$, and ± 2. The five boundary surfaces of the orbitals (which are similar for the d orbitals of other shells) are shown in Fig. 8.34.

8.10 Electron spin

To complete the description of the state of a hydrogen atom, we need to introduce one more concept, that of electron **spin**. The spin of an electron is an intrinsic angular momentum that every electron possesses and which cannot be changed or eliminated (and in this respect resembles its mass or its charge). The name 'spin' is evocative of a ball spinning on its axis, and (so long as it is treated with caution) this classical interpretation can be used to help to visualize the motion.

323

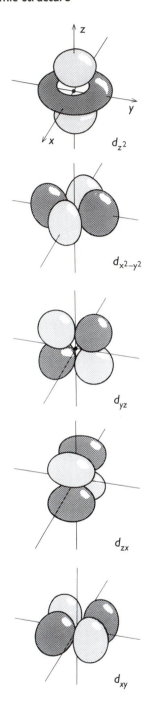

d_{z^2}

$d_{x^2-y^2}$

d_{yz}

d_{zx}

d_{xy}

Fig. 8.34 The boundary surfaces of d orbitals. Two nodal planes in each orbital intersect at the nucleus and separate the four lobes of each orbital. The dark and pale areas denote regions of opposite sign of the wavefunction.

However, spin is in fact *a purely quantum mechanical phenomenon and has no classical counterpart,* so the analogy must be used with care.

We shall make use of two properties of electron spin:

1. Electron spin is described by a quantum number, s (the analogue of l for orbital angular momentum), but s is fixed at the single value $\frac{1}{2}$ for all electrons at all times.

2. The direction of the spin angular momentum can be clockwise or counterclockwise; these two states are distinguished by the quantum number m_s, which can take the values $-\frac{1}{2}$ or $+\frac{1}{2}$, but no other values.

An electron with $m_s = +\frac{1}{2}$ is called an **α electron** and is denoted by the symbol ↑. An electron with $m_s = -\frac{1}{2}$ is called a **β electron** and is denoted by the symbol ↓.

The existence of electron spin was confirmed by an experiment performed by Otto Stern and Walther Gerlach in 1921, who shot a beam of silver atoms through an inhomogeneous magnetic field (Fig. 8.35). A silver atom has 47 electrons, and (for reasons that will become clear later), 23 of the spins are ↑ and 23 spins are ↓; the one remaining spin may be either ↑ or ↓. Because the angular momenta of the ↑ and ↓ spins cancel each other, the atom behaves as if it had the spin of a single electron. The idea behind the Stern–Gerlach experiment was that a rotating, charged body behaves like a magnet and interacts with the applied field. The applied magnetic field pushes or pulls the electron according to the orientation of the electron's spin, and so the initial beam of atoms should split into two beams, one corresponding to atoms with ↑ spin and the other to atoms with ↓ spin. This result was observed.

Other fundamental particles also have characteristic spins. For example, protons and neutrons are spin-$\frac{1}{2}$ particles (that is, for them $s = \frac{1}{2}$) and so invariably spin with a single, irremovable angular momentum. Because the masses of a proton and a neutron are so much greater than the mass of an electron, yet they all have the same spin angular momentum, the classical picture of proton and neutron spin would be of particles spinning much more slowly than an electron. Some elementary particles have $s = 1$, and so have a higher intrinsic angular momentum than an electron. For our purposes the most important spin-1 particle is the photon (the role of the photon spin will be explained shortly). It is a very deep feature of nature that the particles from which matter is built have half-integral spin (such as electrons and quarks, all of which have $s = \frac{1}{2}$) whereas the particles that transmit forces between these particles, and so bind them together into entities like nuclei, atoms, and planets, all have integral spin (such as $s = 1$ for the photon, which transmits the electromagnetic interaction between charged particles).

8.11 Structure, transitions, and selection rules

We can now summarize the structure of a hydrogenic atom as follows:

1. The atom consists of a nucleus and an electron.

2. The distribution of the electron around the nucleus is described by a wavefunction that is specified by three quantum numbers, n, l, and m_l. Different sets of quantum numbers correspond to different distributions of the electron. The spin of the electron is specified by a fourth quantum number m_s, which may have the values $+\frac{1}{2}$ or $-\frac{1}{2}$.

3. The state of lowest energy consists of an electron in a $1s$ orbital; the electron has zero orbital angular momentum around the nucleus.

4. The electron may occupy orbitals of higher energy (in shells of higher principal quantum number), with varying amounts of orbital angular momentum (as determined by the value of l of its orbital). Whatever its orbital angular momentum, it has a constant spin angular momentum.

The connection between observable properties and the quantum numbers of an orbital are summarized in Box 8.1. An electron that is

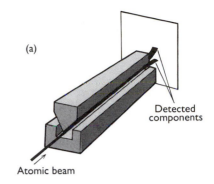

(a)

Detected components

Atomic beam

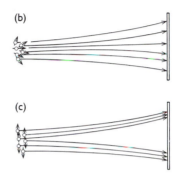

(b)

(c)

Fig. 8.35 (a) The experimental arrangement for the Stern–Gerlach experiment: the magnet is the source of an inhomogeneous field. (b) The classically expected result, in which the electron spin can lie in any orientation. (c) The observed outcome using silver atoms, in which the electron spins can adopt only two orientations ($\uparrow$ and $\downarrow$).

Box 8.1 Hydrogenic atoms

The *wavefunctions* of hydrogenic atoms depend on three quantum numbers:

principal quantum number, $n = 1, 2, 3, \ldots$
angular momentum quantum number, $l = 0, 1, 2, \ldots, n-1$
magnetic quantum number, $m_l = l, l-1, l-2, \ldots, -l$

The *energy* is related to n by eqn (21).

The magnitude of the *orbital angular momentum* of the electron is $\{l(l+1)\}^{1/2}\hbar$ and its component on an arbitrary axis is $m_l\hbar$. Each energy level is n^2-fold degenerate.

The *selection rules* for spectroscopic transitions are

$$\Delta n \text{ unrestricted}, \quad \Delta l = \pm 1 \qquad \Delta m_l = 0 \text{ or } \pm 1$$

described by a wavefunction with $n = 1$ and $l = 0$ is said to **occupy** a $1s$ orbital and to be a **1s electron**. Similarly, we can speak of $2p$, $3d$, ... electrons according to the orbitals they occupy.

A change in the distribution of an electron around the nucleus is called a **transition** between orbitals. Transitions are responsible for the appearance of atomic spectra, as we have already seen. Thus, when an electron makes a transition from an orbital in a shell with a principal quantum number n_2 into an orbital of a shell with principal quantum number n_1, the excess energy is emitted as a photon and contributes to one of the spectral lines given by the Rydberg formula, eqn (18). We can think of the sudden change in the distribution of the electron around the nucleus as jolting the electromagnetic field into oscillation, and that oscillation corresponds to the generation of a photon of light.

Although the Rydberg formula gives the wavenumbers of the transitions correctly, it is not the full story, for it turns out that not all transitions between all available orbitals are possible. For example, it is not possible for an electron in a $3d$ orbital to make a transition to a $1s$ orbital. The reason for the existence of **forbidden transitions**, transitions between orbitals that are not observed spectroscopically, as distinct from **allowed transitions**, which are observed, is the possession by the photon of its intrinsic spin angular momentum with $s = 1$. When a photon, with its one unit of angular momentum, is generated in a transition, the angular momentum of the electron must change by one unit to compensate for the angular momentum carried away by the photon (that is, the angular momentum must be conserved—neither created nor destroyed—just as linear momentum is conserved in collisions). Thus, an electron in a d orbital with $l = 2$ cannot make a transition into an s orbital with $l = 0$ because the photon cannot carry away enough angular momentum. Similarly, an s electron cannot make a transition to another s orbital, because then there is no change in the electron's angular momentum to make up for the angular momentum carried away by the photon.

A **selection rule** is a statement about which transitions are allowed. They are derived (for atoms) by identifying the transitions that conserve angular momentum when a photon is emitted or absorbed. The selection rules for hydrogenic atoms are

$$\Delta l = \pm 1 \qquad \Delta m_l = 0 \text{ or } \pm 1$$

The principal quantum number n can change by any amount consistent with the Δl for the transition because it does not relate directly to the angular momentum.

Example Using the selection rules
To what orbitals may a $4d$ electron make radiative transitions?
Answer Because $l = 2$, the final orbital must have $l = 1$ or 3.

Thus, an electron may make a transition from a 4*d* orbital to any *np* orbital (subject to $\Delta m_l = 0$ or ± 1) and to any *nf* orbital (subject to the same rule). However, it cannot undergo a transition to any other orbital, so a transition to any *ns* orbital or another *nd* orbital is forbidden.

Exercise E8.19 To what orbitals may a 4*s* electron make radiative transitions?

[*np* orbitals only]

Selection rules enable us to construct a **Grotrian diagram** (Fig. 8.36), which is a diagram that summarizes the energies of the states and the allowed transitions between them. The densities of the transition lines in the diagram denote their relative intensities in the spectrum. The intensities may also be calculated from the wavefunctions of the two states, but we shall not deal with this aspect here.

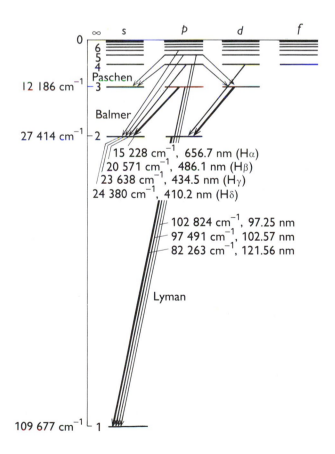

Fig. 8.36 A Grotrian diagram that summarizes the appearance and analysis of the spectrum of atomic hydrogen. The thicker the line, the more intense the transition.

327

The structures of many-electron atoms

The Schrödinger equations for many-electron atoms are extremely complicated because all the electrons interact with one another. Even for a He atom, with its two electrons, no mathematical expression for the orbitals and energies can be given, and we are forced to make approximations. We shall adopt a simple approach based on what we already know about the structure of hydrogenic atoms.

8.12 The orbital approximation

The actual wavefunction of a many-electron atom, if it could be calculated, would be a very complicated function of the coordinates of all the electrons. However, in the **orbital approximation** we suppose that a reasonable first approximation to this unknown exact wavefunction is obtained by thinking of each electron as occupying its 'own' orbital, and writing

$$\psi = \psi(1)\psi(2)\cdots \qquad (24)$$

where $\psi(1)$ is the wavefunction of electron 1, $\psi(2)$ that of electron 2, and so on. We can think of the individual orbitals as resembling the hydrogenic orbitals, but with nuclear charges that are modified by the presence of all other electrons in the atom. This description is only approximate, but it is a useful model discussing the chemical properties of atoms, and is the starting point for more sophisticated descriptions of atomic structure.

The helium atom

The orbital approximation allows us to express the electronic structure of an atom by reporting its **configuration**, the list of occupied orbitals (usually, but not necessarily, in its ground state). Thus, as the ground state of a hydrogenic atom consists of the single electron in a $1s$ orbital, we report its configuration as $1s^1$ (read 'one-s-one').

A He atom has two electrons. We can imagine forming the atom by adding the electrons in succession to the orbitals of the bare nucleus (of charge $2e$). The first electron occupies a $1s$ hydrogenic orbital, but because $Z = 2$, the orbital is more compact than in H itself. The second electron joins the first in the $1s$ orbital, and so the electron configuration of the ground state of He is $1s^2$ (read 'one-s-two').

The Pauli principle

Lithium, with $Z = 3$, has three electrons. Two of its electrons occupy a $1s$ orbital drawn even more closely than in He around the more highly charged nucleus. The third electron, however, does not join the first two in the 1s orbital because that configuration is forbidden by the Pauli exclusion principle:

> The **Pauli exclusion principle** states that no more than two electrons may occupy any given orbital, and if two do occupy one orbital, their spins must be paired.

Electrons with **paired spins**, denoted ↑↓, have zero net spin angular momentum because the spin angular momentum of one electron is cancelled by the spin of the other (one electron has $m_s = +\frac{1}{2}$ and the other has $m_s = -\frac{1}{2}$, and their sum is 0). The exclusion principle is the key to the structure of complex atoms, to chemical periodicity, and to molecular structure. It was proposed by the Austrian, Wolfgang Pauli, in 1924 when he was trying to account for the absence of some lines in the spectrum of helium.

Lithium's third electron cannot enter the 1s orbital because that orbital is already full: we say that the K shell is **complete** and that the two electrons form a **closed shell**. Because a similar closed shell occurs in the He atom, we denote it [He]. The third electron is excluded from the K shell and must occupy the next available orbital, which is one with $n = 2$ and hence belonging to the L shell. However, we now have to decide whether the next available orbital is the 2s orbital or a 2p orbital, and therefore whether the lowest energy configuration of the atom is $[He]2s^1$ or $[He]2p^1$.

Penetration and shielding

Unlike in hydrogenic atoms, the 2s and 2p orbitals (and, in general, all subshells of a given shell) are not degenerate in many-electron atoms. For reasons we shall now explain, s electrons generally lie lower in energy than p electrons of a given shell, and p electrons lie lower than d electrons.

An electron in a many-electron atom experiences a Coulombic repulsion from all the other electrons present. If the electron of interest is at an average distance r from the nucleus, it experiences an average repulsion that can be modelled by placing a point negative charge at the nucleus. This point charge should be equal in magnitude to the total charge of the electrons within a sphere of radius r (Fig. 8.37). The effect of the point negative charge is to reduce the full charge of the nucleus from Ze to $Z_{eff}e$, where Z_{eff} is the **effective atomic number**. We say that the electron experiences a **shielded** nuclear charge, and that the true atomic number is reduced to Z_{eff} by an amount called the **shielding constant**, σ:

$$Z_{eff} = Z - \sigma$$

The electrons do not actually 'block' the full Coulombic attraction of the nucleus: the effective charge is simply a way of expressing the net

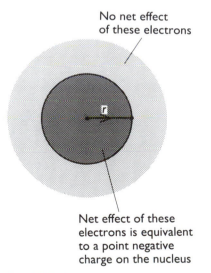

No net effect of these electrons

Net effect of these electrons is equivalent to a point negative charge on the nucleus

Fig. 8.37 An electron at a distance r from the nucleus experiences a Coulombic repulsion from all the electrons within a sphere of radius r. This repulsion is equivalent to a point negative charge located on the nucleus. The negative charge reduces the effective nuclear charge of the nucleus from Ze to $Z_{eff}e$.

329

outcome of the nuclear attraction and the electronic repulsions in terms of a single equivalent charge at the centre of the atom.

Exercise E8.20 What would be a plausible first estimate of the effective nuclear charge experienced by a 2s electron in Li? [slightly higher than 1]

The effective atomic number is different for s and p electrons because they have different wavefunctions (Fig. 8.38). An s electron has a greater **penetration** through inner shells than a p electron in the sense that it is more likely to be found close to the nucleus than a p electron of the same shell (the p orbital, remember, is zero at the nucleus). Because only electrons inside the sphere defined by the location of the electron (in effect, the core electrons) contribute to shielding, an s electron experiences less shielding than a p electron and it therefore experiences a larger Z_{eff}. Consequently, by the combined effects of penetration and shielding, an s electron is more tightly bound than a p electron of the same shell. Similarly, a d electron penetrates less than a p electron of the same shell, and it therefore experiences more shielding and an even smaller Z_{eff}.

Effective atomic numbers for different types of electrons in atoms have been calculated (Table 8.1). We see that in general s electrons do experience higher effective atomic numbers than p electrons, although there are some discrepancies.

The consequence of penetration and shielding is that the energies of subshells in a many-electron atom in general lie in the order

$$s < p < d < f$$

The individual orbitals of a given subshell (such as the three p orbitals of a p subshell) remain degenerate because they all have the same radial characteristics and so experience the same effective nuclear charge.

Fig. 8.38 An electron in an s orbital (here a 3s orbital) is more likely to be found close to the nucleus than an electron in a p orbital of the same shell. Hence it experiences less shielding and is more tightly bound.

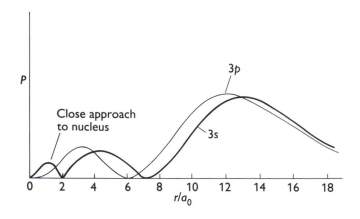

Table 8.1   Effective atomic numbers for valence electrons

	H							He
$1s$	1							1.69
	Li	Be	B	C	N	O	F	Ne
$2s$	1.28	1.91	2.58	3.22	3.85	4.49	5.13	5.76
$2p$			2.42	3.14	3.83	4.45	5.10	5.76
	Na	Mg	Al	Si	P	S	Cl	Ar
$3s$	2.51	3.31	4.12	4.90	5.64	6.37	7.07	7.76
$3p$			4.07	4.29	4.89	5.48	6.12	6.76

We can now complete the Li story. Because the shell with $n = 2$ consists of two nondegenerate subshells, with the $2s$ orbital lower in energy than the three $2p$ orbitals, the third electron occupies the $2s$ orbital. This arrangement results in the ground state configuration $1s^2 2s^1$, or $[\text{He}]2s^1$. It follows that we can think of the structure of the atom as consisting of a central nucleus surrounded by a complete helium-like shell of two $1s$ electrons, and around that a more diffuse $2s$ electron. The electrons in the outermost shell of an atom in its ground state are called the **valence electrons** because they are largely responsible for the chemical bonds that the atom forms (and, as we shall see, the extent to which an atom can form bonds is called its 'valence'). Thus, the valence electron in Li is a $2s$ electron, and lithium's other two electrons belong to its inner shells and take little part in bond formation.

The building-up principle

The extension of the procedure used for H, He, and Li to other atoms is called the **building-up principle** (or the *Aufbau* principle, from the German word for 'building up'). The building-up principle sets down an order of occupation of atomic orbitals that results in the ground state configuration of the neutral atom.

We imagine the bare nucleus of atomic number Z, and then feed into the orbitals Z electrons in succession. The first rule of the building-up principle is as follows:

The order of occupation of subshells is

$$1s \quad 2s \quad 2p \quad 3s \quad 3p \quad 4s \quad 3d \quad 4p \quad 5s \quad 4d \quad 5p \quad 6s$$

and each orbital of each subshell may accommodate up to two electrons.

This order of occupation is approximately the order of energies of the individual orbitals, because in general the lower the energy of the orbital, the lower the total energy of the atom as a whole when that orbital is occupied. However, there are complicating effects arising from electron–electron repulsions that are important when the orbitals have very similar energies (such as the 4s and 3d orbitals near Ca and Sr), and we must then take special care.

We feed the Z electrons in succession into the orbitals subject to the demand of the exclusion principle that no more than two can occupy any single orbital. Because an s subshell consists of only one orbital, up to two electrons may occupy it. Because a p subshell consists of three orbitals, it can accommodate up to six electrons; a d subshell, which consists of five orbitals, can accommodate up to ten electrons.

As an example, consider a carbon atom. Because $Z = 6$ for carbon, there are six electrons to accommodate. Two enter and fill the 1s orbital, two enter and fill the 2s orbital, leaving two electrons to occupy the orbitals of the 2p subshell. Hence its ground configuration is $1s^2 2s^2 2p^2$, or more succinctly $[He]2s^2 2p^2$, with $[He]$ the helium-like $1s^2$ core. However, it is possible to be more precise. On electrostatic grounds, we can expect the last two electrons to occupy different 2p orbitals for they will then be further apart on average and repel each other less than if they were in the same orbital. Thus, one electron can be thought of as occupying the $2p_x$ orbital and the other as occupying the $2p_y$ orbital, and the lowest energy configuration of the atom is $[He]2s^2 2p_x^1 2p_y^1$. The same rule applies whenever degenerate orbitals of a subshell are available for occupation. Thus, another rule of the building-up principle is that

> Electrons occupy different orbitals of a given subshell before doubly occupying any one of them.

Thus, a nitrogen atom ($Z = 7$) has the configuration $[He]2s^2 2p_x^1 2p_y^1 2p_z^1$, and only when we get to oxygen ($Z = 8$) is a 2p orbital doubly occupied, giving the configuration $[He]2s^2 2p_x^2 2p_y^1 2p_z^1$.

An additional point arises when electrons occupy degenerate orbitals (such as the three 2p orbitals) singly, as they do in C, N, and O, for there is then no requirement that their spins should be paired. We need to know whether a lower energy is achieved when the electron spins are the same (both α, for instance, denoted $\uparrow\uparrow$, if there are two electrons in question, as in C) or when they are paired ($\uparrow\downarrow$). This problem is resolved by Hund's rule:

> **Hund's rule** states that, in its ground state, an atom adopts a configuration with the greatest number of unpaired electrons.

The explanation of Hund's rule is complicated, but it reflects the quantum mechanical property of **spin correlation**, that electrons with parallel spins have a tendency to stay well apart and hence repel each other less.†

We can now conclude that in the ground state of a C atom, the two $2p$ electrons have the same spin, that all three $2p$ electrons in an N atom have the same spin, and that the two electrons that singly occupy different $2p$ orbitals in an O atom have the same spin (the two in the $2p_x$ orbital are necessarily paired).

Neon, with $Z = 10$, has the configuration $[He]2s^2 2p^6$, which completes the L shell. This closed-shell configuration is denoted $[Ne]$, and acts as a core for subsequent elements. The next electron must enter the $3s$ orbital and begin a new shell, and so an Na atom, with $Z = 11$, has the configuration $[Ne]3s^1$. Like lithium with the configuration $[He]2s^1$, sodium has a single s electron outside a complete core.

Exercise E8.21 Deduce the ground state electron configuration of sulfur.

$$[[Ne]3s^2 3p_x^2 3p_y^1 3p_z^1]$$

This analysis has brought us to the origin of chemically periodicity. The L shell is completed by eight electrons, and so the element with $Z = 3$ (Li) should have similar properties to the element with $Z = 11$ (Na). Likewise, Be ($Z = 4$) should be similar to $Z = 12$ (Mg), and so on up to the noble gases He ($Z = 2$), Ne ($Z = 10$), and Ar ($Z = 18$).

The occupation of d orbitals

Argon has complete $3s$ and $3p$ subshells, and as the $3d$ orbitals are high in energy, the atom effectively has a closed-shell configuration. Indeed, the $4s$ orbitals are so lowered in energy by their ability to penetrate close to the nucleus that the next electron (for K) occupies a $4s$ orbital rather than a $3d$ orbital and the K atom (the element following Ar) resembles an Na atom. The same is true of a Ca atom, which has the configuration $[Ar]4s^2$. However, at this point, the $3d$ orbitals become comparable in energy to the $4s$ orbitals (Fig. 8.39), and they start to be filled.

Ten electrons can be accommodated in the five $3d$ orbitals, which accounts for the electron configurations of scandium to zinc. However, the building-up principle has less clear-cut predictions about the ground state configurations of these elements because electron–electron repulsions are comparable to the energy difference between the $4s$ and $3d$ orbitals, and a simple analysis no longer works. At gallium, the energy of the $3d$ orbitals has fallen so far below those of

† The effect of spin correlation is to allow the atom to shrink slightly, so the electron–nucleus interaction is improved when the spins are parallel.

Fig. 8.39 Energy levels of many-electron atoms in the periodic table. The inset shows a magnified view close to $Z = 20$.

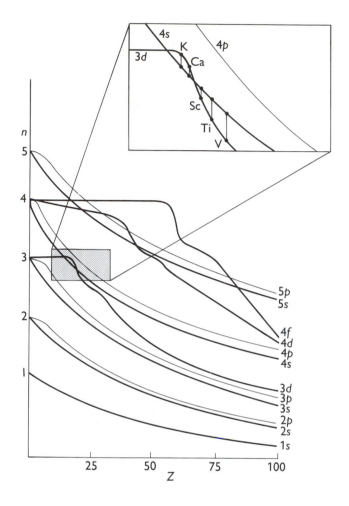

the $4s$ and $4p$ orbitals that they (the $3d$ orbitals) can be largely ignored, and the building-up principle can be used in the same way as in preceding periods. Now the $4s$ and $4p$ subshells constitute the valence shell, and the period terminates with krypton. Because 18 electrons have intervened since argon, this period is the first **long period** of the periodic table. The existence of the **d-block elements** (the 'transition metals') reflects the stepwise occupation of the $3d$ orbitals, and the subtle shades of energy differences along this series gives rise to the rich complexity of inorganic (and bioinorganic) d-metal chemistry. A similar intrusion of the f orbitals in Periods 6 and 7 accounts for the existence of the f block of the periodic table (the lanthanides and actinides).

The configurations of cations and anions

The configurations of cations of elements in the s, p, and d blocks of the periodic table are derived by removing electrons from the ground-state configuration of the neutral atom in a specific order. First, we remove p electrons (if any are present), then s electrons, and then as many d electrons as are necessary to achieve the stated charge. For instance, because the configuration of Fe is $[Ar]3d^64s^2$, an Fe^{3+} cation has the configuration $[Ar]3d^5$. The configurations of anions are derived by continuing the building-up procedure and adding electrons to the neutral atom until the configuration of the next noble gas has been reached. Thus, the configuration of an O^{2-} ion is achieved by adding two electrons to $[He]2s^22p^4$, giving $[He]2s^22p^6$, the same as the configuration of Ne.

Exercise E8.22 Give the electron configurations of (a) a Cu^{2+} ion, and (b) a S^{2-} ion.

$$[(a)\ [Ar]3d^9,\ (b)\ [Ne]3s^23p^6]$$

8.13 Periodic trends in atomic properties

The periodic recurrence of analogous ground state electron configurations as the atomic number increases accounts for the periodic variation in the properties of atoms. Here we concentrate on two aspects of atomic periodicity: atomic radius and ionization energy.

Atomic radius

Atomic radii are of great significance in chemistry, for the size of an atom is one of the most important properties for determining how many chemical bonds an element can form. Moreover, the size and shape of a molecule depends on the sizes of the atoms of which it is composed, and molecular shape and size is a crucial aspect of a molecule's biological function. Atomic radius also has important technological implications, because the similarity of the atomic radii of the d-block elements is the main reason why they can be mixed together to form so many different alloys, particularly varieties of steel. If there is one *single* attribute of an element that determines its chemical properties (either directly, or indirectly through the variation of other properties), then it is atomic radius.

In general, *atomic radii decrease from left to right across a period and increase down each group* (Fig. 8.40). The decrease across a period can be traced to the increase in nuclear charge, which draws the electrons in closer to the nucleus. The increase in nuclear charge is partly cancelled by the increase in the number of electrons, but one electron does not fully shield one nuclear charge, so the increase in nuclear charge dominates. The increase in atomic radius down a group (despite the increase in nuclear charge) is explained by the fact that the valence shells of successive periods correspond to higher principal

Fig. 8.40 The variation of radius through the periodic table. Note the contraction of radii following the lanthanides in Period 6 (following Yb, ytterbium).

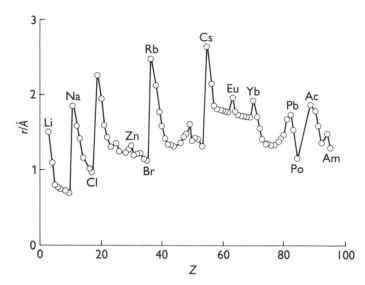

quantum numbers. That is, successive periods correspond to the start and then completion of successive shells of the atom that surround each other like the successive shells of an onion, and the need to occupy an outer shell leads to a larger atom despite the increased nuclear charge.

A modification of the increase down a group is encountered in Period 6, for the radii of the atoms late in the *d* block and in the subsequent *p* block are not as large as would be expected by simple extrapolation down the group. The reason can be traced to the fact that in Period 6 the *f* orbitals are occupied. An *f* electron is a very inefficient shielder of nuclear charge (for reasons connected with its radial extension), and so as the atomic number increases from La to Yb, there is a considerable contraction in radius. By the time the *d* block resumes (at lutetium, Lu), the poorly shielded but considerably increased nuclear charge has drawn in the surrounding electrons, and the atoms are compact. They are so compact that the metals in this region of the periodic table (iridium to lead) are very dense. The reduction in radius below that expected by extrapolation from preceding periods is called the **lanthanide contraction**.

Ionization energy

The minimum energy necessary to remove an electron from a many-electron atom is its **first ionization energy**, I_1. The **second ionization energy**, I_2 is the minimum energy needed to remove a second electron (from the singly charged cation). The variation of the first ionization energy through the periodic table is shown in Fig. 8.41 and some numerical values are given in Table 2.3. The ionization energy of an element plays a central role in its ability to participate in bond formation (since bond formation, as we shall see in Chapter 9, is a consequence of the relocation of electrons from one atom to another) and, after atomic radius, is the most important property for determining an element's chemical characteristics.

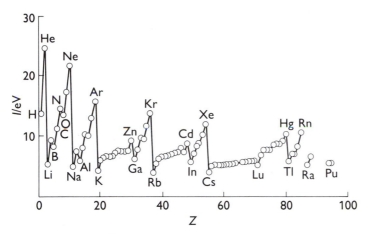

Fig. 8.41 The periodic variation of the first ionization energies of the elements.

Lithium has a low first ionization energy: its outermost electron is well shielded from the nucleus by the core ($Z_{eff} = 1.28$ compared with $Z = 3$) and it is easily removed. Beryllium has a higher nuclear charge than lithium, and its outermost electron (one of the two $2s$ electrons) is more difficult to remove: its ionization energy is larger. The ionization energy decreases between beryllium and boron because in the latter the outermost electron occupies a $2p$ orbital and is less strongly bound than if it had been a $2s$ electron. The ionization energy increases between boron and carbon because the latter's outermost electron is also $2p$ and the nuclear charge has increased. Nitrogen has a still higher ionization energy because of the further increase in nuclear charge.

There is now a kink in the curve which reduces the ionization energy of oxygen below what would be expected by simple extrapolation. At oxygen a $2p$ orbital must become doubly occupied, and the electron–electron repulsions are increased above what would be expected by simple extrapolation along the row. (The kink is less pronounced in the next row, between phosphorus and sulfur, because their orbitals are more diffuse.) The values for oxygen, fluorine, and neon fall roughly on the same line, the increase of their ionization energies reflecting the increasing attraction of the nucleus for the outermost electrons.

The outermost electron in sodium is $3s$. It is far from the nucleus, and the latter's charge is shielded by the compact, complete neon-like core. As a result, the ionization energy of sodium is substantially lower than that of neon. The periodic cycle starts again along this row, and the variation of the ionization energy can be traced to similar reasons.

EXERCISES

8.1 Calculate the power radiated by a $20\,cm \times 3.0\,cm$ section of the surface of a hot body at $1500\,K$.

8.2 The power delivered to a photodetector that collects 8.0×10^7 photons in $3.8\,ms$ from monochromatic light is $0.72\,\mu W$. What is the frequency of the radiation?

8.3 A diffraction experiment requires the use of

electrons of wavelength 0.45 nm. Calculate the velocity of the electrons.

8.4 Calculate the linear momentum of photons of wavelength (a) 750 nm, (b) 70 pm, (c) 19 m.

8.5 The energy required for the ionization of a certain atom is 3.44×10^{-18} J. The absorption of a photon of unknown wavelength ionizes the atom and ejects an electron with velocity 1.03×10^6 m s^{-1}. Calculate the wavelength of the incident radiation.

8.6 The speed of a certain proton is 4.5×10^5 m s^{-1}. If the uncertainty in its momentum is to be reduced to 0.0100 per cent, what uncertainty in its location must be tolerated?

8.7 A particle of mass 6.65×10^{-27} kg is confined to an infinite square well of width L. The energy of the level with $n = 3$ is 2.00×10^{-24} J. Calculate the width of the box.

8.8 Calculate the location of a point in a box of length L at which the probability of a particle being found is 25 per cent of its maximum probability when $n = 1$.

8.9 Calculate the spacing between the levels with $n = 4$ and $n = 5$ of a deuterium atom in a one-dimensional box of length 5.0 nm.

8.10 Calculate the energy per photon and the energy per mole of photons for radiation of wavelength (a) 600 nm (red), (b) 550 nm (yellow), (c) 400 nm (violet), (d) 200 nm (ultraviolet), (e) 150 pm (X-ray), (f) 1 cm (microwave).

8.11 Calculate the linear momenta of the photons specified in Exercise 8.10.

8.12 A sodium lamp emits yellow light (550 nm). How many photons does it emit each second if its power is (a) 1.0 W, (b) 100 W?

8.13 The peak of the Sun's emission occurs at about 480 nm; estimate the temperature of its surface.

8.14 The work function for metallic caesium is 2.14 eV. Calculate the kinetic energy and the speed of the electrons ejected by light of wavelength (a) 700 nm, (b) 300 nm.

8.15 The ionization energy of an H atom is 13.6 eV. Suppose an H atom in interstellar space moves through a region of 100-nm ultraviolet radiation. Can the atom be ionized?

8.16 Calculate the size of the quantum involved in the excitation of (a) an electronic motion of period 10^{-15} s, (b) a molecular vibration of period 10^{-14} s, (c) a pendulum of period 1 s. Express the results in joules and kilojoules per mole.

8.17 Calculate the de Broglie wavelength of (a) a mass of 1.0 g travelling at 1.0 cm s^{-1}, (b) the same, travelling at 100 km s^{-1}, (c) an He atom travelling at 1000 m s^{-1} (a typical speed at room temperature).

8.18 Calculate the de Broglie wavelength of an electron accelerated from rest through a potential difference of (a) 100 V, (b) 1.0 kV, (c) 100 kV.

8.19 (a) Calculate the minimum uncertainty in the speed of a ball of mass 500 g that is known to be within 1.0 μm of a certain point on a bat. (b) What is the minimum uncertainty in the position of a bullet of mass 5.0 g that is known to have a speed somewhere between 350.00 001 m s^{-1} and 350.00 000 m s^{-1}?

8.20 An electron is confined to a linear region with a length of the same order as the diameter of an atom (c. 100 pm). Calculate the minimum uncertainties in its momentum and speed.

8.21 In an X-ray photoelectron experiment, a photon of wavelength 150 pm ejects an electron from the inner shell of an atom and it emerges with a speed of 2.14×10^7 m s^{-1}. Calculate the binding energy of the electron.

8.22 The Planck distribution (eqn (5)) gives the energy density in the wavelength range $\delta\lambda$ at the wavelength λ. Calculate the energy density in the range 650 nm to 655 nm inside a cavity of volume 100 cm^3 when its temperature is (a) 25 °C, (b) 3000 °C.

8.23 The wavelength of the emission maximum from a small pinhole in an electrically heated container was determined at a series of temperatures, and the results are given below. Deduce a value for Planck's constant.

θ/°C	1000	1500	2000	2500	3000	3500
λ_{max}/nm	2181	1600	1240	1035	878	763

8.24 Calculate the wavelength of the line with $n = 4$ in the Balmer series of the spectrum of atomic hydrogen.

8.25 The frequency of one of the lines in the Paschen series of the spectrum of atomic hydrogen is 2.7415×10^{15} Hz. Calculate the principal quantum number of the upper state in the transition.

8.26 One of the energy levels of the H atom is at $27\,414$ cm^{-1}. What is the value of the energy level with which it combines to produce light of wavelength 486.1 nm?

8.27 When 58.4 nm ultraviolet radiation from a helium lamp is directed on to a sample of krypton, electrons are ejected at 1.59×10^6 m s^{-1}. Calculate the ionization energy of krypton.

8.28 Locate the radial nodes in the $3s$ orbital of an H atom.

8.29 What is the orbital angular momentum of an electron in the orbitals (a) $1s$, (b) $3s$, (c) $3d$, (d) $2p$, (e) $3p$? Give the numbers of angular and radial nodes in each case.

8.30 State the orbital degeneracy of the levels in the hydrogen atom that have energy (a) $-hcR_H$, (b) $-hcR_H/9$, and (c) $-hcR_H/25$.

8.31 At what radius does the probability of finding an electron at a point in the H atom fall to 50 per cent of its maximum value?

8.32 At what radius in the H atom does the radial distribution function of the ground state have (a) 50 per cent, (b) 75 per cent of its maximum value?

8.33 Which of the following transitions are allowed in the normal electronic emission spectrum of an atom: (a) $2s \rightarrow 1s$, (b) $2p \rightarrow 1s$, (c) $3d \rightarrow 2p$, (d) $5d \rightarrow 2s$, (e) $5p \rightarrow 3s$?

8.34 How many electrons can occupy the following subshells: (a) $1s$, (b) $3p$, (c) $3d$, and (d) $6g$?

8.35 Give the electron configurations of the ground states of the first eighteen elements in the periodic table.

8.36 The 'Humphreys series' is another group of lines in the spectrum of atomic hydrogen. It begins at $12\,368$ nm and has been traced to 3281.4 nm. What are the transitions involved? What are the wavelengths of the intermediate transitions?

8.37 A series of lines in the spectrum of atomic hydrogen lies at 656.46 nm, 486.27 nm, 434.17 nm, and 410.29 nm. What is the wavelength of the next line in the series? What is the ionization energy of the atom when it is in the lower state of the transitions?

8.38 The Li^{2+} ion is hydrogenic and has a Lyman series at $740\,747$ cm^{-1}, $877\,924$ cm^{-1}, $925\,933$ cm^{-1}, and beyond. Show that the energy levels are of the form $-hcR/n^2$ and find the value of R for this ion. Go on to predict the wavenumbers of the two longest wavelength transitions of the Balmer series of the ion and find the ionization energy of the ion.

9

The chemical bond

In this chapter we consider **valence theory**, the theory of the origin of the numbers, strengths, and three-dimensional arrangement of chemical bonds between atoms. Valence theory is of the greatest importance for accounting for the properties of elements and the compounds they form. Its usefulness ranges from the explanation of the properties of the smallest molecules, such as why N_2 is so inert that it acts as a diluent for the aggressive oxidizing power of atmospheric oxygen, to the properties of the most complex molecules known, such as the function of protein molecules as enzymes and the function of DNA as a genetic material. The description of chemical bonding has become highly developed through the use of computers, and it is now possible to consider the structures of molecules of almost any complexity. However, in this chapter we confine our attention to the qualitative concepts that underlie these computational procedures.

Chemists generally distinguish between two types of bond:

- An **ionic bond** is a chemical bond formed by the transfer of electrons from one atom to another and the consequent attraction between the ions that are so formed.

- A **covalent bond** is a chemical bond formed when two atoms share a pair of electrons.

The character of a covalent bond, which we concentrate on in this chapter, was identified by G. N. Lewis (in 1916, before quantum mechanics was fully developed). We shall assume that Lewis's ideas are familiar, but for convenience they are reviewed in *Further information 9* (p. 342). In this chapter we develop the modern theory of chemical bond formation in terms of the quantum mechanical

341

properties of electrons and place Lewis's ideas in a modern context. We shall see that an ionic bond can be regarded as a limiting type of covalent bond in which the sharing of electrons has given way to the possession by one atom of both electrons of the bond. However, there are certain aspects of ionic solids (solids in which ionic bonding is a good description) that we shall treat separately in Chapter 10.

Lewis's original theory was unable to account for the shapes adopted by molecules; and shape, as we have remarked, is a crucial

FURTHER INFORMATION 9:
The Lewis theory of covalent bonding

In his original formulation of a theory of the covalent bond, G. N. Lewis proposed that each bond consisted of one electron pair and that each atom in a molecule shared electrons until it had acquired an octet configuration characteristic of a noble gas atom near it in the periodic table. (Hydrogen is an exception: it acquires a duplet of electrons, the configuration of its neighbour helium.) Thus, to write down a Lewis structure:

(1) Arrange the atoms as they are found in the molecule.

(2) Add one electron pair (represented by dots, :) between each bonded atom.

(3) Use the remaining electron pairs to complete the octets of all the atoms present either by forming lone pairs or by forming multiple bonds.

(4) Replace bonding electron pairs by bond lines (—) but leave lone pairs as dots (:).

A Lewis structure does not (except in very simple cases), portray the actual *geometrical* structure of the molecule; it is a *topological map* of the arrangement of bonds.

As an example, consider the Lewis structure

of methanol, CH_3OH, in which there are $4 \times 1 + 4 + 6 = 14$ valence electrons (and hence seven electron pairs) to accommodate. The first step is to write the atoms in the arrangement

$$\begin{array}{c} H \\ H \ C \ O \ H \\ H \end{array}$$

The next step is to add electron pairs to denote bonds:

$$\begin{array}{c} H \\ H:\overset{..}{C}:O:H \\ H \end{array}$$

The C atom now has a complete octet and all four H atoms have complete duplets. There are two unused electron pairs, which are used as lone pairs to complete the octet of the O atom:

An example of a species with a multiple bond is acetic acid:

When more than one structure can be written, the only difference being the location of multiple bonds, then the molecule is interpreted as a **resonance hybrid** (a quantum mechanical blend) of the individual Lewis structures. Resonance is depicted by a double headed arrow, ↔. The ozone molecule, O_3, is a resonance hybrid of the two structures shown at the top of the next column.

Many molecules cannot be written in a way

aspect of the properties of a molecule and is central to an enzyme's action. The most elementary (but quite successful) explanation of the shapes adopted by molecules is **valence-shell electron-pair repulsion theory** (VSEPR theory) in which it is supposed that the shape of a molecule is determined by the repulsions between electron pairs in the valence shell. We shall suppose that this theory is familiar; however, a brief outline is given in *Further information 10* (p. 344). In this chapter we shall examine some of the contributions that quantum

that conforms to the octet rule. Some, the **hypervalent molecules**, require an expansion of the octet. Although it is often stated that octet expansion requires the involvement of *d* orbitals, and is therefore confined to Period 3 and subsequent elements, there is good evidence to suggest that octet expansion is a consequence of an atom's size, not its intrinsic orbital structure. Whatever the reason, octet expansion is needed to account for the structures of PCl_5 (to 10 electrons), SF_6 (to 12 electrons), and XeO_4 (to 16 electrons):

Octet expansion is also encountered in species that do not necessarily require it, but which, if it is permitted, may acquire a lower energy. Thus, of the following structures of the SO_4^{2-} ion, the second has a lower energy than the first. The

actual structure of the ion is a resonance hybrid of both structures (together with analogous structures with double bonds in different locations), but the latter structure makes the dominant contribution.

Octet completion is not always energetically appropriate. Such is the case with boron trifluoride, BF_3. Two of the possible Lewis structures for this molecule are

In the former, the B atom has an **incomplete octet**. Nevertheless, it has a lower energy than the other structure, for to form the latter structure one F atom has had partially to relinquish an electron pair, which is energetically demanding for such an electronegative element. The actual molecule is a resonance hybrid of the two structures (and of others with the double bond in different locations), but the overwhelming contribution is from the former structure. Consequently, we regard BF_3 as a molecule with an incomplete octet. This feature is responsible for its ability to act as a Lewis acid (an electron pair acceptor).

The Lewis approach fails for a class of **electron-deficient compounds**, which are molecules that have too few electrons for a Lewis structure to be written. The most famous example is diborane, B_2H_6, which requires at least seven pairs of electrons to bind the eight atoms together, but it has only twelve valence electrons in all. The structures of such molecules can be explained in terms of molecular orbital theory and the concept of delocalized electron pairs, in which the influence of an electron pair is distributed over several atoms.

FURTHER INFORMATION 10: VSEPR theory

In **valence-shell electron-pair repulsion theory** (VSEPR theory) we focus on a single 'central' atom and consider the local arrangement of atoms that are linked to it. For example, in considering the H_2O molecule, we concentrate on the electron pairs in the valence shell of the central O atom. This procedure can be extended to molecules in which there is no obvious 'central atom', such as in benzene, C_6H_6, or hydrogen peroxide, H_2O_2, by focusing attention on a group of atoms, such as a C—CH—C fragment of benzene or an H—O—O fragment of hydrogen peroxide, and considering the arrangement of electron pairs around the central atom of the fragment.

The basic assumption of VSEPR theory is that the *valence-shell electron pairs of the central atom adopt positions that maximize their separations*. Thus, if the atom has four electron pairs in its valence shell, then the pairs adopt a tetrahedral arrangement around the atom; if the atom has five pairs, then the arrangement is trigonal bipyramidal. The maximum-separation arrangements of two to seven electron pairs are summarized in Table 1.

Table 1 Electron-pair arrangements

Number of electron pairs	Arrangement
2	Linear
3	Trigonal planar
4	Tetrahedral
5	Trigonal bipyramidal
6	Octahedral
7	Pentagonal bipyramidal

Once the basic shape of the arrangement of electron pairs has been identified, the pairs are identified as bonding or nonbonding pairs. For instance, in the H_2O molecule, two of the tetrahedrally arranged pairs are bonding pairs and two are nonbonding pairs. Then the shape of the molecule is classified by noting the arrangement of the atoms around the central atom. The H_2O molecule, for instance, has an underlying tetrahedral arrangement of electron pairs, but as only two of the pairs are bonding pairs, the molecule is classified as angular (Fig. 1).

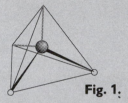

Fig. 1.

It is important to keep in mind the distinction between the arrangement of electron pairs and the shape of the resulting molecule: *the latter is determined by the relative locations of the atoms, not the lone pairs* (Fig. 2).

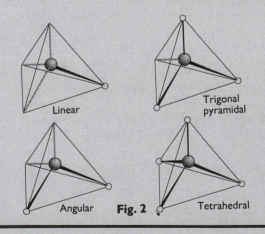

Linear

Trigonal pyramidal

Angular   **Fig. 2**

Tetrahedral

Example Using VSEPR theory to predict molecular shape—I
Predict the shape of an ethane molecule.
Answer We concentrate on one of the C atoms initially. That atom has four electron pairs in its

valence shell (in the molecule), and they adopt a tetrahedral arrangement. All four electron pairs are bonding: three bond H atoms and the fourth bonds the second C atom. Therefore, the arrangement of atoms is tetrahedral around the C atom. The second C atom has the same environment, so we conclude that the ethane molecule consists of two tetrahedral —CH₃ groups (**1**).

1   **2**

Exercise Predict the shape of an H₂O₂ molecule.

[Two angular H—O—O units (**2**)]

The next stage in the application of VSEPR theory is to accommodate the greater repelling effect of lone pairs compared with that of bonding pairs. That is, *bonding pairs tend to move away from lone pairs even though that might bring them closer to other bonding pairs*. The NH₃ molecule provides a simple example. The N atom has four electron pairs in its valence shell and they adopt a tetrahedral arrangement. Three of the pairs are bonding pairs, and the fourth is a lone pair. The basic shape of the molecule is therefore trigonal pyramidal. However, a lower energy is achieved if the three bonding pairs move away from the lone pair, even though they are brought slightly closer together (Fig. 3). We therefore predict an HNH bond angle of slightly less than the

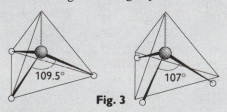

109.5°   107°

Fig. 3

tetrahedral angle of 109.5°, which is consistent with the observed angle of 107°.

Example Using VSEPR theory to predict molecular shape—II
What is the shape of an SF₄ molecule?
Answer The first step is to write a Lewis (electron dot) structure for the molecule to identify the number of electron pairs in the valence shell of the S atom. The Lewis structure shows that there are five electron pairs on the S

atom. Reference to Table 1 shows that the five pairs are arranged as a trigonal bipyramid. Four of the pairs are bonding pairs and one is a lone pair. The repulsions stemming from the lone pair are minimized if the lone pair is placed in an equatorial position: then it is close to the axial pairs (**3**), whereas if it had adopted an axial position it would have been closed to three equatorial pairs (**4**). Finally, the four bonding pairs are allowed to relax away from the single lone pair, to give a distorted see-saw arrangement (**5**).

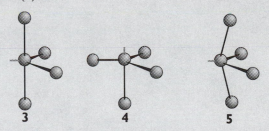

3   **4**   **5**

Exercise Predict the shape of a XeF₄ molecule.

[Square planar (**6**)]

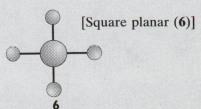

6

345

So far we have not considered how to take into account multiple bonding. It is supposed that the two or three electron pairs, respectively, of double and triple bonds are pinned together in their respective bonding regions, and hence act like a single 'superpair'. For example, each C atom in an ethylene molecule, $CH_2\!=\!CH_2$, is regarded as having *three* pairs (one of them the superpair of two electrons pairs of the double bond); they adopt a trigonal planar arrangement around each atom, so the shape of the molecule

7   **8**

is trigonal planar at each C atom (**7**). Another example is the SO_3^{2-} ion: if we adopt the Lewis

structure

$$:\!\ddot{O}\!:\!\ddot{S}\!:\!\ddot{O}\!:\; \rceil^{2-} \quad \text{or} \quad :\!\ddot{O}\!-\!\ddot{S}\!-\!\ddot{O}\!:\; \rceil^{2-}$$
$$\cdot\ddot{O}\cdot \qquad\qquad\qquad \overset{\|}{:\!\ddot{O}\!:}$$

we see that there are four pairs (one of them a superpair) around the S atom, indicating a tetrahedral arrangement of pairs. One pair is a lone pair, so overall the ion is trigonal pyramidal (**8**). We would reach the same conclusion if we adopted the alternative Lewis structure in which

$$:\!\ddot{O}\!:\!\ddot{S}\!:\!\ddot{O}\!:\; \rceil^{2-} \quad \text{or} \quad :\!\ddot{O}\!-\!\ddot{S}\!-\!\ddot{O}\!:\; \rceil^{2-}$$
$$:\!\ddot{O}\!: \qquad\qquad\qquad \overset{|}{:\!\ddot{O}\!:}$$

there are four electron pairs (none of them a superpair).

theory has made to understanding why a molecule adopts its characteristic shape.

All theories of molecular structure make the same simplification at the outset. Whereas the Schrödinger equation for a hydrogen atom can be solved exactly, an exact solution is not possible for any molecule because even the simplest molecule consists of three particles (two nuclei and one electron). The **Born–Oppenheimer approximation** is therefore adopted, in which it is supposed that the nuclei, being so much heavier than an electron, move relatively slowly and may be treated as stationary while the electrons move around them. We can therefore think of the nuclei as being fixed at an arbitrary separation R, and then solve the Schrödinger equation for the wavefunction of the electrons alone. The approximation is quite good for ground-state molecules, because calculations suggest that the nuclei in H_2 move through only about 1 pm while the electron speeds through 1000 pm, and so the error of assuming that the nuclei are stationary is small.

The Born–Oppenheimer approximation allows us to select an internuclear separation, and (in principle) to solve the Schrödinger equation for the electrons for that nuclear separation. Then we can choose a different separation and repeat the calculation, and so on. In this way we can explore how the energy of the molecule varies with bond length (and, in more complex molecules, with angles too), and obtain a **molecular potential energy curve**. A typical example of such a curve is illustrated in Fig. 9.1. It is called a *potential* energy curve because the kinetic energy of the nuclei (which are supposed

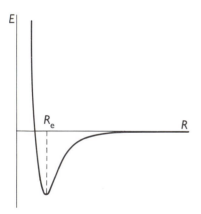

Fig. 9.1 A molecular potential energy curve. The equilibrium bond length, R_e, corresponds to the energy minimum.

stationary) is ignored. Once the curve has been calculated or determined experimentally (by using the spectroscopic techniques described in Chapter 11), we can identify the **equilibrium bond length** (the internuclear separation at the minimum of the curve, R_e) and **bond dissociation energy**, D_e, the depth of the minimum below the energy of the infinitely widely separated atoms.

Valence bond theory

The first step towards a quantum mechanical theory of bonding and molecular shape refines the Lewis electron-dot structures by regarding each electron pair as occupying an orbital that spreads over the two atoms linked by the bond. As in Lewis's pre-quantum mechanical theory, the centre of attention is the electron pair and the bond it forms, but the theory describes the location of the two electrons in terms of a wavefunction. The language we shall introduce, which includes concepts such as spin pairing, σ and π bonds, and hybridization, is widely used throughout chemistry, and is particularly widespread in the description of the properties and reactions of organic compounds.

9.1 Diatomic molecules

In **valence bond theory**, a bond is formed when an electron in one atomic orbital pairs its spin with that of an electron supplied by another atomic orbital (Fig. 9.2). Thus, the formation of a bond in H_2 can be pictured as the pairing of two electrons, one from each atom, to form a **σ bond**. A σ bond has cylindrical symmetry around the internuclear axis, and is so called because, when viewed along the bond, it resembles a pair of electrons in an s orbital. The joint wavefunction for the two electrons is written

$$\psi(\text{H—H}) = \psi_{\text{H1sA}}(1)\psi_{\text{H1sB}}(2) + \psi_{\text{H1sA}}(2)\psi_{\text{H1sB}}(1) \qquad (1)$$

The first term on the right represents a molecule in which electron 1 is in the $1s$ orbital on atom A and electron 2 is in the analogous orbital on atom B. The second term represents the case in which electron 2 is on atom A and electron 1 is on atom B. We have to include both terms in the description of the molecule, because it is impossible to keep track of the two electrons: when the atoms are within bonding distance, either electron may be on either atom with equal probability. *The essential feature of valence bond theory is the pairing of the electrons and the formation of a wavefunction that allows both electrons to be found on either atom.* In general, for orbitals that we can symbolize A and B on atoms A and B respectively, a valence-bond wavefunction for an A—B bond is

$$\psi(\text{A—B}) = A(1)B(2) + A(2)B(1) \qquad (2)$$

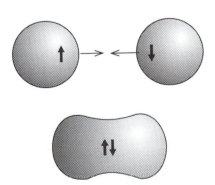

Fig. 9.2 In valence bond theory, a σ bond is formed when two electrons in orbitals on neighbouring atoms pair and the orbitals merge to form a cylindrical electron cloud.

When the energy of the H_2 molecule is calculated as the internuclear separation, R, is changed, a curve like that shown in Fig. 9.1 is obtained. The energy falls below that of two separated H atoms as the two atoms are brought within bonding distance, and each electron is free to migrate to the other atom. However, the energy reduction that follows from this process is counteracted by an increase in energy from the Coulombic repulsion between the two positively charged nuclei, which has the form

$$V_{\text{nuc,nuc}} = +\frac{e^2}{4\pi\varepsilon_0 R}$$

This positive contribution to the energy becomes large as R becomes small, and the total energy curve passes through a minimum and then climbs to a strongly positive value as the two nuclei are pressed together.

A similar description can be applied to more complex molecules, such as the Period 2 **homonuclear diatomic molecules**, which are diatomic molecules in which both atoms belong to the same element. Nitrogen, N_2, is an example. To construct the valence bond description of N_2, we consider the *valence* electron configuration of each atom:

$$N\ 2s^2 2p_x^1 2p_y^1 2p_z^1$$

It is conventional to take the z axis to be the internuclear axis, so we can imagine each atom as having a $2p_z$ orbital pointing towards a $2p_z$ orbital on the other atom, with the $2p_x$ and $2p_y$ orbitals perpendicular to the axis (Fig. 9.3). A σ bond can be formed by spin pairing the two electrons in the neighbouring $2p_z$ orbitals. However, the remaining p orbitals cannot merge to give σ bonds as they do not have cylindrical symmetry around the internuclear axis. Instead, the electrons in them pair to form two **π bonds**. A π bond arises from the spin pairing of electrons in two p orbitals that approach side-by-side, and is so called because, viewed along the internuclear axis, a π bond resembles a pair of electrons in a p orbital. In N_2, there are two π bonds: one is formed from spin pairing in two neighbouring $2p_x$ orbitals and the other from spin pairing in two neighbouring $2p_y$ orbitals. The overall bonding pattern in N_2 is therefore a σ bond plus two π bonds (Fig. 9.4), which is consistent with the Lewis structure :N≡N:.

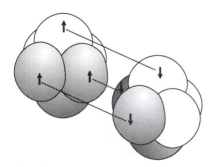

Fig. 9.3 The bonds in N_2 are built by allowing the electrons in the N2p to pair. However, only one orbital on each atom can form a σ bond: the orbitals perpendicular to the axis form π bonds.

Exercise E9.1 Describe the valence-bond ground state of the Cl_2 molecule.
[One $\sigma(Cl3p_z, Cl3p_z)$ bond formed by spin pairing of the two electrons]

9.2 Polyatomic molecules

The concept of bond formation by electron pairing and the merging of orbitals to form a σ bond can be extended readily to polyatomic species. Each σ bond in a polyatomic molecule is formed by the spin pairing of electrons in any atomic orbitals with cylindrical symmetry about the internuclear axis. Likewise, π bonds are formed by pairing electrons that occupy atomic orbitals of the appropriate symmetry. A simple description of the electronic structure of H_2O should make this clear.

The valence electron configuration of the O atom is $2s^2 2p_x^2 2p_y^1 2p_z^1$. The two unpaired electrons in the O2p orbitals can each pair with an electron in an H1s orbital, and each combination results in the formation of a σ bond (each bond has cylindrical symmetry about the respective O—H internuclear distance). Because the $2p_y$ and $2p_z$ orbitals lie at 90° to each other, the two σ bonds also lie at 90° to each other (Fig. 9.5). We predict, therefore that H_2O should be an angular molecule, which it is. However, the model predicts a bond angle of 90°, whereas the actual bond angle is 104°.

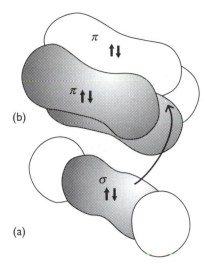

Fig. 9.4 The electrons in the 2p orbitals of two neighbouring N atoms merge to form σ and π bonds. (a) The electrons in the N2p_z orbitals pair to form a bond of cylindrical symmetry (shown separately). (b) Electrons in the N2p orbitals that lie perpendicular to the axis also pair to form *two* π bonds.

Exercise E9.2 Give a valence bond description of NH_3, and predict the bond angle of the molecule on the basis of this description. The experimental bond angle is 107°.

[Three σ(N2p, H1s) bonds; 90°]

While broadly correct, valence bond theory seems to have two deficiencies. One is the poor estimate it provides for the bond angle in H_2O (and other molecules, such as NH_3). Indeed, the theory appears to make worse predictions than VSEPR theory, which predicts HOH and HNH bond angles of slightly less than 109° in H_2O and NH_3, respectively. The second major deficiency is its apparent inability to account for carbon's tetravalence: the ground state configuration of C is $2s^2 2p_x^1 2p_y^1$, which suggests that a carbon atom should be capable of forming only *two* bonds, not four. However, both deficiencies are repaired by the combined consequences of **promotion**, the excitation of an electron to an orbital of higher energy, and **hybridization**, the blending together of the orbitals of the excited atom.

Promotion

The promotion of an electron is its excitation to an orbital of higher energy as a bond is formed. The promotion is worthwhile if the energy it requires can be more than recovered in the greater strength or number of bonds that the promotion allows to be formed. In carbon, for example, the promotion of a 2s electron to a 2p orbital leads to the configuration $2s^1 2p_x^1 2p_y^1 2p_z^1$, with *four* unpaired electrons in separate orbitals. These electrons may pair with four electrons in orbitals provided by four other atoms (such as four H1s orbitals if the molecule is CH_4), and hence form four electron-pair σ bonds. Although energy was required to promote the electron, it is more than recovered by the atom's ability to form four bonds in place of two bonds of the

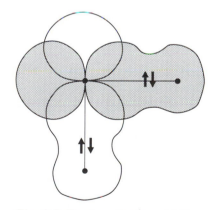

Fig. 9.5 The bonding in an H_2O molecule can be pictured in terms of the pairing of an electron belonging to one H atom with an electron in an O2p orbital; the other bond is formed likewise, but using a perpendicular O2p orbital. The predicted bond angle is 90°, which is in poor agreement with the experimental bond angle (104°).

349

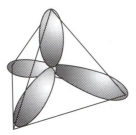

Fig. 9.6 The 2s and three 2p orbitals of a carbon atom hybridize, and the resulting hybrid orbitals point towards the corners of a regular tetrahedron.

unpromoted atom. Promotion and the formation of four bonds is a characteristic feature of carbon because the promoted electron leaves a doubly occupied 2s orbital and enters a vacant 2p orbital, hence significantly relieving the electron–electron repulsion it experiences in the former. Promotion also leads to a doubling of the number of bonds that can be formed.

Hybridization

Nevertheless, our description of the bonding in CH_4 (and its homologues) is still incomplete because it appears to imply the presence of three σ bonds of one type (formed from H1s and C2p orbitals) and a fourth σ bond of a distinctly different character (formed from H1s and C2s). This problem is overcome by realizing that the electron density distribution in the promoted atom is equivalent to the electron density in which each electron occupies a **hybrid orbital** formed by the interference between the C2s and C2p orbitals. The origin of the hybridization can be appreciated by thinking of the four atomic orbitals, which are waves centred on a nucleus, as being like ripples spreading from a single point on the surface of a lake: the waves interfere destructively and constructively in different regions, and give rise to four new shapes. The specific linear combinations that give rise to four equivalent hybrid orbitals are

$$h_1 = s + p_x + p_y + p_z \qquad h_2 = s - p_x - p_y + p_z$$
$$h_3 = s - p_x + p_y - p_z \qquad h_4 = s + p_x - p_y - p_z$$

As a result of the constructive and destructive interference between the component orbitals, each hybrid orbital has a large lobe pointing in the direction of one corner of a regular tetrahedron (Fig. 9.6). Because each hybrid is built from one s orbital and three p orbitals, it is called an sp^3 **hybrid orbital**.

It is now easy to see how the valence bond description of the methane molecule leads to a tetrahedral molecule containing four equivalent C—H bonds. Each hybrid orbital of the promoted atom contains a single unpaired electron; a hydrogen 1s electron can pair with each one, giving rise to a σ bond with lobes pointing to the four corners of tetrahedron. Because each sp^3 hybrid orbital has the same composition, all four σ bonds are identical apart from their orientation in space (Fig. 9.7).

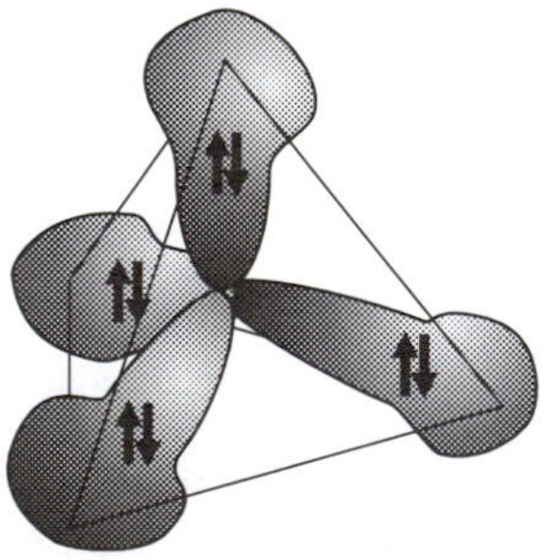

Hybridization can also be used to describe the structure of the ethylene molecule and the torsional rigidity of double bonds. The ethylene molecule is planar, with HCH and HCC bond angles close to 120°. To reproduce the σ-bonding structure that the arrangement of electron pairs requires, we promote each C atom to a $2s^1 2p^3$ configuration, but instead of using all four orbitals to form hybrids, we form sp^2 **hybrid orbitals** by allowing the s orbital and *two* p orbitals to interfere constructively and destructively. As shown in Fig. 9.8, these three hybrid orbitals lie in a plane and point towards the corners of an

Fig. 9.7 The valence bond description of the structure of CH_4. Each σ bond is formed by the pairing of an electron in an H1s orbital with an electron in one of the hybrid orbitals shown in Fig. 9.6. The resulting molecule is regular tetrahedral.

equilateral triangle.† The third $2p$ orbital ($2p_x$) is not included in the hybridization, and its axis is perpendicular to the plane in which the hybrids lie.

The structure of the ethylene molecule can now be described as follows. The sp^2-hybridized C atoms each form three σ bonds by spin pairing with either the h_1 hybrid of the other C atom or with H1s orbitals. The σ framework therefore consists of bonds at 120° to each other. When the two CH$_2$ groups lie in the same plane, the two electrons in the unhybridized p orbitals can pair and form a π bond (Fig. 9.9). The formation of this π bond locks the framework into the planar arrangement, because any rotation of one CH$_2$ group relative to the other leads to a weakening of the π bond (and consequently an increase in energy of the molecule).

A similar description applies to the linear acetylene molecule, H—C≡C—H, but now the carbon atoms are **sp hybridized**, and the σ bonds are formed from hybrid atomic orbitals of the form

$$h_1 = s + p_z \qquad h_2 = s - p_z$$

These two orbitals lie along the z axis. The electrons in them pair either with an electron in the corresponding hybrid orbital on the other C atom or with an electron in the H1s orbitals. Electrons in the two remaining p orbitals on each atom, which are perpendicular to the molecular axis, pair to form two perpendicular π bonds (as in Fig. 9.10).

Other hybridization schemes, particularly those involving d orbitals, are often invoked to account for (or at least be consistent with) other molecular geometries. For example sp^3d^2 hybridization results in six equivalent hybrid orbitals pointing towards the corners of a regular octahedron. (The hybridization of N atomic orbitals always results in the formation of N hybrid orbitals.) This octahedral hybridization scheme is sometimes invoked to account for the structure of octahedral molecules, such as SF$_6$. These schemes are summarized in Table 9.1.

Exercise E9.3 Describe the bonding in a PCl$_5$ molecule in valence-bond terms.

[Five σ bonds formed from sp^3d hybrids]

The 'pure' schemes in Table 9.1 are not the only possibilities: it is possible to form hybrid orbitals with intermediate proportions of atomic orbitals. For example, as more p-orbital character is included

† The explicit combinations we take to form the three hybrid orbitals are

$$h_1 = s + \sqrt{2}p_x \quad h_2 = s + \sqrt{\tfrac{3}{2}}p_x - \sqrt{\tfrac{1}{2}}p_y \quad h_3 = s - \sqrt{\tfrac{3}{2}}p_x - \sqrt{\tfrac{1}{2}}p_y$$

The coefficients are chosen so that the constructive interference occurs in the three directions desired.

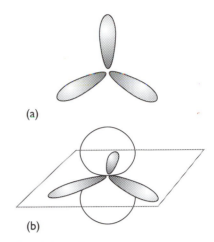

(a)

(b)

Fig. 9.8 (a) Trigonal planar hybridization is obtained when an s and two p orbitals are hybridized. The three lobes lie in a plane and make an angle of 120° with each other. (b) The remaining p orbital in the valence shell of an sp^2-hybridized atom lies perpendicular to the plane of the three hybrids.

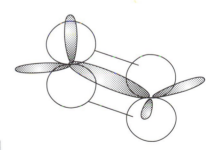

Fig. 9.9 The valence bond description of the structure of a carbon–carbon double bond, as in ethylene (ethene). The electrons in the two sp^2 hybrids that point towards each other pair and form a σ bond (shown stylized here). Electrons in the two p orbitals that are perpendicular to the plane of the hybrids pair, and form a π bond. The electrons in the remaining hybrid orbitals are used to form bonds to other atoms (to H atoms in ethylene).

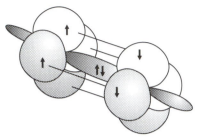

Fig. 9.10 The electronic structure of acetylene (ethyne). The electrons in the two *sp* hybrids on each atom pair to form σ bonds either with the other C atom or with an H atom. The remaining two unhybridized 2*p* orbitals on each atom are perpendicular to the axis: the electrons in corresponding orbitals on each atom pair to form two π bonds.

Table 9.1   Hybrid orbitals

Coordination number	Shape	Hybridization
2	Linear	sp
3	Trigonal planar	sp^2
4	Tetrahedral	sp^3
5	Bipyramidal	sp^3d
6	Octahedral	sp^3d^2

in an *sp* hybridization scheme, the hybridization changes towards sp^2 and the angle between the hybrids changes continuously from 180°, for pure *sp* hybridization, to 120° for pure sp^2 hybridization. If the proportion of *p* character continues to be increased (by reducing the admixture of *s* orbital), then the hybrids eventually become pure *p* orbitals making 90° to each other. Now we can account for the structure of H_2O, with its bond angle of 104°. Each O—H σ bond is formed from an O atom hybrid orbital with a composition that lies between pure *p* (which would lead to a bond angle of 90°) and pure sp^2 (which would lead to a bond angle of 120°). The equilibrium bond angle and hybridization are obtained by calculating the energy of the molecule as the bond angle is varied and searching for the minimum energy of the molecule.

Molecular orbitals

Valence bond theory focuses its attention on individual bonds in molecules. In an alternative approach, that of **molecular orbital theory**, it is accepted that electrons should not be regarded as belonging to particular bonds but should be treated as spreading throughout the entire molecule. This theory has been more fully developed than valence bond theory and provides the language that is widely used in modern discussions of bonding in small inorganic molecules, *d*-metal complexes, and solids. To introduce it, we shall follow the same strategy as in Chapter 8 where the hydrogen atom was taken as the fundamental species for discussing atomic structure, and then developed into a description of complex atoms. In this chapter we use the simplest molecule of all, the hydrogen molecule-ion, H_2^+, to introduce the essential features of bonding, and then use it as a guide to the structures of more complex systems. These applications include diatomic molecules, polyatomic molecules, and finally solids consisting of effectively infinite numbers of atoms.

9.3 The hydrogen molecule-ion

The wavefunctions obtained by solving the Schrödinger equation for the H_2^+ ion are called **molecular orbitals**. A molecular orbital is like an atomic orbital, but spreads throughout the molecule. As for an atomic orbital, the probability of finding an electron at a particular location is proportional to the square of the wavefunction at that point, so where the molecular orbital has a large amplitude, the electron that occupies it has a high probability of being found. Where the molecular orbital is zero (at one of its nodes), there is zero probability of finding the electron.

Exact mathematical formulas for the molecular orbitals for H_2^+ may be obtained (within the Born–Oppenheimer approximation), but they are very complicated functions and do not give much insight into the form of the orbitals, the contributions to the energy, or the form that orbitals are likely to take for more complex polyatomic molecules. Therefore, we shall adopt a simpler procedure that, while more approximate, gives more insight. The approximation we shall describe is used almost universally in modern molecular-orbital computations.

9.4 Linear combinations of atomic orbitals

The approximation we shall adopt is based on the fact that the wavefunction of the electron in H_2^+ resembles a $1s$ orbital centred on one nucleus when it is close to that nucleus, and it resembles a $1s$ orbital centred on the second nucleus when it is close to that nucleus. Then an approximate form for the molecular orbital is

$$\psi = \psi_{1s}(A) + \psi_{1s}(B) \tag{3}$$

where $\psi_{1s}(A)$ is a $1s$ orbital centred on nucleus A and $\psi_{1s}(B)$ the corresponding orbital on nucleus B. When the electron is close to A its distance from B is large, the wavefunction $\psi_{1s}(B)$ is small, and therefore ψ is almost pure $\psi_{1s}(A)$, as we require. Similarly, ψ is almost pure $\psi_{1s}(B)$ close to B. The technical term for a sum of the kind shown in eqn (3) is a **linear combination of atomic orbitals** (LCAO), and we shall use that name from now on. An approximate molecular orbital formed from a linear combination of atomic orbitals is called an LCAO-MO.

The form of the orbital constructed in eqn (3) is shown in Fig. 9.11. It is called a **σ orbital** because it resembles an s orbital when viewed along the axis:† in general, *any molecular orbital with cylindrical symmetry about the internuclear axis is called a σ orbital*. Because it is the σ orbital of lowest energy (as we shall see), it is labelled 1σ. An electron that occupies a σ orbital is called a **σ electron**. In the ground state of the H_2^+ ion, there is a single 1σ electron, and so we report the configuration of the molecule-ion as $1\sigma^1$.

According to the Born interpretation, the probability of finding a 1σ electron at any point in the molecule is proportional to the square of the wavefunction. In the present case,

$$\psi^2 = \{\psi_{1s}(A) + \psi_{1s}(B)\}^2 = \psi_{1s}(A)^2 + \psi_{1s}(B)^2 + 2\psi_{1s}(A)\psi_{1s}(B)$$

According to this expression, the total probability at any point is proportional to the sum of the following three contributions:

1. $\psi_{1s}(A)^2$, the probability that would be obtained if the electron were confined to the orbital on A.

---

† More precisely, it is so called because an electron that occupies a σ orbital has zero orbital angular momentum around the internuclear axis.

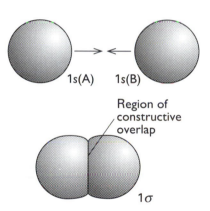

Fig. 9.11 The formation of a bonding molecular orbital (a σ orbital). Two H1s orbitals come together, and where they overlap, interfere constructively and give rise to an enhanced amplitude in the internuclear region. (The vertical line where the orbitals overlap is not a node.) The resulting orbital has cylindrical symmetry about the internuclear axis. When it is occupied by two paired electrons, to give the configuration σ^2, we have a σ bond.

2. $\psi_{1s}(B)^2$, the probability that would be obtained if the electron were confined to the orbital on B.

3. $2\psi_{1s}(A)\psi_{1s}(B)$, an additional contribution to the probability.

The third contribution, which represents an increase in the probability of finding the electron in the internuclear region, arises from the electron's ability to spread over both nuclei when the atoms are within bonding distance of one another. The enhancement can be traced to the **constructive interference** of the two atomic $1s$ orbitals. Each atomic orbital has a positive amplitude in the internuclear region, and so the total amplitude is greater in that region than if the electron were confined to a single atomic orbital.

We shall frequently use the result that *electrons accumulate in regions where atomic orbitals overlap and interfere constructively*. The enhanced probability of finding the electron between the nuclei means that the electron is in a good position to interact strongly with both nuclei. Hence the energy of the molecule is lower than that of the separate atoms, where each electron can interact strongly with only one nucleus.

9.5 Bonding and antibonding orbitals

A 1σ orbital is an example of a bonding molecular orbital:

> A **bonding orbital** is a molecular orbital that, if occupied, contributes to a lowering of the energy of a molecule.

The energy of the molecule decreases as R is decreased from large values because the electron is increasingly likely to be found in the internuclear region as the two atomic orbitals interfere more effectively. However, at small separations, there is too little space between the nuclei for significant accumulation of electron density there. In addition, the nucleus–nucleus repulsion $V_{\text{nuc,nuc}}$ given in eqn (3) becomes large as the distance R is decreased. As a result, after an initial decrease, at small internuclear separations the energy of the molecule rises and the potential energy curve passes through a minimum. Calculations on H_2^+ give the equilibrium bond length as $R_e = 130\,\text{pm}$ and the bond dissociation energy as $D_e = 171\,\text{kJ mol}^{-1}$. The experimental values are $106\,\text{pm}$ and $250\,\text{kJ mol}^{-1}$, and so this simple LCAO-MO description of the molecule, while inaccurate, is not absurdly wrong.

The argument that led to the expression of ψ as the sum of two atomic orbitals is equally well satisfied by writing a molecular orbital as the difference

$$\psi' = \psi_{1s}(A) - \psi_{1s}(B) \tag{4}$$

This linear combination also resembles one or other of the atomic orbitals close to the two nuclei. However, the resulting molecular orbital corresponds to a higher energy than that of 1σ, and it is in fact a good approximation to the next-higher exact solutions of the Schrödinger equation for H_2^+.

Exercise E9.4 Show that the molecular orbital written above is zero on a plane cutting through and perpendicular to the internuclear axis at its mid point. Take each atomic orbital to be of the form e^{-r}.

Because ψ' is cylindrically symmetrical around the internuclear axis it is also a σ orbital, and is denoted 2σ (Fig. 9.12; throughout this chapter we shall label orbitals of a particular type in order of increasing energy). The origin of the higher energy of 2σ relative to that of 1σ can be traced to the nodal plane half way between the nuclei: on this plane the $\psi_{1s}(A)$ and $\psi_{1s}(B)$ wavefunctions cancel as a result of **destructive interference** between their amplitudes, which are of opposite sign. There is zero probability of finding the electron on this nodal plane. In the drawings like that in Figs 9.11 and 9.12, we represent overlap of orbitals with the same sign (as in the formation of 1σ) by shading of the same tint; the overlap of orbitals of opposite sign (as in the formation of 2σ) is represented by one orbital of a light tint (or white) and another orbital of a dark tint.

The 2σ orbital is an example of an **antibonding orbital**, an orbital that, if occupied, raises the energy of the molecule relative to the separated atoms. (Antibonding orbitals are sometimes marked with an asterisk, as in $2\sigma^*$, as a sign that they are of high energy.) The antibonding character of the $2\sigma^*$ orbital is partly a result of the exclusion of the electron from the internuclear region and its relocation outside the bonding region. Because of its location, the electron helps to pull the nuclei apart rather than helping to pull them together. An antibonding orbital is often slightly more strongly antibonding than the corresponding bonding orbital is bonding, and for this reason the **molecular orbital energy-level diagram** (Fig. 9.13), the schematic diagram of the energies of the molecular orbitals and the atomic orbitals from which they are formed, is not quite symmetrical about the atomic energy levels.

9.6 The structures of diatomic molecules

In Chapter 8 we used the hydrogenic atomic orbitals and the building-up principle to deduce the ground electronic configurations of

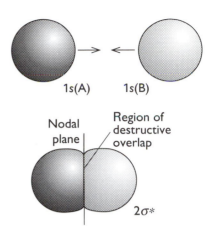

Fig. 9.12 The formation of an antibonding molecular orbital (a σ^* orbital). Two H1s orbitals come together, where they overlap with opposite signs (as depicted by a different density of shading), interfere destructively and give rise to a decreased amplitude in the internuclear region. There is a nodal plane exactly half way between the nuclei, on which any electrons that occupy the orbital will not be found.

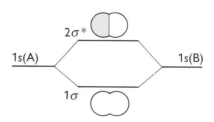

Fig. 9.13 A molecular orbital energy-level diagram for orbitals constructed from (1s, 1s)-overlap, the separation of the levels corresponding to the equilibrium bond length.

355

many-electron atoms. We can adopt an analogous procedure for many-electron diatomic molecules (such as H_2 with two electrons and even Br_2 with 70), but using the H_2^+ molecular orbitals instead. As we shall illustrate in the following sections, first with H_2 and then with heavier molecules, the general procedure is to construct molecular orbitals by forming linear combinations of all suitable atomic orbitals supplied by the atoms (the meaning of 'suitable' will be explained shortly). The electrons supplied by the atoms are then accommodated in the orbitals so as to achieve the lowest overall energy subject to the constraint of the Pauli exclusion principle, that no more than two electrons may occupy a single orbital (and then must be paired). As for atoms, if several molecular orbitals of the same energy are available, we add the electrons to each individual orbital before doubly occupying any one orbital (because that minimizes electron–electron repulsions). We also take note of Hund's rule (Section 8.12), that if electrons do occupy different degenerate orbitals, then they do so with parallel spins.

The hydrogen and helium molecules

We shall illustrate the general procedure by considering H_2, the simplest many-electron diatomic molecule. First, we need to build the molecular orbitals. Because each H atom of H_2 contributes a $1s$ orbital (as in H_2^+), we can form the 1σ and $2\sigma^*$ bonding and antibonding orbitals from them, as we have seen already. At the equilibrium internuclear separation these orbitals will have the energies that can be represented by the horizontal lines in Fig. 9.14. Note that from two atomic orbitals we have built two molecular orbitals. This observation is a special case of the general rule that *from N atomic orbitals we can build N molecular orbitals* (orbitals cannot simply be 'lost' when they are added and subtracted to form linear combinations).

> **Exercise E9.5** How many molecular orbitals can be built from the valence shell orbitals in O_2?
>
> [8]

There are two electrons to accommodate (one supplied from each atom), and both can enter 1σ by pairing their spins. The ground state configuration is therefore $1\sigma^2$, and the atoms are joined by a bond consisting of an electron pair in a bonding σ orbital. These two electrons bind the two nuclei together more strongly and closely than the single electron in H_2^+, and the bond length is reduced from 106 pm to 74 pm. A pair of electrons in a σ orbital is called a **σ bond**, and is very similar to the σ bond of valence bond theory (the two differ in certain details of the electron distribution between the two atoms joined by the bond). We can conclude that *the importance of an*

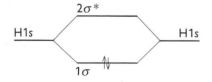

Fig. 9.14 The ground electronic configuration of H_2 is obtained by accommodating the two electrons in the lowest available orbital (the bonding orbital).

electron pair in bonding stems from the fact that two is the maximum number of electrons that can enter each bonding molecular orbital.

The same argument shows why helium is a monatomic gas. We consider a hypothetical He_2 molecule. Each He atom contributes a $1s$ orbital to the linear combination used to form the molecular orbitals, and so 1σ and $2\sigma^*$ molecular orbitals can be constructed. (They differ in detail from those in H_2 because the $He1s$ orbitals are more compact, but the general shape is the same, and we can use qualitatively the same molecular orbital energy-level diagram as for H_2.) Because each atom provides two electrons, there are four electrons to accommodate. Two can enter the 1σ orbital, but then it is full (by the Pauli principle) and the next two must enter the antibonding $2\sigma^*$ orbital (Fig. 9.15). The ground electronic configuration of He_2 is therefore

$$He_2\ 1\sigma^2 2\sigma^{*2}$$

Because an antibonding orbital is slightly more antibonding than a bonding orbital is bonding, the He_2 molecule has a higher energy than the separated atoms, and so does not form. Hence, two ground-state He atoms do not form bonds to each other, and helium is a monatomic gas.

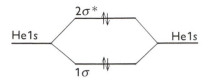

Fig. 9.15 The ground electronic configuration of the four-electron molecule He_2 has two bonding electrons and two antibonding electrons. It has a higher energy than the separated atoms, and so He_2 is unstable relative to two He atoms.

Example Judging the stability of diatomic molecules
Decide whether Li_2 is likely to exist on the assumption that only the valence s orbitals contribute to its molecular orbitals.
Answer Each molecular orbital is built from $2s$ atomic orbitals, which give one bonding and one antibonding combination (1σ and $2\sigma^*$, respectively; we are numbering the orbitals in order of increasing energy, starting with the lowest-energy bonding orbital that can be formed from the valence atomic orbitals). Each Li atom supplies one valence electron, which fills the 1σ orbital, to give the configuration $1\sigma^2$, which is bonding.

Exercise E9.6 Is LiH likely to exist if the Li atom uses only its $2s$ orbital for bonding?
[Yes, $(Li2s, H1s)\sigma^2$]

Period 2 diatomic molecules

We shall now see how the concepts we have introduced apply to other **homonuclear diatomic molecules**, which are diatomic molecules formed from identical atoms, such as N_2 and Cl_2, and ions such as O_2^{2-}. In line with the building-up procedure, we first consider the molecular orbitals that may be formed and do not (at this stage) trouble about how many electrons are available.

357

The atomic orbitals available for the construction of linear combinations are as follows:

> • The **core orbitals**, the orbitals of the inner, closed shells.
>
> • The **valence orbitals**, the orbitals of the valence shell.
>
> • The **virtual orbitals**, the orbitals of the atom that are unoccupied in its ground state.

In elementary treatments (but not in modern sophisticated treatments) the core orbitals are ignored as being too compact to have significant overlap with orbitals on other atoms. The virtual orbitals are ignored on the grounds that they are too high in energy to participate in bonding. Therefore, molecular orbitals are formed using only the valence orbitals.

In Period 2, the valence orbitals are $2s$ and $2p$. Suppose first that we consider these two types of orbital separately. Then the $2s$ orbitals on each atom overlap to form bonding and antibonding combinations that we shall denote 1σ and $2\sigma^*$, respectively. Likewise, the two $2p_z$ orbitals (as remarked earlier, by convention, the internuclear axis is denoted the z axis) have cylindrical symmetry around the internuclear axis, and hence may participate in σ-orbital formation to form bonding and antibonding combinations 3σ and $4\sigma^*$, respectively (Fig. 9.16). The resulting energy levels of the σ orbitals are shown in the molecular orbital energy-level diagram in Fig. 9.17. Now, strictly we should not consider the s and p_z orbitals separately, because they *both* have cylindrical symmetry, and *both* of them can contribute to σ-orbital formation. Therefore, we should combine all four orbitals together to form four σ molecular orbitals, each one of the form

$$\psi = c_1 \psi_{2s}(A) + c_2 \psi_{2s}(B) + c_3 \psi_{2p_z}(A) + c_4 \psi_{2p_z}(B)$$

The four coefficients, c, which represent the different contributions that each atomic orbital makes to the overall molecular orbital, must be found by solving the Schrödinger equation. However, in practice, the two lowest energy combinations of this kind are very similar to the combination 1σ and $2\sigma^*$ of $2s$ orbitals that we have described, and the two highest energy combinations are very similar to the 3σ and $4\sigma^*$ combinations of $2p_z$ orbitals. In each case there will be small differences: the $2\sigma^*$ orbital, for instance, will be contaminated by some $2p_z$ character and the 3σ orbital will be contaminated by some $2s$ character, and their energies will be slightly shifted from where they would be if we considered only the 'pure' combinations. Nevertheless, the changes are not great, and we can continue to think of 1σ and $2\sigma^*$ as forming one bonding and antibonding pair, and of 3σ and $4\sigma^*$ as forming another pair. The four orbitals are shown in the centre column of Fig. 9.17. There is no guarantee that $2\sigma^*$ and 3σ will be in

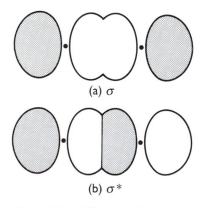

(a) σ

(b) σ^*

Fig. 9.16 (a) The interference leading to the formation of a σ-bonding orbital and (b) the corresponding antibonding orbital when two p orbitals overlap along an internuclear axis.

Atom   Molecule   Atom

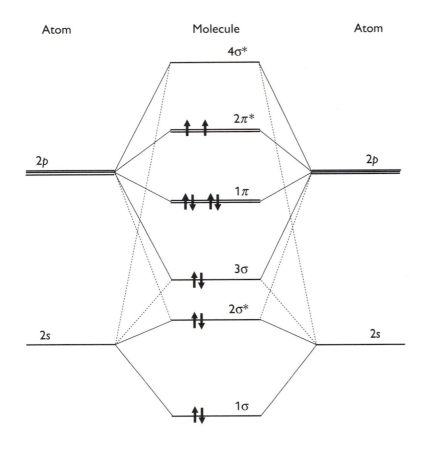

Fig. 9.17 A typical molecular orbital energy-level diagram for Period 2 homonuclear diatomic molecules. The valence atomic orbitals are drawn in the columns on the left and the right; the molecular orbitals are shown in the middle. Note that the π orbitals form doubly degenerate pairs. The solid line joining the molecular orbitals to the atomic orbitals show the principal composition of the molecular orbitals; the dotted lines show that a small admixture of other atomic orbitals is also present. Note the labelling scheme for the molecular orbitals: the scheme is explained in the text. This diagram is suitable for O_2 and F_2.

the exact location shown in the illustration, and the locations shown in Fig. 9.18 are found in some molecules.

Now consider the $2p_x$ and $2p_y$ orbitals of each atom, which are perpendicular to the internuclear axis and may overlap broadside-on. This overlap may be constructive or destructive, and results in a bonding and an antibonding **π orbital**, which we label 1π and $2\pi^*$, respectively. The notation π is the analogue of p in atoms, for when viewed along the axis of the molecule, a π orbital looks like a p orbital (Fig. 9.19).† The two $2p_x$ orbitals overlap to give a bonding and an antibonding π orbital, as do the two $2p_y$ orbitals too. The two bonding combinations have the same energy; likewise, the two antibonding combinations have the same energy. Hence, each π energy level is doubly degenerate and consists of two distinct orbitals. Two electrons in a π orbital constitute a **π bond**: such a bond resembles a π bond of valence bond theory, but the details of the electron distribution are slightly different.

† More precisely, an electron in a π orbital has one unit of orbital angular momentum about the internuclear axis.

Fig. 9.18 A typical molecular orbital energy-level diagram for Period 2 homonuclear diatomic molecules up to and including N_2. The difference from the diagram in Fig. 9.17 stems largely from the amount of admixture represented by the dotted lines.

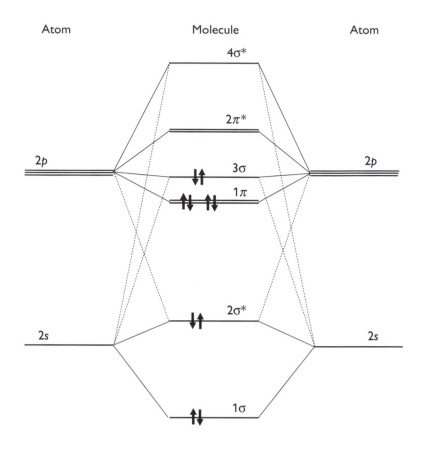

The relative order of the σ and π orbitals in a molecule cannot be predicted readily (it varies with the energy separation between the $2s$ and $2p$ orbitals of the atom), and in some molecules the order shown in Fig. 9.17 applies whereas others have the order shown in Fig. 9.18. The change in order can be seen in Fig. 9.20, which shows the calculated energy levels of the Period 2 homonuclear diatomic molecules. A useful rule is that, for neutral molecules, the order shown in Fig. 9.17 is valid for O_2 and F_2, whereas the order shown in Fig. 9.18 is valid for the preceding elements of the period.

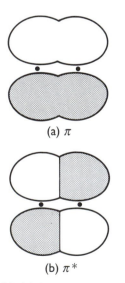

(a) π

(b) π^*

Fig. 9.19 (a) The interference leading to the formation of a π-bonding orbital and (b) the corresponding antibonding orbital.

Example Assessing the contribution of d orbitals
Can d orbitals contribute to σ and π orbitals in diatomic molecules?
Answer We need to assess the symmetry of d orbitals with respect to the internuclear z axis. A d_{z^2} orbital has cylindrical symmetry around z and so can contribute to σ orbitals. The d_{xz} and d_{yz} orbitals have π symmetry with respect to the axis (Fig. 9.21), and so can contribute to π orbitals.

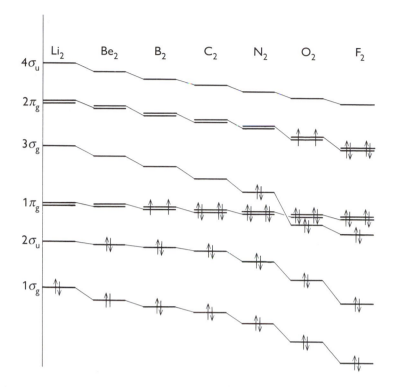

Fig. 9.20 The variation of the orbital energies of Period 2 homonuclear diatomic molecules. The labels g and u are explained on p. 366.

Exercise E9.7 Sketch the 'δ orbitals' (orbitals that resemble *d* orbitals when viewed along the internuclear axis) that may be formed by the remaining two *d* orbitals (and which contribute to bonding in some *d*-metal cluster compounds).

[Fig. 9.21]

Symmetry and overlap

One central feature of molecular orbital theory can now be addressed. We have seen that *s* and p_z orbitals may contribute to the formation of σ orbitals, and that p_x and p_y orbitals may contribute to π orbitals. However, we never have to consider orbitals formed by the overlap of *s* and p_x orbitals (or p_y orbitals). *When building molecular orbitals, it is necessary to consider linear combinations only of atomic orbitals of the same symmetry with respect to the internuclear axis.* Because an *s* orbital has cylindrical symmetry around the internuclear axis, but a p_x orbital does not, the two atomic orbitals cannot contribute to the same molecular orbital. The reason for this distinction based on symmetry can be understood by considering the overlap between an *s* orbital and a p_x orbital (Fig. 9.22): although there is constructive interference between the two orbitals on one side of the axis, there is an exactly compensating amount of destructive interference on the other side of the axis, and the net overlap (and hence the net bonding or antibonding effect) is zero.

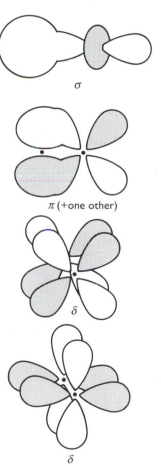

Fig. 9.21 The types of molecular orbital to which *d* orbitals can contribute. The σ and π combinations can be formed with *s*, *p*, and *d* orbitals of the appropriate symmetry, but the δ orbitals can be formed only by the *d* orbitals of the two atoms.

Fig. 9.22 Overlapping s and p orbitals. (a) End-on overlap leads to non-zero overlap and to the formation of an axially symmetric σ orbital. (b) Broadside overlap leads to no net accumulation or reduction of electron density and does not contribute to bonding.

(a)

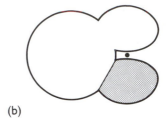

(b)

CALCULUS

The extent to which two orbitals overlap is measured by the **overlap integral**, S:

$$S = \int \psi(A)^* \psi(B)\, d\tau$$

where the integration is over all space and the asterisk denotes a complex conjugate wavefunction. If the atomic orbital $\psi(A)$ on A is small wherever the orbital $\psi(B)$ on B is large, or vice versa, then the product of their amplitudes is everywhere small and the integral—the sum of these products—is small (Fig. 9.23(a)). If $\psi(A)$ and $\psi(B)$ are simultaneously large in some region of space, then S may be large (Fig. 9.23(b)). If the two atomic orbitals are identical (e.g. $1s$ orbitals on the same

nucleus), then $S = 1$. For two $1s$ orbitals at the equilibrium bond length in H_2^+, $S = 0.59$, which is an unusually large value. Typical values for orbitals with $n = 2$ are in the range 0.2 to 0.3.

Now consider the arrangement in Fig. 9.23(c) in which an s orbital overlaps a p_x orbital of a different atom. At some point r the product $\psi(A)^* \psi(B)$ may be large. However, there is a point r' where $\psi(A)^* \psi(B)$ has exactly the same magnitude but an opposite sign. When the integral is evaluated, these two contributions are added together and cancel. For every point in the upper half of the diagram, there is a point in the lower half that cancels, and so $S = 0$. Therefore, there is no net overlap between the s and p orbitals in this arrangement.

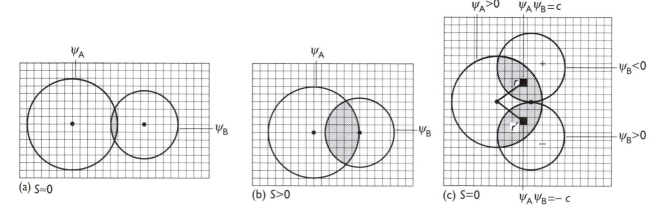

Fig. 9.23 A schematic representation of the contributions to the overlap integral. (a) $S \approx 0$ because the orbitals are far apart and their product is always small. (b) S is large (but less than 1) because the product $\psi_A \psi_B$ is large over a substantial region. (c) $S = 0$ because the positive region of overlap is exactly cancelled by the negative region.

We now have the criteria for selecting atomic orbitals from which molecular orbitals are to be built:

1. Use all available valence orbitals from both atoms (in polyatomic molecules, from all the atoms).

2. Classify the atomic orbitals as having σ and π symmetry with respect to the internuclear axis, and build σ and π orbitals from *all* atomic orbitals of a given symmetry.

3. If there are N atomic orbitals of σ symmetry, then N σ molecular orbitals can be built with progressively higher energy from strongly bonding to strongly antibonding.

4. If there are N' atomic orbitals of π symmetry, then N' π molecular orbitals can be built with progressively higher energy from strongly bonding to strongly antibonding. The π orbitals occur in doubly degenerate pairs.

As a general rule, the energy of each species of orbital (σ or π) increases with the number of internuclear nodes. The lowest energy orbital of a given species has no internuclear nodes and the highest energy orbital has a nodal plane between each pair of neighbouring atoms (Fig. 9.24).

The electronic structures of homonuclear diatomic molecules

Figures 9.17 and 9.18 show the general layout of the valence-shell atomic orbitals of Period 2 atoms on the left and right of the molecular orbital energy-level diagram. The lines in the middle are an indication of the energies of the molecular orbitals that can be formed by overlap of atomic orbitals. From the eight valence-shell orbitals (four from each atom), we can form eight molecular orbitals: four are σ orbitals and four, in two pairs, are doubly degenerate π orbitals. With the orbitals established, the ground state electron configurations of the molecules are derived by adding the appropriate number of electrons to the orbitals and following the building-up rules. Charged species (such as the peroxide ion, O_2^{2-}, and C_2^+) need either more or fewer electrons (for anions and cations, respectively) than the neutral molecules.

We shall illustrate the procedure with N_2, which has ten valence electrons; for this molecule we use Fig. 9.18. The first two electrons pair, enter, and fill the 1σ orbital; the next two enter and fill the $2\sigma^*$ orbital. Six electrons remain. There are two 1π orbitals, and so four electrons can be accommodated in them. The two remaining electrons

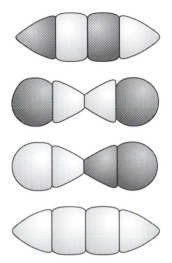

Fig. 9.24 A schematic representation of the four molecular orbitals that can be formed from four s orbitals in a chain of four atoms. The lowest energy combination (the bottom diagram) is formed from atomic orbitals with the same sign, and there are no internuclear nodes (the perpendicular lines in the diagram are not nodes, but a part of the depiction of the shape of the orbital). The next higher orbital has one node (at the centre of the molecule). The next higher orbital has two internuclear nodes, and the uppermost, highest energy orbital has three internuclear nodes, one between each neighbouring pair of atoms, and is fully antibonding. The size of the spheres reflect the contributions of each atom to the molecular orbital; dark and light shading represents different signs.

enter the 3σ orbital. The ground state configuration of N_2 is therefore

$$N_2\,1\sigma^2 2\sigma^{*2} 1\pi^4 3\sigma^2$$

This configuration is also depicted in Fig. 9.20.

The strength of a bond in a molecule is the net outcome of the bonding and antibonding effects of the occupied orbitals. The **bond order**, b, is defined as the difference between the number of occupied bonding orbitals between the atoms of interest and the number of occupied antibonding orbitals. In a diatomic molecule

$$b = \tfrac{1}{2}(n - n^*) \tag{5}$$

where n is the number of electrons in bonding orbitals and n^* is the number of electrons in antibonding orbitals. Each electron pair in a bonding orbital increases the bond order by 1 and each pair in an antibonding orbital decreases it by 1. For H_2, $b = 1$, corresponding to a single bond between the two atoms: this bond order is consistent with the Lewis structure H—H for the molecule. In He_2, which has equal numbers of bonding and antibonding electrons (with $n = 2$ and $n^* = 2$), the bond order is $b = 0$, and there is no bond. In N_2, the 1σ, 3σ, and 1π orbitals are bonding, and $n = 2 + 2 + 4 = 8$; however, $2\sigma^*$ (the antibonding partner of 1σ) is antibonding, so $n^* = 2$ and the bond order of N_2 is

$$b = \tfrac{1}{2}(8 - 2) = 3$$

This value is consistent with the Lewis structure :N≡N:, in which there is a triple bond between the two atoms.

The bond order is a useful parameter for discussing the characteristics of bonds, because it correlates with bond length, and *the greater the bond order between atoms of a given pair of atoms, the shorter the bond*. The bond order also correlates with bond strength, and *the greater the bond order, the greater the strength*. The high bond order of N_2 is consistent with its high dissociation energy ($942\,\text{kJ mol}^{-1}$).

Example Writing the electron configuration of a diatomic molecule

Write the ground state electron configuration of O_2 and calculate the bond order.

Answer The molecular orbital energy-level diagram in Fig. 9.17 should be used, and there are 12 valence electrons to place in the orbitals shown there. The first 10 electrons recreate the N_2 configuration (with a reversal of the order of the 3σ and 1π orbitals); the remaining two electrons must occupy the $2\pi^*$ orbitals. The configuration and bond order are therefore

$$O_2\,1\sigma^2 2\sigma^{*2} 3\sigma^2 1\pi^4 2\pi^{*2} \qquad b = 2$$

This configuration is depicted in Fig. 9.20. Because 1σ, $3\sigma^*$, and 1π are bonding and $2\sigma^*$ and $2\pi^*$ are antibonding, the bond order is

$$b = \tfrac{1}{2}(8 - 4) = 2$$

The bond order of 2 accords with the classical view that oxygen has a double bond and is written O=O.

Exercise E9.8 Write the electron configuration of F_2 and deduce its bond order.

$$[1\sigma^2 2\sigma^{*2} 3\sigma^2 1\pi^4 2\pi^{*4}, \ b = 1]$$

We see from the example above that the electron configuration of O_2 is

$$O_2 \ 1\sigma^2 2\sigma^{*2} 3\sigma^2 1\pi^4 2\pi^{*2}$$

According to the building-up principle, the two $2\pi^*$ electrons in O_2 will occupy different orbitals: one will enter the $2\pi^*$ orbital formed by overlap of $2p_x$ and the other will enter its degenerate partner, the $2\pi^*$ orbital formed from overlap of the $2p_y$ orbitals. Because the two electrons occupy different orbitals, they may have parallel spins ($\uparrow\uparrow$). Therefore, we can predict that an O_2 molecule will be magnetic because the magnetic effects of the two unpaired spins do not cancel. Specifically, oxygen is predicted to be a **paramagnetic** substance, a substance that is drawn into a magnetic field. Most substances (those with paired electron spins) are **diamagnetic** and are pushed out of a magnetic field. That O_2 is in fact a paramagnetic gas is a striking confirmation of the superiority of the molecular orbital description of the molecule over the Lewis description (which requires all the electrons to be paired). The property of paramagnetism is utilized to monitor the oxygen content of incubators by measuring the magnetism of the gases they contain.

An F_2 molecule has two more electrons than an O_2 molecule and the configuration and bond order are

$$F_2 \ 1\sigma^2 2\sigma^{*2} 3\sigma^2 1\pi^4 2\pi^{*4}, \qquad b = 1$$

We conclude that F_2 is a singly bonded molecule, in agreement with the structure F—F. The low bond order is consistent with the low dissociation energy of F_2 (154 kJ mol^{-1}). An Ne_2 molecule has two further electrons:

$$Ne_2 \ 1\sigma^2 2\sigma^{*2} 3\sigma^2 1\pi^4 2\pi^{*4} 4\sigma^{*2}, \qquad b = 0$$

(The $4\sigma^*$ orbital is the antibonding partner of 3σ.) The bond order of zero agrees with the monatomic character of neon.

Example Judging the relative bond strengths of molecules and ions
Judge whether N_2^+ is likely to have a larger or smaller dissociation energy than N_2.

365

Centre of inversion

σ_g

σ_u

π_u

π_g

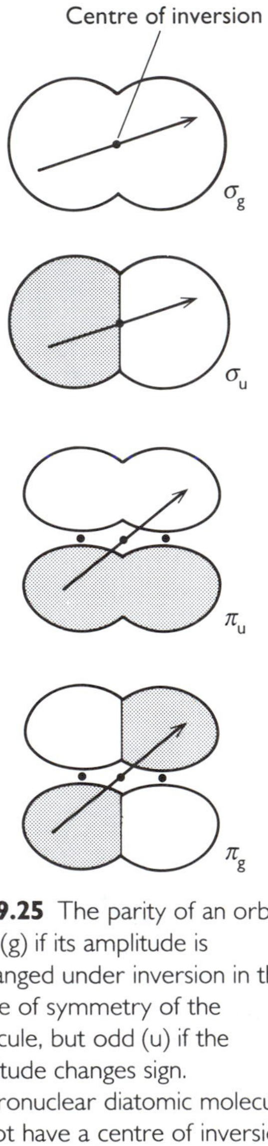

Fig. 9.25 The parity of an orbital is even (g) if its amplitude is unchanged under inversion in the centre of symmetry of the molecule, but odd (u) if the amplitude changes sign. Heteronuclear diatomic molecules do not have a centre of inversion, and so the g, u classification is irrelevant.

Answer Because a species with the larger bond order is likely to have the larger dissociation energy, we should compare their electronic configurations, and assess their bond orders. From Fig. 9.18:

$$N_2 \qquad 1\sigma^2 2\sigma^{*2} 1\pi^4 3\sigma^2, \qquad b = 3$$
$$N_2^+ \qquad 1\sigma^2 2\sigma^{*2} 1\pi^4 3\sigma^1, \qquad b = 2.5$$

Because the cation has the lower bond order, we expect it to have the smaller dissociation energy. The experimental dissociation energies are 945 kJ mol^{-1} for N_2 and 842 kJ mol^{-1} for N_2^+.

Exercise E9.9 Which can be expected to have the higher dissociation energy, F_2 or F_2^+?

$[F_2^+]$

Parity

We have seen a little of the importance of the symmetry of atomic orbitals in the construction of molecular orbitals. Symmetry plays a role in many discussions of molecular orbitals themselves, particularly in the interpretation of the electronic transitions that give rise to molecular spectra (Chapter 11).

The classification of molecular orbitals as σ and π is one aspect of *molecular* (as distinct from *atomic*) symmetry that we have already seen, because σ and π orbitals differ in their symmetry about the internuclear axis. However, there is another aspect of molecular symmetry that applies in *homonuclear* diatomic molecules rather than to diatomic molecules in general. In homonuclear diatomic molecules, orbitals may also be classified according to their **parity**, their behaviour under the process called **inversion** (Fig. 9.25). To decide on the parity of an orbital, we consider an arbitrary point in a homonuclear diatomic molecule, and note the sign of the orbital (we are denoting sign by light or dark shading). Then we imagine travelling through the centre of the molecule and out to the corresponding point on the other side. If the orbital has the same sign (the same shading) at that point, it has **even parity** and is denoted g (from *gerade*, the German word for 'even'). If the orbital has opposite sign, it has **odd parity** and is denoted u (from *ungerade*, 'uneven'). The parity designation applies only to homonuclear diatomic molecules, because heteronuclear diatomic molecules (such as HCl) do not have inversion symmetry of the kind that we have described: such molecules do not have a 'centre'.

Figure 9.25 shows that a bonding σ orbital has even parity; so it is written σ_g; an antibonding σ orbital has odd parity and is written σ_u. A bonding π orbital has odd parity and is denoted π_u and an antibonding π orbital has even parity, denoted π_g. It follows that the full dress version of the electron configuration of the ground state of N_2 is

$$N_2 \, 1\sigma_g^2 2\sigma_u^{*2} 1\pi_u^4 3\sigma_g^2$$

As we have indicated, this detailed specification of a configuration is needed (in the main) only when discussing electronic transitions and molecular selection rules.

Exercise E9.10 Write the full electron configuration of the ground state of the F_2 molecule.

$$[1\sigma_g^2 2\sigma_u^{*2} 3\sigma_g^2 1\pi_u^4 2\pi_g^{*4}]$$

Heteronuclear diatomic molecules

A **heteronuclear diatomic molecule** is a diatomic molecule formed from atoms of two different elements, such as CO and HCl. The electron distribution in the covalent bond between the atoms is not symmetrical between the atoms because it is energetically favourable for the electron pair to be found closer to one atom rather than the other. This imbalance results in a **polar bond**, which is a covalent bond in which the electron pair is shared unequally by the two atoms.

The atom that draws the bonding electron pair to it more strongly is called the more **electronegative** atom. A numerical scale of electronegativity was originally formulated by Linus Pauling, who based it on considerations of bond strength and the degree of admixture of ionic bonding into a covalent description of the bond. A somewhat simpler definition was proposed by Robert Mulliken, who related the electronegativity, χ, to the ionization energy and the electron affinity of the element:

$$\chi = \tfrac{1}{2}(I + E_{ea}) \tag{6}$$

This relation is plausible, because an atom that has a high electronegativity is likely to be one that has a high ionization energy (so that it is unlikely to lose electrons to another atom in the molecule) and a high electron affinity (so that it is energetically favourable for an electron to move towards it). The Mulliken electronegativities are broadly in line with the Pauling electronegativities, and a list of the latter is given in Table 9.2. Electronegativities show a periodicity, and the elements with the highest electronegativities are those close to fluorine in the periodic table. An element with a low electronegativity (such as an alkali metal) is said to be **electropositive**. The most electropositive elements are found close to caesium in the periodic table.

The location of the bonding electron pair close to one atom in a heteronuclear molecule results in that atom having a net negative charge, which is called a **partial negative charge** and denoted $\delta-$. There is a compensating partial positive charge $\delta+$ on the other atom. In a typical heteronuclear diatomic molecule, the more electronegative

Table 9.2 Electronegativities of the main-group elements

| | | | H | | | |
			2.1			
Li	**Be**	**B**	**C**	**N**	**O**	**F**
1.0	1.5	2.0	2.5	3.0	3.5	4.0
Na	**Mg**	**Al**	**Si**	**P**	**S**	**Cl**
0.9	1.2	1.5	1.8	2.1	2.5	3.0
K	**Ca**	**Ga**	**Ge**	**As**	**Se**	**Br**
0.8	1.0	1.6	1.8	2.0	2.4	2.8
Rb	**Sr**	**In**	**Sn**	**Sb**	**Te**	**I**
0.8	1.0	1.7	1.8	1.9	2.1	2.5
Cs	**Ba**	**Tl**	**Pb**	**Bi**	**Po**	**At**
0.7	0.9	1.8	1.8	1.9	2.0	2.2

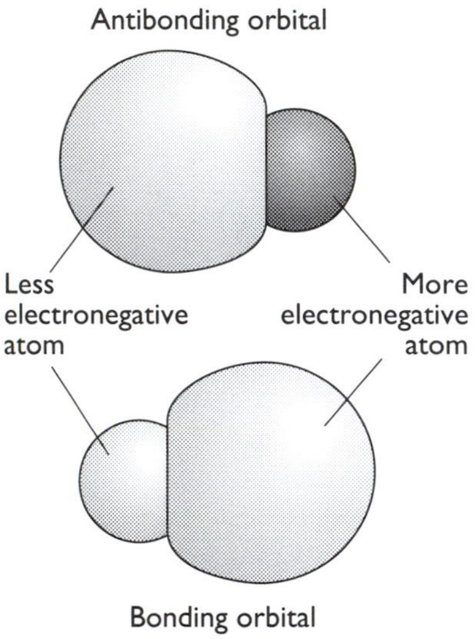

element has the partial negative charge and the more electropositive element has the partial positive charge.

> **Exercise E9.11** Predict the (weak) polarity of a C—H bond.
> $$[^{\delta-}C—H^{\delta+}]$$

Antibonding orbital

Less electronegative atom — More electronegative atom

Bonding orbital

Fig. 9.26 A schematic representation of the relative contributions of atoms of different electronegativities to bonding and antibonding molecular orbitals. In the bonding orbital, the more electronegative atom makes the greater contribution (represented by the larger sphere), and the electrons of the bond are more likely to be found on that atom. The opposite is true of an antibonding orbital. A part of the reason why an antibonding orbital is of high energy is that the electrons that occupy it are likely to be found on the more electropositive atom.

Polar covalent bonds

A **polar covalent bond** consists of two electrons in an orbital of the form

$$\psi = c_A \psi(A) + c_B \psi(B) \qquad (7)$$

with unequal coefficients. The proportion of the atomic orbital $\psi(A)$ in the bond is c_A^2 and that of $\psi(B)$ is c_B^2. A **nonpolar bond**, a covalent bond in which the electron pair is shared equally between the two atoms and there are zero partial charges on each atom, has $c_A^2 = c_B^2$. A pure ionic bond, in which one atom has obtained virtually sole possession of the electron pair (as in Cs^+F^-), has one coefficient equal to zero (so that A^+B^- would have $c_A = 0$ and $c_B = 1$). A general feature of molecular orbitals between dissimilar atoms is that *the atomic orbital with the lower energy* (that belonging to the more electronegative atom) *makes the larger contribution to the bonding molecular orbital*. The opposite is true of the antibonding orbital, for which the principal contribution comes from the atomic orbital with higher energy (the less electronegative atom). A schematic representation of this point is shown in Fig. 9.26.

These features of polar bonds can be illustrated by considering HF. The general form of the molecular orbitals of HF is

$$\psi = c_H\psi(H) + c_F\psi(F) \qquad (8)$$

where $\psi(H)$ is an H1s orbital and $\psi(F)$ is an F2p_z orbital. Because the ionization energy of a hydrogen atom is 13.6 eV, we know that the energy of the H1s orbital is -13.6 eV (as usual, the zero of energy is the infinitely separated electron and proton, Fig. 9.27). Similarly, from the ionization energy of fluorine, which is 18.6 eV, we know that the energy of the F2p_z orbital is -18.6 eV, about 5 eV lower than the H1s orbital. It follows that the bonding σ orbital in HF is mainly F2p_z and the antibonding σ orbital is mainly H1s orbital in character. The two electrons in the bonding orbital are most likely to be found in the F2p_z orbital, so there is a partial negative charge on the F atom and a partial positive charge on the H atom.

A systematic way of finding the coefficients in the linear combinations used to build molecular orbitals is provided by the variation principle:

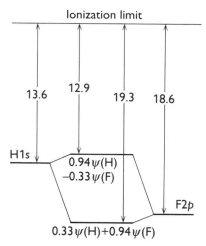

Fig. 9.27 The atomic orbital energy levels of H and F atoms and the molecular orbitals they form. The bonding orbital has predominantly F atom character and the antibonding orbital has predominantly H atom character.

> The **variation principle** states that if an arbitrary wavefunction is used to calculate the energy, then the value calculated is never less than the true energy.

The arbitrary wavefunction is called the **trial wavefunction**. The principle implies that if we vary the coefficients in the trial wavefunction until we achieve the lowest energy, then those coefficients will be the best available for a molecular orbital of the form selected (that is, for one built from the given atomic orbitals). For example, when the variation principle is applied to an H_2 molecule, the lowest energy is obtained when the two H1s orbitals contribute equally to a bonding orbital. However, when the principle is applied to an HF molecule, the lowest energy is obtained for the orbital

$$\psi = 0.33\psi_{1s}(H) + 0.94\psi_{2p_z}(F)$$

We see that indeed the F2p_z orbital does make the greater contribution to the bonding σ orbital.

Exercise E9.12 What percentage of its time does a σ electron in HF spend in a F2p_z orbital?
[88 per cent ($=0.94^2 \times 100$ per cent)]

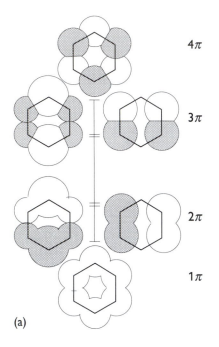

4π

3π

2π

1π

(a)

(b)

Fig. 9.28 (a) The π molecular orbitals of benzene. The lowest energy orbital is fully bonding between neighbouring atoms but the uppermost orbital is fully antibonding. The two pairs of doubly degenerate molecular orbitals have an intermediate number of internuclear nodes. As usual, light and dark shading represents different signs of the wavefunction. When six electrons occupy the three lowest orbitals, the resulting *overall* electron density is the same at each atom: this uniformity is represented by the double-doughnut shape in (b).

9.7 The structures of polyatomic molecules

The bonds in polyatomic molecules are built in the same way as in diatomic molecules, the only difference being that we use more atomic orbitals to construct the molecular orbitals, and these molecular orbitals spread over the entire molecule. In general, a molecular orbital is a linear combination of *all* the atomic orbitals of all the atoms in the molecule. In H_2O, for instance, the atomic orbitals are the two H1s orbitals, the O2s orbital, and the three O2p orbitals (if we consider only the valence shell). From these six atomic orbitals we can construct six molecular orbitals that spread over all three atoms. The molecular orbitals differ in energy—the lowest energy, most strongly bonding orbitals have the least number of nodes between adjacent atoms, and the highest energy, most strongly antibonding orbitals have the greatest numbers of nodes between neighbouring atoms.

The bonding influence of a single electron pair is distributed over all the atoms, and each electron pair (the maximum number of electrons that can occupy any single molecular orbital) helps to bind *all* the atoms together. In the LCAO-MO approximation, each molecular orbital is modelled as a linear combination of atomic orbitals, with atomic orbitals contributed by all the atoms in the molecule. No attempt is made to construct bonds between individual pairs of atoms, and all s and p orbitals are included in the construction of molecular orbitals without resorting to the artifice of hybridization. Thus, a typical molecular orbital in H_2O is constructed from H1s orbitals (denoted s_A and s_B) and O2s and O2p orbitals (denoted s_O and p_O) and has the composition

$$\psi = c_1 s_A + c_2 s_O + c_3 p_O + c_4 s_B \tag{9}$$

Because four atomic orbitals are being used to form the molecular orbital, there will be four possible molecular orbitals: the lowest energy (most-bonding) orbital will have no internuclear nodes and the highest energy (most-antibonding) orbital will have a node between each pair of neighbouring nuclei.

As an illustration, consider the molecular orbitals that may be formed from the p orbitals perpendicular to the molecular plane of benzene, C_6H_6. Because there are six such orbitals, it is possible to form six molecular orbitals of the form

$$\psi = c_1 p_1 + c_2 p_2 + c_3 p_3 + c_4 p_4 + c_5 p_5 + c_6 p_6 \tag{10}$$

The lowest energy, most strongly bonding orbital has no internuclear nodes, and has the form

$$\psi = p_1 + p_2 + p_3 + p_4 + p_5 + p_6$$

and is illustrated in Fig. 9.28. It is strongly bonding because the constructive interference between neighbouring p orbitals results in a good accumulation of electron density between the nuclei (but slightly off the internuclear axis, as in the π bonds of diatomic molecules).

The most antibonding orbital has the form

$$\psi = p_1 - p_2 + p_3 - p_4 + p_5 - p_6$$

The alternation of signs in the linear combination results in destructive interference between neighbours, and the molecular orbital has a nodal plane between each pair of neighbours, as shown in the illustration. The remaining four molecular orbitals are more difficult to establish by qualitative arguments, but they have the form shown in Fig. 9.28, and lie in energy between the most bonding and most antibonding orbitals. Note that the four intermediate orbitals form two doubly degenerate pairs, one net bonding and the other net antibonding.

The energies of the six orbitals may be calculated by solving the Schrödinger equation, and are shown in the molecular orbital energy-level diagram (Fig. 9.29). There are six electrons to be accommodated (one is supplied by each C atom), and they occupy the lowest three orbitals. It is an important feature of the configuration that *the only molecular orbitals occupied have a net bonding character*, for this is one contribution to the stability (in the sense of low energy) of the benzene molecule. It is helpful to note the similarity between the molecular orbital energy-level diagram for benzene and that for N_2 (see Fig. 9.18): the strong bonding in benzene is echoed in the strong bonding in nitrogen.

A feature of the molecular-orbital description of benzene is that each molecular orbital spreads either all round or partially round the C_6 ring. That is, π bonding is **delocalized**, and each electron pair helps to bind together several or all of the C atoms. The delocalization of bonding influence is a primary feature of molecular orbital theory, and we encounter it in its extreme form in the electronic structures of solids.

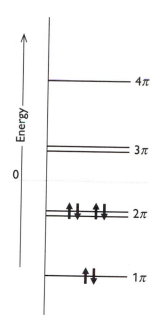

Fig. 9.29 The molecular orbital energy-level diagram for benzene, and the configuration in its ground state.

The band theory of solids

Modern chemistry is deeply concerned with the properties of solids. Apart from their intrinsic usefulness for construction, modern solids have made possible the semiconductor revolution and recent advances in ceramics have given rise to the hope that we may be on the verge of a superconductor revolution. Solids are used widely in chemical industry as catalysts, where the details of their action often depends on the details of their electronic structure, particularly at the surfaces where the reaction takes place. Advances in understanding of electron mobility in solids is also useful in biology, where electron transport processes are responsible for many biochemical processes, particularly photosynthesis and respiration.

For our purposes, we shall concentrate on the simplest types of solid. We shall distinguish two types of solid according to the variation of their electrical conductivity with temperature. A **metallic conductor** is a substance with a conductivity that decreases as the temperature is raised. Metallic conductors include the metallic elements, their alloys, and graphite. A **semiconductor** is a substance with a conductivity that increases as the temperature is raised. Semiconductors include silicon, diamond, and gallium arsenide. A semiconductor generally has a lower conductivity than that typical of metals, but the magnitude of the conductivity is not the criterion of the distinction. It is conventional to classify semiconductors with very low electrical conductivities as **insulators**. We shall use the latter term, but it should be appreciated that it is one of convenience rather than one of fundamental significance. We shall not consider **superconductors**, which are substances that conduct electricity with zero resistance.

9.8 The formation of bands

The electronic structures of solids can be treated in a uniform way in terms of molecular orbital theory. Indeed, it is possible to regard a solid as just another system in which electron delocalization (like that in benzene, but on a much greater scale) is responsible for bonding and properties. In a solid, atom after atom lies in a three-dimensional array and electrons take part in bonding that spreads throughout the sample. We shall consider a one-dimensional solid initially, which consists of a single, infinitely long line of atoms, each one having one s orbital available for forming molecular orbitals. We can construct the LCAO-MOs of the solid by adding atoms to a line and then find the electronic structure by using the building-up principle.

One atom contributes one s orbital at a certain energy (Fig. 9.30(a)). When a second atom is brought up it overlaps the first, and forms a bonding and antibonding orbital (Fig. 9.30(b)). The third atom overlaps its nearest neighbour (and only slightly the next-nearest), and three molecular orbitals are formed from these three atomic orbitals (Fig. 9.30(c)). The fourth atom leads to the formation of a fourth molecular orbital (Fig. 9.30(d)). At this stage we can begin to see that the general effect of bringing up successive atoms is to spread the range of energies covered by the molecular orbitals, and also to fill in the range of energies with more and more orbitals (one more for each atom). When N atoms have been added to the line, there are N molecular orbitals covering a band of finite width. When N is infinitely large, the difference between neighbouring energy levels is infinitely small, but the band still has finite width. This band consists of N different molecular orbitals, the lowest energy orbital being fully bonding, and the highest energy orbital being fully antibonding between adjacent atoms (Fig. 9.31).

The band formed from the overlap of s orbitals is called the **s band**. If the atoms have p orbitals available, the same procedure leads to a **p band** (as in the upper half of Fig. 9.31). If the atomic p orbitals lie higher in energy than the s orbitals, then the p band lies higher than the s band, and there may be a **band gap**, a range of energies for which no orbitals exist.

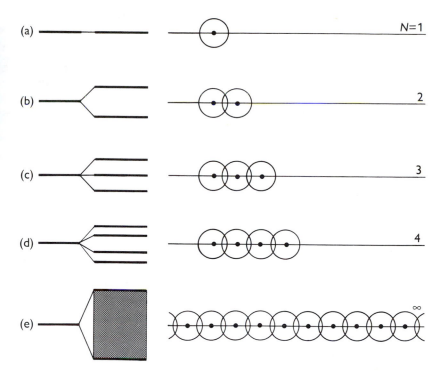

Fig. 9.30 The formation of a band of N molecular orbitals by successive addition of N atoms to a line. Note that the band remains of finite width, and although it looks continuous, it consist of N different orbitals even when N approaches infinity.

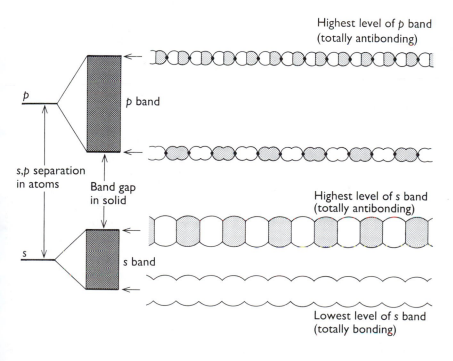

Fig. 9.31 The overlap of s orbitals gives rise to an s band, and the overlap of p orbitals gives rise to a p band. In this case the s and p orbitals of the atoms are so widely spaced that there is a band gap. In many cases the separation is less, and the bands overlap.

Highest level of p band (totally antibonding)

p band

Highest level of s band (totally antibonding)

s,p separation in atoms

Band gap in solid

s band

Lowest level of s band (totally bonding)

373

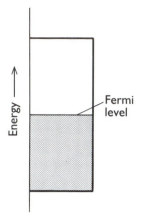

Fig. 9.32 When N electrons occupy a band of N orbitals, it is only half full and the electrons near the Fermi level (the top of the filled levels) are mobile.

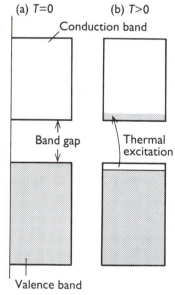

Fig. 9.33 (a) When 2N electrons are present, the band is full and the material is an insulator at $T = 0$. (b) At temperatures above $T = 0$, electrons populate the levels of the conduction band at the expense of the valence band, and the solid is a semiconductor.

9.9 The occupation of orbitals

Now consider the electronic structure of a solid formed from atoms each able to contribute one electron (for example, the alkali metals). There are N atomic orbitals and therefore N molecular orbitals squashed into an apparently continuous band. There are N electrons to accommodate. Because the orbitals in the bands are so close together in energy, electrons can be excited out of them by thermal motion of the atoms. The possibility of thermal excitation is a complication we can avoid by considering the solid at $T = 0$ when there is no such motion and all the electrons occupy the lowest available orbitals.

At $T = 0$, only the lowest $\frac{1}{2}N$ molecular orbitals are occupied (Fig. 9.32), and the highest occupied molecular orbital is called the **Fermi level**. However, unlike in the discrete molecules we have considered so far, there are empty orbitals very close in energy to the Fermi level, and so it requires hardly any energy to excite the uppermost electrons. Some of the electrons are therefore very mobile, and give rise to electrical conductivity.

The electrical conductivity of a metallic solid decreases with increasing temperature even though more electrons are excited into empty orbitals. This apparent paradox is resolved by noting that the increase in temperature causes more vigorous thermal motion of the atoms, so collisions between the moving electrons and an atom are more likely. That is, the electrons are scattered out of their paths through the solid, and are less efficient at transporting charge.

9.10 Insulators and semiconductors

When each atom provides two electrons, the 2N electrons fill the N orbitals of the s band. The Fermi level now lies at the top of the band (at $T = 0$), and there is a gap before the next band begins (Fig. 9.33(a)). As the temperature is increased, electrons populate the empty orbitals of the upper band (Fig. 9.33(b)). They are now mobile, and the solid is an electric conductor. In fact, it is a semiconductor, because the electrical conductivity depends on the number of electrons that are promoted across the gap, and that number increases as the temperature is raised. If the gap is large, though, very few electrons will be promoted at ordinary temperatures, and the conductivity will remain close to zero, giving an insulator. Thus, the conventional distinction between an insulator and a semiconductor is related to the size of the band gap, and is not absolute like the distinction between a metal (incomplete bands at $T = 0$) and a semiconductor (full bands at $T = 0$).

Another method of increasing the number of charge carriers and enhancing the semiconductivity of a solid is to implant foreign atoms into an otherwise pure material. If these dopants can trap electrons (as when Ge atoms are introduced into arsenic), they withdraw electrons from the filled band, leaving holes which allow the remaining electrons

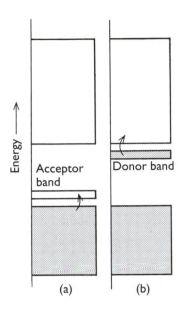

Fig. 9.34 (a) A dopant with fewer electrons than its host can form a narrow band that accepts electrons from the valence band. The holes in the valence band are mobile, and the substance is a p-type semiconductor. (b) A dopant with more electrons than its host forms a narrow band that can supply electrons to the conduction band. The electrons it supplies are mobile, and the substance is an n-type semiconductor.

to move (Fig. 9.34(a)). This doping procedure gives rise to **p-type semiconductivity**, the p indicating that the holes are positive relative to the electrons in the band. Alternatively, a dopant might carry excess electrons (for example, P atoms introduced into germanium), and these additional electrons occupy otherwise empty bands, giving **n-type semiconductivity** (Fig. 9.34(b)), where n denotes the negative charge of the carriers. The preparation of doped but otherwise ultrapure materials was described in Section 4.9.

EXERCISES

9.1 Give the ground state electron configurations and bond orders of (a) Li_2, (b) Be_2, and (c) C_2.

9.2 Give the ground state electron configurations of (a) H_2^-, (b) N_2, and (c) O_2.

9.3 Give the ground state electron configurations of (a) CO, (b) NO, and (c) CN^-. For heteronuclear diatomic molecules, a good first approximation is that the energy-level diagram is much the same as for homonuclear diatomic molecules.

9.4 From the ground state electron configurations of B_2 and C_2, predict which molecule should have the greater bond dissociation energy.

9.5 Which of the molecules N_2, NO, O_2, C_2, F_2, and CN would you expect to be stabilized by (a) the addition of an electron to form AB^-, (b) the removal of an electron to form AB^+?

9.6 Sketch the molecular orbital energy-level diagrams for (a) CO, (b) XeF and deduce their ground state electron configurations. Is XeF likely to have a shorter bond length than XeF^+?

9.7 Where it is appropriate, give the parity of (a) $2\pi^*$ in F_2, (b) 3σ in NO, (c) 1δ in Tl_2, (d) $2\delta^*$ in Fe_2.

9.8 Give the parities of the six π molecular orbitals of benzene.

9.9 Use the electron configurations of NO and N_2 to predict which is likely to have the shorter bond length.

9.10 Show that the sp^2 hybrid orbital $(s + \sqrt{2}p)/\sqrt{3}$ is normalized to 1 if the s and p orbitals are normalized to 1.

9.11 Normalize the molecular orbital $\psi_s(A) +$

$\lambda\psi_s(B)$ in terms of the parameter λ and the overlap integral S.

9.12 Which of the following triatomic species are expected to be linear: (a) CO_2, (b) NO_2, (c) NO_2^+, (d) NO_2^-, (e) SO_2, (f) H_2O, (g) H_2O^{2+}? Give reasons in each case.

9.13 Use VSEPR theory to predict the shapes of (a) H_2S, (b) SF_6, (c) XeF_4, (d) SF_4.

9.14 Construct the molecular orbital energy-level diagrams of (a) ethene (ethylene) and (b) ethyne (acetylene) on the basis that the molecules are formed from the appropriately hybridized CH_2 or CH fragments.

9.15 Predict the electronic configurations of (a) the benzene anion, (b) the benzene cation.

9.16 In the free-electron molecular orbital (FEMO) theory, the electrons in a conjugated molecule are treated as independent particles in a box of length L. Sketch the form of the two occupied orbitals in butadiene predicted by this model and predict the minimum excitation energy of the molecule. The tetraene $CH_2{=}CHCH{=}CHCH{=}CHCH{=}CH_2$ can be treated as a box of length $8R$, where $R \approx 140$ pm (as in this case, an extra half bond-length is often added at each end of the box). Calculate the minimum excitation energy of the molecule and sketch the appearance of the highest occupied and lowest unoccupied molecular orbitals.

10

Cohesion and structure

CONTENTS

- The origin of cohesion
- Fluids
- Crystal structure
- Natural biopolymers
- Exercises

The next level of structural organization that we shall consider arises from the forces that are responsible for the cohesion of particles and the condensation of gases into liquids and solids. The same forces are responsible for the shapes that proteins adopt, for they twist their long polypeptide chains into their characteristic shapes and then pin them together in the arrangement essential to their function. In some cases the cohesive forces can be identified without difficulty: for example, the cohesion of ionic solids is largely a result of the Coulomb interaction between the ions. However, the interactions that are responsible for the condensation and solidification of molecular substances are more subtle, and we shall need to explore their origin more fully.

The origin of cohesion

In this section we examine the forces that bind closed-shell species together. First, we consider the energetics of ionic solids, and then turn to the weaker interactions that are responsible for the cohesion of species with zero net charge, such as the forces responsible for the structure of solid argon and of ice. For the latter interactions, we need to understand some of the electric properties of molecules, in particular their electric dipole moments and polarizabilities.

The term **van der Waals forces** denotes the interactions between closed-shell molecules. The attractive contributions to these forces include the interactions between the partial electric charges of polar molecules. Van der Waals forces also include the repulsive interactions that are responsible for the prevention of the complete collapse of matter to nuclear densities. The repulsive interactions arise from the Pauli principle and the exclusion of electrons from regions of space where the orbitals of closed-shell species overlap.

10.1 Lattice enthalpy

A measure of the strength of binding in a solid, the analogue of the dissociation energy of individual molecules, is the **lattice enthalpy**, ΔH_L. The lattice enthalpy is defined as the enthalpy change accompanying the complete separation of the entities that compose the solid (such as ions if the solid is ionic, and molecules if the solid is molecular). For example, the lattice enthalpy of an ionic solid such as calcium chloride, $CaCl_2$, is the enthalpy change accompanying the process

$$CaCl_2(s) \rightarrow Ca^{2+}(g) + 2Cl^-(g) \quad \Delta H_L$$

and the lattice enthalpy of a molecular solid, such as ice, is

$$H_2O(s) \rightarrow H_2O(g) \quad \Delta H_L$$

The lattice enthalpy is a *positive* quantity, and the greater its value the stronger the bonding within the solid. Some experimental lattice enthalpies are given in Table 10.1. Most lattice enthalpies of ionic solids are of the order of $10^3 \, kJ \, mol^{-1}$, but there are considerable variations. The lattice enthalpies of ionic solids are significantly larger than those of most molecular solids. A feature to note from the data in the table is that ionic solids composed of small, highly charged ions (such as Al_2O_3) have high lattice enthalpies. A feature not shown in the table is that molecules capable of forming hydrogen bonds give rise to higher lattice enthalpies than molecules that cannot form such bonds.

The determination of lattice enthalpy

Lattice enthalpies are determined experimentally by using a **Born–Haber cycle**. In a Born–Haber cycle, a cycle (or closed path) of steps is formed which includes lattice formation as one stage. A typical cycle for an ionic compound has the form

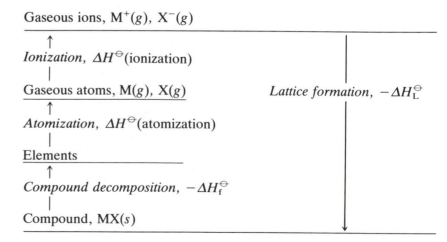

Table 10.1 Lattice enthalpies, $\Delta H_L^\ominus/kJ\,mol^{-1}$

LiF	1037	LiCl	852	LiBr	815	LiI	761
NaF	926	NaCl	786	NaBr	752	NaI	705
KF	821	KCl	717	KBr	689	KI	649
MgO	3850	CaO	3461	SrO	3283	BaO	3114
MgS	3406	CaS	3119	SrS	2974	BaS	2832

For a solid element (such as a metal like sodium), the standard enthalpy of atomization, $\Delta H^\ominus$(atomization), is the standard enthalpy of sublimation; for a gaseous element (such as chlorine), it is the standard enthalpy of dissociation. The standard enthalpy of ionization is the ionization enthalpy (for the formation of cations) and the electron-gain enthalpy (for the formation of anions). The value of the lattice enthalpy—the only unknown in a well-chosen cycle—is found from the requirement that the sum of the enthalpy changes around a complete cycle is zero (because enthalpy is a state property):

$$\Delta H^\ominus(\text{atomization}) + \Delta H^\ominus(\text{ionization}) + (-\Delta H_L^\ominus) + (-\Delta H_f^\ominus) = 0$$

Therefore,

$$\Delta H_L^\ominus = \Delta H^\ominus(\text{atomization}) + \Delta H^\ominus(\text{ionization}) - \Delta H_f^\ominus$$

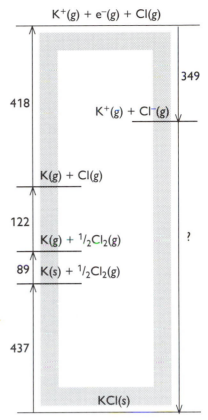

Fig. 10.1 The Born–Haber cycle for the calculation of the lattice enthalpy of potassium chloride. The sum of the enthalpy changes around the cycle is zero. Upward pointing arrows denote positive changes in enthalpy; downward pointing arrows denote negative enthalpy changes. All the steps in the cycle correspond to the same temperature.

Example Using a Born–Haber cycle to determine a lattice enthalpy
Calculate the lattice enthalpy of KCl(s) using a Born–Haber cycle and the information given in the margin.

Answer The required cycle is shown in Fig. 10.1. The first step is the sublimation of solid potassium

	$\Delta H^\ominus/kJ$
$K(s) \rightarrow K(g)$	+89

followed by its ionization

$K(g) \rightarrow K^+(g) + e^-(g)$	+418

Chloride ions are formed by dissociation of Cl_2:

$\frac{1}{2}Cl_2(g) \rightarrow Cl(g)$	+122

followed by electron gain

$Cl(g) + e^-(g) \rightarrow Cl^-(g)$	-349

The solid is now formed

$K^+(g) + Cl^-(g) \rightarrow KCl(s)$	$-\Delta H_L$

Step	$\Delta H^\ominus/kJ\,mol^{-1}$
Sublimation of K(s)	+89
Ionization of K(g)	+418
Dissociation of $Cl_2(g)$	+244
Electron attachment to Cl(g)	-349
Formation of KCl(s)	-438

and the cycle is completed by decomposing KCl(s) into its elements

$$KCl(s) \rightarrow K(s) + \tfrac{1}{2}Cl_2(g) \qquad +437$$

(the reverse of its formation). The sum of the enthalpies is $-\Delta H_L + 717 \text{ kJ}$; however, the sum must be equal to zero, so $\Delta H_L = 717 \text{ kJ}$.

Exercise E10.1 Calculate the lattice enthalpy by magnesium bromide from the following data and the information in Appendix 1

Step	$\Delta H^{\ominus}/\text{kJ}$
Sublimation of Mg(s)	+148
Ionization of Mg(g) to $Mg^{2+}(g)$	+2187
Vaporization of $Br_2(l)$	+31

[2421 kJ]

Coulombic contributions to lattice enthalpies

The total Coulombic potential energy of a crystal is the sum of the individual Coulomb potential energy terms

$$V_{AB} = \frac{(z_A e) \times (z_B e)}{4\pi\varepsilon_0 r_{AB}}$$

This expression applies to ions of charge numbers z_A and z_B (with cations having positive charge numbers and anions negative charge numbers) separated by a distance r_{AB} (ε_0 is the vacuum permittivity, see inside front cover). The sum over all the pairs of ions in the solid may be carried out for any crystal structure, although in practice it is lengthy because nearest neighbours contribute a large negative term, second-nearest neighbours an only slightly weaker positive term, and so on. The overall result, however, is that there is a net attraction between the cations and anions and a favourable (negative) contribution to the energy of the solid.

For example, in a uniformly spaced one-dimensional line of alternating cations and anions of separation d with $z_A = +z$ and $z_B = -z$, the interaction of one ion with all the others is equal to

$$V = -\frac{z^2 e^2}{4\pi\varepsilon_0 d} \times 2\ln 2$$

JUSTIFICATION

Consider a line of alternating cations and anions extending in an infinite direction to the left and right of the ion of interest. The Coulombic energy of interaction with the ions on the right is the following sum of terms, where the negative terms represent attractions between ions of charge opposite to that of the ion of interest and the positive terms represent repulsions between ions of the same charge:

$$V = \frac{1}{4\pi\varepsilon_0} \times \left(-\frac{z^2 e^2}{d} + \frac{z^2 e^2}{2d} - \frac{z^2 e^2}{3d} + \frac{z^2 e^2}{4d} - \cdots \right)$$

$$= -\frac{z^2 e^2}{4\pi\varepsilon_0 d} \times (1 - \tfrac{1}{2} + \tfrac{1}{3} - \tfrac{1}{4} + \cdots)$$

$$= -\frac{z^2 e^2}{4\pi\varepsilon_0 d} \times \ln 2$$

This calculation has used the relation

$$\ln 2 = 1 - \tfrac{1}{2} + \tfrac{1}{3} - \tfrac{1}{4} + \cdots$$

The interaction of the ion of interest with the ions to its left is the same, so the *total* potential energy of interaction is twice this expression, as given in the text.

As in this case of a line of ions, it is found that, apart from the explicit appearance of the ion charge numbers, the value of the sum depends only on the type of the lattice and on a single scale parameter d, the separation of the centres of nearest neighbours. Therefore, in general, for the potential energy per mole of formula units we may write

$$V = N_A \frac{e^2}{4\pi\varepsilon_0} \left(\frac{z_A z_B}{d} \right) \times A \qquad (1)$$

where A is called a **Madelung constant**, N_A is Avogadro's constant, and z_A and z_B are the charge numbers of the ions. Because the charge number of cations is positive and that of anions is negative, V is negative overall, corresponding to a lowering in potential energy relative to the gas of widely separated ions. The value of the Madelung constant for a single line of ions is $2 \ln 2 = 1.386$, as we have already seen. The computed values of the Madelung constant for some common lattices are given in Table 10.2.

The potential energy of interaction between the ions of an ionic crystal is the principal attractive contribution to the lattice enthalpy. However, the ions also repel each other when they are pressed together and their wavefunctions overlap, and these repulsions reduce the strength of the net attraction between ions. It follows that the lattice enthalpy is smaller than V itself. The effect of repulsions is taken into account by the **Born–Meyer equation** for the lattice enthalpy:

$$\Delta H_L = |z_A z_B| \frac{N_A e^2}{4\pi\varepsilon_0 d} \left(1 - \frac{d^*}{d} \right) A \qquad (2)$$

where d^* is an empirical parameter that is often taken as 34.5 pm. The

Table 10.2 Madelung constants

Structural type	A
Caesium chloride	1.763
Fluorite	2.519
Rock salt	1.748
Rutile	2.408

details of this expression are not important: what is important is that the expression confirms the experimental results that:

1. Lattice enthalpy increases with charge number of the ions.

2. Lattice enthalpy increases as the ionic radii decrease.

The second conclusion follows from the fact that the smaller the ionic radii, the smaller the value of d.

Exercise E10.2 Which can be expected to have the greater lattice enthalpy, magnesium oxide or strontium oxide?

[MgO]

10.2 Permanent and induced electric dipole moments

To understand the origin of the attractive intermolecular forces between uncharged species, we need to understand the origin of the electric dipole moment of a molecule. An **electric dipole** consists of two charges q and $-q$ separated by a distance R. A dipole is represented by a vector $\boldsymbol{\mu}$ that points from the negative charge to the positive charge. The magnitude of this vector is qR and is called the **electric dipole moment**, μ. An electric dipole is often represented by the arrow $\leftrightarrow$ added to the Lewis structure for the molecule, with the $+$ marking the positive end, as in $\overrightarrow{\text{HCl}}$ (note that the direction of $\leftrightarrow$ is opposite to that of the vector $\boldsymbol{\mu}$). Dipole moments are generally reported in **debye**, D, where

$$1\,\text{D} = 3.336 \times 10^{-30}\,\text{C m}$$

The debye is named after Peter Debye, the Dutch pioneer of the study of the dipole moments of molecules. The dipole moment of a pair of charges e and $-e$ separated by 100 pm (1 Å) is $1.6 \times 10^{-29}\,\text{C m}$, corresponding to 4.8 D. Dipole moments of small molecules are typically about 1 D.

A **polar molecule** is a molecule with an electric dipole moment that arises from the partial charges on atoms linked by polar bonds (Section 9.6). **Nonpolar molecules**, molecules that have zero electric dipole moment, may acquire a dipole moment in an electric field as a result of the distortion the field causes in their electronic distributions and nuclear positions. Similarly, polar molecules may have their existing dipole moments modified by the applied field.

The partial charge of a dipole lowers the potential energy of an ion of opposite charge or the partial charge of opposite sign in another dipole. As a result, permanent and induced dipole moments are important in chemistry through their role in intermolecular forces and in their contribution to the ability of a substance to act as a solvent for ionic solids. The latter ability stems from the fact that the partial charge at one end of a dipole may be attracted to an ion of opposite charge and hence contribute an exothermic term to the enthalpy of solution (Fig. 10.2).

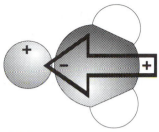

Fig. 10.2 An electric dipole of a solvent molecule can interact with an ion and adhere to it. Here the negative partial charge of the dipole of H_2O is attracted to the positive charge of a cation; the positive partial charge of H_2O (which is largely on the H atoms) is further away from the cation, and repels it less strongly than the negative partial charge, which is closer, and attracts it. The net interaction is attractive. This process contributes to the hydration of cations in aqueous solution.

Polar molecules

All heteronuclear diatomic molecules are polar because the difference in electronegativities of their two atoms results in nonzero partial charges. Typical dipole moments include 1.08 D for HCl and 0.42 for HI (Table 10.3). A very approximate relation between the dipole moment and the difference in Pauling electronegativities, $\Delta\chi$ (Table 9.2), of the two atoms is

$$\mu/D = \Delta\chi \qquad (3)$$

Exercise E10.3 Estimate the electric dipole moment of an HBr molecule.

[0.8 D; experimental value 0.80 D]

The more electronegative atom is usually the negative end of the dipole, but there are exceptions, particularly when antibonding orbitals are occupied. Thus the dipole moment of CO is very small (0.12 D) but the negative end of the dipole is on the C atom even though the O atom is more electronegative. It should be recalled from the discussion in Section 9.6 that electrons in antibonding orbitals tend to be found closer to the less electronegative atom, so they can give rise to a negative partial charge on that atom.

Whether or not a polyatomic molecule is polar depends on its symmetry. Indeed, molecular symmetry is more important than the question of whether or not the atoms in the molecule are of the same element. Thus the homonuclear triatomic molecule O_3 (which is angular) is polar because the electron density on the central O atom is different from that on the two outer O atoms. The heteronuclear triatomic molecule CO_2 (which is linear) is nonpolar because the two bond dipoles point in opposite directions and cancel.

Exercise E10.4 Use VSEPR theory to judge whether ClF_3 is polar or nonpolar.

[Polar]

Table 10.3 Dipole moments (μ) and polarizability volumes (α')†

	μ/D	$\alpha'/10^{-24}\ cm^3$
Ar	0	1.66
CCl_4	0	10.5
C_6H_6	0	10.4
H_2	0	0.819
H_2O	1.85	1.48
NH_3	1.47	2.22
HCl	1.08	2.63
HBr	0.80	3.61
HI	0.42	5.45

† The polarizability volumes are mean values over all orientations of the molecule.

383

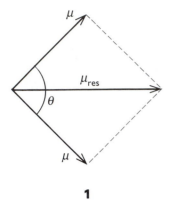

1

To some extent it is possible to resolve the dipole moment of a polyatomic molecule into contributions from various groups of atoms in the molecule (Fig. 10.3) and the directions in which these individual contributions lie. Thus, *p*-dichlorobenzene is nonpolar by symmetry on account of the cancellation of two equal but opposing C—Cl moments. *o*-Dichlorobenzene has a dipole moment which is approximately the resultant of two monochlorobenzene dipole moments arranged at 60°. The technique of vector addition can be applied with fair success to other series of related molecules, and the resultant μ_{res} of two equal dipole moments, μ, that make an angle θ to each other (as in **1**) is approximately

$$\mu_{res} = 2\mu \cos \tfrac{1}{2}\theta \qquad (4)$$

Exercise E10.5 Estimate the relative electric dipole moments of *ortho* and *meta* disubstituted benzenes.
$$[\mu(\text{ortho})/\mu(\text{meta}) = 1.7]$$

Interaction between dipoles

Most of the work in this section will be based in one way or the other on the Coulomb potential energy of interaction between two charges

Fig. 10.3 The dipole moments of the dichlorobenzene isomers can be obtained approximately by vectorial addition of two chlorobenzene dipole moments (1.57 D).

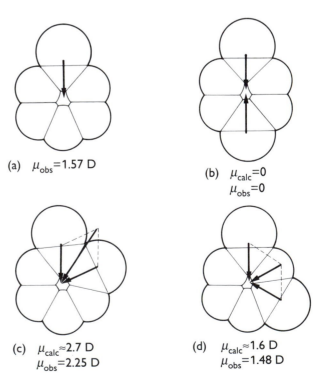

(a) $\mu_{obs} = 1.57$ D

(b) $\mu_{calc} = 0$
     $\mu_{obs} = 0$

(c) $\mu_{calc} \approx 2.7$ D
     $\mu_{obs} = 2.25$ D

(d) $\mu_{calc} \approx 1.6$ D
     $\mu_{obs} = 1.48$ D

q_1 and q_2 separated by a distance r:

$$V = \frac{q_1 q_2}{4\pi\varepsilon_0 r} \qquad (5)$$

It is easy to adapt this expression to give the interaction between a charge and a dipole and to extend it to the interaction between two dipoles. For example, it follows from eqn (5) that the potential energy of a dipole μ_1 in the presence of a charge q_2 in the orientation shown in (2) is

$$V = -\frac{\mu_1 q_2}{4\pi\varepsilon_0 r^2} \qquad (6)$$

The interaction energy decreases more rapidly with distance than that between two point charges because, from the viewpoint of the single charge q_2, the partial charges of the dipole seem to merge and cancel as the distance r increases. Likewise, the interaction energy between two dipoles μ_1 and μ_2 in the orientation shown in (3) is

$$V = -2 \times \frac{\mu_1 \mu_2}{4\pi\varepsilon_0 r^3} \qquad (7)$$

The potential energy of interaction between two polar molecules is a complicated function of the angle between them. However, when the two dipoles are parallel, as in (4), the energy is simply

$$V = \frac{\mu_1 \mu_2 f}{4\pi\varepsilon_0 r^3} \quad \text{with} \quad f = 1 - 3\cos^2\theta \qquad (8)$$

This expression applies to polar molecules in a fixed orientation in a solid. In a fluid of completely freely rotating molecules the interaction between dipoles would average to zero because the attractions and repulsions cancel. However, because the potential energy of a dipole near another dipole depends on their relative orientations, the molecules do not in fact rotate completely freely even in a gas. As a result, the lower energy orientations are marginally favoured, and so there is a nonzero interaction between polar molecules (Fig. 10.4).

The detailed calculation of the interaction energy is quite complicated, but the form of the final answer is very simple:

$$V = -\frac{C}{r^6} \quad \text{where} \quad C \propto (\mu_1 \mu_2)^2 / T \qquad (9)$$

The important features of this expression are the dependence of the average interaction energy on the inverse *sixth* power of the separation and its inverse dependence on the temperature. The latter reflects the way in which the greater thermal motion at higher temperatures overcomes the mutual orienting effects of the dipoles. At 25 °C the average interaction energy for pairs of molecules with $\mu = 1$ D is about $-1.4\,\mathrm{kJ\,mol^{-1}}$ when the separation is 0.3 nm. This energy should be compared with the average molar kinetic energy of $\frac{3}{2}RT = 3.7\,\mathrm{kJ\,mol^{-1}}$ at the same temperature. The two energies are not very dissimilar but they are both much less than the energies involved in the making and breaking of chemical bonds.

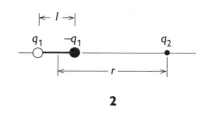

2

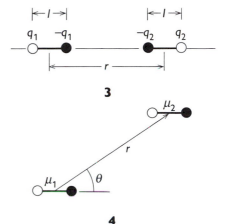

3

4

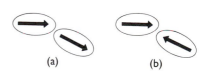

(a)   (b)

Fig. 10.4 When a pair of molecules can adopt all relative orientations with equal probability, the favourable orientations (a) and the unfavourable ones (b) cancel, and the average interaction is zero. In an actual fluid, the interactions in (a) slightly predominate.

385

Induced dipole moments

So far, we have considered interactions between charged species (an ion interacting with another ion) and between species that have no net charge but which have nonzero dipole moments (two polar molecules interacting, or a polar molecule like water interacting with an ion). Now we move on to consider the interaction between molecules at least one of which has neither a net charge nor a dipole moment. An example might be a polar group on a polypeptide interacting with a Xe atom that is being used as an anaesthetic.

Although a molecule might not have a *permanent* electric dipole moment (methane, for example, does not) it may acquire an **induced dipole moment**, μ^*, as a result of the influence of an electric field (perhaps a field generated by a nearby ion or polar molecule). The charge of the ion or the partial charge of the dipole distorts the electron distribution of the other nearby molecule, and gives rise to an electric dipole in it. The magnitude of this induced dipole moment is proportional to the strength of the applied electric field, E, and we write

$$\mu^* = \alpha E \tag{10}$$

where the constant α is the **polarizability** of the molecule. The polarizability has the units $J^{-1}C^2m^2$ when the field strength is expressed in volts per metre ($V\,m^{-1}$) and μ is expressed in coulomb-metres ($C\,m$). Because that collection of units is awkward, α is usually converted to **polarizability volume**, α', by using the relation

$$\alpha' = \frac{\alpha}{4\pi\varepsilon_0} \tag{11}$$

The units of $4\pi\varepsilon_0$ are $J^{-1}C^2m^{-1}$, so α' has the dimensions of volume (hence its name). Values are usually quoted in cm^3 (and often in $Å^3$, where $1\,Å^3 = 10^{-24}\,cm^3$, for then values come out close to 1).† The larger the polarizability volume, the larger the polarizability of the molecule and the greater the distortion that is caused by a given electric field.

Exercise E10.6 What strength of electric field (in $kV\,cm^{-1}$) is required to induce an electric dipole moment of $1.0\,\mu D$ in a molecule of polarizability volume $0.11\,Å^3$ (like CCl_4)?
[$2.7\,kV\,cm^{-1}$]

Some experimental polarizability volumes of molecules are given in Table 10.3. The values reflect the strengths with which the nuclear

† When using older compilations of data, it is useful to note that polarizability volumes have the same numerical values as the 'polarizabilities' reported using c.g.s. electrical units, and so the tabulated values previously called 'polarizabilities' can be used directly.

charges control the electron distributions and prevent their distortion by the applied field. If the molecule has few electrons, they are tightly controlled by the nuclear charges and the polarizability of the molecule is low. If the molecule contains large atoms with electrons some distance from the nucleus, the nuclear control is smaller, the electron distribution is flabbier, and the polarizability is greater. The polarizability depends on the orientation of the molecule with respect to the field unless the molecule is tetrahedral (such as CCl_4) or octahedral (such as SF_6).

A polar molecule with dipole moment μ_1 can induce a dipole moment in a polarizable molecule (which may itself be either polar or nonpolar) because the partial charges of the polar molecule give rise to an electric field that distorts the second molecule. That induced dipole interacts with the permanent dipole of the first molecule, and the two are attracted to each other. The formula for the interaction energy is

$$V = -\frac{C}{r^6} \quad \text{with} \quad C \propto \mu_1^2 \alpha_2 \tag{12}$$

Here, α_2 is the polarizability of molecule 2 and μ_1 is the permanent dipole moment of molecule 1. For a molecule with $\mu = 1\,\mathrm{D}$ (such as HCl) near a molecule of polarizability volume $\alpha' = 10 \times 10^{-24}\,\mathrm{cm}^3$ (such as benzene, Table 10.3) the average interaction energy is about $-0.8\,\mathrm{kJ\,mol^{-1}}$ when the separation is 0.3 nm.

Dispersion interactions

Finally, we consider the interactions between species neither of which has a net charge or an electric dipole moment (such as two Xe atoms in a gas). Nonpolar molecules can interact even though neither has a permanent dipole moment. The evidence for the existence of interactions between them is the formation of condensed phases of nonpolar substances, such as the condensation of molecular hydrogen and of gaseous xenon.

The interaction between nonpolar molecules arises from the transient dipoles that all molecules possess as a result of fluctuations in the instantaneous positions of their electrons (Fig. 10.5). Suppose, for instance, the electrons in one molecule flicker into an arrangement that results in partial positive and negative charges and thus gives it an instantaneous dipole moment μ_1. This dipole polarizes the other molecule, and induces in it an instantaneous dipole moment μ_2. The two dipoles attract each other and the potential energy of the pair is lowered. Although the first molecule will go on to change the size and direction of its dipole, the second will follow it; that is, the two dipoles are correlated in direction like two meshing gears, with a positive partial charge on one molecule appearing close to a negative partial charge on the other molecule, and vice versa. Because of this correlation of the relative positions of the partial charges, and their resulting attractive interaction, the attraction between the two instantaneous dipoles does not average to zero. Instead, it gives rise to a **dispersion interaction**, which is also known as the **London interaction**

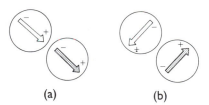

(a)   (b)

Fig. 10.5 In the dispersion interaction, an instantaneous dipole (shaded arrow) on one molecule induces a dipole (light arrow) on another molecule, and the two dipoles then interact to lower the energy. The directions of the two instantaneous dipoles are correlated, and although they occur in different orientations at different instants, the interaction does not average to zero.

(the physicist Fritz London first described it). Polar molecules also interact by a dispersion interaction: such molecules also possess instantaneous dipoles, the only difference beting that the time average of each fluctuating dipole does not vanish but corresponds to the permanent dipole.

The strength of the dispersion interaction depends on the polarizability of the first molecule because the instantaneous dipole moment μ_1 depends on the looseness of the nuclear charge's control over the outer electrons. The strength also depends on the polarizability of the second molecule, for the polarizability determines how readily a dipole can be induced by another molecule. The actual calculation of the dispersion interaction is quite involved, but a reasonable approximation to the interaction energy is given by the **London formula**:

$$V = -\frac{C}{r^6} \quad \text{with} \quad C \propto \alpha_1 \alpha_2 \left(\frac{I_1 I_2}{I_1 + I_2} \right) \tag{13}$$

Here I_1 and I_2 are the ionization energies of the two molecules (Table 2.3). Once again, the interaction turns out to be proportional to the inverse sixth power of the separation. For two CH_4 molecules, $V = -5\,\text{kJ}\,\text{mol}^{-1}$ when $r = 0.3\,\text{nm}$.

Hydrogen bonding

The strongest intermolecular interaction arises from the formation of a **hydrogen bond**, in which a hydrogen atom lies between two strongly electronegative atoms and binds them together. The bond is normally denoted $X\text{—H}\cdots Y$, with X and Y nitrogen, oxygen, or fluorine. Unlike the other interactions we have considered, hydrogen bonding is not universal but is restricted to molecules that contain these atoms.

The most elementary description of the formation of a hydrogen bond is that it is the result of a Coulombic interaction between the partly exposed positive charge of a proton bound to an electron-withdrawing X atom (in the fragment X—H) and the negative charge of a lone pair on the second atom Y:

$$^{\delta-}X\text{—H}^{\delta+} \cdots \,^{\delta-}{:}Y$$

A slightly more sophisticated version of this description is to regard hydrogen bond formation as the formation of a Lewis acid–base complex in which the partly exposed proton of the X—H group is the Lewis acid and :Y, with its lone pair, is the Lewis base:

$$X\text{—H} + :Y \rightarrow X\text{—H}\cdots Y$$

Molecular orbital theory provides an alternative description that is more in line with the concept of delocalized bonding and the ability of an electron pair to bind more than one pair of atoms (Section 9.7). Thus, if the X—H bond is regarded as being formed from the overlap of an orbital on X (which we denote $X\sigma$) and a hydrogen $1s$ orbital (H1s), and the lone pair on Y occupies an orbital on Y (the orbital $Y\sigma$), then when the two molecules are close together, we can build molecular orbitals from all three orbitals:

$$\psi = c_1 X\sigma + c_2 H 1s + c_3 Y\sigma$$

From these three atomic orbitals we can build three molecular orbitals (Fig. 10.6): one is bonding, one almost nonbonding, and the third antibonding. These three orbitals need to accommodate four electrons (two from the original X—H bond and two from the lone pair of Y), and they may do so if two enter the bonding orbital and two enter the nonbonding orbital. Because the antibonding orbital remains empty, the net effect is a lowering of energy. The coefficients of the orbitals are not equal, and the charge distribution arising from the electrons can be interpreted as about 80 per cent ionic and 20 per cent covalent.

Hydrogen bond formation, which has a typical strength of the order of $20\ \text{kJ mol}^{-1}$, dominates all other van der Waals interactions (if the molecules have the appropriate constitution). It accounts for the rigidity of molecular solids such as sucrose and ice, the secondary structure of proteins (the formation of helices and sheets of polypeptide chains), the low vapour pressure of liquids such as water, and their high viscosities and surface tensions. Hydrogen bonding also contributes to the solubility in water of species such as ammonia and compounds containing hydroxyl groups, and to the hydration of anions. In this last case, even ions such as Cl^- and HS^- can participate in hydrogen bond formation with water, for their charge enables them to interact with the hydroxylic protons of H_2O even though uncharged chlorine and sulfur do not participate in hydrogen bonding.

The strengths and distance dependences of the attractive forces that we have considered so far are summarized in Table 10.4.

Table 10.4 Interaction potential energies

Interaction type	Distance dependence of potential energy	Typical energy†/ kJ mol^{-1}	Comment
Ion–ion	$1/r$	250	Only between ions
Ion–dipole	$1/r^2$	15	
Dipole–dipole	$1/r^3$	2	Between stationary polar molecules
	$1/r^6$	0.3	Between rotating polar molecules
London (dispersion)	$1/r^6$	2	Between all types of molecule

† The energy of a hydrogen bond A—H $\cdots$ B is typically $20\ \text{kJ mol}^{-1}$ and occurs on contact for A or B = N, O, or F.

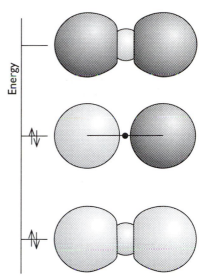

Fig. 10.6 A schematic portrayal of the molecular orbitals that can be formed from X, H, and Y orbitals and which gives rise to an X—H $\cdots$ Y hydrogen bond. The lowest energy combination is fully bonding, the next nonbonding, and the uppermost is antibonding. The antibonding orbital is not occupied by the electrons provided by the X—H bond and the :Y lone pair, so the configuration shown may result in a net lowering of energy in certain cases (namely when the X and Y atoms are N, O, or F).

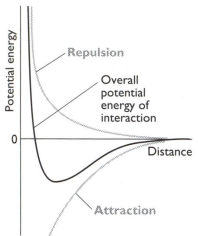

Fig. 10.7 The general form of an intermolecular potential energy curve (the graph of the potential energy of two closed shell species as the distance between them is changed). The attractive (negative) contribution has a long range, but the repulsive (positive) interaction increases more sharply once the molecules come into contact. The overall potential energy is shown by the heavy line.

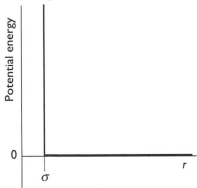

Fig. 10.8 The true intermolecular potential can be modelled in a variety of ways. One of simplest is this hard-sphere potential, in which there is no potential energy of interaction until the two molecules are separated by a distance σ, when the potential energy rises abruptly to infinity as the impenetrable hard spheres repel each other.

390

10.3 The total interaction

The total attractive interaction energy between rotating molecules that cannot participate in hydrogen bonding is the sum of the contributions from the dipole–dipole, dipole–induced dipole, and dispersion interactions. However, only the dispersion interaction contributes if both molecules are nonpolar. All three interactions vary as the inverse sixth power of the separation, and so we may write

$$V = -\frac{C}{r^6} \tag{14}$$

where C is a coefficient that depends on the identity of the molecules.

When molecules are squeezed together (for instance, during the impact of a collision or under the force exerted by a weight pressing on a substance), the nuclear and electronic repulsion and the rising kinetic energy of the confined electrons begin to dominate the attractive forces (Fig. 10.7). The repulsions increase steeply with decreasing separation in a way that can be deduced only by very extensive, complicated molecular-structure calculations. In many cases, however, progress can be made by using a greatly simplified representation of the potential energy, where the details are ignored and the general features expressed by a few adjustable parameters.

One such approximation is the **hard-sphere potential**, in which it is assumed that the potential energy rises abruptly to infinity as soon as the particles come within some separation, σ, called the **collision diameter** (Fig. 10.8):

$$V = \infty \quad \text{for} \quad r \leq \sigma, \qquad V = 0 \quad \text{for} \quad r > \sigma \tag{15}$$

This very simple potential is surprisingly useful for assessing a number of properties.

Another widely used approximation is to express the short-range repulsive potential energy as inversely proportional to a high power of r:

$$V = +\frac{C^*}{r^n} \tag{16}$$

where C^* is another constant (the asterisk signifies repulsion). Typically n is taken as 12, in which case the repulsion dominates the $1/r^6$ attractions strongly at short separation because then $C^*/r^{12} \gg C/r^6$. The sum of the repulsive interaction and the attractive interaction is called the **Lennard–Jones potential**. It is normally written in the form

$$V = 4\varepsilon \left\{ \left(\frac{\sigma}{r} \right)^{12} - \left(\frac{\sigma}{r} \right)^6 \right\} \tag{17}$$

and is drawn in Fig. 10.9. The two parameters are now ε, the depth of the well, and σ, the separation at which $V = 0$. Some typical values are listed in Table 10.5. Although the (12, 6)-potential has been used in many calculations, there is plenty of evidence to show that $1/r^{12}$ is a very poor representation of the repulsive potential, and that the exponential form $e^{-r/\sigma}$ is superior. An exponential function is more faithful to the exponential decay of atomic wavefunctions at large distances, and hence to the distance-dependence of the overlap that is responsible for repulsion.

Exercise E10.7 At what separation does the minimum of the potential-energy curve occur for a Lennard–Jones potential?

$$[r = 2^{1/6}\sigma]$$

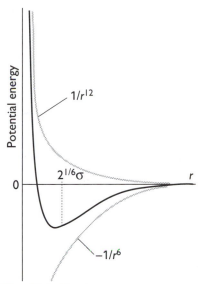

Fig. 10.9 The Lennard–Jones potential is another approximation to the true intermolecular potential-energy curves. It models the attractive component by a contribution that is proportional to $1/r^6$, and the repulsive component by a contribution that is proportional to $1/r^{12}$. (Specifically, these choices result in the Lennard–Jones 12, 6 potential.) Although there are good theoretical reasons for the former, there is plenty of evidence to show that $1/r^{12}$ is only a very poor approximation to the repulsive part of the curve.

Fluids

Intermolecular forces are responsible for the imperfections of real gases and for the formation of condensed phases of molecular compounds; they are responsible for the surface tensions and viscosities of liquids. In this and the following sections we shall explore how intermolecular forces influence the properties of fluids. Broadly speaking, a fluid phase is a gas if the kinetic energies of the molecules greatly exceed the potential energy of their interaction (so the molecules are in almost free flight), and the fluid phase is a liquid if the kinetic energies and potential energies of interaction are comparable, so the molecules can move relative to each other.

10.4 Real gases

Initially we consider real gases in which the interactions result in only a small deviation of the properties from those of a perfect gas. Real gases differ from perfect gases in their equations of state (Section 1.9), exhibiting a Joule–Thomson cooling effect when they expand (Section 1.10), and in their viscosities. The meeting place of theory and experiment is one of the equations of state, particularly the virial equation of state (eqn (22) of Chapter 1):

$$\frac{pV_m}{RT} = 1 + \frac{B}{V_m} + \frac{C}{V_m^2} + \cdots \tag{18}$$

where B is the second virial coefficient and C is the third virial coefficient; both coefficients depend on the temperature. The virial coefficients are consequences of the intermolecular forces, and so by measuring them we have a route to the experimental determination of the strengths of the forces acting between the gas molecules.

When the intermolecular potential V depends only on the separation r of the particles, as in the dispersion interaction of closed-shell atoms and spherical molecules, the second virial coefficient can be calculated quite easily. For example, if the intermolecular potential is

Table 10.5 Lennard–Jones parameters for the (12, 6) potential

	$\varepsilon/\text{kJ mol}^{-1}$	σ/pm
Ar	128	342
Br_2	536	427
C_6H_6	454	527
Cl_2	368	412
H_2	34	297
He	11	258
Xe	236	406

modelled by the hard-sphere potential specified in eqn (15), then it turns out that

$$B = \tfrac{2}{3}\pi N_A \sigma^3 \qquad (19)$$

and the larger the molecules (the larger the value of σ), the greater the second virial coefficient, and the greater the deviation of the gas from perfect behaviour.

JUSTIFICATION

The relation between the intermolecular potential V and the second virial coefficient B is

$$B = -2\pi N_A \int_0^\infty (e^{-V/RT} - 1)r^2 \, dr$$

If we substitute the values of V into the integral, the term in parentheses is equal to -1 from $r = 0$ up to $r = \sigma$ because $e^{-V/RT} = 0$ when $V = \infty$ (because $e^{-\infty} = 0$). For $r > \sigma$, when $V = 0$, $e^{-V/RT} = $

1 (because $e^0 = 1$), so the term in parentheses is zero. Therefore, the value of the integral is

$$B = 2\pi N_A \int_0^\sigma r^2 \, dr = \tfrac{2}{3}\pi N_A \sigma^3$$

because

$$\int x^2 \, dx = \frac{x^3}{3}$$

Exercise E10.8 The second virial coefficient of helium is $12.0 \, \text{cm}^3 \, \text{mol}^{-1}$ at $0\,°C$; estimate the radius of the atom assuming that a hard-sphere potential describes its interactions.

[212 pm]

The evaluation of B for a Lennard–Jones potential is more difficult, but nevertheless can be carried out, so the magnitude of the intermolecular forces can be related directly to departures from perfect behaviour. The third virial coefficient C arises in part from the mutual effect of three-particle contributions to the potential energy, and is even more difficult to calculate.

10.5 The structures of liquids

The starting point for the discussion of solids is the well ordered structure of perfect crystals. The starting point for the discussion of gases is the totally chaotic distribution of the molecules of a perfect gas. The liquid state is intermediate between these two extremes: there is some structure and some chaos.

The pair distribution function

The particles of a liquid are held together by intermolecular forces, but their kinetic energies are comparable to their potential energies. As a result, the whole structure is very mobile. The best description of the average locations of the particles is in terms of the **pair distribution function**, g. The pair distribution function is defined so that $g\delta r$ is the probability that a molecule will be found in the range δr at a distance r from another. It follows that if g passes through a maximum at a radius of, for instance, 0.5 nm, then the most probable distance (regardless of direction) at which a second molecule will be found will be at 0.5 nm from the first molecule.

In a crystal, g is a periodic array of sharp spikes, representing the certainty (in the absence of defects and thermal motion) that particles lie at definite locations. This regularity continues out to large distances, and so we say that crystals have **long-range order**. When the crystal melts, the long-range order is lost and wherever we look at long distances from a given particle there is equal probability of finding a second particle. Close to the first particle, though, there may be a remnant of order (Fig. 10.10). Its nearest neighbours might still adopt approximately their original positions, and even if they are displaced by newcomers, the new particles might adopt their vacated positions. It may still be possible to detect a sphere of nearest neighbours at a distance r_1, and perhaps beyond them a sphere of next-nearest neighbours at r_2. The existence of this **short-range order** means that g can be expected to have a broad peak at r_1, a smaller peak at r_2, and perhaps some more structure beyond that.

The shape of the pair distribution function can be determined by X-ray diffraction, for g can be extracted from the diffuse diffraction pattern characteristic of liquid samples in much the same way as a crystal structure is obtained from X-ray diffraction of crystals (see Section 10.11). The peaks corresponding to the shells of local structure shown in the example in Fig. 10.11 (for water) are distinguishable. Closer analysis shows that any given H_2O molecule is surrounded by

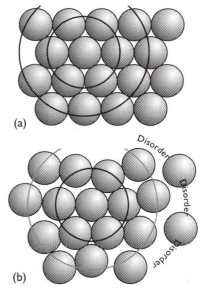

Fig. 10.10 (a) In a perfect crystal at $T = 0$, the distribution of molecules (or ions) is highly regular, and the pair distribution function shows a series of sharp peaks showing the regular organization of rings of neighbours around any selected central ion. (b) In a liquid, there remain some elements of structure close to each molecule, but the further the distance, the less the correlation. The pair distribution function now shows a pronounced (but broadened) peak corresponding to the nearest neighbours of the molecule of interest (which are only slightly more disordered than in the solid), and a suggestion of a peak for the next ring of molecules, but little structure at greater distances.

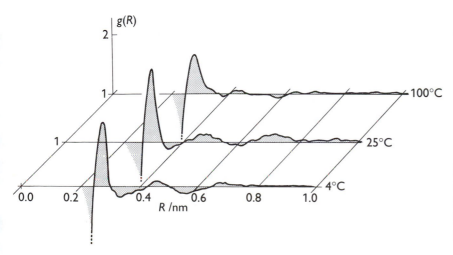

Fig. 10.11 The pair distribution function of the oxygen atoms in liquid water at three temperatures. Note the expansion as the temperature is raised.

Cohesion and structure

Fig. 10.12 A fragment of the crystal structure of ice. Each O atom is at the centre of a tetrahedron of four O atoms at a distance of 276 pm. The central O atom is attached by two short O—H bonds to two H atoms and by two relatively long O $\cdots$ H bonds to two neighbouring H_2O molecules. Overall, the structure consists of planes of hexagonal puckered rings of H_2O molecules.

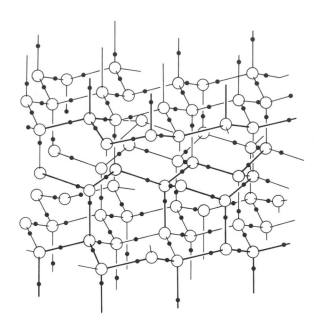

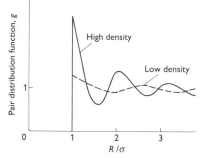

Fig. 10.13 The radial distribution function for a simulation of a liquid using impenetrable hard spheres (ball-bearings).

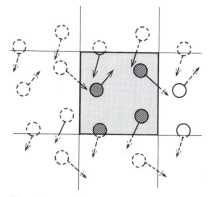

Fig. 10.14 In a two-dimensional molecular dynamics simulation that uses 'periodic boundary conditions', when one particle leaves the cell its mirror image enters through the opposite face.

other molecules at the corners of a tetrahedron, similar to the arrangement in ice (Fig. 10.12). The form of g at 100 °C (Fig. 10.11) shows that the intermolecular forces (in this case, largely hydrogen bonds) are strong enough to affect the local structure right up to the boiling point.

The calculation of g

Because the pair distribution function can be calculated by making assumptions about the intermolecular forces, it can be used to test theories of liquid structure. However, even a fluid of hard spheres without attractive interactions (a box of ball-bearings) gives a function that oscillates near the origin (Fig. 10.13), and one of the factors influencing, and sometimes dominating, the structure of a liquid is the geometrical problem of stacking together reasonably hard spheres. The attractive part of the potential modifies this basic structure, but sometimes only quite weakly, by **gathering and trapping particles into one another's vicinity**.

There are several ways of building the intermolecular potential into the calculation of g. Numerical methods take a box of between 10^2 and 10^3 particles (the number increases as computers grow more powerful), and the rest of the liquid is simulated by surrounding the box with replications (Fig. 10.14). Then, whenever a particle leaves through a face, its image arrives through the opposite face so that the total number remains constant. In the **molecular dynamics** approach, the history of an initial arrangement is followed by calculating the trajectories of all the particles under the influence of the intermolecular potentials. Newton's laws are used to predict where each particle will be after a short time interval (about 10^{-15} s, which is shorter than the time between collisions), and then the calculation is repeated for tens of thousands of such steps. The procedure gives a series of snapshots of the liquid and g can be calculated as before.

10.6 Molecular motion in liquids

The motion of molecules in liquids can be studied theoretically using the molecular dynamics simulation, because the detailed information it gives includes the velocities of the particles. Their motion may be studied experimentally by a variety of methods, and the results indicate that some types of molecule rotate in a series of small (about 5°) steps, but that others jump through about 1 radian (57°) in each step. Another technique is **inelastic neutron scattering**, in which the energy neutrons collect or discard as they pass through a sample is interpreted in terms of the motion of its particles, and very detailed information can be obtained.

The net rate at which a molecule moves through a liquid is expressed in terms of the **diffusion coefficient**, D, which is large if the molecule is very mobile. Some values are given in Table 10.6. To use the diffusion coefficient, we think of the motion of the molecule as occurring in a series of short jumps in random directions, a so-called **random walk**. Although the randomly walking molecule may take many steps in a given time, it has only a small probability of being found far from its starting point because some of the steps lead it away from the starting point but others lead it back. The net distance travelled in a time t from the starting point is measured by the **root-mean square distance**, d, with

$$d^2 = 2Dt \qquad (20)$$

Thus, the net distance increases only as the square root of the time.

Table 10.6 Diffusion coefficients at 25 °C, $D/10^{-5} \, cm^2 s^{-1}$

H_2O in water	2.26
Air in carbon tetrachloride	3.63
CH_3OH in water	1.58
$C_{12}H_{22}O_{11}$ (sucrose) in water	0.522
NH_2CH_2COOH in water	0.673
O_2 in carbon tetrachloride	3.82

Exercise E10.9 The self-diffusion coefficient of H_2O in water is $2.26 \times 10^{-5} \, cm^2 \, s^{-1}$ at 25 °C. How long does it take for an H_2O molecule to travel (a) 1.0 cm, (b) 2.0 cm from its starting point in a sample of unstirred water?

[(a) 6.1 h, (b) 25 h]

Although the net effect of diffusion is very slow in liquids (and much slower in solids), each step takes only a very short time. If it is supposed that each step is of length λ and takes a time τ, then the **Einstein–Smoluchowski equation** states that

$$D = \frac{\lambda^2}{2\tau} \qquad (21)$$

Therefore, the longer the step and the shorter the time it takes for the steps to occur, the greater the diffusion coefficient.

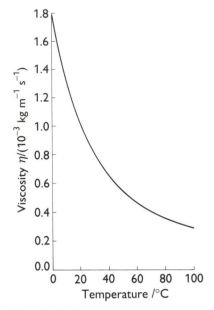

Fig. 10.15 The experimental temperature dependence of the viscosity of water. As the temperature is increased, more molecules are able to escape from the potential wells provided by their neighbours, and so the liquid becomes more fluid.

Exercise E10.10 Suppose an H_2O molecule moves through one molecular diameter (about 200 pm) each time it takes a step in a random walk. What is the time for each step at 25 °C?

[9 ps]

The diffusion coefficient increases with temperature because an increase in temperature enables a molecule to escape more easily from the attractive forces exerted by its neighbours. The speed of the random walk is increased, which corresponds to a decrease in the time τ in the Einstein–Smoluchowski equation. If we suppose that the rate of the random walk follows an Arrhenius temperature dependence with an activation energy E_a, then the diffusion coefficient can be expected to follow the relation

$$D = D_0 e^{-E_a/RT} \qquad (22)$$

The rate at which particles diffuse through a liquid is related to the viscosity, and so the viscosity should show the same temperature dependence as in eqn (22). This decreasing exponential dependence is observed, at least over reasonably small temperature ranges (Fig. 10.15). One problem is that the density of the liquid changes as it is heated, and makes a pronounced contribution to the diffusion coefficient and the viscosity. Thus, the temperature dependence of viscosity at constant volume, when the density is constant, is much less than that at constant pressure. The intermolecular potentials govern the magnitude of E_a, but the problem of calculating it is immensely difficult and still largely unsolved.

Exercise E10.11 Estimate the activation energy for the viscosity of water from the graph in Fig. 10.15, using the viscosities at 40 °C and 80 °C.
[*Hint:* use an equation like eqn (22) to formulate an expression for the logarithm of the ratio of the two viscosities.]

[4 kJ mol^{-1}]

10.7 Liquid crystals

A **mesophase** may arise when molecules have highly anisotropic shapes, such as being long and thin, as in (**5**), or disc-like. When the solid melts, some aspects of the long-range order characteristic of the solid may be retained, and the new phase may be a **liquid crystal**, a substance having liquid-like imperfect long-range order but some crystal-like aspects of short-range order. One type of retained long-range order gives rise to a **smectic phase** (from the Greek word for 'soapy'), in which the molecules align themselves in layers (Fig. 10.16(a)). Other materials, and some smectic liquid crystals at higher temperatures, lack the layered structure but retain a parallel alignment (Fig. 10.16(b)): this mesophase is the **nematic phase** (from the Greek for 'thread'). The strongly anisotropic optical properties of nematic lquid crystals, and their response to electric fields, is the basis of their use as data displays in calculators and watches. In the **cholesteric phase** (from the Greek for 'bile solid') the molecules lie in sheets at angles that change slightly between each sheet. As a result, they form helical structures with a pitch that depends on the temperature. These cholesteric liquid crystals diffract light and, because the pitch depends on temperature, give rise to colours that depend on the temperature.

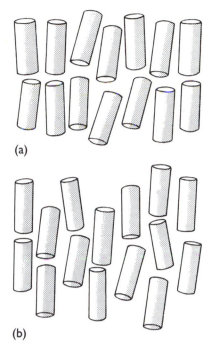

(a)

(b)

Fig. 10.16 The arrangement of molecules in (a) the smectic phase and (b) the nematic phase of a liquid crystal. They both have elements of order, but it is different in different directions.

10.8 Colloids

A **colloid** is a dispersion of small particles of one material in another. In this context, 'small' means something less than about 500 nm in diameter (about the wavelength of light). In general they are aggregates of numerous atoms or molecules, but are too small to be seen with an ordinary optical microscope. They pass through most filter papers, but can be detected by light scattering, sedimentation, and osmosis.

The name given to the colloid depends on the two phases involved. **Sols** are dispersions of solids in liquids (such as clusters of gold atoms in water) or of solids in solids (such as ruby glass, which is a gold-in-glass sol, and achieves its colour by scattering). **Aerosols** are dispersions of liquids in gases (like fog and many sprays) and of solids in gases (such as smoke): the particles are often large enough to be seen with a microscope. **Emulsions** are dispersions of liquids in liquids (such as milk).

A further classification of colloids is as **lyophilic** (solvent attracting) and **lyophobic** (solvent repelling); in the case of water as solvent, the terms **hydrophilic** and **hydrophobic** are used instead. Lyophobic colloids include the metal sols. Lyophilic colloids generally have some chemical similarity to the solvent, such as —OH groups able to form hydrogen bonds. A **gel** is a semirigid mass of a lyophilic sol in which all the dispersion medium has been absorbed by the sol particles.

The preparation of aerosols can be as simple as sneezing (which produces an aerosol). Laboratory and commercial methods make use of several techniques. Material (e.g. quartz) may be ground in the presence of the dispersion medium. Passing a heavy electric current through a cell may lead to the crumbling of an electrode into colloidal particles; arcing between electrodes immersed in the support medium also produces a colloid. Chemical precipitation sometimes results in a colloid. A precipitate (for example, silver iodide) already formed may be converted to a colloid by the addition of **peptizing agent**, a

397

substance that disperses a colloid (for example, potassium iodide, which provides ions that adhere to the colloidal particles). Clays may be peptized by alkalis, the OH^- ion being the active agent.

Emulsions are normally prepared by shaking the two components together, although some kind of **emulsifying agent** has to be used in order to stabilize the product. This emulsifier may be a soap (a salt of a long-chain carboxylic acid), a surfactant, or a lyophilic sol that forms a protective film around the dispersed phase. In milk, which is an emulsion of fats in water, the emulsifying agent is casein, a protein containing phosphate groups. That casein is not completely successful in stabilizing milk is apparent from the formation of cream on the surface: the dispersed fats coalesce into oily droplets which float to the surface. This coalescence may be prevented by ensuring that the emulsion is dispersed very finely initially: violent agitation with ultrasonics brings this dispersal about, the product being 'homogenized' milk.

Aerosols are formed when a spray of liquid is torn apart by a jet of gas. The dispersal is aided if a charge is applied to the liquid, for then the electrostatic repulsions blast the jet apart into droplets. This procedure may also be used to produce emulsions, for the charged liquid phase may be squirted into another liquid.

Colloids are often purified by **dialysis**, a form of osmosis. The aim is to remove much (but not all, for reasons explained later) of the ionic material that may have accompanied their formation. A membrane (e.g. cellulose) is selected which is permeable to solvent and ions, but not to the colloid particles. Dialysis is very slow, and is normally accelerated by applying an electric field and making use of the charge carried by many colloids; the technique is then called **electrodialysis**.

Surface, structure, and stability

The principal feature of colloids is the very great surface area of the dispersed phase in comparison with the same amount of ordinary material. For example, a 1 cm cube of material has a surface area of $6 \, cm^2$, but when it is dispersed as 10^{18} little 10 nm cubes the total surface area is $6 \times 10^6 \, cm^2$ (about the size of a tennis court). This dramatic increase in area means that surface effects are of dominating importance in colloid chemistry.

As a result of their great surface area, colloids are thermodynamically unstable with respect to the bulk: that is, colloids have at thermodynamic tendency to reduce their surface area (like a liquid). Their apparent stability must therefore be a consequence of the kinetics of collapse: colloids are kinetically, not thermodynamically, stable (just as diamond is a kinetically stable solid, because it takes a very long time to convert to graphite). At first sight, though, even the kinetic argument seems to fail: colloidal particles attract each other over large distances, and so there is a long-range force tending to collapse them down into a single blob.

Several factors oppose the long-range dispersion attraction. For example, there may be a protective film at the surface of the colloid particles that stabilizes the interface and cannot be penetrated when two particles touch. Thus the surface atoms of a platinum sol in water react chemically and are turned into $—Pt(OH)_3H_3$, and this layer encases the particle like a shell. A fat can be emulsified by a soap because the long hydrocarbon tails penetrate the oil droplet but the $—CO_2^-$ head groups (or other hydrophilic groups in detergents) surround the surface, form hydrogen bonds with water, and give rise to a shell of negative charge that repels a possible approach from another similarly charged particle.

Micelle formation and the hydrophobic interaction

By a **surfactant** we mean a species that accumulates at the interface of two fluids (one of which may be air) and modifies the properties of the surface. A typical surfactant molecule consists of a long hydrocarbon tail that dissolves in hydrocarbon and other nonpolar materials, and a hydrophilic **head group**, such as a carboxylate group, $—CO_2^-$, that dissolves in a polar solvent (typically water). Soaps, for example, consist of the alkali metal salts of long-chain carboxylic acids, and the surfactant in detergents is typically a long-chain benzenesulfonic acid $(R—C_6H_4SO_3H)$. The mode of action of a surfactant in a detergent, and of soap, is to dissolve in both the aqueous phase and the hydrocarbon phase where their surfaces are in contact, and hence to solubilize the hydrocarbon phase so that it can be washed away (Fig. 10.17).

Surfactant molecules can group together as **micelles**, colloid-sized clusters of molecules, even in the absence of grease droplets, for their hydrophobic tails tend to congregate, and their hydrophilic heads provide protection (Fig. 10.18). Micelles form only above the **critical micelle concentration** (CMC) and above the **Krafft temperature**. Nonionic surfactant molecules may cluster together in swarms of 1000 or more, but ionic species tend to be disrupted by the Coulomb repulsions between head groups and are normally limited to groups of between 10 and 100 molecules. The shapes of the individual micelles vary with concentration. Although spherical micelles do occur, close to the CMC they are, more commonly, flattened spheres, and rod-like at higher concentrations. The interior of a micelle is like a droplet of oil, and magnetic resonance shows that the hydrocarbon tails are mobile, but slightly more restricted than in the bulk.

Micelles are important in industry and biology on account of their solubilizing function: matter can be transported by water after it has been dissolved in their hydrocarbon interiors. For this reason, micellar systems are used as detergents and drug carriers, and for organic synthesis, froth flotation, and petroleum recovery.

The thermodynamics of micelle formation shows that the enthalpy of formation in aqueous systems is probably positive (that is, that they are endothermic) with $\Delta H = 1–2$ kJ per mole of surfactant. That they do form above the CMC indicates that the entropy change accompanying their formation must then be positive (in order for the free

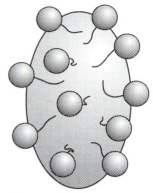

Fig. 10.17 A surfactant molecule in a detergent or soap acts by sinking its hydrophobic hydrocarbon tail into the grease, so leaving its hydrophilic head groups on the surface of the grease where they can interact attractively with the surrounding water.

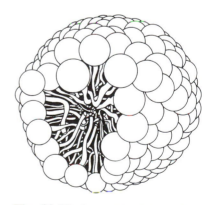

Fig. 10.18 A representation of a spherical micelle. The hydrophilic groups are represented by spheres, and the hydrophobic hydrocarbon chains are represented by the stalks: the latter are mobile.

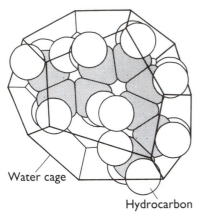

Water cage

Hydrocarbon

Fig. 10.19 When a hydrocarbon molecule is surrounded by water, the water molecules form a clathrate cage. As a result of this acquisition of structure, the entropy of the water decreases, and so the dispersal of the hydrocarbon into water is entropy-opposed; its coalescence is entropy-favoured.

energy of formation to be negative), and measurement suggest a value of about $+140\,\mathrm{J\,K^{-1}\,mol^{-1}}$ at room temperature. That the entropy change is positive even though the molecules are clustering together shows that there must be a contribution to the entropy from the solvent and that its molecules must be more free to move once the solute molecules have herded into small clusters. This is plausible, because each individual solute molecule is held in an organized solvent cage (Fig. 10.19), but once the micelle has formed the solvent molecules need form only a single (admittedly larger) cage. The increase in entropy when hydrophobic groups cluster together and reduce their structural demands on the solvent is the origin of the **hydrophobic interaction** that tends to stabilize groupings of hydrophobic groups in biological macromolecules. The hydrophobic interaction is an example of an ordering process that is stabilized by a tendency toward greater disorder of the solvent.

The electric double layer

Apart from the physical stabilization of colloid particles, a major source of kinetic stability is the existence of an electric charge on their surfaces. On account of this charge, ions of opposite charge tend to cluster nearby.

Two regions of charge must be distinguished. First, there is a fairly immobile layer of ions that stick tightly to the surface of the colloidal particle, and which may include water molecules (if that is the support medium). The radius of the sphere that captures this rigid layer is called the **radius of shear**, and is the major factor determining the mobility of the particles. The electric potential at the radius of shear relative to its value in the distant, bulk medium is called the **zeta potential**, ζ, or **electrokinetic potential**. The charged unit attracts an oppositely charged ionic atmosphere. The inner shell of charge and the outer atmosphere is called the **electric double layer**.

At high concentrations of ions of high charge number, the atmosphere is dense and the potential falls to its bulk value within a short distance. In this case there is little electrostatic repulsion to hinder the close approach of two colloid particles. As a result, **flocculation**, the coagulation of the colloid, readily occurs as a consequence of the van der Waals forces. This is the basis of the empirical **Schultze–Hardy rule**, that hydrophobic colloids are flocculated most efficiently by ions of opposite charge type and high charge number. The Al^{3+} ions in alum are very effective, and are used to induce the congealing of blood. When river water containing colloidal clay flows into the sea, the brine induces coagulation and is a major cause of silting in estuaries.

Metal oxide sols tend to be positively charged whereas sulfur and the noble metals tend to be negatively charged. Naturally occurring macromolecules also acquire a charge when dispersed in water, and an important feature of proteins and other natural macromolecules is that their overall charge depends on the pH of the medium. For instance, in acid environments protons attach to basic groups, and the net charge of the macromolecule is positive; in basic media the net charge is negative as a result of proton loss. At the **isoelectric point** the pH is such that there is no net charge on the macromolecule.

Example Determining the isoelectric point
The speed with which bovine serum albumin (BSA) moves through water under the influence of an electric field was monitored at several values of pH, and the data are listed below. What is the isoelectric point of the protein;

pH	4.20	4.56	5.20	5.65	6.30	7.00
Speed/μm s^{-1}	0.50	0.18	−0.25	−0.65	−0.90	−1.25

Answer The macromolecule is not influenced by the electric field when it is uncharged. Therefore, the isoelectric point is the pH at which it does not migrate in an electric field. We should therefore plot speed against pH and find by interpolation the pH at which the speed is zero. The data are plotted in Fig. 10.20. The speed is zero at pH = 4.8; hence pH = 4.8 is the isoelectric point.

Exercise E10.12 The following data were obtained for another protein:

pH	4.5	5.0	5.5	6.0
Speed/μm s^{-1}	−0.10	−0.20	−0.30	−0.35

Estimate the pH of the isoelectric point.

[4.3]

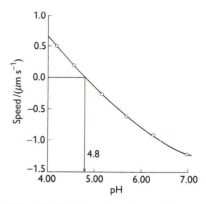

Fig. 10.20 The plot of speed against pH allows the isoelectric point to be detected as the pH at which the speed is zero.

The primary role of the electric double layer is to confer kinetic stability. Colliding colloidal particles break through the double layer and coalesce only if the collision is sufficiently energetic to disrupt the layers of ions and solvating molecules, or if thermal motion has stirred away the surface accumulation of charge. This may happen at high temperatures, which is one reason why sols precipitate when they are heated. The protective role of the double layer is the reason why it is important not to remove all the ions when a colloid is being purified by dialysis, and why proteins coagulate most readily at their isoelectric point.

The presence of charge on colloidal particles and natural macromolecules also permits us to control their motion, such as in dialysis and electrophoresis. Apart from its application to the determination of molar mass, electrophoresis has several analytical and technological applications. One analytical application is to the separation of different macromolecules, and a typical apparatus is illustrated in Fig. 10.21. Technical applications include silent ink-jet printers, the painting of objects by airborne charged paint droplets, and electrophoretic rubber

Fig. 10.21 The layout of a simple electrophoresis apparatus. The sample is introduced into the trough in the gel, and the different components form separated bands under the influence of the potential difference.

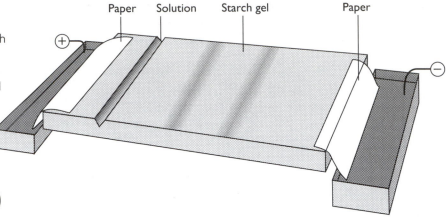

Paper   Solution   Starch gel   Paper

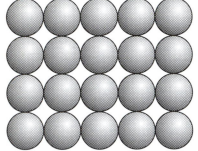

(a)

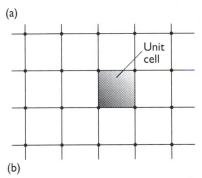

(b)

Fig. 10.22 (a) A crystal consists of a uniform array of atoms, molecules, or ions, as represented by these spheres (in many cases, the components of the crystal are far from spherical, but this diagram illustrates the general idea). (b) The location of each atom, molecule, or ion can be represented by a single point; here (for convenience only), the locations are denoted by a point at the centre of the sphere. The unit cell, which is shown shaded, is the smallest block from which the entire array of points can be constructed without rotating or otherwise modifying the block.

forming by deposition of charged rubber molecules on anodes formed into the shape of the desired product (for example, surgical gloves).

Crystal structure

Now we turn to the structures of solids. We shall concentrate on **crystalline solids**, which are characterized by regular arrays of atoms, molecules, or ions. Early in the history of modern science it was suggested that the regular external form of crystals implied an internal regularity. It is that internal regularity that we explore here.

10.9 Unit cells

The pattern that atoms, ions, or molecules adopt in a crystal is expressed in terms of an array of points which represents the location of the individual species (Fig. 10.22). A **unit cell** of a crystal is the small volume containing an array of these points and which may be used to construct the entire crystal by purely translational displacements, much as a wall may be constructed from bricks (Fig. 10.23). An infinite number of different unit cells can describe the same structure, but it is conventional to choose the cell with sides that have the shortest lengths and which are most nearly perpendicular to each other.

Unit cells are classified into one of seven **crystal systems** according to the symmetry they possess under rotations about different axes. The **cubic system**, for example, has four three-fold rotation axes. (A three-fold rotation is a rotation that returns the unit cell to the same appearance three times during a complete revolution, after rotations through 120°, 240°, and 360°, Fig. 10.24(a).) The four axes lie in a tetrahedral array. The **monoclinic system** has one two-fold axis. (A

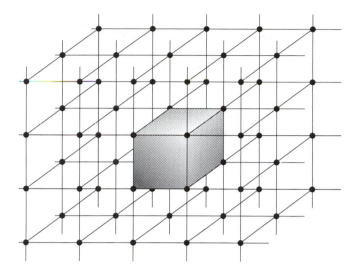

Fig. 10.23 A unit cell, here shown in three dimensions, is like a brick used to construct a wall. Once again, only pure translations are allowed in the construction of the crystal. (Some bonding patterns for actual walls use rotations of bricks, so for these a single brick is not a unit cell.)

Table 10.7 The essential symmetries of the seven crystal systems

The systems	Essential symmetries
Triclinic (1)	None
Monoclinic (2)	One two-fold axis
Orthorhombic (3)	Three perpendicular two-fold axes
Rhombohedral (4)	One three-fold axis
Tetragonal (5)	One four-fold axis
Hexagonal (6)	One six-fold axis
Cubic (7)	Four three-fold axes in a tetrahedral arrangement

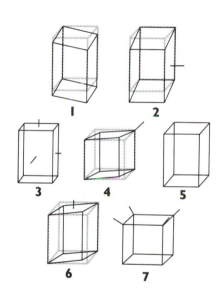

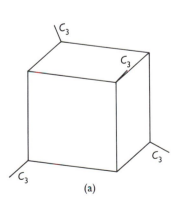

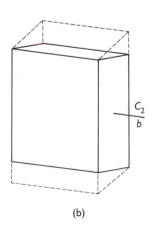

Fig. 10.24 (a) A unit cell belonging to the cubic system has four three-fold axes (denoted C_3) arranged tetrahedrally. (b) A unit cell belonging to the monoclinic system has one two-fold (denoted C_2) axis (along b).

two-fold rotation is one that leaves the cell apparently unchanged twice during a complete revolution, after rotations through 180° and 360°, Fig. 10.24(b).) The **essential symmetries**, the elements that must be present for the unit cell to belong to a particular system, are listed in Table 10.7.

The points that denote the positions of atoms and ions in unit cells may take different positions inside the unit cells, so each crystal system can occur in a number of different varieties. For example, points may occur only at the corners of the unit cells, or they may also occur on

Fig. 10.25 The fourteen Bravais lattices. The letter P denotes a primitive unit cell, I a body-centred unit cell, F a face-centred unit cell, and C (or A or B) a cell with lattice points on two opposite faces.

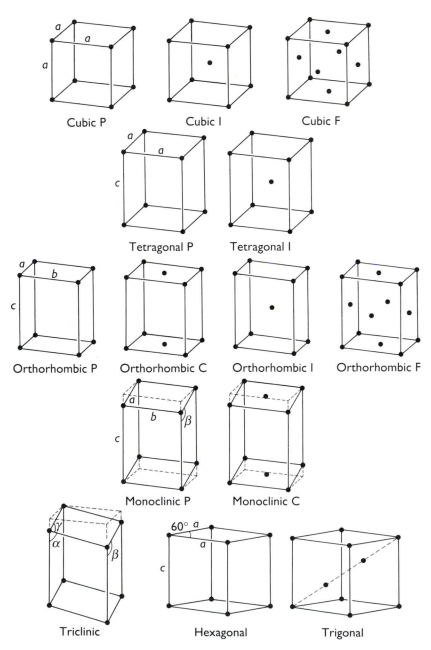

the faces and in the body of the cell without destroying the cell's essential symmetry. These different possibilities give rise to fourteen distinct types of unit cell, which are called **Bravais lattices** (Fig. 10.25). Unit cells with points only at the corners are called **primitive** and denoted P. A **body-centred unit cell** (I) also has a point at its centre. A **face-centred unit cell** (F) has points at its corners and also at the centres of its six faces. A **side-centred unit cell** (A, B, or C, depending on which faces are occupied) has points at its corners and at the centres of two opposite faces.

10.10 The identification of crystal planes

The spacing of the points in a lattice is an important quantitative aspect of a crystal's structure. Because two-dimensional arrays of points are easier to visualize than three-dimensional arrays, we shall introduce the concepts by referring to two-dimensional structures initially, and then extend the conclusions by analogy to three dimensions.

Consider the two-dimensional rectangular array of points formed from a unit cell of sides a, b (Fig. 10.26). We can distinguish the four sets of planes in the illustration by the distances along the axes at which the planes intersect the axes. One way of labelling the planes would therefore be to denote each set by the smallest intersection distances. For example, we could denote the four sets in the illustration as $(1a, 1b)$, $(3a, 2b)$, $(-1a, 1b)$, and $(\infty a, 1b)$. If, however, we agreed always to quote distances along the axes as multiples of the lengths of the unit cell, then we could label the planes more simply as $(1, 1)$, $(3, 2)$, $(-1, 1)$, and $(\infty, 1)$. If the array in Fig. 10.26 is the top view of a three-dimensional rectangular lattice in which the unit cell

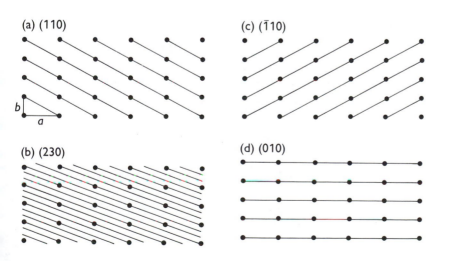

(a) (1̄10)

(b) (230)

(c) (1̄10)

(d) (010)

Fig. 10.26 Some of the planes that can be drawn through the points of the space lattice and their corresponding Miller indices (*hkl*).

405

Fig. 10.27 Some representative planes in three dimensions and their Miller indices. Note that a zero indicates that a plane is parallel to the corresponding axis, and that the indexing may also be used for unit cells with nonorthogonal axes.

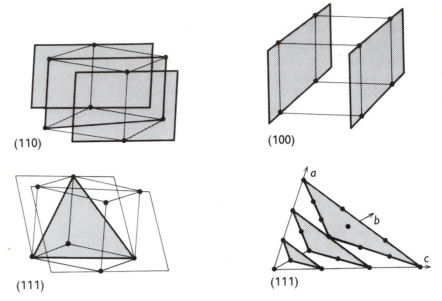

(110) (100)

(111) (111)

has a length c in the z direction, all four sets of planes intersect the z axis at infinity, and so the full labels of the sets of planes of lattice points would be $(1, 1, \infty)$, $(3, 2, \infty)$, $(-1, 1, \infty)$, and $(\infty, 1, \infty)$.

The presence of ∞ in the labels is inconvenient, and can be eliminated by taking the reciprocals of the labels (this step turns out to have further advantages, as we shall see). The resulting **Miller indices** are the reciprocals of the numbers in the parentheses with fractions cleared. For example, the $(1, 1, \infty)$ planes in Fig. 10.26(a) are the (110) planes in the Miller notation. Similarly, the $(3, 2, \infty)$ planes in Fig. 10.26b become first $(\frac{1}{3}, \frac{1}{2}, 0)$ and then $(2, 3, 0)$ when fractions are cleared, and so they are referred to as the (230) planes. Negative indices are written with a bar over the number, and Fig. 10.26(c) shows the $(\bar{1}10)$ planes. Figure 10.27 shows some planes in three dimensions, including an example of an array with axes that are not mutually perpendicular.

Exercise E10.13 A representative member of a set of planes in a crystal intersects the axes at $3a$, $2b$, and $2c$; what are the Miller indices of the planes?

[(233)]

It is helpful to keep in mind the fact, as illustrated in Fig. 10.26, that the smaller the value of h in the Miller index (hkl), the more nearly parallel the plane is to the a axis. The same is true of k and the b axis and l and the c axis. When $h = 0$, the planes intersect the a axis at infinity, and so the $(0kl)$ planes are parallel to the a axis. Similarly, the $(h0l)$ planes are parallel to b and the $(hk0)$ planes are parallel to c.

The Miller indices are very useful for calculating the separation of planes. The expression for a rectangular array built from a unit cell of sides of lengths a, b, and c is

$$\frac{1}{d^2} = \frac{h^2}{a^2} + \frac{k^2}{b^2} + \frac{l^2}{c^2} \tag{23}$$

Example Using the Miller indices
Calculate the separation of (a) the (123) planes and (b) the (246) planes of an orthorhombic cell with $a = 0.82$ nm, $b = 0.94$ nm, and $c = 0.75$ nm.
Answer For the first part, we simply substitute the information into eqn (23):

$$\frac{1}{d^2} = \frac{1^2}{(0.82 \text{ nm})^2} + \frac{2^2}{(0.94 \text{ nm})^2} + \frac{3^2}{(0.75 \text{ nm})^2} = 22 \text{ nm}^{-2}$$

Hence, $d = 0.21$ nm. For the second part, instead of repeating the calculation, we note that if all three Miller indices are multiplied by 2,

$$\frac{1}{d'^2} = \frac{(2h)^2}{a^2} + \frac{(2k)^2}{b^2} + \frac{(2l)^2}{c^2} = \frac{2^2}{d^2}$$

where d is the separation given by eqn (23). That is,

$$d' = \tfrac{1}{2}d$$

Hence, in the present case, $d' = 0.11$ nm. In general, increasing the indices uniformly by a factor n decreases the separation of the planes by the same factor n.

Exercise E10.14 Calculate the separation of the (133) and (399) planes in the same lattice.
[0.20 nm, 0.065 nm]

10.11 The determination of structure

One of the most important techniques for the determination of the structures of crystals is **X-ray diffraction**. In its simplest form, the technique can be used to identify the lattice type and the separation of the planes of lattice points (and hence the distance between the nuclei of atoms and ions). In its most sophisticated version, X-ray diffraction provides detailed information about the location of all the atoms in molecules as complex as proteins and nucleic acids. The current considerable success of modern molecular biology has stemmed from X-ray diffraction techniques that have grown in sensitivity and scope as

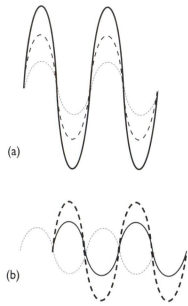

Fig. 10.28 When two waves (drawn as broken lines) are in the same region of space they interfere. Depending on their relative phase, they may interfere either (a) constructively, to give an enhanced amplitude, or (b) destructively, to give a smaller amplitude.

computing techniques have become more powerful. Here we shall confine attention to the principles of the technique, and illustrate how it may be used to determine the spacing of atoms in a crystal.

Diffraction

A characteristic property of waves is that they interfere with one another, giving a greater amplitude where their displacements add and a smaller amplitude where their displacements subtract (Fig. 10.28). Because the intensity of electromagnetic radiation is proportional to the square of the amplitude of the waves, the regions of constructive and destructive interference show up as regions of enhanced and diminished intensities. The phenomenon of **diffraction** is the interference that is caused by an object in the path of waves, and the pattern of varying intensity that results is called the **diffraction pattern** (Fig. 10.29; the phenomenon was first mentioned in Section 8.3 in connection with the wave properties of electrons). Diffraction occurs when the dimensions of the diffracting object are comparable to the wavelength of the radiation: sound waves (with wavelengths of the order of 1 m) are diffracted by macroscopic objects about 1 m across, and light waves (with wavelengths of the order of 500 nm) are diffracted by narrow slits of widths of the order of 10^2 nm.

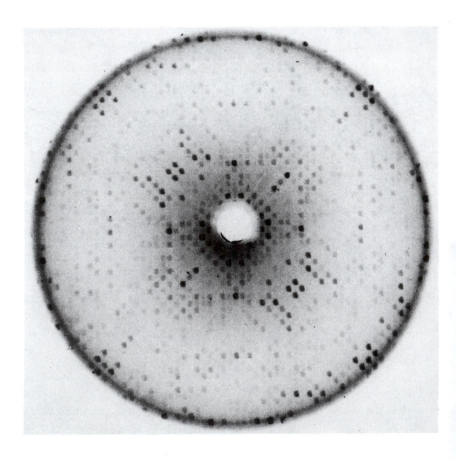

Fig. 10.29 A typical diffraction pattern obtained in a version of the X-ray diffraction technique. The black dots are the reflections, the points of maximum constructive interference, that are used to determine the structure of the crystal. The sample is ribulose bis-phosphate carboxylase. (Photograph supplied by Dr M. Adams, University of Oxford.)

X-rays have wavelengths comparable to bond lengths in molecules and the spacing of atoms in crystals (about 100 pm), and so they are diffracted by them. By analysing the diffraction pattern, it is possible to draw up a detailed picture of the location of atoms. Electrons moving at about 2×10^4 km s^{-1} (after acceleration through about 4 kV) have wavelengths of about 20 pm, and may also be diffracted by molecules. Neutrons generated in a nuclear reactor, and then slowed to thermal velocities, have similar wavelengths and may also be used for diffraction studies. We shall confine our attention to X-ray diffraction, the most widely used of the three techniques.

Example Estimating the wavelength of accelerated electrons
Calculate the wavelength of electrons that have been accelerated from rest through a potential difference of 4.0 kV.
Answer The key to the calculation is the de Broglie relation $\lambda = h/p$ between the wavelength λ and the linear momentum p. After acceleration through a potential difference V an electron has an energy eV, which is stored entirely as kinetic energy. Because the kinetic energy of a particle of mass m_e and momentum p is equal to $p^2/2m_e$, the two expressions for the energy may be equated, and from

$$\frac{p^2}{2m_e} = eV$$

we can obtain first an expression for the momentum

$$p = \sqrt{2m_e eV}$$

and then, from the de Broglie relation, an expression for the wavelength of the electron:

$$\lambda = \frac{h}{\sqrt{2m_e eV}}$$

Substitution of the data and the fundamental constants (from the tables inside the front cover) then gives

$$\lambda = \frac{6.626 \times 10^{-34} \text{ J s}}{[2 \times (9.109 \times 10^{-31} \text{ kg}) \times (1.602 \times 10^{-19} \text{ C}) \times (4.0 \times 10^3 \text{ V})]^{1/2}}$$
$$= 1.9 \times 10^{-11} \text{ m, or 19 pm}$$

We have used 1 C V = 1 J and 1 J = 1 kg m^2 s^{-2}.

Exercise E10.15 Calculate the wavelength of a neutron that is travelling with a kinetic energy kT, where k is Boltzmann's constant in a reactor at a temperature of 500 °C.
[110 pm]

The short wavelength electromagnetic radiation known as X-rays are produced by bombarding a metal with high-energy electrons. The electrons decelerate as they plunge into the metal and generate radiation with a continuous range of wavelengths. This radiation is called **bremsstrahlung** (*Bremse* is German for brake, *Strahlung* for ray). Superimposed on the continuum are a few high-intensity, sharp peaks. These peaks arise from the interaction of the incoming electrons with the electrons in the inner shells of the atoms. A collision expels an electron, and an electron of higher energy drops into the vacancy, emitting the excess energy as an X-ray photon. An example of the process is the expulsion of an electron from the K shell (the shell with $n = 1$) of a copper atom, followed by the transition of an outer electron into the vacancy, which gives rise to copper's K_α radiation of wavelength 154 pm.

In 1923, 17 years after X-rays had been discovered by Wilhelm Röntgen, the German physicist Max von Laue suggested that they might be diffracted when passed through a crystal, for their wavelengths are comparable to the separation of atoms. Laue's suggestion was confirmed almost immediately by Walter Friedrich and Paul Knipping, and has grown since then into a technique of extraordinary power.

The Bragg law

The earliest approach to the analysis of diffraction patterns produced by crystals was to regard a plane of atoms as a semitransparent mirror, and to model a crystal as stacks of reflecting planes of separation d (Fig. 10.30). The model makes it easy to calculate the angle the crystal must make to the incoming beam of X-rays for constructive interference to occur. It has also given rise to the name **reflection** to denote an intense spot arising from constructive interference.

The path-length difference of the two rays shown in the illustration is

$$AB + BC = 2d \sin \theta$$

where θ is the **glancing angle**. For many glancing angles the path-length difference is not an integral number of wavelengths, and the waves interfere destructively. However, when the path-length difference is an integral number of wavelengths ($AB + BC = n\lambda$, with n an integer), the reflected waves are in phase and interfere constructively. It follows that a reflection should be observed when the glancing angle satisfies the **Bragg law**

$$n\lambda = 2d \sin \theta$$

Reflections with $n = 2, 3, \ldots$ are called second order, third order, and so on; they correspond to path-length differences of 2λ, 3λ, and so on. In modern work it is normal to absorb the n into d, to write the Bragg law as

$$\lambda = 2d \sin \theta \tag{24}$$

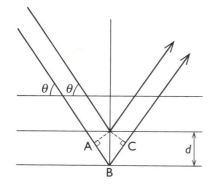

Fig. 10.30 The derivation of the Bragg law treats each lattice plane as reflecting the incident radiation. The path lengths differ by AB + BC, which depends on the glancing angle θ. Constructive interference (a 'reflection') occurs when AB + BC is equal to an integral number of wavelengths.

and to regard the nth order reflection as arising from the (nh, nk, nl) planes of the crystal (see the example on p. 407).

The primary use of the Bragg law is in the determination of the spacing between the layers of atoms, for once the angle θ corresponding to a reflection has been determined, d may readily be calculated.

Example Using the Bragg law

A reflection from the (111) planes of a cubic crystal was observed at a glancing angle of 11.2° when Cu K_α X-rays of wavelength 154 pm were used. What is the length of the side of the unit cell?

Answer According to the Bragg Law, the (111) planes responsible for the diffraction have separation

$$d = \frac{\lambda}{2 \sin \theta} = \frac{154 \text{ pm}}{2 \sin 11.2°} = 396 \text{ pm}$$

The separation of the (111) planes of a cubic lattice of side a is given by eqn (23) as

$$\frac{1}{d^2} = \frac{1^2}{a^2} + \frac{1^2}{a^2} + \frac{1^2}{a^2} = \frac{3}{a^2}$$

Therefore,

$$a = 3^{1/2} \times d = 686 \text{ pm}$$

Exercise E10.16 Calculate the angle at which the same lattice will give a reflection from the (123) planes.

[24.8°]

The powder method

Laue's original method consisted of passing a broad-band beam of X-rays into a single crystal, and recording the diffraction pattern photographically. The idea behind the approach was that a crystal might not be suitably orientated to act as a diffraction grating for a single wavelength, but whatever its orientation the Bragg law would be satisfied for at least one of the wavelengths if a range of wavelengths was present in the beam. An alternative technique was developed by Peter Debye and Paul Scherrer and independently by Albert Hull. They used monochromatic radiation and a powdered sample.

When the sample is a powder, some of the crystallites will always be orientated so as to satisfy the Bragg condition. For example, some of them will be oriented so that their (111) planes, of spacing d, give rise to a diffracted intensity at the glancing angle θ (Fig. 10.31). The (111) planes of some other crystallites may be at the angle θ to the beam, but at an arbitrary angle about the line of its approach. It follows that

Fig. 10.31 The same set of planes in two microcrystallites with different orientations around the direction of the incident beam give diffracted rays that lie on a cone. The full powder diffraction pattern is formed by cones corresponding to reflections from all the sets of (*hkl*) planes that satisfy the Bragg law. (A reflection at a glancing angle θ gives rise to a reflection at an angle 2θ to the direction of the incident beam; see inset.)

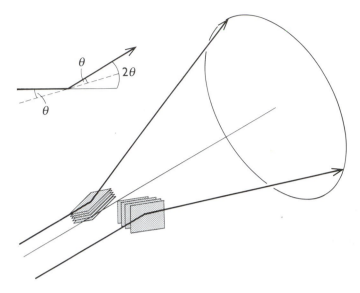

the diffracted beams will lie on a cone around the incident beam of half-angle 2θ. Other crystallites will be orientated with different planes satisfying the Bragg law and give rise to a cone of diffracted intensity with a different half-angle. In principle, each set of (*hkl*) planes gives rise to a different diffraction cone because some of the randomly orientated crystallites will have the correct angle to diffract the incident beam.

The **Debye–Scherrer method**, which makes use of this approach, is illustrated in Fig. 10.32. The sample is in a capillary tube which is rotated to ensure that the crystallites are randomly orientated. The diffraction cones are photographed as arcs of circles where they cut the strip of film, and some typical patterns are shown in Fig. 10.33. In modern diffractometers the sample is spread on a flat plate and the diffraction pattern is monitored electronically. The major application is now for qualitative analysis because the diffraction pattern is a kind of fingerprint and may be recognizable. The technique is also used for

Fig. 10.32 In the Debye–Scherrer method, a monochromatic X-ray beam is diffracted by a powder sample. The crystallites give rise to cones of intensity which are detected by a photographic film wrapped round the circumferencee of the camera.

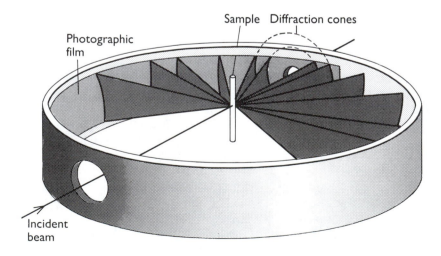

the initial determination of the dimensions and symmetries of unit cells.

Modern X-ray diffraction, which utilizes an **X-ray diffractometer** (Fig. 10.34), is now a highly sophisticated technique. By far the most detailed information comes from developments of the techniques pioneered by the Braggs (William and his son Lawrence), in which a single crystal is employed as the diffracting grating and a monochromatic beam of X-rays is used to generate the diffraction pattern. The single crystal (which may be only a fraction of a millimetre in length) is rotated relative to the beam, and the diffraction pattern is monitored and recorded electronically for each crystal orientation. The analysis of the diffraction pattern is then carried out on a computer that is an intrinsic part of the diffractometer, and the results presented as a detailed structural map of the unit cell of the crystal, showing the relative locations of all the atoms it contains.

10.12 Information from X-ray analysis

The bonding within a solid, the origin of its cohesion, may be of various kinds. Simplest of all (in principle) is the bonding in **metallic solids**, in which electrons are delocalized over arrays of identical cations and bind the whole together into a rigid but malleable structure. The origin of this type of bonding in terms of molecular orbitals was described in Section 9.8. In many cases, because the delocalized molecular orbitals can accommodate bonding patterns with very little directional character, the crystal structures of metals are determined largely by the manner in which spherical metal cations can pack together into an orderly array. In **covalent solids**, covalent bonds in a definite spatial orientation link the atoms in a network extending

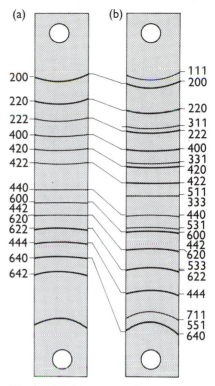

Fig. 10.33 X-ray powder photographs of (a) KCl, (b) NaCl. The numbers are the Miller indices of the sets of planes responsible for the diffraction. Note that the film covers one half of the circumference of the powder camera, the lines at the top corresponding to the smallest diffraction angles.

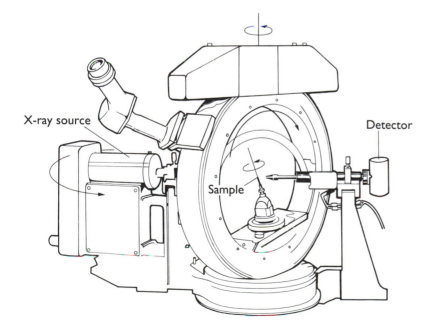

Fig. 10.34 A four-circle diffractometer. The setting of the orientations of the components is controlled by computer, each reflection is monitored in turn, and their intensities are recorded.

413

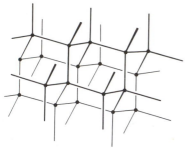

Fig. 10.35 A fragment of the structure of diamond. Each C atom is tetrahedrally bonded to four neighbours. This framework-like structure results in a rigid crystal with a high thermal conductivity.

through the crystal. The stereochemical demands of valence now override the geometrical problem of packing spheres together, and elaborate and extensive structures may be formed. A famous example of a covalent solid is diamond (Fig. 10.35), in which each sp^3-hybridized carbon atom is bonded tetrahedrally by σ bonds to its four neighbours. Covalent solids are often hard and unreactive. **Molecular solids**, which are the subject of the overwhelming majority of modern structural determinations, are bonded together by the van der Waals interactions described earlier in the chapter.

In each case, the observed crystal structure is nature's solution to the problem of condensing objects of various shapes into an aggregate of minimum energy (actually, for temperatures above zero, of minimum Gibbs free energy). Exactly the same is true of the structures of proteins, nucleic acids, and other biological macromolecules, but for them the packing pattern is generally more complex because the individual units (the peptide groups in proteins) are more varied, they are held together by chemical bonds (the peptide links) as well as van der Waals forces, and the surrounding medium (typically water and the ions it contains) may play an important structural role.

10.13 The packing of identical spheres: metal crystals

Most metallic elements crystallize in one of three simple forms, two of which can be explained in terms of organizing spheres into the closest possible packing. In such **close-packed structures** the spheres representing the atoms are packed together with least waste of space and each sphere has the greatest possible number of nearest neighbours.

Close packing

A **close-packed layer** of identical spheres, one with maximum utilization of space, can be formed as shown in Fig. 10.36(a). A second close-packed layer can be formed by placing spheres in the depressions of the first layer (Fig. 10.36(b)). The third layer may be added in either of two ways, both of which result in the same degree of close packing. In one, the spheres are placed so that they reproduce the first layer (Fig. 10.36(c)), to give an ABA pattern of layers. Alternatively, the spheres may be placed over the gaps in the first layer (Fig. 10.36(d)), so giving an ABC pattern.

Two structures are formed if the two stacking patterns are repeated in the vertical direction. If the ABA pattern is repeated, to give the sequence of layers ABABAB . . . , the spheres are **hexagonally close-packed** (hcp); the name reflects the symmetry of the unit cell (Fig. 10.37). Metals with hcp structures include beryllium, cadmium, cobalt, manganese, titanium, and zinc. Solid helium (which forms only under pressure) also adopts this arrangement of atoms. Alternatively, if the ABC pattern is repeated to give the sequence of layers ABCABC . . . , the spheres are **cubic close packed** (ccp); once again, the name reflects

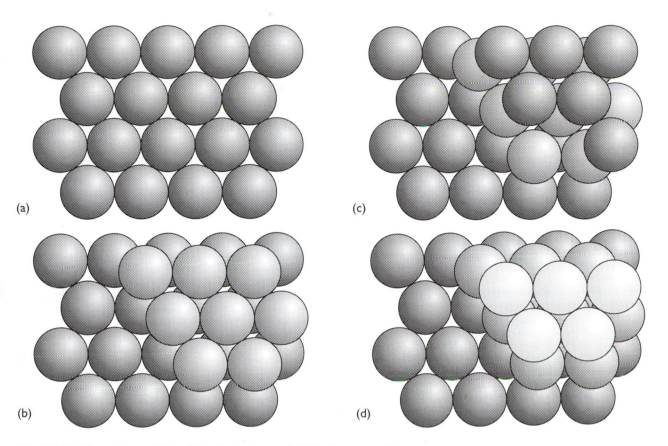

(a)

(b)

(c)

(d)

Fig. 10.36 The close packing of identical spheres. (a) The first layer of close-packed spheres. (b) The second layer of close-packed spheres occupies the dips of the first layer. The two layers are the AB component of the structure. (c) The third layer of close-packed spheres might occupy the dips lying directly above the spheres in the first layer, resulting in an ABA structure. (d) Alternatively, the third layer might lie in the dips that are not above the spheres in the first layer, resulting in an ABC structure.

the symmetry of the unit cell (Fig. 10.38). Metals with this structure include silver, aluminium, gold, calcium, copper, nickel, lead, and platinum. The noble gases other than helium also adopt a ccp structure.

The compactness of the ccp and hcp structures is indicated by their **coordination number**, the number of atoms immediately surrounding any selected atom, which is 12 in both cases. Another measure of their compactness is the **packing fraction**, the fraction of space occupied by the spheres, which is 0.740. That is, in a close-packed solid of identical spheres, 74.0 per cent of the available space is occupied and only 26.0 per cent of the total volume is empty space. The fact that many metals are close packed accounts for one of their common characteristics, their high densities.

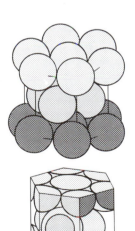

Fig. 10.37 A hexagonal close-packed structure. The tinting of the spheres (denoting the three layers of atoms) is the same as in Fig. 10.36.

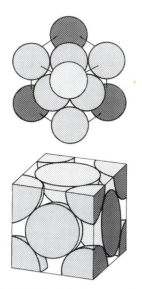

Fig. 10.38 A cubic close-packed structure. The tinting of the spheres is the same as in Fig. 10.36.

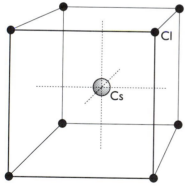

Fig. 10.39 The caesium-chloride structure consists of two interpenetrating simple cubic lattices, one of cations and the other of anions, so that each cube of ions of one kind has a counter-ion at its centre. This illustration shows a single unit cell with a Cs^+ ion at the centre. By imagining eight of these unit cells stacked together to form a bigger cube, it should be possible to imagine an alternative form of the unit cell with Cs^+ at the corners and a Cl^- ion at the centre.

416

Less closely packed structures

A number of common metals adopt structures that are not close packed, which suggests that directional covalent bonding between neighbouring atoms is beginning to influence the structure and impose a specific geometrical arrangement. One such arrangement results in a **body-centred cubic** (bcc) lattice, with one sphere at the centre of a cube formed by eight others. The bcc structure is adopted by a number of common metals, including barium, caesium, chromium, iron, potassium, and tungsten. The coordination number of a bcc lattice is only 8 and its packing fraction is only 0.68, showing that only about two-thirds of the available space is actually occupied.

Exercise E10.17 What is the coordination number of a primitive cubic lattice in which there is a lattice point at each corner of a cube?

[6]

10.14 Ionic crystals

When the structures of ionic crystals are modelled by stacks of spheres we must allow for the fact that the two or more types of ion present in the compound have different radii (generally with the cations smaller than the anions) and different charges. The coordination number of ionic lattice is the number of nearest neighbours of opposite charge, and cations and anions may have different environments in the same crystal.

Even if, by chance, the ions have the same size, the problem of ensuring that the unit cells are electrically neutral makes it impossible to achieve 12-coordinate close-packed structures (which is a reason why ionic solids are generally less dense than metals). The closest packing that can be achieved is the 8-coordination of the **caesium-chloride structure** (Fig. 10.39) in which each cation is surrounded by eight anions and each anion is surrounded by eight cations. In the caesium-chloride structure, an ion of one charge occupies the centre of a cubic unit cell with eight counter ions at its corners. The caesium-chloride structure is adopted by caesium chloride itself and by calcium sulfide, caesium cyanide (with some distortion), and one form of brass (CuZn).

When the radii of the ions differ by more than in caesium chloride, even 8-coordinate packing cannot be achieved. One common structure adopted is the 6-coordinated **rock-salt structure** (Fig. 10.40) typified by sodium chloride (rock salt is a mineral form of sodium chloride) in which each cation is surrounded by six anions and each anion is surrounded by six cations. The rock-salt structure can be interpreted as the interpenetration of two slightly expanded fcc lattices, one of cations and the other of anions. It is the structure of sodium chloride itself and of several other compounds of formula MX, including potassium bromide, silver chloride, and magnesium oxide.

The switch from the adoption of the caesium-chloride structure to the adoption of the rock-salt structure occurs (in a number of examples) in accord with the **radius-ratio rule**, which is based on the value of the radius ratio

$$\rho = \frac{r_{smaller}}{r_{larger}} \qquad (25)$$

The two radii are those of the larger and smaller ions in the crystal. The radius-ratio rule states that the caesium-chloride structure should be expected when

$$\rho > \sqrt{3} - 1 = 0.732$$

and thus the rock-salt structure should be expected when

$$\sqrt{2} - 1 = 0.414 < \rho < 0.732$$

For $\rho < 0.414$, when the two types of ions have markedly different radii (like water melons and grapefruit), the most efficient packing leads to 4-coordination of the type exhibited by the sphalerite (or zinc blende) form of zinc sulfide, ZnS. The radius-ratio rule, which is based on geometrical considerations of sphere packing, is moderately well supported by observation. The deviation of a structure from the prediction is often taken to be an indication of a shift from ionic towards covalent bonding.

The ionic radii that are used to calculate ρ, and wherever else it is important to know the sizes of ions, are derived from the internuclear separation of adjacent ions in a crystal. However, it is necessary to apportion the total distance between the two ions by defining the radius of one ion and reporting all others on that basis. One scale that is widely used is based on the value 140 pm for the radius of the O^{2-} ion (Table 10.8). Other scales are also available (such as one based on

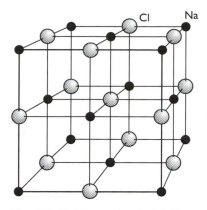

Fig. 10.40 The rock-salt (NaCl) structure consists of two mutually interpenetrating slightly expanded face-centred cubic lattices. Each ion is surrounded by six ions of opposite charge.

Table 10.8   Ionic radii, r/pm

Li⁺	Be²⁺	B³⁺	N³⁻	O²⁻	F⁻
59	27	12	171	140	133
Na⁺	**Mg²⁺**	**Al³⁺**	**P³⁻**	**S²⁻**	**Cl⁻**
102	72	53	212	184	181
K⁺	**Ca²⁺**	**Ga³⁺**	**As³⁻**	**Se²⁻**	**Br⁻**
138	100	62	222	198	196
Rb⁺	**Sr²⁺**				**I⁻**
149	116				220
Cs⁺	**Ba²⁺**				
170	136				

F^- for discussing halides), and it is essential not to mix values from different scales. Because ionic radii are so arbitrary, predictions based on them (such as with the radius-ratio rule) must be viewed cautiously.

Exercise 10.18 Is sodium iodide likely to have a rock-salt or caesium-chloride structure?

[rock salt]

Natural biopolymers

Natural macromolecules need a precisely maintained conformation in order to function. The achievement of a specfic conformation is the major remaining problem in protein synthesis, for although primary structures can be built, the product is inactive because the secondary structure cannot yet be produced. The conformation is sustained by a variety of intermolecular forces of the kind we have encountered in this chapter, including hydrogen bonding, the hydrophobic effect, and dipolar and dispersion interactions.

10.15 The secondary structure

The **primary structure** of a biopolymer is the sequence of its monomer units. For polypeptides, which we consider here, the primary structure is an ordered list of the amino acid residues. The **secondary structure** of a polypeptide is the spatial arrangement of the polypeptide chain under the influence of hydrogen bonding between the various peptide residues (the amino acid groups).

The origin of the secondary structures of proteins is found in the rules formulated by Linus Pauling and Robert Corey. The essential feature is the stabilization of structures by hydrogen bonds involving the —CO—NH— **peptide link**. The latter can act both as a donor of the H atom (the NH part of the link) and as an acceptor (the CO part). The **Corey–Pauling rules** are then as follows:

1. The atoms of the peptide link lie in a plane (Fig. 10.41).

2. The N, H, and O atoms of a hydrogen bond lie in a straight line (with displacements of H tolerated up to not more than 30° from the N—O bond).

3. All NH and CO groups are engaged in bonding.

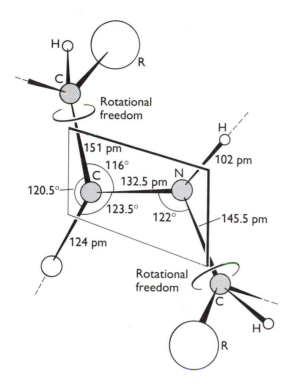

Fig. 10.41 The dimensions that characterize the peptide link. The C—CO—NH—C atoms define a plane (the C—N bond has partial double-bond character), but there is rotational freedom around the C—CO and N—C bonds.

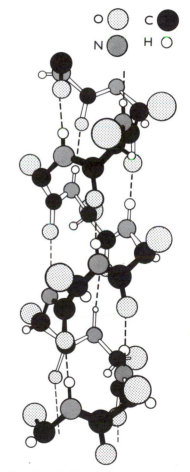

These rules are satisfied by two structures. One, in which hydrogen bonding occurs between peptide links of the same chain, is the *α* **helix**. The other, in which hydrogen bonding links different chains, is the *β*-**pleated sheet**; this form is the secondary structure of the protein fibroin, the constituent of silk.

The *α* helix is illustrated in Fig. 10.42. Each turn of the helix contains 3.6 amino acid residues, and so the period of the helix corresponds to 5 turns (18 residues). The pitch of a single turn is 544 pm. The N—H · · · O bonds lie parallel to the axis and link every fifth group (so residue *i* is linked to residues *i* − 4 and *i* + 4). There is freedom for the helix to be arranged as either a right- or a left-handed screw, but the overwhelming majority of natural polypeptides are right-handed on account of the preponderance of the L-configuration of the naturally occurring amino acids, as we explain below. The reason for their preponderance is uncertain, but it may be related to the symmetries of fundamental particles and the nonconservation of parity (the fact that this universe behaves differently from its hypothetical mirror image).

The stabilities of different polypeptide geometries can be investigated by calculating the total potential energy of all the interactions between nonbonded atoms and looking for a minimum. It turns out, in agreement with experience, that a right-handed *α* helix of L-amino acids is marginally more stable than a left-handed helix of the same acids.

The geometry of the chain can be specified by two angles, ϕ (the torsional angle for the N—C bond) and ψ (the torsional angle for the

Fig. 10.42 The polypeptide *α* helix. There are 3.6 residues per turn, and a translation along the helix of 150 pm per residue, giving a pitch of 544 pm. The diameter (ignoring side chains) is about 600 pm.

Fig. 10.43 The definition of the torsional angles, ψ and ϕ, between two peptide units. In this case (an α-L-polypeptide) the chain has been drawn in its all-*trans* form, with $\psi = \phi = 180°$.

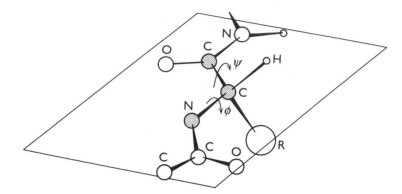

C—C bond). The illustration in Fig. 10.43 defines these angles.† It also shows the all-*trans* form of the chain, in which all ϕ and ψ are 180°. A helix is obtained when all the ϕ are equal and when all the ψ are equal. For a right-handed α helix, all $\phi = -57°$ and all $\psi = -47°$. For a left-handed α helix, both angles are positive. Because only two angles are needed to specify the conformation of a helix, and they range from $-180°$ to $+180°$, the potential energy of the entire molecule can be represented by a point on a plane in which one axis represents ϕ and the other ψ.

The potential energy of a given conformation (ϕ, ψ) can be calculated using the expressions developed in Section 10.2. For example, the interaction energy of two atoms separated by a distance r (which we know once ϕ and ψ are specified) can be given the Lennard–Jones formula, eqn (17). If the partial charges on the atoms (arising from ionic character in the bonds) are known, we can include a Coulomb contribution of the form $1/r$. This is sometimes done by ascribing charges $-0.28e$ and $+0.28e$ to N and H respectively, and $-0.39e$ and $+0.39e$ to O and C respectively. There is also a torsional contribution arising from the barrier to internal rotation of one bond relative to another and which is normally expressed as

$$V = A(1 + \cos 3\phi) + B(1 + \cos 3\psi)$$

A and B are constants of the order of $1\ \mathrm{kJ\ mol^{-1}}$.

The potential energy contours for helical forms of polypeptide chains formed from the nonchiral amino acid glycine (R = H) and alanine (R = CH$_3$) are shown in Fig. 10.44(a) and 10.44(b), respectively. They were computed by summing all the contributions described above for each choice of angles, and then plotting contours of equal potential energy. The glycine map is symmetrical, with minima of equal depth at $\phi = -80°$, $\psi = +90°$ and at $\phi = +80°$, $\psi = -90°$. In contrast, the map for L-alanine is unsymmetrical, and there are three distinct low-energy conformations (marked I, II, III). The minima of regions I and II lie close to the angles typical of right- and left-handed α helices, but the former has a lower minimum, which is consistent

† A positive angle means that the front atom must be rotated clockwise to bring it into an eclipsed position relative to the rear atom.

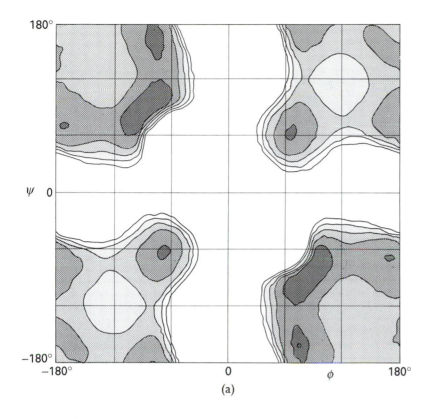

(a)

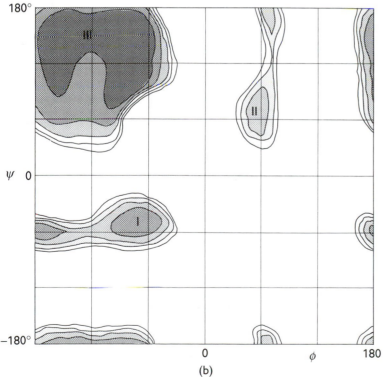

(b)

Fig. 10.44 Energy contour diagrams for (a) a glycyl residue and (b) an alanyl residue of a polypeptide chain. The darker the shading the lower the potential energy. The glycine diagram is symmetrical, but regions I and II in the alanine diagram, which correspond to right- and left-handed helices, are unsymmetrical, and the minimum in region I lies lower than that in region II. (After D. A. Brant and P. J. Flory (1967), *J. Mol. Biol.* **23,** 47.)

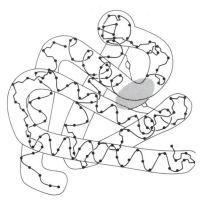

Fig. 10.45 The structure of myoglobin. Only the α-carbon atom positions are shown. The haem group, the oxygen-binding group, is shown as a shaded region. (Based on M. F. Perutz (1964), copyright *Scientific American;* with permission.)

with the formation of right-handed helices from the naturally occurring L-amino acids.

10.16 Higher-order structure

Helical polypeptide chains are folded into a tertiary structure if there are other bonding influences between the residues of the chain that are strong enough to overcome the interactions responsible for the secondary structure. The folding influences include —S—S— links, ionic interactions (which depend on the pH), and strong H-bonds (such as O—H $\cdots$ O—), and are illustrated by the structure of myoglobin (Fig. 10.45), the full structure (of 2600 atoms) having been determined by X-ray diffraction. The folding of the basic α helix caused by —S—S— links can be seen in the structure; about 77 per cent of the structure is α helix, the rest being involved in the folds.

Proteins with $M > 50\,\mathrm{kg\,mol^{-1}}$ are often found to be aggregates of two or more polypeptide chains. The possibility of such quaternary structure often confuses the determination of their molar masses, for different techniques might give values differing by factors of 2 or more. Haemoglobin, which consists of four myoglobin-like chains, is an example.

Protein denaturation can be caused by several means, and different aspects of structure may be affected. The permanent waving of hair, for example, is a reorganization at the quaternary level. Hair is a form of the protein keratin, and its quaternary structure is thought to be a multiple helix, with the α helices bound together by —S—S— links and hydrogen bonds (although there is some dispute about its precise structure). The process of permanent waving consists of disrupting these links, unravelling the keratin quaternary structure, and then reforming it into a more fashionable disposition. The 'permanence' is only temporary, however, because the structure of the newly formed hair is genetically controlled. Incidentally, normal hair grows at a rate that requires at least ten twists of the keratin helix to be produced each second, and so very close inspection of the human scalp would show it to be literally writhing with activity.

Denaturation at the secondary level is brought about by agents that destroy hydrogen bonds. Thermal motion may be sufficient, in which case denaturation is a kind of intramolecular melting. When eggs are cooked the albumin is denatured irreversibly, and the protein collapses into a structure resembling a random coil. The **helix–coil transition** is sharp, like ordinary melting, because it is a cooperative process: when one hydrogen bond has been broken it is easier to break its neighbours, and then even easier to break theirs, and so on. The disruption cascades down the helix, and the transition occurs sharply. Denaturation may also be brought about chemically. For instance, a solvent that forms stronger hydrogen bonds than those within the helix will compete successfully for the NH and CO groups. Acids and bases can cause denaturation by protonation or deprotonation of various groups.

EXERCISES

10.1 Estimate the lattice enthalpy of magnesium oxide from the data in Appendix 1 and the ionization and electron gain enthalpies in Chapter 2.

10.2 Estimate the ratio of the lattice enthalpies of MgO and CaO from the Born–Haber equation by using the ionic radii in Table 10.8.

10.3 The ClF_3 molecule has five electron pairs around the central Cl atom, and so may have either three equatorial F atoms or two axial and one equatorial F atoms. Since the molecule is polar, which structure does it have?

10.4 The electric dipole moment of toluene (methylbenzene) is 0.4 D. Estimate the dipole moments of the three xylenes (dimethylbenzenes). Which answer can you be sure about?

10.5 Calculate the resultant of two dipoles of magnitude 1.5 D and 0.80 D that make an angle 109.5° to each other.

10.6 Find an expression for the potential energy of interaction of a dipole moment μ with a point charge q representing an ion that is at a distance R from the dipole and collinear with it. Base your calculation on the expression for the Coulombic potential energy of interaction of two point charges, and suppose that the two partial charges $+q'$ and $-q'$ of the dipole are separated by a distance l from one another (with $\mu = q'l$).

10.7 Repeat the calculation in Exercise 10.6 for two collinear, parallel, identical point dipoles separated by a distance R.

10.8 The Lennard–Jones potential gives the potential energy of interaction between molecules. Given that the force is the negative slope of the potential, calculate the distance-dependence of the force acting between the molecules. What is the separation at which the force is zero?

10.9 Acetic acid vapour contains a proportion of planar, hydrogen-bonded dimers. The apparent dipole moment of molecules in pure gaseous acetic acid increases with increasing temperature. Suggest an interpretation of the latter observation.

10.10 Suppose the repulsive term in a Lennard–Jones (12, 6) potential is replaced by an exponential function of the form $\exp(-r/\sigma)$. Sketch the form of the potential energy and locate the distance at which it is a minimum.

10.11 Suppose that at a particular temperature it was found that the third virial coefficient C was approximately equal to B^2. Show that under these circumstances, $V_m = RT/p + B$.

10.12 The relation between the second virial coefficient and the intermolecular potential energy of interaction is given on p. 392. Suppose that V represents a hard sphere of radius σ and a shallow attractive potential of magnitude ε that is constant out to a radius σ'. Sketch the form of the potential and calculate B on the assumption that $\sigma' \gg \sigma$ and that $\varepsilon \ll kT$. Is it possible to identify the van der Waals parameters a and b in terms of σ and ε? (See Exercise 1.16 for a clue.)

10.13 Sketch the form of the radial distribution function for a liquid that locally resembles (a) a cubic close-packed structure and (b) a body-centred cubic structure. In each case, show only the first two spheres of neighbours (the nearest and the next nearest).

10.14 (a) Estimate the diffusion coefficient for a molecule that leaps 100 pm each 1.5 ps. (b) What would be the diffusion coefficient if the molecule travelled only half as far on each step?

10.15 (a) Show that the net distance travelled in a time t by a molecule in a liquid is proportional to the step length and the square root of the time interval. (b) How many individual steps must a molecule take to be 1000 step lengths away from its origin?

10.16 Calculate the packing fraction of (a) a stack of cylinders and (b) a primitive cubic lattice.

10.17 Draw a set of points as a rectangular array based on unit cells of side a and b, and mark the planes with Miller indices (10), (01), (11), (12), (23), (41), (4$\bar{1}$).

10.18 Repeat Exercise 10.17 for an array of points in which the a and b axes make 60° to each other.

10.19 In a certain unit cell, planes cut through the crystal axes at $(2a, 3b, c)$, (a, b, c), $(6a, 3b, 3c)$,

423

(2a, −3b, −3c). Identify the Miller indices of the planes.

10.20 Draw an orthorhombic unit cell and mark on it the (100), (010), (001), (011), (101), and (101) planes.

10.21 Draw a triclinic unit cell and mark on it the (100), (010), (001), (011), (101), and (101) planes.

10.22 Calculate the separation of the (111), (211), and (100) planes in a crystal in which the cubic unit cell has side 432 pm.

10.23 The glancing angle of a Bragg reflection from a set of crystal planes separated by 99.3 pm is 20.85°. Calculate the wavelength of the X-rays.

10.24 Copper K_α radiation consists of two components of wavelengths 154.433 pm and 154.051 pm. Calculate the separation of the diffraction lines arising from the two components in a powder diffraction pattern recorded in a camera of radius 5.74 cm from planes of separation 77.8 pm.

10.25 The compound Rb_3TlF_6 has a tetragonal unit cell with dimensions $a = 651$ pm and $c = 934$ pm. Calculate the volume of the unit cell.

10.26 The orthorhombic unit cell of $NiSO_4$ has the dimensions $a = 634$ pm, $b = 784$ pm, and $c = 516$ pm, and the density of the solid is estimated as $3.9\,g\,cm^{-3}$. Determine the number of formula units per unit cell and calculate a more precise value of the density.

10.27 The unit cells of $SbCl_3$ are orthorhombic with dimensions $a = 812$ pm, $b = 947$ pm, and $c = 637$ pm. Calculate the spacing of the (411) planes.

10.28 The separation of (100) planes of lithium metal is 350 pm and its density is $0.53\,g\,cm^{-3}$. Is the structure of lithium fcc or bcc?

10.29 Copper crystallizes in an fcc structure with unit cells of side 361 pm. Predict the appearance of the powder diffraction pattern using 154 pm radiation. What is the density of copper?

11

Molecular spectroscopy

One of the most important methods for identifying a species, determining its structure, and exploring its physical properties is spectroscopy. We have already seen that photons act as messengers from inside atoms, and that atomic spectra can be used to obtain detailed information about the electronic structure of atoms. The same is true of the spectroscopic techniques that depend on the interaction of electromagnetic radiation with molecules: they too reveal remarkably detailed information about molecules and are of immense help in the identification of unknown species. Spectral lines in molecular spectroscopy originate from the emission or absorption of a photon when the energy of a molecule changes. The difference from atomic spectroscopy is that the energy of a molecule can change not only as a result of electronic transitions but also because it can undergo transitions between its rotational and its vibrational states.

General features of spectroscopy

All types of spectra have some features in common, and we examine these characteristics first.

11.1 Experimental techniques

In **emission spectroscopy**, a molecule undergoes a transition from a state of high energy E_1 to a state of lower energy E_2 and emits the excess energy as a photon (Fig. 11.1(a)). In **absorption spectroscopy**, the absorption of nearly monochromatic (single-frequency) incident

425

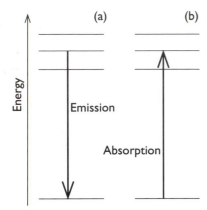

Fig. 11.1 (a) In emission spectroscopy, a molecule returns to a lower state (typically the ground state) from an excited state, and emits the excess energy as a photon. (b) The same transition may be observed in absorption, when the incident radiation supplies a photon that can excite the molecule from its ground state to an excited state.

radiation is monitored as it is swept over a range of frequencies (Fig. 11.1(b)). In **Raman spectroscopy**, a monochromatic incident beam—typically in the visible region of the spectrum and generated by a laser—is passed through the sample and the scattered radiation is analysed. Some of the incident photons collide with the molecules, giving up some of their energy or gaining energy from the molecule, and emerge with a different frequency (Fig. 11.2).

The energy $h\nu$ of the photon emitted or absorbed, and therefore the frequency, ν, of the radiation emitted or absorbed, is given by the Bohr frequency condition (Section 8.7):

$$hv = |E_1 - E_2| \qquad (1)$$

This relation is often expressed in terms of the **wavelength**, λ of the radiation (see *Further information 8* on p. 290) by using the relation

$$\lambda = \frac{c}{v} \qquad (2a)$$

or the **wavenumber**, $\bar{v}$, which is defined as

$$\bar{v} = \frac{v}{c} = \frac{1}{\lambda} \qquad (2b)$$

The units of wavenumber are almost always chosen as reciprocal centimetres, cm^{-1}. Figure 11.3 summarizes the frequencies, wavelengths, and wavenumbers of the various regions of the electromagnetic spectrum.

Emission, absorption, and Raman spectroscpy give the same information about energy-level separations, but practical considerations generally determine which technique is employed. In practice, emission spectroscopy, if it is used at all, is used mainly for the study of energetically excited reaction products. Absorption spectroscopy is much more widely used, and we shall concentrate on it. Raman spectroscopy is becoming increasingly widely used for a variety of investigations (of the kind described later) now that intense, monochromatic laser radiation is readily available.

The apparatus

The radiation source generally produces radiation spanning a range of frequencies, but in a few cases (including lasers) it generates nearly monochromatic radiation. For the far infrared, the source is a mercury arc inside a quartz envelope, most of the radiation being generated by the hot quartz. A **Nernst filament** is used to generate radiation in the near infrared: it consists of a heated ceramic filament containing rare-earth oxides, and emits radiation closely resembling that of a hot black body (of the kind considered in Section 8.1). For the visible region of the spectrum, a tungsten–iodine lamp is used, which gives out intense white light. A discharge through deuterium gas or xenon in quartz is still widely used for the near ultraviolet. A device called a **klystron** (which is also used in radar installations) is used to generate microwaves.

In all but specialized techniques using monochromatic microwave radiation and certain types of laser radiation, spectrometers include a component for separating the frequencies of the radiation so that the variation of the absorption with frequency can be monitored. In conventional spectrometers, this component is a **dispersing element** that separates different frequencies into different spatial directions. The simplest dispersing element is a glass or quartz prism. (Fig. 11.4), but a **diffraction grating** is also widely used. A diffraction grating consists of a glass or ceramic plate into which fine grooves have been cut about 1000 nm apart (a spacing comparable to the wavelength of visible light) and covered with a reflective aluminium coating. The grating causes interference between waves reflected from its surface, and constructive interference occurs at specific angles that depend on the frequency of the radiation being used. That is, each wavelength of light is directed into a specific direction.

The third component of a spectrometer is the **detector**, the device that converts incident radiation into an electric current for signal processing or plotting. Radiation-sensitive semiconductor devices are increasingly used for this purpose in infrared spectroscopy. However, in the optical and ultraviolet region, photographic recording or a **photomultiplier** are widely used. In the latter device, an incident photon ejects an electron from a photosensitive surface, the electron is accelerated by a potential difference and ejects a shower of electrons where it strikes a screen. The latter electrons are accelerated, and each one releases a further shower on impact with another screen. Thus the impact of the initial photon is converted into a cascade of electrons, which is converted into a current in an external circuit.

Although semiconductor detectors are increasingly being used in the infrared, thermocouples are still widely used. A thermocouple detector consists typically of a blackened gold foil to which are attached thermoelectric alloys which generate an electric current when the temperature is raised. A **thermistor bolometer** is essentially a resistance thermometer, and is typically formed from a mixture of oxides deposited on quartz. In each case the radiation is chopped by a shutter that rotates in the beam so that an alternating signal is obtained from the detector (an oscillating signal is easier to amplify than a steady signal). A microwave detector is typically a crystal diode consisting of a tungsten tip in contact with a semiconductor, such as germanium, silicon, or gallium arsenide.

The highest resolution is obtained when the sample is gaseous and of such low pressure that collisions between the molecules are infrequent. Gaseous samples are essential for rotational (microwave) spectroscopy, for only in that phase can molecules rotate freely. To achieve sufficient absorption, the path lengths of gaseous samples must be very long, of the order of metres. Long path lengths are achieved

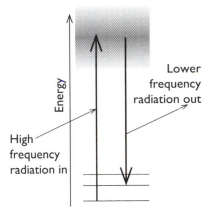

Fig. 11.2 In Raman spectroscopy, an incident photon is scattered from a molecule with either an increase in frequency (if the radiation collects energy from the molecule) or—as shown here—with a lower frequency if it loses energy to the molecule. The process can be regarded as taking place by an excitation of the molecule to a wide range of states (represented by the shaded band), and the subsequent return of the molecule to a lower state; the net energy change is then carried away by the photon.

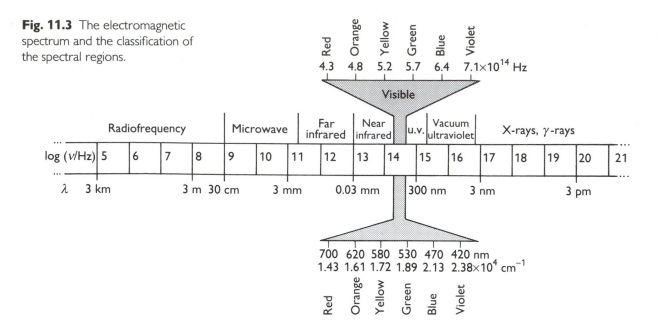

Fig. 11.3 The electromagnetic spectrum and the classification of the spectral regions.

by multiple passage of the beam between two parallel mirrors at each end of the sample cavity.

For infrared spectroscopy, the sample is typically a liquid held between windows of sodium chloride (which is transparent down to $700\ \text{cm}^{-1}$) or potassium bromide (down to $400\ \text{cm}^{-1}$). Other ways of preparing the sample include grinding into a paste with 'Nujol', a hydrocarbon oil, or pressing it into a solid disc, perhaps with powdered potassium bromide.

Measures of intensity

The intensity with which radiation is absorbed depends on the identity of the absorbing species (which we denote J), the frequency, ν, of the radiation, the molar concentration, $[J]$, of the species in the sample, and the path length, l, of the radiation through the sample. It is found experimentally that the **transmittance**, T, the ratio of the emerging intensity I to the incident intensity I_0, is given by the **Beer–Lambert law**:

$$\log T = -\varepsilon[J]l \tag{3}$$

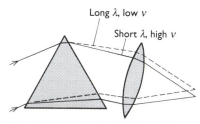

Fig. 11.4 One simple dispersing element is a prism, which separates frequencies spatially by making use of the higher refractive index for high-frequency radiation. The shortest wavelengths for which a glass prism can be used is about 400 nm, but quartz can be used down to 180 nm.

The coefficient ε is the **molar absorption coefficient** (formerly the 'extinction coefficient') of the species, and it depends on the frequency of the incident light. Its units are those of $1/(\text{concentration} \times \text{length})$, and it is normally convenient† to express it in $\text{mol}^{-1}\,\text{L}\,\text{cm}^{-1}$. The dimensionless product $A = \varepsilon[J]l$ is called the **absorbance** (formerly the 'optical density') of the sample. The

† The alternative units $\text{cm}^2\,\text{mol}^{-1}$ bring out the point that ε is a molar cross-section for absorption, and the greater the cross-section of the molecule for absorption, the greater the reduction in the intensity of the beam.

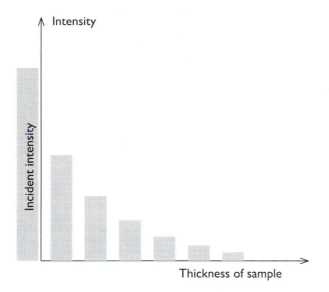

Beer–Lambert law is sometimes written in the form

$$I = I_0 \times 10^{-\varepsilon[J]l} \tag{4}$$

This expression shows that the transmitted intensity decreases rapidly with the length of the sample (Fig. 11.5).

JUSTIFICATION

The Beer–Lambert law can be derived by noting that the change in intensity dI that occurs when electromagnetic radiation passes through a layer of thickness dx is proportional to the thickness of the layer, the concentration, and the incident intensity in the region of the absorbing species. Because dI is negative (the intensity is reduced by absorption), we can write

$$dI = -\kappa[J]I \, dx$$

where κ is the proportionality coefficient, or equivalently

$$\frac{dI}{I} = -\kappa[J] \, dx$$

These expressions apply to each successive layer into which the sample can be regarded as being divided. Therefore, to obtain the intensity, I, that emerges from a sample of thickness l when the incident intensity is I_0, we add (that is, integrate) all the successive changes:

$$\int_{I_0}^{I} \frac{dI}{I} = -\kappa \int_0^l [J] \, dx$$

If the concentration is uniform, $[J]$ is independent of x, and the expression integrates to

$$\ln \frac{I}{I_0} = -\kappa[J]l$$

The relation between natural and common logarithms is

$$\ln x = 2.303 \log x$$

Therefore, by writing $\kappa = 2.303\varepsilon$, we obtain

$$\log \frac{I}{I_0} = -\varepsilon[J]l$$

which, on substituting $T = I/I_0$, is the Beer–Lambert law.

Example Calculating the molar absorption coefficient

Radiation of wavelength 256 nm was passed through 1.0 mm of a solution that contained C_6H_6 at a concentration of 0.050 mol L^{-1}. The light intensity is reduced to 16 per cent of its initial value (so $T = 0.16$). Calculate the absorbance and the molar absorption coefficient of the sample. What would be the transmittance through a 2.0 mm cell?

Answer For ε we use eqn (3) with $T = 0.16$:

$$\varepsilon = \frac{-1}{[C_6H_6]l} \times \log \frac{I}{I_0}$$

$$= \frac{-1}{0.050 \text{ mol L}^{-1} \times 0.10 \text{ cm}} \times \log 0.16 = 160 \text{ mol}^{-1} \text{ L cm}^{-1}$$

The absorbance is therefore

$$A = \varepsilon[C_6H_6]l = 160 \text{ mol}^{-1} \text{ L cm}^{-1} \times 0.050 \text{ mol L}^{-1} \times 0.10 \text{ cm}$$
$$= 0.80$$

The transmittance through a 2.0 mm (0.20 cm) sample is given by

$$T = 10^{-\varepsilon[C_6H_6]l} = 10^{-160 \times 0.050 \times 0.20} = 0.025$$

That is, the emergent light is reduced to 2.5 per cent of its incident intensity.

Exercise E11.1 The transmittance of an aqueous solution that contained Cu^{2+} ions at a concentration of 0.10 mol L^{-1} at 600 nm was measured as 0.30 in a 5.0 mm cell. Calculate the molar absorption coefficient of $Cu^{2+}(aq)$ at that wavelength and the absorbance of the solution. What would be the transmittance through a 1.0 mm cell?

[10 mol^{-1} L cm^{-1}, $A = 0.50$, $T = 0.79$]

The maximum value of the molar absorption coefficient, ε_{max}, is an indication of the intensity of a transition. Typical values for strong transitions are of the order of 10^4–10^5 mol^{-1} L cm^{-1}, indicating that in a 0.01 mol L^{-1} solution the intensity of light (of the appropriate frequency) falls to 10 per cent of its initial value after passing through about 0.1 mm of solution.

11.2 Intensities and linewidths

The intensity of a spectral line depends on the number of molecules that are in the initial state and the strength with which individual molecules are able to interact with the electromagnetic field and generate or absorb photons. If we confine our attention to vibrational and electronic spectroscopy, then the situation is very simple: almost all vibrational absorptions and all electronic absorptions occur from the ground state of a molecule, because that is the only state populated at room temperature. However, molecules can be prepared in short-lived excited states as a result of chemical reaction, electric discharge, or photolysis. In these cases the populations may be quite different from those at thermal equilibrium, and absorption and emission spectra—if they can be recorded quickly enough—then arise from transitions from all the populated levels.

Spectral lines are not infinitely narrow, and in condensed media may spread over several thousand reciprocal centimetres. One important broadening process in gaseous samples is the **Doppler effect**, in which radiation is shifted in frequency when the source is moving towards or away from the observer. When a source emitting radiation of frequency v recedes with a speed s, the observer detects radiation of frequency

$$v' = \frac{v}{1 + \dfrac{s}{c}} \qquad (5a)$$

where c is the speed of the radiation (the speed of light for electromagnetic radiation, the speed of sound for sound waves). A source approaching the observer appears to be emitting radiation of frequency

$$v' = \frac{v}{1 - \dfrac{s}{c}} \qquad (5b)$$

Exercise E11.2 A laser line occurs at 628.443 cm^{-1}. What wavenumber will an observer detect when approaching the laser at (a) 1 m s^{-1}, (b) 1000 m s^{-1}?
[(a) 628.443 cm^{-1}, (b) 628.445 cm^{-1}]

Molecules reach high speeds in all directions in a gas, and a static observer detects the corresponding Doppler-shifted range of frequencies. Some molecules approach the observer, some move away; some move quickly, others slowly. The detected spectral 'line' is the absorption or emission profile arising from all the resulting Doppler shifts. The profile reflects the Maxwell distribution of molecular speeds (Section 1.5), and we observe a bell-shaped Gaussian curve (a shape of the form e$^{-x^2}$). The Doppler line shape is therefore also a Gaussian curve (Fig. 11.6), and calculation shows that when the temperature is T and the molar mass of the molecule is M, the width

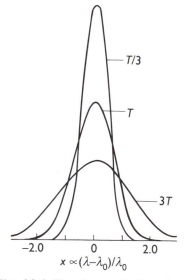

Fig. 11.6 The shape of a Doppler-broadened spectral line reflects the Maxwell distribution of speeds in the sample at the temperature of the experiment. Notice that the line broadens as the temperature is increased. The width at half height, $\delta\lambda$, is given by eqn (6).

$x \propto (\lambda - \lambda_0)/\lambda_0$

431

of the line at half-height is

$$\delta\lambda = \frac{2\lambda}{c} \times \sqrt{\frac{2RT\ln 2}{M}} \qquad (6)$$

The Doppler width increases with temperature because the molecules acquire a wider range of speeds. Therefore, to obtain spectra of maximum sharpness, it is best to work with cold samples.

Exercise E11.3 The sun emits a spectral line at 677.4 nm which has been identified as arising from a transition in highly ionized ^{57}Fe. Its width at half-height is 5.3 pm. What is the temperature of the Sun's surface?

$$[T = 6.8 \times 10^3 \text{ K}]$$

Another source of line broadening is related to the lifetimes of the states involved in the transition. When the Schrödinger equation is solved for a system that is changing with time, it is found that it is impossible to specify the energy levels exactly. If, on average, a system survives in a state for a time τ, the lifetime of the state, then its energy levels are blurred by δE, where the extent of this **lifetime broadening** is given by the expression

$$\delta E \approx \frac{\hbar}{\tau} \qquad (7a)$$

Expressing the energy spread in wavenumbers through $\delta E = hc\delta\tilde{\nu}$ and using the values of the fundamental constants gives the practical form of the relation as

$$\delta\tilde{\nu} \approx \frac{5.3 \text{ cm}^{-1}}{\tau/\text{ps}} \qquad (7b)$$

Only if τ is infinite can the energy of a state be specified exactly. However, no excited state has an infinite lifetime; therefore, all states are subjected to some lifetime broadening, and the shorter the lifetimes of the states involved in a transition, the broader the spectral lines.

Exercise E11.4 What is the uncertainty in wavenumber of a transition from a state with a lifetime of 5.0 ps?

$$[1.0 \text{ cm}^{-1}]$$

Two processes are principally responsible for the finite lifetimes of excited states and hence for their widths. The dominant one is

collisional deactivation, which arises from collisions between molecules or with the walls of the container. If the **collisional lifetime** is τ_{col}, then the resulting collisional linewidth is $\delta E_{col} \approx \hbar/\tau_{col}$. The collisional lifetime can be lengthened, and the broadening minimized, by working at low pressures.

Because the rate of spontaneous emission cannot be changed, it is a natural limit to the lifetime of an excited state, and the resulting lifetime broadening is the **natural linewidth** of the transition. The natural linewidth cannot be changed by modifying the conditions. Natural linewidths depend strongly on the transition frequency (they increase as v^3), and so low-frequency transitions (such as the microwave transitions of rotational spectroscopy) have very small natural linewidths, and collisional and Doppler line-broadening processes are dominant. The natural lifetimes of electronic transitions are very much shorter than for vibrational transitions, and so the natural linewidths of electronic transitions are much greater than those of vibrational and rotational transitions. For example, a typical electronic excited state natural lifetime is about 10^{-8} s (10^4 ps), corresponding to a natural linewidth of about 5×10^{-4} cm^{-1} (15 MHz).

Vibrational spectra

All molecules are capable of vibrating, and complex molecules may do so in a large number of different modes. Even benzene, with 12 atoms, can vibrate in 30 different modes, some of which involve the periodic enlarging and contracting of the ring and others its buckling into various shapes. A molecule as big as a protein can vibrate in tens of thousands of different ways. However, according to quantum mechanics, no vibration can be excited unless the molecule has been provided with a certain minimum energy. Vibrations can be excited by the absorption of electromagnetic radiation, and observing the frequencies at which absorption occurs gives very valuable structural information and tells us about the flexibility of molecules. We shall see in a moment that vibrational transitions occur in the infrared region of the electromagnetic spectrum. Vibrational spectroscopy is therefore also called **infrared spectroscopy**.

11.3 The vibrations of molecules

We shall base our discussion on Fig. 11.7, which shows a typical potential energy curve (it is a reproduction of Fig. 9.1) of a diatomic molecule as its bond is lengthened by pulling one atom away from the other or shortened by pushing the atoms together. In regions close to the equilibrium bond length R_e (at the minimum of the curve) the potential energy can be approximated by a parabola (a curve of the form $y = x^2$), and we can write

$$V = \tfrac{1}{2}k(R - R_e)^2 \tag{8}$$

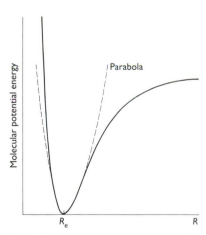

Fig. 11.7 The molecular potential energy curve can be approximated by a parabola near the bottom of the well. A parabolic potential results in harmonic oscillation. At high vibrational excitation energies the parabolic approximation is poor.

433

where k is the **force constant** of the bond. The name derives from the relation between the restoring force experienced by the atoms when a bond is stretched and the extent of the stretching:

$$\text{restoring force} = -k(R - R_e)$$

That is, k is the constant of proportionality between the displacement from equilibrium and the strength of the resulting restoring force. The steeper the walls of the potential (the stiffer the bond), the greater the force constant.

Exercise E11.5 What are the SI units of k? Confirm that V is expressed in joules.

$$[\text{N m}^{-1}; (1 \text{ N m}^{-1}) \times (1 \text{ m})^2 = 1 \text{ J}]$$

Vibrational energy levels

The permitted energy of a vibrating molecule are obtained by solving the Schrödinger equation for the motion of the two atoms of masses m_1 and m_2 with the potential energy in eqn (8) and are

$$E_v = (v + \tfrac{1}{2})\hbar\omega, \quad \text{with} \quad v = 0, 1, 2, \ldots \quad \text{and} \quad \omega = 2\pi\nu \quad (9)$$

where

$$\omega = \sqrt{\frac{k}{\mu}}, \quad \text{with} \quad \mu = \frac{m_1 m_2}{m_1 + m_2} \quad (10)$$

These energy levels are illustrated in Fig. 11.8, which shows that they form a uniformly spaced ladder, with separation $\hbar\omega$ between neighbouring levels. It is important to note that the vibrational energy levels depend on the **effective mass**, μ, of the molecule, not its total mass. If atom 1 were as heavy as a brick wall, we could neglect m_2 in the sum $m_1 + m_2$ in the denominator of μ, and so would find $\mu \approx m_2$, the mass of the lighter atom. The vibration would then be that of a light atom relative to a stationary wall (this is approximately the case in HI, for example, where the I atom barely moves and $\mu \approx m_H$). In the case of a homonuclear diatomic molecule, for which $m_1 = m_2$, the effective mass is half the mass of one atom: $\mu = \tfrac{1}{2}m$.

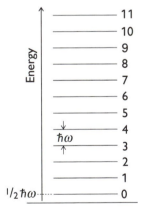

Fig. 11.8 The energy levels of a harmonic oscillator. The quantum number v ranges from 0 to infinity, and the permitted energy levels form a uniform ladder with spacing $\hbar\omega$. The molecule has a zero point energy of $\tfrac{1}{2}\hbar\omega$, which can never be removed.

Exercise E11.6 An HCl molecule has a force constant of 516 N m^{-1}, a reasonably typical value. Calculate the vibrational frequency, ω, of the molecule and the energy separation between any two neighbouring vibrational energy levels.

$$[5.63 \times 10^{14} \text{ s}^{-1}, 5.94 \times 10^{-20} \text{ J}]$$

Incident electromagnetic radiation can excite a molecular vibration if its photons carry enough energy. Because a typical vibrational excitation energy is of the order of 10^{-20}–10^{-19} J (see Exercise E11.6), the frequency of the radiation should be of the order of 10^{13}–10^{14} Hz (from $\Delta E = h\nu$). This frequency corresponds to infrared radiation, so vibrational transitions are observed by **infrared spectroscopy.**

A significant feature of eqn (9) is that, because the lowest vibrational state is when $v = 0$, corresponding to an energy of $\frac{1}{2}\hbar\omega$, *a molecular vibration cannot be stopped completely.* That is, each vibrational mode of a molecule has an irremovable **zero-point energy.** We can think of the molecule as perpetually quivering about its equilibrium bond lengths and angles (this quivering would occur even at $T = 0$). For a diatomic molecule, with only one mode of vibration, the energy locked up as zero-point vibration is quite small, but for a macromolecule with tens of thousands of vibrations, the energy stored in this way may be considerable.

The allowed transitions

We met the concept of a **selection rule** in Section 8.11 as a statement about which transitions are forbidden and which are allowed. Selection rules also apply to molecular spectra, and the form they take depends on the type of transition. The underlying classical idea is that, *if the molecule is to be able to interact with the electromagnetic field and absorb or create a photon of frequency ν, then it must possess, at least transiently, a dipole oscillating at that frequency.* This transient dipole moment associated with the transition—a sudden emergence and then loss of positive and negative partial charges—is called a **transition moment.**

A **gross selection rule** specifies the general features a molecule must have if it is to have a spectrum of a given kind. For a vibration of a molecule to give rise to an absorption spectrum, *the electric dipole moment of the molecule must change during the vibration.* The classical basis of this rule is that the molecule can shake the electromagnetic field into oscillation if its (the molecule's) dipole changes as it vibrates (Fig. 11.9). The molecule need not have a *permanent* dipole: the rule requires only a *change* in dipole moment, possibly from zero. Some vibrations do not affect the molecule's dipole moment (for example, the stretching motion of a homonuclear diatomic molecule), and so they neither absorb nor generate radiation. Homonuclear diatomic molecules are therefore **infrared inactive**, because their dipole moments remain zero however long the bond, but heteronuclear diatomic molecules are **infrared active**.

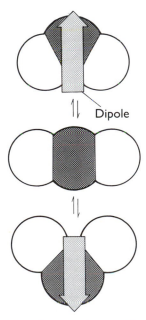

Dipole

Fig. 11.9 The oscillation of a molecule, even if it is nonpolar, may result in an oscillating dipole that can interact with the electromagnetic field.

Example Using the gross selection rules
State which of the following molecules have vibrational absorption spectra: N_2, CO_2, OCS, H_2O, $CH_2{=}CH_2$, C_6H_6.

Answer Molecules that give rise to vibrational spectra have dipole moments that change during the course of a vibration. Therefore, we must judge whether a distortion of the molecule can change its dipole moment (including changing it from zero). All the molecules except N_2 possess at least one vibrational mode that results in a change of dipole moment, and so all except N_2 can show a vibrational absorption spectrum. It should be noted that not all the modes of complex molecules are vibrationally active. For example, a vibration of CO_2 in which the O—C—O bonds stretch and contract symmetrically is inactive because it leaves the dipole moment unchanged (at zero).

Exercise E11.7 Repeat the question for H_2, NO, N_2O, CH_4.
[NO, N_2O, CH_4]

A more detailed study of the transition moment leads to the **specific selection rules** that express the allowed transitions in terms of the changes in quantum numbers (as in the rule $\Delta l = \pm 1$ for atoms). For a vibrational transition, the selection rule is $\Delta v = \pm 1$. It follows that the change in energy for the transition from a state with quantum number v to one with quantum number $v + 1$ is

$$\Delta E = (v + \tfrac{3}{2})\hbar\omega - (v + \tfrac{1}{2})\hbar\omega = \hbar\omega$$

The frequency v (in hertz) of the photon absorbed is $\omega/2\pi$ and its wavenumber is $\omega/2\pi c$. At room temperature almost all the molecules will be in their vibrational ground states initially. Hence, the dominant spectral transition will be from $v = 0$ to $v = 1$. It follows from the calculation of ω for HCl (in Exercise E11.6), that $v = 8.95 \times 10^{13}$ Hz, so the infrared spectrum of the molecule will be an absorption at that frequency. The corresponding wavenumber and wavelength are 2990 cm^{-1} and 3.35 μm, respectively.

Exercise E11.8 The force constant of the C=O group in a peptide link is approximately 1.2 kN m^{-1}. At what wavenumber would you expect it to absorb?
[*Hint:* for the effective mass, treat the group as a CO molecule.]

[at approximately 1700 cm^{-1}]

Vibrational Raman spectra of diatomic molecules

The gross selection rule for vibrational Raman transitions is that the molecular *polarizability*, its responsiveness to an applied electric field

Table 11.1 Properties of diatomic molecules

	$\bar{v}/\text{cm}^{-1}$	r/pm	$k/\text{N m}^{-1}$	$D_e/\text{kJ mol}^{-1}$
$^1\text{H}_2^+$	2333	106	160	256
$^1\text{H}_2$	4400	74	575	432
$^2\text{H}_2$	3118	74	577	440
$^1\text{H}^{19}\text{F}$	4138	92	955	564
$^1\text{H}^{35}\text{Cl}$	2991	127	516	428
$^1\text{H}^{81}\text{Br}$	2649	141	412	363
$^1\text{H}^{127}\text{I}$	2308	161	314	295
$^{14}\text{N}_2$	2358	110	2294	942
$^{16}\text{O}_2$	1580	121	1177	494
$^{19}\text{F}_2$	892	142	445	154
$^{35}\text{Cl}_2$	560	199	323	239

(Section 10.2), should change as the molecule vibrates. The polarizability plays a role in Raman spectroscopy because the molecule must be squeezed and stretched by the incident radiation in order that a vibrational excitation may occur during the photon–molecule collision. Both homonuclear and heteronuclear diatomic molecules swell and contract during a vibration, and the control of the nuclei over the electrons, and hence the molecular polarizability, changes too. Both types of diatomic molecule are therefore vibrationally Raman active.

The specific selection rule for vibrational Raman transitions is the same as for an infrared absorption ($\Delta v = 1$). The photons that are scattered with a lower wavenumber than that of the incident light, and which contribute to the **Stokes lines**, are those for which $\Delta v = +1$. These lines are stronger than the **anti-Stokes lines**, which correspond to photons that are scattered with a higher wavenumber and arise from transitions in which $\Delta v = -1$, because very few molecules are in an excited vibrational state initially.

The information available from vibrational Raman spectra adds to that from infrared spectroscopy because homonuclear diatomic molecules can also be studied. The spectra can be interpreted in terms of the force constants, dissociation energies, and bond lengths, and some of the information obtained is included in Table 11.1.

11.4 The vibrations of polyatomic molecules

There is only one mode of vibration in a diatomic molecule (the periodic extension and compression of the bond between the two atoms). In a nonlinear polyatomic molecule of N atoms there are $3N - 6$ modes because bonds may stretch and angles may bend. For example, H_2O is a triatomic nonlinear molecule, and has three modes of vibration. Naphthalene, $C_{10}H_8$, has 48 distinct modes of vibration.

JUSTIFICATION

Each atom may change its location by varying one of its coordinates, and so the total number of such displacements that are available is $3N$. Three of these coordinates are needed to specify the location of the centre of mass of the molecule, and so three of these displacements correspond to the translational motion of the molecule as a whole. The remaining $3N - 3$ displacements are 'internal' modes of the molecule that leave its centre of mass unchanged. Three angles are needed to specify the orientation of a nonlinear molecule in space (Fig. 11.10). Therefore three of the $3N - 3$ internal displacements are rotational, so leaving $3N - 6$ displacements of the atoms relative to each other. The latter displacements are the vibrational modes. A similar calculation for linear molecules, which require only two angles to specify their orientation in space, shows that such molecules have $3N - 5$ vibrational modes.

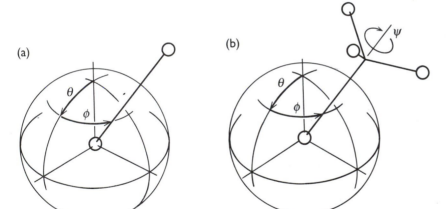

Fig. 11.10 (a) The orientation of a linear molecule requires the specification of two angles (the latitude and longitude of its axis). (b) The orientation of a nonlinear molecule requires the specification of three angles (the latitude and longitude of its axis and the angle of twist—the azimuthal angle—around that axis).

The description of the vibrational motion of a polyatomic molecule is much simpler if combinations of the stretching and bending motions of individual bonds are taken. For example, although we could describe two of the four vibrations of a CO_2 molecule as individual carbon–oxygen bond stretches, ν_L and ν_R in Fig. 11.11, the description of the motion becomes much simpler if two combinations of these vibrations are used instead. One combination is ν_1 in Fig. 11.12: this combination is the **symmetric stretch**. Another combination is ν_3, the **antisymmetric stretch**, in which the two O atoms always move in the same direction. Both modes are independent in the sense that if one is excited, then it does not excite the other. They are two of the

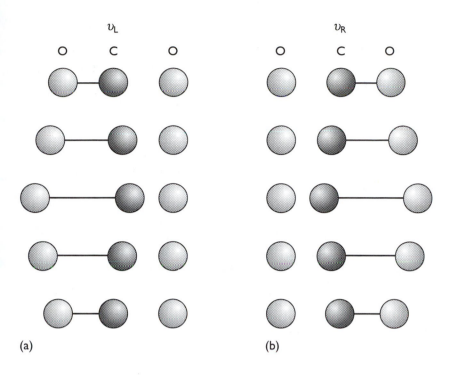

ν_L ν_R

O   C   O   O   C   O

(a)   (b)

Fig. 11.11 The stretching vibrations of a CO_2 molecule can be represented in a number of ways. In this representation, (a) one O═C bond vibrates and the remaining O atom is stationary, and (b) the C═O bond vibrates while the other O atom is stationary. Because the stationary atom is linked to the C atom, it does not remain stationary for long. That is, if one vibration begins, it rapidly stimulates the other to occur.

normal modes of the molecule, its independent, collective vibrational displacements. The two other normal modes are the **bending modes**, ν_2. In general, a normal mode is an independent, synchronous motion of atoms or groups of atoms that may be excited without leading to the excitation of any other normal mode.

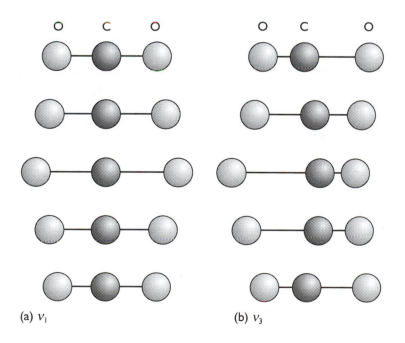

O   C   O   O   C   O

(a) ν_1 (b) ν_3

Fig. 11.12 Alternatively, linear combinations of the two modes can be taken to give these two normal modes of the molecule. The mode in (a) is the *symmetric stretch* and that in (b) is the *antisymmetric stretch*. The two modes are independent, and if either of them is stimulated, the other remains unexcited. A normal mode description greatly simplifies the description of the vibrations of the molecule.

439

The four normal modes of CO_2 and the N_{vib} normal modes of polyatomic molecules in general are the key to the description of molecular vibrations. Each normal mode behaves like an independent harmonic oscillator and the energies of the vibrational levels are given by the same expression as in eqn (10), but with an effective mass, μ, that depends on the extent to which each of the atoms moves during the vibration. (Atoms that do not move, such as the C atom in the symmetric stretch of CO_2, do not contribute to the effective mass.) The force constant also depends in a complicated way on the extent to which bonds bend and stretch during a vibration. Typically, a normal mode that is largely a bending motion has a lower force constant (and hence a lower frequency) than a normal mode that is largely a stretching motion.

The gross selection rule for infrared activity is that the motion corresponding to a normal mode should give rise to a changing dipole moment. Deciding whether this is so can sometimes be done by inspection. For example, the symmetric stretch of CO_2 leaves the dipole moment unchanged (at zero), and so this mode is infrared inactive and makes no contribution to the molecule's infrared spectrum. The antisymmetric stretch, however, changes the dipole moment because the molecule becomes unsymmetrical as it vibrates, and so this mode is infrared active. The fact that the mode does absorb infrared radiation enables carbon dioxide to act as a 'greenhouse gas' by absorbing infrared radiation emitted from the surface of the Earth. Because the dipole moment change is parallel to the molecular axis in the antisymmetric stretching mode, the transitions arising from this mode are classified as **parallel bands** in the spectrum. Both bending modes are also infrared active: they are accompanied by a changing dipole perpendicular to the molecular axis (Fig. 11.13), and so transitions involving them lead to a **perpendicular band** in the spectrum.

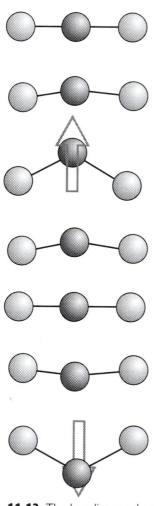

Fig. 11.13 The bending mode of CO_2 results in the formation of an electric dipole (depicted by the arrow at the stage in the vibration when it is a maximum), which oscillates as the molecule vibrates.

Some of the normal modes of organic molecules can be regarded as motions of individual functional groups whereas others cannot be regarded as localized in this way and are motions of the molecule as a whole. The latter are generally of relatively low frequency, and occur below about 1500 cm^{-1} in the spectrum. The resulting whole-molecule region of the absorption spectrum is called the **fingerprint region** of the spectrum, for it is characteristic of the molecule. The matching of the fingerprint region with a spectrum of a known compound in a library of infrared spectra is a very powerful way of confirming the identity of a particular substance.

The characteristic vibrations of functional groups that occur outside the fingerprint region are very useful for the identification of an unknown compound. Most of these vibrations can be regarded as stretching modes, for the lower frequency bending modes usually occur in the fingerprint region and so are less readily identified. The characteristic wavenumbers of some functional groups are listed in Table 11.2.

Table 11.2 Typical vibration wavenumbers

Vibration type	$\tilde{\nu}/cm^{-1}$
C—H stretch	2850–2960
C—H bend	1340–1465
C—C stretch, bend	700–1250
C=C stretch	1620–1680
C≡C stretch	2100–2260
O—H stretch	3590–3650
C=O stretch	1640–1780
C≡N stretch	2215–2275
N—H stretch	3200–3500
Hydrogen bonds	3200–3570

Example Interpreting an infrared spectrum
The infrared spectrum of an organic compound is shown in Fig. 11.14. Suggest an identification.
Answer Some of the features at wavenumbers above 1500 cm^{-1} can be identified by comparison with the data in Table 11.2: (a) C—H stretch of a benzene ring, indicating a substituted benzene; (b) carboxylic acid O—H stretch, indicating a carboxylic acid; (c) the strong absorption of a conjugated C≡C group, indicating a substituted alkyne; (d) this strong absorption is also characteristic of a carboxylic acid that is conjugated to a carbon–carbon multiple bond; (e) a characteristic vibration of a benzene ring, confirming the deduction drawn from (a); (f) a characteristic absorption of a nitro group (—NO$_2$) connected to a multiply bonded carbon–carbon system, suggesting a nitro-substituted benzene. The molecule contains as components a benzene ring, a C≡C bond, a —COOH group, and a —NO$_2$ group. The molecule is in fact O_2N—C_6H_4—C≡C—COOH (and more detailed analysis and comparison of the fingerprint region shows it to be the *para* isomer).

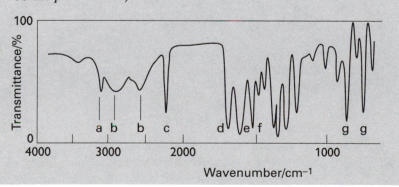

Fig. 11.14 A typical infrared absorption spectrum taken by forming a sample into a disk with potassium bromide. As explained in the example, the substance can be identified as $O_2NC_6H_4C≡CCOOH$.

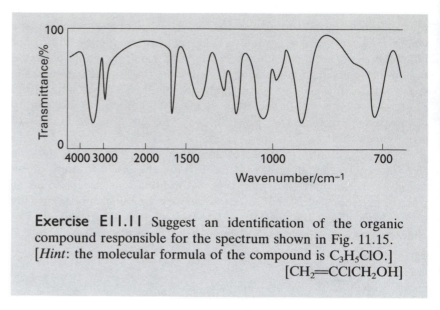

Fig. 11.15 The spectrum considered in Exercise E11.11.

Exercise E11.11 Suggest an identification of the organic compound responsible for the spectrum shown in Fig. 11.15. [*Hint*: the molecular formula of the compound is C_3H_5ClO.]

$[CH_2{=}CClCH_2OH]$

11.5 Vibrational Raman spectra of polyatomic molecules

When a photon strikes a polyatomic molecule it may deposit energy in it as a vibrational motion or acquire additional energy from a vibration that is already excited. As a result, the incident photon may undergo Raman scattering and travel away from the molecule with a reduced or an increased frequency. The normal modes of vibration of molecules are Raman active and participate in this kind of scattering if they are accompanied by a changing polarizability. However, it is often quite difficult to judge by inspection when this is so. The symmetric stretch of CO_2, for example, alternately swells and contracts the molecule: this motion changes its polarizability, and so the mode is Raman active. The other modes of CO_2 leave the polarizability unchanged, and so they are Raman inactive.

In particular cases it is possible to make use of the following very general rule about the infrared and Raman activity of vibrational modes:

The **exclusion rule** states that if the molecule is symmetrical under inversion, then no modes can be both infrared and Raman active.

(A mode may be inactive in both.) A molecule is symmetric under inversion if it looks unchanged when each atom is projected through a single point, the **centre of inversion**, and out an equal distance on the

442

other side. Because we can often judge intuitively when a mode changes the molecular dipole moment, we can use this rule to identify modes that are not Raman active. The rule applies to CO_2 but to neither H_2O nor CH_4 because they have no centre of symmetry.

Exercise E11.12 One vibrational mode of benzene is a 'breathing mode' in which the ring alternately expands and contracts. May it be vibrationally Raman active?

[Yes]

One application of vibrational Raman spectroscopy is to the determination of the structures of nonpolar molecules such as XeF_4 and SF_6. Another application makes use of the fact that the intensity characteristics of Raman transitions, which depend on molecular polarizabilities, are more readily transferred from molecule to molecule than the intensities of infrared spectra, which depend on dipole moments and are more sensitive to the other groups present in a molecule and to the solvent. Hence, Raman spectra are useful in the identification of organic and inorganic species in solution. An example of the technique is shown in Fig. 11.16, which shows the vibrational Raman spectrum of an aqueous solution of lysozyme and, for comparison, a superposition of the Raman spectra of the constituent amino acids. The differences are indications of the effects of conformation, environment, and specific interactions (such as S—S linking) in the enzyme molecule.

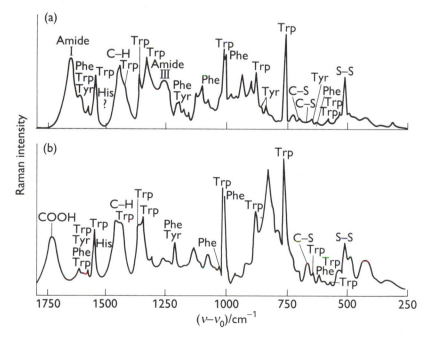

Fig. 11.16 The vibrational Raman spectrum of lysozyme in water and the superposition of the Raman spectra of the constituent amino acids. (From D. A. Long (1977), *Raman spectroscopy*. Copyright McGraw-Hill Inc. Used with the permission of the McGraw-Hill Book Company.)

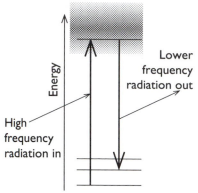

Fig. 11.17 In the resonance Raman effect, the incident radiation has a frequency corresponding to an actual electronic excitation of the molecule. A photon is emitted when the excited state returns to a state close to the ground state.

A modification of the basic Raman effect involves using incident radiation that coincides with the frequency of an electronic transition of the sample (Fig. 11.17; compare Fig. 11.2, where the incident radiation does not coincide with an actual electronic transition of the sample). The technique is then called **resonance Raman spectroscopy**. It is characterized by a much greater intensity in the scattered radiation and, because only a few vibrational modes contribute to the scattering, to a greatly simplified Raman spectrum. The resonance Raman spectrum shown in Fig. 11.18, for example, is of solid potassium chromate. The nine peaks that are identified are the Stokes lines that correspond to the excitation of the symmetric breathing mode of the tetrahedral CrO_4^{2-} ion and the transfer of up to nine vibrational quanta during the photon–ion collision. The high intensity of the resonance Raman transitions is used to examine the metal ions in biological macromolecules (such as the iron in haemoglobin and cytochromes or the cobalt in vitamin B), which are present in such low abundances that conventional Raman spectroscopy cannot detect them.

Electronic transitions: ultraviolet and visible spectra

The energies needed to change the distributions of electrons in molecules are of the order of several electronvolts (1 eV is about $8000\ cm^{-1}$). Consequently, the photons emitted or absorbed when such changes occur lie in the visible and ultraviolet regions of the spectrum, which spread from about $14\,000\ cm^{-1}$ for red light to $21\,000\ cm^{-1}$ for blue, and on to $50\,000\ cm^{-1}$ for ultraviolet radiation (Table 11.3). Indeed, many of the colours of the objects in the world around us, including the green of vegetation, the colours of flowers and of synthetic dyes, and the colours of pigments and minerals, stem from transitions in which electrons are shifted from one orbital of a molecule or ion into another. The migration of electrons that takes place when chlorophyll absorbs red and blue light (leaving green to be

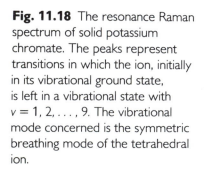

Fig. 11.18 The resonance Raman spectrum of solid potassium chromate. The peaks represent transitions in which the ion, initially in its vibrational ground state, is left in a vibrational state with $v = 1, 2, \ldots, 9$. The vibrational mode concerned is the symmetric breathing mode of the tetrahedral ion.

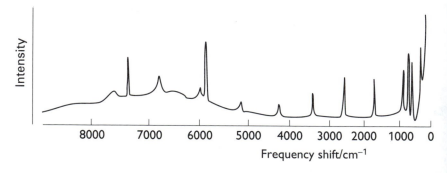

Table 11.3 Colour, frequency, and energy of light

Colour	λ/nm	$\nu/10^{14}$ Hz	$\bar{\nu}/10^4$ cm^{-1}	E/eV	E/kJ mol^{-1}
Infrared	1000	3.00	1.00	1.24	120
Red	700	4.28	1.43	1.77	171
Orange	620	4.84	1.61	2.00	193
Yellow	580	5.17	1.72	2.14	206
Green	530	5.66	1.89	2.34	226
Blue	470	6.38	2.13	2.64	254
Violet	420	7.14	2.38	2.95	285
Near ultraviolet	300	10.0	3.33	4.15	400
Far ultraviolet	200	15.0	5.00	6.20	598

reflected) is the primary energy-harvesting step by which our planet captures energy from the Sun and uses it to drive the nonspontaneous reactions of photosynthesis. In some cases the relocation of electrons may be so extensive that it results in the breaking of a bond and the dissociation of the molecule: such processes give rise to the numerous reactions of photochemistry, including the reactions that both sustain and damage the atmosphere.

11.6 The Franck–Condon principle

The nuclei in a molecule are subjected to different Coulomb forces from the surrounding electrons after an electronic transition has occurred and the molecule may respond by bursting into vibration. As a result, some of the energy used to redistribute an electron is in fact used to stimulate the vibrations of the absorbing molecules, and instead of a single purely electronic absorption line being observed, the absorption spectrum consists of many lines. Such **vibrational structure** of an electronic transition can be resolved if the sample is gaseous, but in a liquid or solid the lines usually merge together and result in a broad, almost featureless band of absorption (Fig. 11.19). Superimposed on the vibrational transitions that accompany the electronic transition is an additional line structure that arises from rotational transitions. The electronic spectra of gaseous samples are, in comparison, very complicated, but rich in information.

The details of the appearance of the vibrational structure of a band are explained by the following principle:

The **Franck–Condon principle** states that, because the nuclei are so much more massive than the electrons, an electronic transition takes place faster than the nuclei can respond.

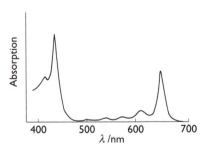

Fig. 11.19 The absorption spectrum of chlorophyll in the visible region. Note that it absorbs in the red and blue regions, and that green light is not absorbed.

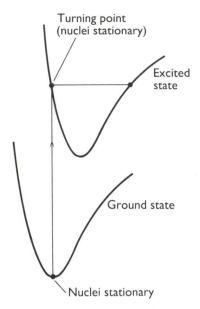

Fig. 11.20 According to the Franck–Condon principle, the most intense vibronic transition is from the ground vibrational state to the vibrational state lying vertically above it. Transitions to other vibrational levels also occur, but with lower intensity.

As a result of the electronic transition, electronic density is rapidly built up in new regions of the molecule and removed from others, and the initially stationary nuclei suddenly experience a new force field. They respond to the new force by beginning to vibrate, and (in classical terms) swing backwards and forwards from their original separation (which was maintained during the rapid electronic excitation). The stationary equilibrium separation of the nuclei in the initial electronic state therefore becomes a turning point, the points of a vibration when the nuclei are at the end points of their swing, in the final electronic state (Fig. 11.20). In practice, the electronically excited molecule may be formed in one of several excited vibrational states, so the absorption occurs at several different wavenumbers. As remarked above, in a condensed medium, the individual transitions are so broad that they merge together to give a broad band of absorption.

11.7 Specific types of transition

The absorption of a photon can often be traced to the excitation of electrons that belong to a small group of atoms. For example, when a carbonyl group is present, an absorption at about 290 nm is normally observed. Groups with characteristic optical absorptions are called **chromophores** (from the Greek for 'colour bringer'), and their presence often accounts for the colours of substances.

A *d*-metal complex may absorb light as a result of the transfer of an electron from the ligands into the *d* orbitals of the central atom, or vice versa (Fig. 11.21). In such **charge-transfer transitions** the electron moves through a considerable distance, which means that the redistribution of charge may be large and the absorption correspondingly intense. This mode of chromophore activity is shown by the permanganate ion, MnO_4^-: the charge redistribution that accompanies the migration of an electron from the O atoms to the central Mn atom accounts for its intense violet colour (resulting from absorption in the range 420 to 700 nm).

The transition responsible for absorption in carbonyl compounds can be traced to the lone pairs of electrons on the O atom. One of these electrons may be excited into an empty π^* orbital of the carbonyl group (Fig. 11.22), which gives rise to an ***n*-to-π^* transition**. Typical absorption energies are about 4 eV.

Fig. 11.21 A schematic representation of a charge transfer transition in which an electron initially on the outer atoms (represented by the grey tint in a) migrate to the central ion in (b).

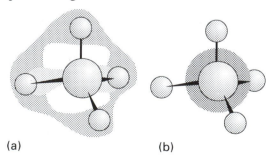

(a)   (b)

A C=C double bond acts as a chromophore because the absorption of a photon excites a π electron into an antibonding π^* orbital (Fig. 11.23). The chromophore activity is therefore due to a **π-to-π^* transition**. Its energy is around 7 eV for an unconjugated double bond, which corresponds to an absorption at 180 nm (in the ultraviolet). When the double bond is part of a conjugated chain, the energies of the molecular orbitals lie closer together and the transition shifts into the visible region of the spectrum. Many of the reds and yellows of vegetation are a consequence of transitions of this kind. For example, the carotenes that are present in green leaves (but are concealed by the intense absorption of the chlorophyll until the latter decays in the autumn) collect some of the solar radiation incident on the leaf by a π-to-π^* transition in their long conjugated hydrocarbon chains.

An important example of a π-to-π^* transition is provided by the photochemical mechanism of vision. The retina of the eye contains 'visual purple', which is a protein in combination with 11-*cis*-retinal (**1**). The 11-*cis*-retinal acts as a chromophore, and is the primary receptor for photons entering the eye. A solution of 11-*cis*-retinal absorbs at about 380 nm, but in combination with the protein (a link which might involve the elimination of the terminal carbonyl group) the absorption maximum shifts to about 500 nm and tails into the blue. The conjugated double bonds are responsible for the ability of the molecule to absorb over the entire visible region, but they also play another important role. In its electronically excited state the conjugated chain can isomerize, one half of the chain being able to twist about an excited C=C bond, which is now no longer torsionally rigid, and forming all-*trans*-retinal (**2**). On account of its different shape, the new isomer cannot fit into the protein. The primary step in vision therefore appears to be photon absorption followed by isomerization: the uncoiling of the molecule then triggers a nerve impulse to the brain.

11.8 Radiative decay

One process by which an electronically excited molecule can discard its excess energy is by **radiative decay**, in which an electron collapses back into a lower energy orbital and in the process generates a photon. As a result, an observer sees the sample glowing (if the emitted radiation is in the visible region of the spectrum).

There are two principal modes of radiative decay, fluorescence and phosphorescence. In **fluorescence**, the spontaneously emitted radiation ceases immediately after the exciting radiation is extinguished. In

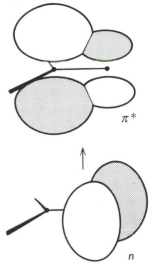

Fig. 11.22 The $>$C=O group acts as a chromophore primarily on account of the excitation of a nonbonding O lone-pair electron to an antibonding C=O π^* orbital.

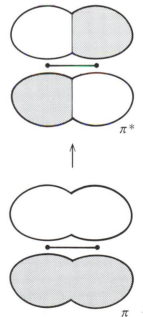

Fig. 11.23 The C=C double bond acts as a chromophore. One of its important transitions is the π^*-to-π transition illustrated here, in which an electron is promoted from a π orbital to the corresponding antibonding orbital.

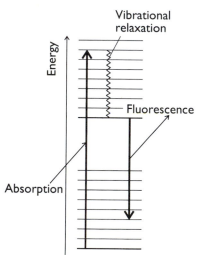

Fig. 11.24 The sequence of steps leading to fluorescence. After the initial absorption the upper vibrational states undergo radiationless decay—the process of vibrational relaxation—by giving up energy to the surroundings. A radiative transition then occurs from the ground state of the upper electronic state. In practice, the separation of the ground states of the electronic states (the heavy horizontal lines) is 10 to 100 times greater than the separation of the vibrational levels.

phosphorescence, the spontaneous emission may persist for long periods (even hours, but characteristically seconds or fractions of seconds). The difference suggests that fluorescence is an immediate conversion of absorbed light into re-emitted energy and that phosphorescence involves the storage of energy in a reservoir from which it slowly leaks. One very important application of radiative decay is to the operation of lasers, and we shall consider the steps involved.

Fluorescence and phosphorescence

Figure 11.24, which is a simple example of a **Jablonski diagram**, a schematic portrayal of molecular electronic and vibrational energy levels, shows the sequence of steps involved in fluorescence. The initial absorption takes the molecule to an excited electronic state, and if the absorption spectrum were monitored it would look like the one shown in Fig. 11.25(a). The excited molecule is subjected to collisions with the surrounding molecules, and as it gives up energy it steps down the ladder of vibrational levels. The surrounding molecules, however, might be unable to accept the larger energy difference needed to lower the molecule to the ground state. It might therefore survive long enough to generate a photon and emit the remaining excess energy as radiation. The downward electronic transition is vertical (in accord with the Franck–Condon principle) and the fluorescence spectrum (Fig. 11.25(b)) has a vibrational structure characteristic of the *lower* electronic state.

Fluorescence occurs at a lower frequency than that of the incident radiation because the fluorescence radiation is emitted after some vibrational energy has been discarded into the surroundings. The vivid oranges and greens of fluorescent dyes are an everyday manifestation of this effect: they absorb in the ultraviolet and blue, and fluoresce in the visible. The mechanism also suggests that the intensity of the fluorescence ought to depend on the ability of the solvent molecules to accept the electronic and vibrational quanta. It is indeed found that a solvent composed of molecules with widely spaced vibrational levels (such as water) may be able to accept the large quantum of electronic energy and so quench the fluorescence.

Figure 11.26 is a Jablonski diagram that shows the sequence of events leading to phosphorescence. The first steps are the same as in fluorescence, but the presence of a **triplet state** plays a decisive role. A triplet state is one in which two electrons in different orbitals have *parallel* spins: the ground state of O_2 which was discussed in Section 9.6 is an example. The name 'triplet' reflects the (quantum mechanical) fact that two parallel spins can adopt only three orientations with respect to an external magnetic field ($\uparrow\uparrow$, $\rightrightarrows$, and $\downarrow\downarrow$); an ordinary spin-paired state ($\uparrow\downarrow$) is called a **singlet state** because there is only one arrangement in space for such a pair of spins.

The ground state of a typical phosphorescent molecule is a singlet because its electrons are all paired, and the excited state to which the absorption excites the molecule is also a singlet. The peculiar feature of a phosphorescent molecule, however, is that it possesses an excited

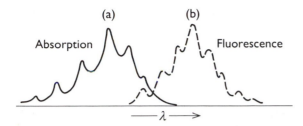

Fig. 11.25 The absorption spectrum (a) shows a vibrational structure characteristic of the upper state. The fluorescence spectrum (b) shows a structure characteristic of the lower state; it is also displaced to lower frequencies and resembles a mirror image of the absorption.

triplet state of an energy similar to that of the excited singlet state, and into which the excited singlet state may convert. Hence, if there is a mechanism for unpairing two electron spins (and so converting ↑↓ into ↑↑), the molecule may undergo **intersystem crossing** and become a triplet state. The unpairing of electron spins is possible if the molecule contains atoms of a heavy element (such as sulfur), because their nuclei can exert such strong magnetic fields on a nearby electron that the spin may be reversed.†

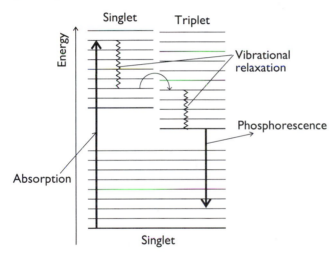

Fig. 11.26 The sequence of steps leading to phosphorescence. The important step is the crossing from an excited singlet to an excited triplet state. The triplet state acts as a slowly radiating reservoir because the return to the ground state is very slow.

If an excited singlet molecule crosses into a triplet state, it continues to deposit energy into the surroundings and to step down the ladder of vibrational states. However, it is now stepping down the triplet's ladder, and at the lowest vibrational energy level it is trapped. The solvent cannot extract the final, large quantum of electronic excitation energy, and the molecule cannot radiate its energy because return to the ground state is forbidden: a triplet state cannot convert into a singlet state because an electron spin cannot reverse in a transition. The radiative transition, however, is not totally forbidden because the mechanism that was responsible for the intersystem crossing also

† This argument is greatly abbreviated. The phenomenon responsible for the reversal of an electron spin is called *spin–orbit coupling*, in which the magnetic field arising from an electron's orbital motion around the nucleus interacts with the spin magnetic moment of the electron, and flips it into a new orientation. The strength of the orbitally generated magnetic field increases as the nuclear charge increases.

breaks the selection rule. The molecules are therefore able to emit weakly, and the emission may continue long after the original excited state was formed.

The mechanism of phosphorescence summarized in Fig. 11.26 accounts for the observation that the excitation energy seems to become trapped in a slowly leaking reservoir. It also suggests (as is confirmed experimentally) that phosphorescence should be most intense from solid samples: energy transfer is then less efficient and the intersystem crossing has time to occur as the singlet excited state loses vibrational energy. The mechanism also suggests that the phosphorescence efficiency should depend on the presence of a moderately heavy atom (with its ability to flip electron spins, see the footnote), which is in fact the case.

Another fate for an electronically excited molecule is **dissociation**, or fragmentation (Fig. 11.27). The onset of dissociation can be detected in an absorption spectrum by seeing that the vibrational structure of a band terminates at a certain energy. Absorption occurs in a continuous band above this **dissociation limit** (the highest frequency before the onset of continuous absorption) because the final state is unquantized translational motion of the fragments. Locating the dissociation limit is a valuable way of determining the bond dissociation energy.

Lasers

Lasers have transformed experimental chemistry as much as they have the everyday world. The word **laser** is an acronym formed from **l**ight **a**mplification by **s**timulated **e**mission of **r**adiation. As this name suggests, it is a process that depends on stimulated emission as distinct from the spontaneous emission processes characteristic of fluorescence and phosphorescence. In **stimulated emission**, an excited state is stimulated to emit a photon by the presence of radiation of the same frequency, and the more photons there are present, the greater the probability of the emission. To picture the process, we can think of the oscillations of the electromagnetic field as periodically (at the frequency at which the field oscillates) distorting the excited molecule at the frequency of the transition and hence encouraging the molecule to generate a photon of the same frequency. The essential feature of laser action is the strong **gain**, or growth of intensity, that results: the more photons present of the appropriate frequency, the more photons of that frequency the excited molecules will be stimulated to form, and so the laser medium fills with photons.

One requirement of laser action is the existence of an excited state that has a long enough lifetime for it to participate in stimulated emission. Another requirement is the existence of a greater population in the upper state than in the lower state where the transition terminates, for otherwise there would be a net absorption rather than the net emission required. Because at thermal equilibrium the opposite is true, it is necessary to achieve a **population inversion** in

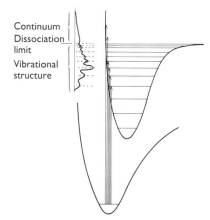

Fig. 11.27 When absorption occurs to unbound states of the upper electronic state, the molecule dissociates and the absorption is a continuum. Below the dissociation limit the electronic spectrum shows a normal vibrational structure.

which there are more molecules in the upper state than in the lower. One way of achieving population inversion is illustrated in Fig. 11.28. The inversion is achieved indirectly through an intermediate state, I. Thus, the molecule is excited to I, which then gives up some of its energy nonradiatively (by passing energy on to vibrations of the surroundings) and changes into a lower state B; the laser transition is the return of B to a lower state A. Because four levels are involved overall, this arrangement leads to a **four-level laser**. The transition from X to I is caused by an intense flash of light in the process, called **pumping**. In some cases the pumping flash is achieved with an electric discharge through xenon or with the radiation from another laser.

Laser radiation has a number of advantages for applications in chemistry. One advantage is its highly monochromatic character, which enables very precise spectroscopic observations to be made. Another advantage is the ability of laser radiation to be produced in very short pulses (currently, as brief as about 1 fs, where 1 fs, 1 femtosecond, is 10^{-15} s): as a result, very fast chemical events, such as the individual transfer of atoms during a chemical reaction, can be followed. Laser radiation is also very intense, which reduces the time needed for spectroscopic observations: this characteristic is particularly useful in Raman spectroscopy, where the scattered radiation is of very low intensity.

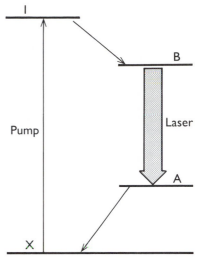

Fig. 11.28 The transitions involved in a four-level laser. Because the laser transition terminates in an excited state (A), the population inverson between B and A is much easier to achieve than when the lower state of the laser transition is the ground state.

11.9 Photoelectron spectroscopy

The exposure of a molecule to high-frequency radiation can result in the ejection of an electron. This **photoejection** is the basis of another type of spectroscopy in which the energies of the ejected **photoelectrons** are monitored. The principle of the technique is that an incoming photon of frequency ν has an energy $h\nu$; that energy may be transferred to an electron to remove it from its orbital (which requires an energy I, where I is the ionization energy of the electron from the orbital it occupies), leaving the remainder of the original energy of the photon to appear as kinetic energy of the electron, $\frac{1}{2}m_e v^2$, where v is the velocity of the photoelectron after it has been ejected. From the conservation of energy, we can write

$$h\nu = I + \tfrac{1}{2}m_e v^2 \tag{11}$$

Therefore, by monitoring the velocity of the photoelectron, and knowing the frequency of the incident radiation, it is possible to deduce the ionization energy of the electron and hence the strength with which the electron was bound in the molecule (Fig. 11.29): the slower the ejected electron, the deeper the orbital from which it was ejected. The apparatus is a modification of a mass spectrometer (Fig. 11.30), in which the velocity of the photoelectrons is measured by determining the strength of the electric field required to bend their paths on to the detector.

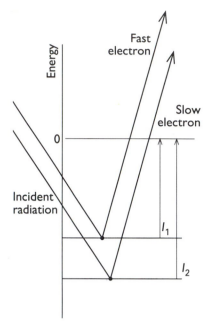

Fig. 11.29 The basic principle of photoelectron spectroscopy. An incoming photon of known energy collides with an electron in one of the orbitals and expels it with a kinetic energy that is equal to the difference between the energy supplied by the photon and the ionization energy from the occupied orbital. An electron from an orbital with a low ionization energy will emerge with a high kinetic energy (and high speed) whereas an electron from an orbital with a high ionization energy will be ejected with a low kinetic energy (and low speed).

Exercise E11.14 What is the velocity of photoelectrons that are ejected from a molecule with 21 eV radiation (from a helium discharge lamp) and are known to come from an orbital of ionization energy 12 eV?

$$[1.8 \times 10^3 \text{ km s}^{-1}]$$

A typical **photoelectron spectrum** (of HBr) is shown in Fig. 11.31. If we disregard the fine structure, we see that the HBr lines fall into two main groups. The least tightly bound electrons (with the lowest ionization energies and hence highest kinetic energies when ejected) are the lone pairs of the Br atom. The next ionization energy lies at 15.2 eV, and corresponds to the removal of an electron from the H—Br σ bond.

The HBr spectrum shows that ejection of a σ electron is accompanied by a considerable amount of vibrational excitation. The Franck–Condon principle would account for this observation if ejection were accompanied by an appreciable change of equilibrium bond length between HBr and HBr$^+$: if that is so, then the ion is formed in a bond-compressed state, which is consistent with the important bonding effect of the σ electrons. The lack of much vibrational structure in the other band is consistent with the nonbonding role of the Br $4p_x$ and $4p_y$ lone-pair electrons, for the equilibrium bond length is little changed when one is removed.

Example Interpreting a UV photoelectron spectrum
The highest kinetic-energy electrons in the spectrum of H_2O using 21.22 eV He radiation are at about 9 eV and show a large vibrational spacing of 0.41 eV. The symmetric stretching mode of the neutral H_2O molecule lies at 3652 cm^{-1}. What conclusions can be drawn about the nature of the orbital from which the electron is ejected? Note that $1 \text{ eV} \approx 8000 \text{ cm}^{-1}$.
Answer Because 0.41 eV corresponds to 3310 cm^{-1}, which is similar to the 3652 cm^{-1} of the nonionized molecule, we can suspect that the electron is ejected from an orbital that has little influence on the bonding in the molecule. That is, photoejection is from a largely nonbonding orbital.

Exercise E11.15 In the same spectrum of H_2O, the band near 7.0 eV shows a long vibrational series with spacing 0.125 eV. The bending mode of H_2O lies at 1596 cm^{-1}. What conclusions can you draw about the characteristics of the orbital occupied by the photoelectron?
[The electron contributed to long distance H—H bonding across the molecule]

Nuclear magnetic resonance

One of the most widely used and helpful forms of spectroscopy, and a technique that has transformed the practice of chemistry and its dependent disciplines, makes use of an effect that is familiar from classical physics. When two pendulums are joined by the same slightly flexible support and one is set in motion, the other is forced into oscillation by the motion of the common axle, and energy flows between the two. The energy transfer occurs most efficiently when the frequencies of the two oscillators are identical (Fig. 11.32). The condition of strong effective coupling when the frequencies are identical is called **resonance**, and the excitation energy is said to **resonate** between the coupled oscillators.

Resonance is the basis of a number of everyday phenomena, including the response of radios to the weak oscillations of the electromagnetic field generated by a distant transmitter. In this section we explore a spectroscopic application that when originally developed (and in some cases still) depends on matching a set of energy levels to a source of monochromatic radiation and observing the strong absorption that occurs at resonance.

11.10 Principles of magnetic resonance

The application of resonance that we describe here depends on the fact that many nuclei possess spin angular momentum (just as an

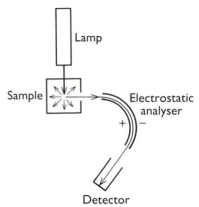

Fig. 11.30 A photoelectron spectrometer consists of a source of ionizing radiation, such as a helium discharge lamp for ultraviolet photon electron spectroscopy (UPS) and an X-ray source for X-ray photoelectron spectroscopy (XPS), an electrostatic analyser, and an electron detector. The deflection of the electron path caused by the analyser depends on their speed.

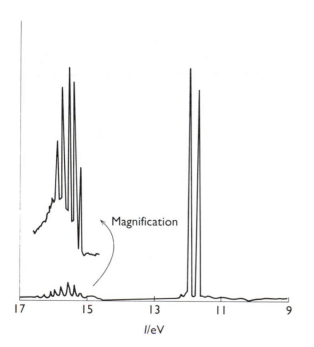

Fig. 11.31 The photoelectron spectrum of HBr. The lowest ionization-energy band corresponds to the ejection of a Br lone-pair electron. The higher ionization-energy band corresponds to the ejection of a bonding electron. The structure of the latter band is due to the vibrational excitation of HBr$^+$ that results from the ionization.

453

Fig. 11.32 Two pendulums hanging from the same support (which helps to convey the motion of one pendulum to the other, and so weakly couples them together) show the phenomenon of resonance when their two natural frequencies (as determined by their lengths) are the same. When their natural frequencies are different ((a) and (c)), the motion of one has only a slight effect on the other. When their natural frequencies are the same (b), if one is set in motion the other begins to swing in time with it.

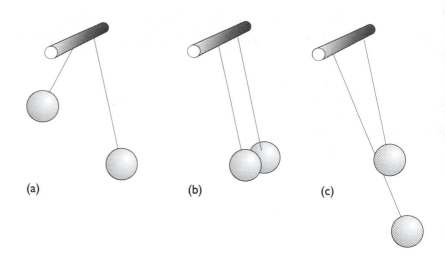

electron does, but nuclei possess a wider range of values). A nucleus with spin quantum number I (the analogue of s for electrons, and which may be an integer or half-integer) may take $2I + 1$ different orientations, relative to an arbitrary axis, which are distinguished by the quantum number m_I:

$$m_I = I, I - 1, \ldots, -I$$

A proton has $I = \frac{1}{2}$ (the same as an electron) and may adopt either of two orientations ($m_I = +\frac{1}{2}$ and $-\frac{1}{2}$). A ^{14}N nucleus has $I = 1$ and may adopt any of three orientations ($m_I = +1, 0, -1$). In this chapter we shall consider only **spin-$\frac{1}{2}$ nuclei**, those with $I = \frac{1}{2}$. Spin-$\frac{1}{2}$ nuclei include protons (^1H), ^{13}C, ^{19}F, and ^{31}P nuclei (Table 11.4). As for electrons, the state with $m_I = +\frac{1}{2}$ ($\uparrow$) is denoted α and that with $m_I = -\frac{1}{2}$ ($\downarrow$) is denoted β. It is worth bearing in mind that two very common nuclei, ^{12}C and ^{16}O, have zero spin and hence are invisible in magnetic resonance.

The energies of nuclei in magnetic fields

A nucleus with nonzero spin behaves like a tiny magnet. The orientation of this magnet is determined by the value of m_I, and in a magnetic field B the $2I + 1$ orientations of the nucleus have different energies, which are given by

$$E_{m_I} = -g_I \mu_N B m_I \tag{12}$$

where g_I is the **nuclear g-factor**, a characteristic of the nucleus, and μ_N is the **nuclear magneton**:

$$\mu_N = \frac{e\hbar}{2m_p} = 5.051 \times 10^{-27} \text{ J T}^{-1} \tag{13}$$

Table 11.4 Nuclear spin properties

Nucleus	Natural abundance, per cent	Spin, I
^1H	99.98	$\frac{1}{2}$
$^1\text{H (D)}$	0.0156	1
^{12}C	98.99	0
^{13}C	1.11	$\frac{1}{2}$
^{14}N	99.64	1
^{16}O	99.96	0
^{17}O	0.037	$\frac{5}{2}$
^{19}F	100	$\frac{1}{2}$
^{31}P	100	$\frac{1}{2}$
^{35}Cl	75.4	$\frac{3}{2}$
^{37}Cl	24.6	$\frac{3}{2}$

(m_p is the mass of the proton and T denotes the unit *tesla* which is used to measure the intensity of a magnetic field.) Nuclear g-factors are numbers of the order of 1: for protons, $g_I = 5.5857$. Positive values of g_I indicate that the nuclear magnet lies in the same direction as the nuclear spin (this is the case for protons) whereas negative values of g_I indicate that the magnet points in the opposite direction. The strength of a nuclear magnet is about 2000 times weaker than that of the magnet associated with electron spin.

The energy separation of the two states of spin-$\frac{1}{2}$ nuclei (Fig. 11.33) is

$$\Delta E = E_\beta - E_\alpha = \tfrac{1}{2}g_I\mu_N B - (-\tfrac{1}{2}g_I\mu_N B) = g_I\mu_N B$$

For most nuclei, g_I is positive, so the β state lies above the α state and there are slightly more α spins than β spins. If the sample is bathed in radiation of frequency ν, the energy separations come into resonance with the radiation when the frequency satisfies the **resonance condition**

$$h\nu = g_I\mu_N B \qquad (14)$$

At resonance there is strong coupling between the nuclear spins and the radiation, and strong absorption occurs as the spins flip from ↑ to ↓.

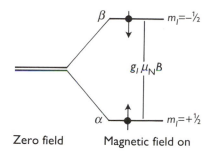

Fig. 11.33 The nuclear spin energy levels of a spin-$\frac{1}{2}$ nucleus (e.g. ^{1}H or ^{13}C) in a magnetic field. Resonance occurs when the energy separation of the levels matches the energy of the photons in the electromagnetic field.

Exercise E11.16 Calculate the frequency (in megahertz) at which radiation comes into resonance with proton spins in a 12 T magnetic field.

[510 MHz]

The technique

In its simplest form, **nuclear magnetic resonance** (NMR) is the study of the properties of molecules containing magnetic nuclei by means of the application of a magnetic field and the observation of the frequency at which they come into resonance with a radiofrequency electromagnetic field. When applied to proton spins, the technique is called **proton magnetic resonance** (^{1}H-NMR). In the early days of the technique the only nuclei that could be studied were protons (which behave like relatively strong magnets because g_I is large), but now a wide variety of nuclei (especially ^{13}C and ^{31}P) are investigated routinely.

An NMR spectrometer consists of a magnet that can produce a uniform, intense field and the appropriate sources of radiofrequency radiation. In simple instruments the magnetic field is provided by an electromagnet; for serious work, a superconducting magnet capable of producing fields of the order of 10 T and more is used. (A magnetic field of 10 T is very strong: a small magnet, for example, gives a magnetic field of only a few millitesla.) The use of high magnetic fields has two advantages. One is that the field increases the energy separation and therefore the population difference between the two spin states, and hence a stronger net absorption is obtained. Secondly, a high field simplifies the appearance of certain spectra.

Exercise E11.17 Calculate the ratio of numbers of ↑ and ↓ protons in a sample that is exposed to (a) a 1.0 T magnetic field, (b) a 10 T magnetic field at 20 °C.
[*Hint*: use the Boltzmann ratio (*Further information* 7, p. 258) and the energy separation calculated from eqn (12).]
[(a) 1.000 007 0, (b) 1.000 070]

The sample (of volume about $1\,cm^3$) is placed on the axis of the cylindrically wound magnet, and is rotated at about 15 cycles per second. This spinning helps to remove magnetic inhomogeneities and ensures that all the magnetic nuclei experience the same field. Although a superconducting magnet operates at the temperature of liquid helium (4 K), the sample itself is normally at room temperature.

11.11 The information in NMR spectra

Nuclear spins interact with the *local* magnetic field. The local field may differ from the applied field either on account of the local electronic structure of the molecule or because there is another magnetic nucleus nearby.

The chemical shift

The applied magnetic field can induce a circulating motion of the electrons in the molecule, and that motion gives rise to a small additional magnetic field δB. This additional field is proportional to the applied field, and it is conventional to express it as

$$\delta B = -\sigma B \tag{15}$$

where σ is the **shielding constant** (which may be positive or negative according to whether the induced field lies in the opposite or same direction as the applied field). The ability to the applied field to induce the circulation of electrons through the nuclear framework of the molecule depends on the details of the electronic structure near the magnetic nucleus of interest, and so nuclei in different chemical groups have different shielding constants.

Because the total local field is

$$B_{loc} = B + \delta B = (1 - \sigma)B$$

the resonance condition is

$$h\nu = g_I\mu_N B_{loc} = g_I\mu_N(1 - \sigma)B \tag{16}$$

and is different for nuclei in different parts of the molecule (because σ varies with the molecular environment). Hence, different nuclei, even

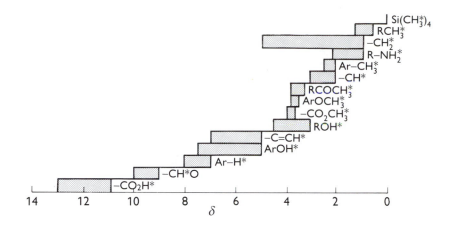

Fig. 11.34 The range of typical chemical shifts for ^{1}H resonances.

those of the same element so long as they are in different parts of a molecule, come into resonance at different frequencies.

The **chemical shift** of a nucleus is the difference between its resonance frequency and that of a reference standard. The standard for protons is the proton resonance in tetramethylsilane, $Si(CH_3)_4$, commonly referred to as TMS, which bristles with protons and dissolves without reaction in many samples. Other references are used for other nuclei. For ^{13}C, the reference frequency is the ^{13}C resonance in TMS and for ^{31}P it is the ^{31}P resonance in 85 per cent $H_3PO_4(aq)$. The separation of the resonance of a particular group of nuclei from the standard increases with the strength of the applied magnetic field because the induced field is proportional to the applied field, and the stronger the latter, the greater the shift.

Chemical shifts are reported on the **δ scale**, which is defined as

$$\delta = \frac{\nu - \nu^{\ominus}}{\nu^{\ominus}} \times 10^6 \tag{17}$$

where $\nu^{\ominus}$ is the resonance frequency of the standard. The advantage of the δ scale is that shifts reported on it are independent of the applied field (because both numerator and denominator are proportional to the applied field). The resonance frequencies themselves, however, do depend on the applied field through

$$\nu - \nu^{\ominus} = \delta \times \nu^{\ominus} \times 10^{-6} \tag{18}$$

A positive δ indicates that the resonance frequency of the group of nuclei in question is higher than that of the standard. Hence $\delta > 0$ indicates that the local magnetic field is stronger than that experienced by the nuclei in the standard under the same conditions. Some typical chemical shifts are given in Fig. 11.34.

Example Using the chemical shift
At what frequency shift from TMS would a group of nuclei with $\delta = 1.00$ resonate in a spectrometer operating at 500 MHz?

457

Answer From eqn (18), the shift in resonance from the standard (TMS) is

$$\nu - \nu^{\ominus} = 1.00 \times 500 \text{ MHz} \times 10^{-6} = 500 \text{ Hz}$$

In a spectrometer operating at 100 MHz, the shift would be only 100 Hz.

Exercise E11.18 What is the shift of the resonance from TMS of a group of nuclei with $\delta = 3.50$ and an operating frequency of 350 MHz?

[1.23 kHz]

The existence of a chemical shift explains the general features of the spectrum of ethanol shown in Fig. 11.35. The CH_3 protons form one group of nuclei with $\delta = 1$. The two CH_2 protons are in a different part of the molecule, experience a different local magnetic field, and hence resonate at $\delta = 3$. Finally, the OH proton is in another environment, and has a chemical shift of $\delta = 4$.

We can use the relative intensities of the signal (the areas under the absorption lines) to help distinguish which group of lines corresponds to which chemical group, and spectrometers can integrate the absorption—determine the areas under the absorption signal—automatically (as is shown in Fig. 11.35). In ethanol the group intensities are in the ratio $3:2:1$ because there are three CH_3 protons, two CH_2 protons, and one OH proton in each molecule. Counting the number of magnetic nuclei as well as noting their chemical shifts is valuable analytically because it helps us identify the compound present in a sample.

Fig. 11.35 An NMR spectrum of ethanol. The bold letters denote the protons giving rise to the resonance peak, and the step-like curve is the integrated signal.

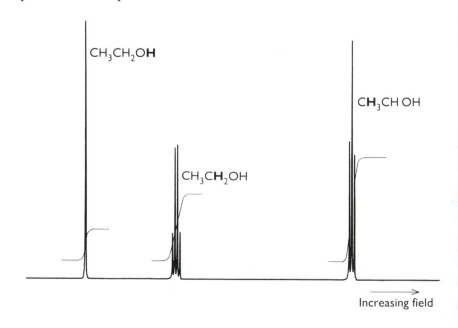

CH$_3$CH$_2$O**H**

CH$_3$C**H**$_2$OH

C**H**$_3$CH OH

Increasing field

If $\sigma > 0$, then the chemical shift is **negative** (to low frequencies) and we say that the nucleus is **shielded**.† If $\sigma < 0$, then the chemical shift is positive and we say that the nucleus is **deshielded**. For some purposes it is convenient to regard σ as the sum of a positive diamagnetic contribution, σ_d, and a negative paramagnetic contribution, σ_p, and to write

$$\sigma = \sigma_d + \sigma_p$$

Then σ is positive if the diamagnetic contribution dominates, and is negative if the paramagnetic contribution dominates.

The diamagnetic contribution arises from the ability of the applied field to generate orbital motion of the electrons in the molecule. The resulting circulation of the charge generates a magnetic field that opposes the applied field and reduces the frequency needed for resonance. The paramagnetic contribution, σ_p, arises from the ability of the applied field to force the electrons to circulate through the molecule by taking advantage of the availability of orbitals that are unoccupied in the ground state. It is zero in free atoms and around the axes of linear molecules (such as H—C≡C—H) where the electrons can circulate freely. Proton shielding constants are often dominated by the diamagnetic contribution, but this is not the case for the nuclei of other elements.

The fine structure

The splitting of resonances into individual lines in Fig. 11.35 is called the **fine structure** of the spectrum. It arises because each magnetic nucleus contributes to the local field experienced by the other nuclei and modifies their resonance frequencies. The strength of the interaction is expressed in terms of the **spin–spin coupling constant**, J and reported in hertz (Hz). Spin coupling constants are an intrinsic property of the molecule and independent of the strength of the applied field.

We shall consider first a molecule that contains two spin-$\frac{1}{2}$ nuclei, A and X. Suppose that the spin of X is α; then A will resonate at a certain frequency as a result of the combined effect of the external field, the shielding constant, and the spin–spin interaction of the nucleus A with X. The spin–spin coupling will result in one line in the spectrum of A being shifted by $\frac{1}{2}J$ from the frequency it would have in the absence of coupling. If the spin of X is β, then A will resonate at a frequency shifted by $-\frac{1}{2}J$. Therefore, instead of a single line from A, we get a doublet of lines separated by a frequency J (Fig. 11.36). The same splitting occurs in the X resonance: instead of a single line it is a doublet with splitting J (the same value as for the splitting of A).

---

† The shielding constant σ and the chemical shift δ indicate similar properties, but whereas the shielding constant is an indication of the shielding or antishielding of a proton in an absolute sense, the value of δ indicates the shielding relative to the standard (TMS).

459

Fig. 11.36 The effect of spin–spin coupling on an NMR spectrum of two spin-$\frac{1}{2}$ nuclei with widely different chemical shifts. Each resonance is split into two lines separated by J. Full circles indicate α spins, open circles indicate β spins.

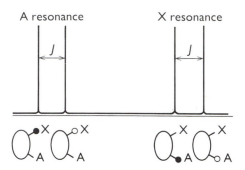

N	Intensity distribution
0	1
1	1 1
2	1 2 1
3	1 3 3 1
4	1 4 6 4 1
:	:

If there is another X nucleus in the molecule with the same chemical shift as the first X (giving an AX_2 species), then the resonance of A is split into a doublet by one X, and each line of the doublet is split again by the same amount (Fig. 11.37) by the second X. This splitting results in three lines in the intensity ratio $1:2:1$ (because the central frequency can be obtained in two ways). As in the AX case discussed above, the X resonance of the AX_2 species is split into a doublet by A.

Three equivalent X nuclei (an AX_3 species) split the resonance of A into four lines of intensity ratio $1:3:3:1$ (Fig. 11.38). The X resonance remains a doublet as a result of the splitting caused by A. In general, N equivalent spin-$\frac{1}{2}$ nuclei split the resonance of a nearby spin or group of equivalent spins into $N+1$ lines with an intensity distribution given by Pascal's triangle (shown in the margin). Subsequent rows of the triangle are formed by adding together the two adjacent numbers in the line above.

> **Exercise E11.19** Complete the next line of the triangle, the pattern arising from five equivalent protons.
>
> $[1:5:10:10:5:1]$

> **Example** Accounting for the fine structure in a spectrum
> Account for the fine structure in the ^{1}H-NMR spectrum of the C—H protons of ethanol.
> **Answer** The three protons of the CH_3 group split the single resonance of the CH_2 protons into a $1:3:3:1$ quartet with a splitting J. Likewise, the two protons of the CH_2 group split the single resonance of the CH_3 protons into a $1:2:1$ triplet. Each of these lines is split into a doublet to a small extent by the OH proton.

> **Exercise E11.20** What fine structure can be expected for the protons in NH_4^+?
>
> $[1:1:1$ triplet from N$]$

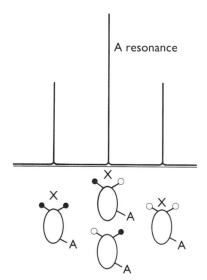

Fig. 11.37 The origin of the $1:2:1$ triplet in the A resonance of an AX_2 species. The two X nuclei may have the $2^2 = 4$ spin arrangements $(\uparrow\uparrow)$; $(\uparrow\downarrow)$, $(\downarrow\uparrow)$; $(\downarrow\downarrow)$. The middle two arrangements are responsible for the coincident resonances of A.

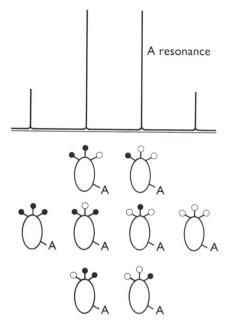

A resonance

Fig. 11.38 The origin of the 1:3:3:1 quartet in the A resonance of an AX_3 species where A and X are spin-$\frac{1}{2}$ nuclei with widely different chemical shifts. There are $2^3 = 8$ arrangements of the spins of the three X nuclei, and their effects on the A nucleus give rise to four groups of resonances.

The spin–spin coupling constant of two nuclei joined by n bonds is normally denoted ^nJ, with subscripts for the types of nuclei involved. Thus, $^1J_{CH}$ is the coupling constant for a proton joined directly to a ^{13}C atom, and $^2J_{CH}$ is the coupling constant when the same two nuclei are separated by two bonds (as in ^{13}C—C—H). A typical value of $^1J_{CH}$ is between 10^2 to 10^3 Hz; $^2J_{CH}$ is about 10 times less, between about 10 and 10^2 Hz. Both ^3J and ^4J give detectable effects in a spectrum, but couplings over larger numbers of bonds can generally be ignored.

Example Interpreting an NMR spectrum
Suggest an interpretation of the 60 MHz ^1H-NMR spectrum in Fig. 11.39.
Answer We need to look for groups with characteristic chemical shifts (Fig. 11.34) and account for the fine structure as was done for ethanol. The resonance at $\delta = 3.4$ corresponds to $>CH_2$ in an ether; that at $\delta = 1.2$ corresponds to CH_3 in CH_3CH_2. The fine structure of the $>CH_2$ group (a 1:3:3:1 quartet) is characteristic of splitting caused by CH_3; the fine structure of the CH_3 resonance is characteristic of splitting caused by CH_2. The fine structure constant is $J = \pm 60$ Hz (the same for each group). The compound is probably $(CH_3CH_2)_2O$.

Exercise E11.21 What changes in the spectrum would be observed on recording it at 300 MHz?
[Groups of lines 5 times further apart in frequency (but the same δ values); no change in spin–spin splitting]

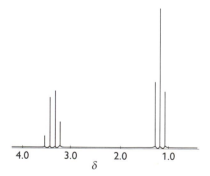

Fig. 11.39 The NMR spectrum considered in the example.

EXERCISES

11.1 Which of the following molecules may show infrared absorption spectra: (a) H_2, (b) HCl, (c) CO_2, (d) H_2O, (e) CH_3CH_3, (f) CH_4, (g) CH_3Cl, (h) N_2?

11.2 The wavenumber of the fundamental vibrational transition of Cl_2 is 565 cm^{-1}. Calculate the force constant of the bond.

11.3 What is the Doppler-shifted wavelength of a red (660 nm) traffic light approached at 50 m.p.h.? At what speed would it appear green (520 nm)?

11.4 A spectral line of $^{48}Ti^{8+}$ in a distant star was found to be shifted from 654.2 nm to 706.5 nm and to be broadened to 61.8 pm. What is the speed of recession and the surface temperature of the star?

11.5 Estimate the lifetime of a state that gives rise to a line of width (a) 0.1 cm^{-1}, (b) 1 cm^{-1}, (c) 100 MHz.

11.6 A molecule in a liquid undergoes about 1×10^{13} collisions in each second. Suppose that (a) every collision is effective in deactivating the molecule vibrationally and (b) that one collision in 100 is effective. Calculate the width (in cm^{-1}) of vibrational transitions in the molecule.

11.7 The hydrogen halides have the following fundamental vibrational wavenumbers:

	HF	HCl	HBr	HI
$\bar{v}/\text{cm}^{-1}$	4141.3	2988.9	2649.7	2309.5

Calculate the force constants of the hydrogen–halogen bonds.

11.8 From the data in Exercise 11.7, predict the fundamental vibrational wavenumbers of the deuterium halides.

11.9 Consider the vibrational mode that corresponds to the uniform expansion of the benzene ring. Is it (a) Raman, (b) infrared active?

11.10 The molar absorption coefficient of a substance dissolved in hexane is known to be $855 \text{ mol}^{-1} \text{ L cm}^{-1}$ at 270 nm. Calculate the percentage reduction in intensity when light of that wavelength passes through 2.5 mm of a solution of concentration $3.25 \times 10^{-3} \text{ mol L}^{-1}$.

11.11 When light of wavelength 400 nm passes through 3.5 mm of a solution of an absorbing substance at a concentration $6.67 \times 10^{-4} \text{ mol L}^{-1}$, the transmission is 65.5 per cent. Calculate the molar absorption coefficient of the solute at this wavelength and express the answer in $\text{cm}^2 \text{ mol}^{-1}$.

11.12 The molar absorption coefficient of a solute at 540 nm is $286 \text{ mol}^{-1} \text{ L cm}^{-1}$. When light of that wavelength passes through a 6.5 mm cell containing a solution of the solute, 46.5 per cent of the light was absorbed. What is the concentration of the solution?

11.13 The compound $CH_3CH\!\!=\!\!CHCHO$ has a strong absorption in the ultraviolet at $46\,950 \text{ cm}^{-1}$ and a weak absorption at $30\,000 \text{ cm}^{-1}$. Justify these features in terms of the structure of the compound.

11.14 The following data were obtained for the absorption by Br_2 in carbon tetrachloride using a 2.0 mm cell. Calculate the molar absorption coefficient (ε) of bromine at the wavelength employed.

$[Br_2]/\text{mol L}^{-1}$	0.0010	0.0050	0.0100	0.0500
T/per cent	81.4	35.6	12.7	3.0×10^{-3}

11.15 A 2.0 mm cell was filled with a solution of benzene in a nonabsorbing solvent. The concentration of the benzene was 0.010 mol L^{-1} and the wavelength of the radiation was 256 nm (where there is a maximum in the absorption). Calculate the molar absorption coefficient of benzene at this wavelength given that the transmission was 48 per cent. What will the transmittance be in a 4.0 mm cell at the same wavelength?

11.16 A swimmer enters a gloomier world (in one sense) on diving to greater depths. Given that the mean molar absorption coefficient of sea water in the visible region is $6.2 \times 10^{-5} \text{ mol}^{-1} \text{ L cm}^{-1}$, calculate the depth at which a diver will experience (a) half the surface intensity of light, (b) one tenth that intensity.

11.17 In a particular photoelectron spectrum using 21.22 eV photons, electrons were ejected with kinetic energies of 11.01 eV, 8.23 eV, and 5.22 eV. Sketch the molecular orbital energy-level diagram for the species, showing the ionization energies of the three identifiable orbitals.

11.18 What would be the nuclear magnetic resonance spectrum for a proton resonance line that was split by interaction with six identical protons?

11.19 What would be the nuclear magnetic resonance spectrum for a proton resonance line that was split by interaction with (a) two, (b) three equivalent nitrogen nuclei (the spin of a nitrogen nucleus is 1)?

11.20 ^{32}S has a nuclear spin of $\frac{3}{2}$ and a nuclear g factor of 0.4289. Calculate the energies of the nuclear spin states in a magnetic field of 7.500 T.

11.21 Calculate the energy difference between neighbouring nuclear spin states of a ^{14}N nucleus ($I = 1$, $g = 0.4036$) in a 15.00 T magnetic field. What frequency is required for resonance?

11.22 Calculate the magnetic field needed to satisfy the resonance condition for unshielded protons in a 150.0 MHz radiofrequency field.

11.23 The chemical shift of the CH_3 protons in acetaldehyde (ethanal) is $\delta = 2.20$ and that of the CHO proton is 9.80. What is the difference in local magnetic field between the two regions of the molecule when the applied field is (a) 1.5 T, (b) 7.0 T?

11.24 Using the information in Fig. 11.34 state the splitting (in Hz) between the methyl and aldehydic proton resonances in a spectrometer operating at (a) 60 MHz, (b) 350 MHz.

11.25 Sketch the appearance of the ^1H-NMR spectrum of acetaldehyde using $J = 2.90$ Hz and the data in Fig. 11.34 in a spectrometer operating at (a) 60 MHz, (b) 350 MHz.

11.26 Sketch the form of the ^{19}F-NMR spectra of a natural sample of $^{10}BF_4^-$ and $^{11}BF_4^-$.

11.27 Sketch the form of an $A_3M_2X_4$ spectrum, where A, M, and X are protons with distinctly different chemical shifts and $J_{AM} > J_{AX} > J_{MX}$.

Appendices

A1 Thermodynamic data at 25 °C (298 K)

A1.1 Inorganic substances

Substance	Molar mass $M/\text{g mol}^{-1}$	Enthalpy of formation $\Delta H_f^{\ominus}/\text{kJ mol}^{-1}$	Free energy of formation $\Delta G_f^{\ominus}/\text{kJ mol}^{-1}$	Entropy $S^{\ominus}/\text{J K}^{-1}\text{mol}^{-1}$
Aluminium				
$Al(s)$	26.98	0	0	28.33
$Al^{3+}(aq)$	26.98	-524.7	-481.2	-321.7
$Al_2O_3(s)$	101.95	-1675.7	-1582.3	50.92
$Al(OH)_3(s)$	78.00	-1276		
$AlCl_3(s)$	133.24	-704.2	-628.8	110.67
Antimony				
$SbH_3(g)$	153.24	$+145.11$	$+147.75$	232.78
$SbCl_3(g)$	228.11	-313.8	-301.2	337.80
$SbCl_5(g)$	299.02	-394.34	-334.29	401.94
Arsenic				
$As(s)$ (grey)	74.92	0	0	35.1
$As_2S_3(s)$	246.04	-169.0	-168.6	163.6
$AsO_4^{3-}(aq)$	138.92	-888.14	-648.41	-162.8
Barium				
$Ba(s)$	137.34	0	0	62.8
$Ba^{2+}(aq)$	137.34	-537.64	-560.77	9.6
$BaO(s)$	153.34	-553.5	-525.1	70.42
$BaCO_3(s)$	197.35	-1216.3	-1137.6	112.1
$BaCO_3(aq)$	197.35	-1214.78	-1088.59	-47.3

A1.1 Inorganic substances (continued)

Substance	Molar mass $M/\text{g mol}^{-1}$	Enthalpy of formation $\Delta H_f^{\ominus}/\text{kJ mol}^{-1}$	Free energy of formation $\Delta G_f^{\ominus}/\text{kJ mol}^{-1}$	Entropy $S^{\ominus}/\text{J K}^{-1}\text{mol}^{-1}$
Boron				
B(s)	10.81	0	0	5.86
$B_2O_3(s)$	69.62	−1272.8	−1193.7	53.97
$BF_3(g)$	67.81	−1137.0	−1120.3	254.12
Bromine				
$Br_2(l)$	159.82	0	0	152.23
$Br_2(g)$	159.82	+30.91	+3.11	245.46
Br(g)	79.91	+111.88	+82.40	175.02
$Br^-(aq)$	79.91	−121.55	−103.96	82.4
HBr(g)	90.92	−36.40	−53.45	198.70
Calcium				
Ca(s)	40.08	0	0	41.42
Ca(g)	40.08	+178.2	+144.3	154.88
$Ca^{2+}(aq)$	40.08	−542.83	−553.58	−53.1
CaO(s)	56.08	−635.09	−604.03	39.75
$Ca(OH)_2(s)$	74.10	−986.09	−898.49	83.39
$Ca(OH)_2(aq)$	74.10	−1002.82	−868.07	−74.5
$CaCO_3(s)$ (calcite)	100.09	−1206.9	−1128.8	92.9
$CaCO_3(s)$ (aragonite)	100.09	−1207.1	−1127.8	88.7
$CaCO_3(aq)$	100.09	−1219.97	−1081.39	−110.0
$CaF_2(s)$	78.08	−1219.6	−1167.3	68.87
$CaF_2(aq)$	78.08	−1208.09	−1111.15	−80.8
$CaCl_2(s)$	110.99	−795.8	−748.1	104.6
$CaBr_2(s)$	199.90	−682.8	−663.6	130
$CaC_2(s)$	64.10	−59.8	−64.9	69.96
$CaSO_4(s)$	136.14	−1434.11	−1321.79	106.7
$CaSO_4(aq)$	136.14	−1452.10	−1298.10	−33.1
Carbon[†]				
C(s) (graphite)	12.011	0	0	5.740
C(s) (diamond)	12.011	+1.895	+2.900	2.377
C(g)	12.011	+716.68	+671.26	158.10
CO(g)	28.01	−110.53	−137.17	197.67
$CO_2(g)$	44.01	−393.51	−394.36	213.74
$CO_3^{2-}(aq)$	60.01	−677.14	−527.81	−56.9
$CCl_4(l)$	153.82	−135.44	−65.21	216.40
$CS_2(l)$	76.4	+89.70	+65.27	151.34
HCN(g)	27.03	+135.1	+124.7	201.78
HCN(l)	27.03	+108.87	+124.97	112.84
Cerium				
Ce(s)	140.12	0	0	72.0
$Ce^{3+}(aq)$	140.12	−696.2	−672.0	−205
$Ce^{4+}(aq)$	140.12	−537.2	−503.8	−301

A1.1 Inorganic substances (continued)

Substance	Molar mass $M/\text{g mol}^{-1}$	Enthalpy of formation $\Delta H_f^{\ominus}/\text{kJ mol}^{-1}$	Free energy of formation $\Delta G_f^{\ominus}/\text{kJ mol}^{-1}$	Entropy $S^{\ominus}/\text{J K}^{-1}\text{mol}^{-1}$
Chlorine				
$Cl_2(g)$	70.91	0	0	223.07
$Cl(g)$	35.45	+121.68	+105.68	165.20
$Cl^-(g)$	35.45	−167.16	−131.23	56.5
$HCl(g)$	36.46	−92.31	−95.30	186.91
$HCl(aq)$	36.46	−167.16	−131.23	56.5
Copper				
$Cu(s)$	63.54	0	0	33.15
$Cu^+(aq)$	63.54	+71.67	+49.98	40.6
$Cu^{2+}(aq)$	63.54	+64.77	+65.49	−99.6
$Cu_2O(s)$	143.08	−168.6	−146.0	93.14
$CuO(s)$	79.54	−157.3	−129.7	42.63
$CuSO_4(s)$	159.60	−771.36	−661.8	109
$CuSO_4 \cdot 5H_2O(s)$	249.68	−2279.7	−1879.7	300.4
Deuterium				
$D_2(g)$	4.028	0	0	144.96
$D_2O(g)$	20.028	−249.20	−234.54	198.34
$D_2O(l)$	20.028	−294.60	−243.44	75.94
Fluorine				
$F_2(g)$	38.00	0	0	202.78
$F^-(aq)$	19.00	−332.63	−278.79	−13.8
$HF(g)$	20.01	−271.1	−273.2	173.78
$HF(aq)$	20.01	−332.63	−278.79	−13.8
Hydrogen (see also deuterium)				
$H_2(g)$	2.016	0	0	130.68
$H(g)$	1.008	+217.97	+203.25	114.71
$H^+(aq)$	1.008	0	0	0
$H_2O(l)$	18.02	−285.83	−237.13	69.91
$H_2O(g)$	18.02	−241.82	−228.57	188.83
$H_2O_2(l)$	34.01	−187.78	−120.35	109.6
$H_2O_2(aq)$	24.01	−191.17	−134.03	143.9
Iodine				
$I_2(s)$	253.81	0	0	116.14
$I_2(g)$	253.81	+62.44	+19.33	260.69
$I^-(aq)$	126.90	−55.19	−51.57	111.3
$HI(g)$	127.91	+26.48	+1.70	206.59
Iron				
$Fe(s)$	55.85	0	0	27.28
$Fe^{2+}(aq)$	55.85	−89.1	−78.90	−137.7
$Fe^{3+}(aq)$	55.85	−48.5	−4.7	−315.9
$Fe_3O_4(s)$ (magnetite)	231.54	−1118.4	−1015.4	146.4

A1.1 Inorganic substances (continued)

Substance	Molar mass $M/\text{g mol}^{-1}$	Enthalpy of formation $\Delta H_f^{\ominus}/\text{kJ mol}^{-1}$	Free energy of formation $\Delta G_f^{\ominus}/\text{kJ mol}^{-1}$	Entropy $S^{\ominus}/\text{J K}^{-1}\text{mol}^{-1}$
Iron (*continued*)				
$Fe_2O_3(s)$ (haematite)	159.69	-824.2	-742.2	87.40
$FeS(s, \alpha)$	87.91	-100.0	-100.4	60.29
$FeS(aq)$	87.91		$+6.9$	
$FeS_2(s)$	119.98	-178.2	-166.9	52.93
Lead				
$Pb(s)$	207.19	0	0	64.81
$Pb^{2+}(aq)$	207.19	-1.7	-24.43	10.5
$PbO_2(s)$	239.19	-277.4	-217.33	68.6
$PbSO_4(s)$	303.25	-919.94	-813.14	148.57
$PbBr_2(s)$	367.01	-278.7	-261.92	161.5
$PbBr_2(aq)$	367.01	-244.8	-232.34	175.3
Magnesium				
$Mg(s)$	24.31	0	0	32.68
$Mg(g)$	24.31	$+147.70$	$+113.10$	148.65
$Mg^{2+}(aq)$	24.31	-466.85	-454.8	-138.1
$MgO(s)$	40.31	-601.70	-569.43	26.94
$MgCO_3(s)$	84.32	-1095.8	-1012.1	65.7
$MgBr_2(s)$	184.13	-524.3	-503.8	117.2
Mercury				
$Hg(l)$	200.59	0	0	76.02
$Hg(g)$	200.59	$+61.32$	$+31.82$	174.96
$HgO(s)$	216.59	-90.83	-58.54	70.29
$Hg_2Cl_2(s)$	472.09	-265.22	-210.75	192.5
$HgCl_2(s)$	271.50	-224.3	-178.6	146.0
Nitrogen				
$N_2(g)$	28.01	0	0	191.61
$NO(g)$	30.01	$+90.25$	$+86.55$	210.76
$N_2O(g)$	44.01	$+82.05$	$+104.20$	219.85
$NO_2(g)$	46.01	$+33.18$	$+51.31$	240.06
$N_2O_4(g)$	92.01	$+9.16$	$+97.89$	304.29
$HNO_3(l)$	63.01	-174.10	-80.71	155.60
$HNO_3(aq)$	63.01	-207.36	-111.25	146.4
$NO_3^-(aq)$	62.01	-205.0	-108.74	146.4
$NH_3(g)$	17.03	-46.11	-16.45	192.45
$NH_3(aq)$	17.03	-80.29	-26.50	111.3
$NH_4^+(aq)$	18.04	-132.51	-79.31	113.4
$NH_2OH(s)$	33.03	-114.2		
$HN_3(g)$	43.03	$+294.1$	$+328.1$	238.97
$N_2H_4(l)$	32.05	$+50.63$	$+149.34$	121.21
$NH_4NO_3(s)$	80.04	-365.56	-183.87	151.08
$NH_4Cl(s)$	53.49	-314.43	-202.87	94.6
$NH_4ClO_4(s)$	117.49	-295.31	-88.75	186.2

A1.1 Inorganic substances (continued)

Substance	Molar mass $M/\text{g mol}^{-1}$	Enthalpy of formation $\Delta H_f^{\ominus}/\text{kJ mol}^{-1}$	Free energy of formation $\Delta G_f^{\ominus}/\text{kJ mol}^{-1}$	Entropy $S^{\ominus}/\text{J K}^{-1}\text{mol}^{-1}$
Oxygen				
$O_2(g)$	32.00	0	0	205.14
$O_3(g)$	48.00	+142.7	+163.2	238.93
$OH^-(aq)$	17.01	−229.99	−157.24	−10.75
Phosphorus				
$P(s)$ (white)	30.97	0	0	41.09
$P_4(g)$	123.90	+58.91	+24.44	297.98
$PH_3(g)$	34.00	+5.4	+13.4	210.23
$P_4O_{10}(s)$	283.89	−2984.0	−2697.0	228.86
$H_3PO_3(aq)$	82.00	−964.8		
$H_3PO_4(l)$	98.00	−1266.9		
$H_3PO_4(aq)$	98.00	−1277.4	−1018.7	
$PCl_3(l)$	137.33	−319.7	−272.3	217.18
$PCl_3(g)$	137.33	−287.0	−267.8	311.78
$PCl_5(g)$	208.24	−374.9	−305.0	364.6
$PCl_5(s)$	208.24	−443.5		
Potassium				
$K(s)$	39.10	0	0	64.18
$K(g)$	39.10	+89.24	+60.59	160.34
$K^+(aq)$	39.10	−252.38	−283.27	102.5
$KOH(s)$	56.11	−424.76	−379.08	78.9
$KOH(aq)$	56.11	−482.37	−440.50	−91.6
$KF(s)$	58.10	−567.27	−537.75	66.57
$KCl(s)$	74.56	−436.75	−409.14	82.59
$KBr(s)$	119.01	−393.80	−380.66	95.90
$KI(s)$	166.01	−327.90	−324.89	106.32
$KClO_3(s)$	122.55	−397.73	−296.25	143.1
$KClO_4(s)$	138.55	−432.75	−303.09	151.0
$K_2S(s)$	110.27	−380.7	−364.0	105
$K_2S(aq)$	110.27	−471.5	−480.7	190.4
Silicon				
$Si(s)$	28.09	0	0	18.83
$SiO_2(s, \alpha)$	60.09	−910.94	−856.64	41.84
Silver				
$Ag(s)$	107.87	0	0	42.55
$Ag^+(aq)$	107.87	+105.58	+77.11	72.68
$Ag_2O(s)$	231.74	−31.05	−11.20	121.3
$AgBr(s)$	187.78	−100.37	−96.90	107.1
$AgBr(aq)$	187.78	−15.98	−26.86	155.2
$AgCl(s)$	143.32	−127.07	−109.79	96.2
$AgCl(aq)$	143.32	−61.58	−54.12	129.3
$AgI(s)$	234.77	−61.84	−66.19	115.5

469

A1.1 Inorganic substances (continued)

Substance	Molar mass $M/\text{g mol}^{-1}$	Enthalpy of formation $\Delta H_f^{\ominus}/\text{kJ mol}^{-1}$	Free energy of formation $\Delta G_f^{\ominus}/\text{kJ mol}^{-1}$	Entropy $S^{\ominus}/\text{J K}^{-1}\text{mol}^{-1}$
Silver (continued)				
$AgI(aq)$	234.77	+50.38	+25.52	184.1
$AgNO_3(s)$	169.88	−124.39	−33.41	140.92
Sodium				
$Na(s)$	22.99	0	0	51.21
$Na(g)$	22.99	+107.32	+76.76	153.71
$Na^+(aq)$	22.99	−240.12	−261.91	59.0
$NaOH(s)$	40.00	−425.61	−379.49	64.46
$NaOH(aq)$	40.00	−470.11	−419.15	48.1
$NaCl(s)$	58.44	−411.15	−384.14	72.13
$NaBr(s)$	102.90	−361.06	−348.98	86.82
$NaI(s)$	149.89	−287.78	−286.06	98.53
Sulfur				
$S(s)$ (rhombic)	32.06	0	0	31.80
$S(s)$ (monoclinic)	32.06	+0.33	+0.1	32.6
$S^{2-}(aq)$	32.06	+33.1	+85.8	−14.6
$SO_2(g)$	64.06	−296.83	−300.19	248.22
$SO_3(g)$	80.06	−395.72	−371.06	256.76
$H_2SO_4(l)$	98.08	−813.99	−690.00	156.90
$H_2SO_4(aq)$	98.08	−909.27	−744.53	20.1
$SO_4^{2-}(aq)$	96.06	−909.27	−744.53	20.1
$H_2S(g)$	34.08	−20.63	−33.56	205.79
$H_2S(aq)$	34.08	−39.7	−27.83	121
$SF_6(g)$	146.05	−1209	−1105.3	291.82
Tin				
$Sn(s)$ (white)	118.69	0	0	51.55
$Sn(s)$ (grey)	118.69	−2.09	0.13	44.14
$SnO(s)$	134.69	−285.8	−256.9	56.5
$SnO_2(s)$	150.69	−580.7	−519.6	52.3
Zinc				
$Zn(s)$	65.37	0	0	41.63
$Zn^{2+}(aq)$	65.37	−153.89	−147.06	−112.1
$ZnO(s)$	81.37	−348.28	−318.30	43.64

† For organic compounds of carbon, see Appendix A1.2.

A1.2 Organic compounds

Substance	Molar mass M/g mol^{-1}	Enthalpy of combustion $\Delta H_c^{\ominus}$/kJ mol^{-1}	Enthalpy of formation $\Delta H_f^{\ominus}$/kJ mol^{-1}	Free energy of formation $\Delta G_f^{\ominus}$/kJ mol^{-1}	Entropy $S^{\ominus}$/J K^{-1} mol^{-1}
Hydrocarbons					
$CH_4(g)$ (methane)	16.04	−890	−74.81	−50.72	186.26
$C_2H_2(g)$ (acetylene)	26.04	−1300	+226.73	+209.20	200.94
$C_2H_4(g)$ (ethylene)	28.05	−1411	+52.26	+68.15	219.56
$C_2H_6(g)$ (ethane)	30.07	−1560	−84.68	−32.82	229.60
$C_3H_6(g)$ (propylene)	42.00	−2058	+20.42	+62.78	266.6
$C_3H_6(g)$ (cyclopropane)	42.08	−2091	+53.30	+104.45	237.4
$C_3H_8(g)$ (propane)	44.01	−2220	−103.85	−23.49	270.2
$C_4H_{10}(g)$ (butane)	58.13	−2878	−126.15	−17.03	310.1
$C_5H_{12}(g)$ (pentane)	72.15	−3537	−146.44	−8.20	349
$C_6H_6(l)$ (benzene)	78.12	−3268	+49.0	+124.3	173.3
$C_6H_6(g)$ (benzene)	78.12	−3302			
$C_7H_8(l)$ (toluene)	92.13	−3910	+12.0	+113.8	221.0
$C_7H_8(g)$ (toluene)	92.13	−3953			
$C_6H_{12}(l)$ (cyclohexane)	84.16	−3920	−156.4	+26.7	204.4
$C_6H_{12}(g)$ (cyclohexane)	84.16	−3953			
$C_8H_{18}(l)$ (octane)	114.23	−5471	−249.9	+6.4	358
$C_{10}H_8(s)$ (naphthalene)	128.18	−5147	+78.53		
Alcohols and phenols					
$CH_3OH(l)$ (methanol)	32.04	−726	−238.86	−166.27	126.8
$CH_3OH(g)$ (methanol)	32.04	−764	−200.66	−161.96	239.81
$C_2H_5OH(l)$ (ethanol)	46.07	−1368	−277.69	−174.78	160.7
$C_2H_5OH(g)$ (ethanol)	46.07	−1409	−235.10	−168.49	282.70
$C_6H_5OH(s)$ (phenol)	94.11	−3054	−164.6	−50.42	144.0
Carboxylic acids					
$HCOOH(l)$ (formic)	46.03	−255	−424.72	−361.35	128.95
$CH_3COOH(l)$ (acetic)	60.05	−875	−484.5	−389.9	159.8
$CH_3COOH(aq)$ (acetic)	60.05		−485.76	−396.46	86.6
$(COOH)_2(s)$ (oxalic)	90.04	−254	−827.2	−697.9	120
$C_6H_5COOH(s)$ (benzoic)	122.13	−3227	−385.1	−245.3	167.6

A1.2 Organic compounds (continued)

Substance	Molar mass $M/\text{g mol}^{-1}$	Enthalpy of combustion $\Delta H_c^{\ominus}/\text{kJ mol}^{-1}$	Enthalpy of formation $\Delta H_f^{\ominus}/\text{kJ mol}^{-1}$	Free energy of formation $\Delta G_f^{\ominus}/\text{kJ mol}^{-1}$	Entropy $S^{\ominus}/\text{J K}^{-1}\text{mol}^{-1}$
Aldehydes and ketones					
$HCHO(g)$ (formaldehyde)	30.03	−571	−108.57	−102.53	218.77
$CH_3CHO(l)$ (acetaldehyde)	44.05	−1166	−192.30	−128.12	160.2
$CH_3CHO(g)$ (acetaldehyde)	44.05	−1192	−166.19	−128.86	250.3
$CH_3COCH_3(l)$ (acetone)	58.08	−1790	−248.1	−155.4	200
Sugars					
$C_6H_{12}O_6(s)$ (glucose)	180.16	−2808	−1268	−910	212
$C_6H_{12}O_6(aq)$ (glucose)	180.16			−917	
$C_6H_{12}O_6(s)$ (fructose)	180.16	−2810	−1266		
$C_{12}H_{22}O_{11}(s)$ (sucrose)	342.30	−5645	−2222	−1545	360
Nitrogen compounds					
$CO(NH_2)_2(s)$ (urea)	60.06	−632	−333.51	−197.33	104.60
$C_6H_5NH_2(l)$ (aniline)	93.13	−3393	+31.6	+149.1	191.3
$NH_2CH_2COOH(s)$ (glycine)	75.07	−969	−532.9	−373.4	103.51
$CH_3NH_2(g)$ (methylamine)	31.06	−1085	−22.97	32.16	243.41

A2   Standard reduction potentials at 25 °C (298 K)

A2.1 Potentials (in electrochemical order)

Reduction half-reaction	$E^{\ominus}/\text{V}$	Reduction half-reaction	$E^{\ominus}/\text{V}$
Strongly oxidizing		$AgF + e^- \rightarrow Ag + F^-$	+0.78
$H_4XeO_6 + 2H^+ + 2e^- \rightarrow XeO_3 + 3H_2O$	+3.0	$Fe^{3+} + e^- \rightarrow Fe^{2+}$	+0.77
$F_2 + 2e^- \rightarrow 2F^-$	+2.87	$BrO^- + H_2O + 2e^- \rightarrow Br^- + 2OH^-$	+0.76
$O_3 + 2H^+ + 2e^- \rightarrow O_2 + H_2O$	+2.07	$MnO_4^{2-} + 2H_2O + 2e^- \rightarrow MnO_2 + 4OH^-$	+0.60
$S_2O_8^{2-} + 2e^- \rightarrow 2SO_4^{2-}$	+2.05	$MnO_4^- + e^- \rightarrow MnO_4^{2-}$	+0.56
$Ag^{2+} + e^- \rightarrow Ag^+$	+1.98	$I_2(s) + e^- \rightarrow 2I^-$	+0.54
$Co^{3+} + e^- \rightarrow Co^{2+}$	+1.81	$Cu^+ + e^- \rightarrow Cu$	+0.52
$H_2O_2 + 2H^+ + 2e^- \rightarrow 2H_2O$	+1.78	$I_3^- + 2e^- \rightarrow 3I^-$	+0.53
$Au^+ + e^- \rightarrow Au$	+1.69	$NiO(OH) + H_2O + e^- \rightarrow Ni(OH)_2 + OH^-$	+0.49
$Pb^{4+} + 2e^- \rightarrow Pb^{2+}$	+1.67	$O_2 + 2H_2O + 4e^- \rightarrow 4OH^-$	+0.40
$2HClO + 2H^+ + 2e^- \rightarrow Cl_2 + 2H_2O$	+1.63	$ClO_4^- + H_2O + 2e^- \rightarrow ClO_3^- + 2OH^-$	+0.36
$Hg^{2+} + 2e^- \rightarrow Hg$	+1.62	$Cu^{2+} + 2e^- \rightarrow Cu$	+0.34
$Ce^{4+} + e^- \rightarrow Ce^{3+}$	+1.61	$Hg_2Cl_2 + 2e^- \rightarrow 2Hg + 2Cl^-$	+0.27
$2HBrO + 2H^+ + 2e^- \rightarrow Br_2 + 2H_2O$	+1.60	$AgCl + e^- \rightarrow Ag + Cl^-$	+0.22
$MnO_4^- + 8H^+ + 5e^- \rightarrow Mn^{2+} + 4H_2O$	+1.51	$Bi^{3+} + 3e^- \rightarrow Bi$	+0.20
$Mn^{3+} + e^- \rightarrow Mn^{2+}$	+1.51	$SO_4^{2-} + 4H^+ + 2e^- \rightarrow H_2SO_3 + H_2O$	+0.17
$Au^{3+} + 3e^- \rightarrow Au$	+1.40	$Cu^{2+} + e^- \rightarrow Cu^+$	+0.15
$Cl_2 + 2e^- \rightarrow 2Cl^-$	+1.36	$Sn^{4+} + 2e^- \rightarrow Sn^{2+}$	+0.15
$Cr_2O_7^{2-} + 14H^+ + 6e^- \rightarrow 2Cr^{3+} + 7H_2O$	+1.33	$AgBr + e^- \rightarrow Ag + Br^-$	+0.07
$O_3 + H_2O + 2e^- \rightarrow O_2 + 2OH^-$	+1.24	$NO_3^- + H_2O + 2e^- \rightarrow NO_2^- + 2OH^-$	+0.01
$O_2 + 4H^+ + 4e^- \rightarrow 2H_2O$	+1.23	$Ti^{4+} + e^- \rightarrow Ti^{3+}$	0.00
$MnO_2 + 4H^+ + 2e^- \rightarrow Mn^{2+} + 2H_2O$	+1.23	$2H^+ + 2e^- \rightarrow H_2$	0, by
$ClO_4^- + 2H^+ + 2e^- \rightarrow ClO_3^- + H_2O$	+1.23		definition
$Pt^{2+} + 2e^- \rightarrow Pt$	+1.20	$Fe^{3+} + 3e^- \rightarrow Fe$	−0.04
$Br_2 + 2e^- \rightarrow 2Br^-$	+1.09	$O_2 + H_2O + 2e^- \rightarrow HO_2^- + OH^-$	−0.08
$Pu^{4+} + e^- \rightarrow Pu^{3+}$	+0.97	$Pb^{2+} + 2e^- \rightarrow Pb$	−0.13
$NO_3^- + 4H^+ + 3e^- \rightarrow NO + 2H_2O$	+0.96	$In^+ + e^- \rightarrow In$	−0.14
$2Hg^{2+} + 2e^- \rightarrow Hg_2^{2+}$	+0.92	$Sn^{2+} + 2e^- \rightarrow Sn$	−0.14
$ClO^- + H_2O + 2e^- \rightarrow Cl^- + 2OH^-$	+0.89	$AgI + e^- \rightarrow Ag + I^-$	−0.15
$NO_3^- + 2H^+ + e^- \rightarrow NO_2 + H_2O$	+0.80	$Ni^{2+} + 2e^- \rightarrow Ni$	−0.23
$Ag^+ + e^- \rightarrow Ag$	+0.80	$V^{3+} + e^- \rightarrow V^{2+}$	−0.26
$Hg_2^{2+} + 2e^- \rightarrow 2Hg$	+0.79	$Co^{2+} + 2e^- \rightarrow Co$	−0.28

473

A2.1 Potentials (continued)

Reduction half-reaction	$E^{\ominus}/V$	Reduction half-reaction	$E^{\ominus}/V$
$In^{3+} + 3e^- \rightarrow In$	-0.34	$Mn^{2+} + 2e^- \rightarrow Mn$	-1.18
$Tl^+ + e^- \rightarrow Tl$	-0.34	$V^{2+} + 2e^- \rightarrow V$	-1.19
$PbSO_4 + 2e^- \rightarrow Pb + SO_4^{2-}$	-0.36	$Ti^{2+} + 2e^- \rightarrow Ti$	-1.63
$Ti^{3+} + e^- \rightarrow Ti^{2+}$	-0.37	$Al^{3+} + 3e^- \rightarrow Al$	-1.66
$In^{2+} + e^- \rightarrow In^+$	-0.40	$U^{3+} + 3e^- \rightarrow U$	-1.79
$Cr^{3+} + e^- \rightarrow Cr^{2+}$	-0.41	$Be^{2+} + 2e^- \rightarrow Be$	-1.85
$Fe^{2+} + 2e^- \rightarrow Fe$	-0.44	$Mg^{2+} + 2e^- \rightarrow Mg$	-2.36
$In^{3+} + 2e^- \rightarrow In^+$	-0.44	$Ce^{3+} + 3e^- \rightarrow Ce$	-2.48
$S + 2e^- \rightarrow S^{2-}$	-0.48	$La^{3+} + 2e^- \rightarrow La$	-2.52
$In^{3+} + e^- \rightarrow In^{2+}$	-0.49	$Na^+ + e^- \rightarrow Na$	-2.71
$Ga^+ + e^- \rightarrow Ga$	-0.53	$Ca^{2+} + 2e^- \rightarrow Ca$	-2.87
$O_2 + e^- \rightarrow O_2^-$	-0.56	$Sr^{2+} + 2e^- \rightarrow Sr$	-2.89
$U^{4+} + e^- \rightarrow U^{3+}$	-0.61	$Ba^{2+} + 2e^- \rightarrow Ba$	-2.91
$Se + 2e^- \rightarrow Se^{2-}$	-0.67	$Ra^{2+} + 2e^- \rightarrow Ra$	-2.92
$Cr^{3+} + 3e^- \rightarrow Cr$	-0.74	$Cs^+ + e^- \rightarrow Cs$	-2.92
$Zn^{2+} + 2e^- \rightarrow Zn$	-0.76	$Rb^+ + e^- \rightarrow Rb$	-2.93
$Cd(OH)_2 + 2e^- \rightarrow Cd + 2OH^-$	-0.81	$K^+ + e^- \rightarrow K$	-2.93
$2H_2O + 2e^- \rightarrow H_2 + 2OH^-$	-0.83	$Li^+ + e^- \rightarrow Li$	-3.05
$Te + 2e^- \rightarrow Te^{2-}$	-0.84		
$Cr^{2+} + 2e^- \rightarrow Cr$	-0.91	*Strongly reducing*	

A2.2 Potentials (in alphabetical order)

Reduction half-reaction	$E^{\ominus}/V$	Reduction half-reaction	$E^{\ominus}/V$
$Ag^+ + e^- \rightarrow Ag$	$+0.80$	$2H^+ + 2e^- \rightarrow H_2$	0, by
$Ag^{2+} + e^- \rightarrow Ag^+$	$+1.98$		definition
$AgBr + e^- \rightarrow Ag + Br^-$	$+0.07$	$2H_2O + 2e^- \rightarrow H_2 + 2OH^-$	-0.83
$AgCl + e^- \rightarrow Ag + Cl^-$	$+0.22$	$2HBrO + 2H^+ + 2e^- \rightarrow Br_2 + 2H_2O$	$+1.60$
$AgF + e^- \rightarrow Ag + F^-$	$+0.78$	$2HClO + 2H^+ + 2e^- \rightarrow Cl_2 + 2H_2O$	$+1.63$
$AgI + e^- \rightarrow Ag + I^-$	-0.15	$H_2O_2 + 2H^+ + 2e^- \rightarrow 2H_2O$	$+1.78$
$Al^{3+} + 3e^- \rightarrow Al$	-1.66	$H_4XeO_6 + 2H^+ + 2e^- \rightarrow XeO_3 + 3H_2O$	$+3.0$
$Au^+ + e^- \rightarrow Au$	$+1.69$	$Hg_2^{2+} + 2e^- \rightarrow 2Hg$	$+0.79$
$Au^{3+} + 3e^- \rightarrow Au$	$+1.40$	$Hg^{2+} + 2e^- \rightarrow Hg$	$+1.62$
$Ba^{2+} + 2e^- \rightarrow Ba$	-2.91	$Hg_2Cl_2 + 2e^- \rightarrow 2Hg + 2Cl^-$	$+0.27$
$Be^{2+} + 2e^- \rightarrow Be$	-1.85	$2Hg^{2+} + 2e^- \rightarrow Hg_2^{2+}$	$+0.92$
$Bi^{3+} + 3e^- \rightarrow Bi$	$+0.20$	$I_2(s) + e^- \rightarrow 2I^-$	$+0.54$
$Br_2 + 2e^- \rightarrow 2Br^-$	$+1.09$	$I_3^- + 2e^- \rightarrow 3I^-$	$+0.53$
$BrO^- + H_2O + 2e^- \rightarrow Br^- + 2OH^-$	$+0.76$	$In^+ + e^- \rightarrow In$	-0.14
$Ca^{2+} + 2e^- \rightarrow Ca$	-2.87	$In^{2+} + e^- \rightarrow In^+$	-0.40
$Cd(OH)_2 + 2e^- \rightarrow Cd + 2OH^-$	-0.81	$In^{3+} + 2e^- \rightarrow In^+$	-0.44
$Cd^{2+} + 2e^- \rightarrow Cd$	-0.40	$In^{3+} + 3e^- \rightarrow In$	-0.34
$Ce^{3+} + 3e^- \rightarrow Ce$	-2.48	$In^{3+} + e^- \rightarrow In^{2+}$	-0.49
$Ce^{4+} + 2e^- \rightarrow Ce^{3+}$	$+1.61$	$K^+ + e^- \rightarrow K$	-2.93
$Cl_2 + 2e^- \rightarrow 2Cl^-$	$+1.36$	$La^{3+} + 3e^- \rightarrow La$	-2.52
$ClO^- + H_2O + 2e^- \rightarrow Cl^- + 2OH^-$	$+0.89$	$Li^+ + e^- \rightarrow Li$	-3.05
$ClO_4^- + 2H^+ + 2e^- \rightarrow ClO_3^- + H_2O$	$+1.23$	$Mg^{2+} + 2e^- \rightarrow Mg$	-2.36
$ClO_4^- + H_2O + 2e^- \rightarrow ClO_3^- + 2OH^-$	$+0.36$	$Mn^{2+} + 2e^- \rightarrow Mn$	-1.18
$Co^{2+} + 2e^- \rightarrow Co$	-0.28	$Mn^{3+} + e^- \rightarrow Mn^{2+}$	$+1.51$
$Co^{3+} + e^- \rightarrow Co^{2+}$	$+1.81$	$MnO_2 + 4H^+ + 2e^- \rightarrow Mn^{2+} + 2H_2O$	$+1.23$
$Cr^{2+} + 2e^- \rightarrow Cr$	-0.91	$MnO_4^- + 8H^+ + 5e^- \rightarrow Mn^{2+} + 4H_2O$	$+1.51$
$Cr_2O_7^{2-} + 14H^+ + 6e^- \rightarrow 2Cr^{3+} + 7H_2O$	$+1.33$	$MnO_4^- + e^- \rightarrow MnO_4^{2-}$	$+0.56$
$Cr^{3+} + 3e^- \rightarrow Cr$	-0.74	$MnO_4^{2-} + 2H_2O + 2e^- \rightarrow MnO_2 + 4OH^-$	$+0.60$
$Cr^{3+} + e^- \rightarrow Cr^{2+}$	-0.41	$Na^+ + e^- \rightarrow Na$	-2.71
$Cs^+ + e^- \rightarrow Cs$	-2.92	$Ni^{2+} + 2e^- \rightarrow Ni$	-0.23
$Cu^+ + e^- \rightarrow Cu$	$+0.52$	$NiO(OH) + H_2O + e^- \rightarrow Ni(OH)_2 + OH^-$	$+0.49$
$Cu^{2+} + 2e^- \rightarrow Cu$	$+0.34$	$NO_3^- + 2H^+ + e^- \rightarrow NO_2 + H_2O$	$+0.80$
$Cu^{2+} + e^- \rightarrow Cu^+$	$+0.15$	$NO_3^- + 4H^+ + 3e^- \rightarrow NO + 2H_2O$	$+0.96$
$F_2 + 2e^- \rightarrow 2F^-$	$+2.87$	$NO_3^- + H_2O + 2e^- \rightarrow NO_2^- + 2OH^-$	$+0.10$
$Fe^{2+} + 2e^- \rightarrow Fe$	-0.44	$O_2 + 2H_2O + 4e^- \rightarrow 4OH^-$	$+0.40$
$Fe^{3+} + 3e^- \rightarrow Fe$	-0.04	$O_2 + 4H^+ + 4e^- \rightarrow 2H_2O$	$+1.23$
$Fe^{3+} + e^- \rightarrow Fe^{2+}$	$+0.77$	$O_2 + e^- \rightarrow O_2^-$	-0.56
$Ga^+ + e^- \rightarrow Ga$	-0.53	$O_2 + H_2O + 2e^- \rightarrow HO_2^- + OH^-$	-0.08

A2.2 Potentials (continued)

Reduction half-reaction	$E^{\ominus}/V$	Reduction half-reaction	$E^{\ominus}/V$
$O_3 + 2H^+ + 2e^- \rightarrow O_2 + H_2O$	+2.07	$Sn^{2+} + 2e^- \rightarrow Sn$	−0.14
$O_3 + H_2O + 2e^- \rightarrow O_2 + 2OH^-$	+1.24	$Sn^{4+} + 2e^- \rightarrow Sn^{2+}$	+0.15
$Pb^{2+} + 2e^- \rightarrow Pb$	−0.13	$Sr^{2+} + 2e^- \rightarrow Sr$	−2.89
$Pb^{4+} + 2e^- \rightarrow Pb^{2+}$	+1.67	$Te + 2e^- \rightarrow Te^{2-}$	−0.84
$PbSO_4 + 2e^- \rightarrow Pb + SO_4^{2-}$	−0.36	$Tl^+ + e^- \rightarrow Tl$	−0.34
$Pt^{2+} + 2e^- \rightarrow Pt$	+1.20	$Ti^{2+} + 2e^- \rightarrow Ti$	−1.63
$Pu^{4+} + e^- \rightarrow Pu^{3+}$	+0.97	$Ti^{3+} + e^- \rightarrow Ti^{2+}$	−0.37
$Ra^{2+} + 2e^- \rightarrow Ra$	−2.92	$Ti^{4+} + e^- \rightarrow Ti^{3+}$	0.00
$Rb^+ + e^- \rightarrow Rb$	−2.93	$U^{3+} + 3e^- \rightarrow U$	−1.79
$S + 2e^- \rightarrow S^{2-}$	−0.48	$U^{4+} + e^- \rightarrow U^{3+}$	−0.61
$SO_4^{2-} + 4H^+ + 2e^- \rightarrow H_2SO_3 + H_2O$	+0.17	$V^{2+} + 2e^- \rightarrow V$	−1.19
$S_2O_8^{2-} + 2e^- \rightarrow 2SO_4^{2-}$	+2.05	$V^{3+} + e^- \rightarrow V^{2+}$	−0.26
$Se + 2e^- \rightarrow Se^{2-}$	−0.67	$Zn^{2+} + 2e^- \rightarrow Zn$	−0.76

Further reading

I The properties of gases

General reference: P. W. Atkins. *Physical Chemistry,* Oxford University Press and W. H. Freeman & Co. (1990). Chapter 1.

The ideal gas law at the center of the Sun. D. B. Clark. *J. Chem. Ed.* **1989,** *66,* 826.

The choice of names and symbols for quantities in chemistry. I. M. Mills. *J. Chem. Ed.* **1989,** *66,* 887.

The many faces of van der Waals equation of state. J. G. Eberhardt. *J. Chem. Ed.* **1989,** *66,* 906.

Graham's law: defining gas velocities. T. Kenney. *J. Chem. Ed.* **1990,** *67,* 871.

Supercritical fluid: liquid, gas, both, or neither? A different approach. E. F. Meyer and T. P. Meyer. *J. Chem. Ed.* **1986,** *63,* 463.

P-V-T isotherms of real gases: Experimental versus calculated values. J. L. Pauley and E. H. Davis. *J. Chem. Ed.* **1986,** *63,* 466.

A gas kinetic explanation of simple thermodynamic processes. B. A. Waite. *J. Chem. Ed.* **1985,** *62,* 2245.

2 Thermodynamics: the first law

General reference: P. W. Atkins. *Physical Chemistry,* Oxford University Press and W. H. Freeman & Co. (1990). Chapter 2.

General definitions of work and heat in thermodynamic processes. E. A. Gislason and N. C. Craig. *J. Chem. Ed.* **1987,** *64,* 660.

Standard enthalpies of formation of ions in solution. T. S. Solomon. *J. Chem. Ed.* **1991,** *68,* 41.

Understanding the language: Problem solving and the first law of thermodynamics. M. Hamby. *J. Chem. Ed.* **1990,** *67,* 923.

Simplification of some thermodynamic calculations. E. R. Boyko and J. F. Belliveau. *J. Chem. Ed.* **1990,** *67,* 743.

Heat, work, and metabolism. J. N. Spencer. *J. Chem. Ed.* **1985,** *62,* 5715.

3 Thermodynamics: the second law

General reference: P. W. Atkins. *Physical Chemistry,* Oxford University Press and W. H. Freeman & Co. (1990). Chapter 4. See also *The second law.* Scientific American Library (1984).

Thermodynamic reversibility. 1. What is it? H. B. Hollinger and M. J. Zensen. *J. Chem. Ed.* **1991,** *68,* 31.

Entropy analysis of four familiar processes. N. C. Craig. *J. Chem. Ed.* **1988,** *65,* 760.

Order and disorder and entropies of fusion. D. F. G. Gilson. *J. Chem. Ed.* **1992,** *69,* 23.

4 Phase equilibria

General reference: P. W. Atkins. *Physical Chemistry,* Oxford University Press and W. H. Freeman & Co. (1990). Chapters 6, 7, and 8.

The critical point and the number of degrees of freedom. R. Battino. *J. Chem. Ed.* **1991,** *68,* 276.

The direct relation between altitude and boiling point. B. L. Earl. *J. Chem. Ed.* **1990,** *67,* 45.

The true meaning of component in the Gibbs phase rule. H. E. Franzen. *J. Chem. Ed.* **1986,** *63,* 948.

The liquid–vapor phase diagram of CO_2 near the critical temperature. A. M. Halpern and M.-F. Lin. *J. Chem. Ed.* **1986,** *63,* 38.

5 Chemical equilibria

General reference: P. W. Atkins. *Physical Chemistry,* Oxford University Press and W. H. Freeman & Co. (1990). Chapter 9. For an elementary introduction, see P. W. Atkins and J. A. Beran, *General chemistry.* Scientific American Books (1992), Chapters 13 to 15.

Le Chatelier's principle, temperature effects, and entropy. J. A. Campbell. *J. Chem. Ed.* **1985,** *62,* 231.

A general approach for calculating polyprotic acid speciation and buffer capacity. D. W. King and D. R. Kester. *J. Chem. Ed.* **1990,** *67,* 932.

Calculation of solubilities of carbonates and phosphates in water as influenced by competitive acid–base reactions. C. Lagier and A. Olivieri. *J. Chem. Ed.* **1990,** *67,* 934.

Equilibria and $\Delta G^{\ominus}$. J. J. MacDonald. *J. Chem. Ed.* **1990,** *67,* 745.

Equilibrium, free energy, and entropy: rates and differences. J. J. MacDonald. *J. Chem. Ed.* **1990,** *67,* 380.

The K_a values of water and the hydronium ion for comparision with other acids. M. J. Campbell and B. A. Waite. *J. Chem. Ed.* **1990,** *67,* 386.

The extent of acid–base reactions. R. J. Thompson. *J. Chem. Ed.* **1990,** *67,* 220.

More about the extent of acid–base reactions. M. E. Cardinalli, C. Giomini, and G. Marrosu. *J. Chem. Ed.* **1991,** *68,* 989.

An efficient method for the treatment of weak acid/base equilibria. J. H. Burness. *J. Chem. Ed.* **1990,** *67,* 224.

Solution of acid–base equilibria by successive approximation. A. C. Oliveri. *J. Chem. Ed.* **1990,** *67,* 229.

The perils of carbonic acid and equilibrium constants. W. P. Jencks and R. A. Altura. *J. Chem. Ed.* **1988,** *65,* 770.

Calculated and observed solubilities of salts with reference to $CaSO_4$. R. B. Martin. *J. Chem. Ed.* **1986,** *63,* 471.

6 Electrochemistry

General reference: P. W. Atkins. *Physical Chemistry,* Oxford University Press and W. H. Freeman & Co. (1990). Chapter 10.

An effective approach to teaching electrochemistry. V. I. Birss and D. R. Truax. *J. Chem. Ed.* **1990,** *67,* 403.

Alleviating the common confusion caused by polarity in electrochemistry. P. J. Moran and E. Gileadi. *J. Chem. Ed.* **1989,** *66,* 912.

7 Chemical kinetics

General reference: P. W. Atkins. *Physical Chemistry,* Oxford University Press and W. H. Freeman & Co. (1990). Chapters 26, 27, and 28. See also K. J. Laidler, *Chemical kinetics.* Harper & Row (1987).

A simple method for analyzing first-order kinetics. B. Borderei, D. Lavabre, G. Levy, and J. C. Micheau. *J. Chem. Ed.* **1990,** *67,* 459.

The kinetics of isotopic exchange reactions. S. R. Logan. *J. Chem. Ed.* **1990,** *67,* 371.

Simplified rate-law integration for reactions that are first-order in each of two reactants. E. Levin and J. G. Eberhart. *J. Chem. Ed.* **1989,** *66,* 705.

Homogeneous, heterogeneous, and enzymatic catalysis. S. T. Oyama and G. A. Somorjai, *J. Chem. Ed.* **1988,** *65,* 765.

The application of physical organic chemistry to biochemical problems. F. Westheimer. *J. Chem. Ed.* **1986,** *63,* 4096.

A new perspective on kinetic and thermodynamic control of reactions. R. B. Snadden. *J. Chem. Ed.* **1985,** *62,* 653.

8 Atomic structure

General reference: P. W. Atkins. *Physical Chemistry,* Oxford University Press and W. H. Freeman & Co. (1990). Chapters 11 and 13.

See also *Quanta: A handbook of concepts*, Oxford University Press (1991) and D. F. Shriver, P. W. Atkins, and C. H. Langford, *Inorganic chemistry*. Oxford University Press and W. H. Freeman & Co. (1990).

Orbital shape representations. O. Kikuchi and K. Suzuki. *J. Chem. Ed.* **1985,** *62,* 2065.

How do electrons get across nodes? A problem in the interpretation of quantum theory. P. G. Nelson. *J. Chem. Ed.* **1990,** *67,* 647.

Very large hydrogen atoms in interstellar space. D. B. Clark. *J. Chem. Ed.* **1991,** *67,* 454.

The use of effective nuclear charge (Z^*) calculations to illustrate the relative energies of ns and $(n-1)d$ orbitals. C. P. Brink. *J. Chem. Ed.* **1991,** *68,* 376.

How to get more from ionization energies in the teaching of atomic structure. P. Mirone. *J. Chem. Ed.* **1991,** *68,* 132.

The periodicity of electron affinity. R. T. Myers. *J. Chem. Ed.* **1990,** *67,* 307.

Periodic contractions among the elements: Or, on being the right size. J. Mason. *J. Chem. Ed.* **1988,** *65,* 17.

Oxidation numbers and their limitations. A. A. Woolf. *J. Chem. Ed.* **1988,** *65,* 45.

9 Molecular structure

General reference: P. W. Atkins. *Physical Chemistry,* Oxford University Press and W. H. Freeman & Co. (1990). Chapter 14. See also *Quanta: A handbook of concepts,* Oxford University Press (1991).

The chemical bond revisited. N. C. Baird. *J. Chem. Ed.* **1986,** *63,* 660.

Lewis structures, formal charge, and oxidation number: A more user-friendly approach. J. E. Packer and S. D. Woodgate. *J. Chem. Ed.* **1991,** *67,* 456.

The relative energies of molecular orbitals for second-row homonuclear diatomic molecules: the effect of *s-p* mixing. H. L. Nyquist. *J. Chem. Ed.* **1991,** *68,* 731.

Describing electron distribution in the hydrogen molecule: A new approach C. J. Willis. *J. Chem. Ed.* **1991,** *68,* 743.

Molecular shape prediction and the lone-pair electrons on the central atom. S. M. Al-Mousawi, *J. Chem. Ed.* **1990,** *67,* 861.

Valence-bond theory and chemical structure. D. J. Klein and N. Trinajstić. *J. Chem. Ed.* **1990,** *67,* 633.

The Lewis electron-pair model, spectroscopy, and the role of the orbital picture in describing the electronic structure of molecules. G. A. Gallup. *J. Chem. Ed.* **1988,** *65,* 671.

Some chemical and electronic considerations of solid state semiconductor crystals. H. J. Hinitz. *J. Chem. Ed.* **1986,** *63,* 956.

10 Cohesion and structure

General reference: P. W. Atkins. *Physical Chemistry*, Oxford University Press and W. H. Freeman & Co. (1990). Chapters 21 and 23.

Lattice enthalpies of ionic halides, hydrides, oxides, and sulfides: second-electron affinities of atomic oxygen and sulfur. J. Holbrook, R. Sabry-Grant, B. C. Smith, and T. V. Tandel. *J. Chem. Ed.* **1990,** *67,* 304.

Coordination and radius ratio: A graphical representation. A. A. Woolf. *J. Chem. Ed.* **1989,** *66,* 509.

A general approach to radius ratios of simple ionic crystals. B. D. Sharma. *J. Chem. Ed.* **1986,** *63,* 504.

Predictions of crystal structure based on radius ratio: how reliable are they? L. C. Nathan. *J. Chem. Ed.* **1985,** *62,* 215.

A two-dimensional model of a liqiud: The pair-correlation function. M. Contrera and J. Valenzuela. *J. Chem. Ed.* **1986,** *63,* 7.

Interstitial solid solutions: Cooperation of energy and geometry. C. G. Lindsay. *J. Chem. Ed.* **1985,** *62,* 675.

11 Molecular spectroscopy

General reference: P. W. Atkins. *Physical Chemistry*, Oxford University Press and W. H. Freeman & Co. (1990). Chapters 16, 17, and 18.

The early history of spectroscopy. N. C. Thomas. *J. Chem. Ed.* **1991,** *68,* 631.

Beer's law revisited. M. M. Berberan-Santos. *J. Chem. Ed.* **1990,** *67,* 757.

A new look at carbonyl electronic transitions. G. Henderson. *J. Chem. Ed.* **1990,** *67,* 392.

C. V. Raman and the discovery of the Raman effect. F. A. Miller and G. B. Kaufman. *J. Chem. Ed.* **1989,** *66,* 795.

Answers to Exercises

1.1 10 atm.

1.2 (a) no, 24 atm; (b) 22 atm.

1.3 (a) 2.57×10^3 Torr; (b) 3.38 atm.

1.4 30 K.

1.5 388 K.

1.6 (a) $4.0\ m^3$; (b) $2.4 \times 10^2\ m^3$.

1.7 4.22×10^{-2} atm.

1.8 (a) 1.83 L; (b) 48.4 kPa.

1.9 $132\ g\ mol^{-1}$.

1.10 $16.4\ g\ mol^{-1}$.

1.11 (a) (i) 1.0 atm, (ii) 8.2×10^2 atm; (b) (i) 0.99 atm, (ii) 1.7×10^3 atm.

1.12 (a) 2.0 atm H_2, 1.0 atm N_2; (b) 3.0 atm.

1.13 $0.5\ m^3$.

1.14 1.5 kPa.

1.15 3.2×10^{-2} atm.

1.16 $B = b - a/RT$, $C = b^2$; $a = 1.26\ L^2\ atm\ mol^{-2}$, $b = 34.6\ cm^3\ mol^{-1}$.

1.17 (a) (i) 693, (ii) 1363, (iii) $2497\ m\ s^{-1}$; (b) (i) 346, (ii) 682, (iii) $1249\ m\ s^{-1}$.

1.18 81 mPa.

1.19 13 MPa.

1.20 $1\ \mu m$.

1.21 (a) $5 \times 10^{10}\ s^{-1}$; (b) $5 \times 10^9\ s^{-1}$; (c) $5 \times 10^4\ s^{-1}$.

1.22 (a) $6.1 \times 10^{33}\ s^{-1}$; (b) $6 \times 10^{31}\ s^{-1}$; (c) $6 \times 10^{21}\ s^{-1}$.

1.23 $4 \times 10^8\ s^{-1}$.

1.24 (a) 6.8 nm; (b) 68 nm; 0.7 cm.

1.25 9.1×10^{-3}.

1.26 λ independent of T.

2.1 (a) 98 J; (b) 16 J.

2.2 2.6 kJ.

2.3 -0.10 kJ.

2.4 (a) -88 J; (b) -167 J.

2.5 $+123$ J.

2.6 $+2.9$ J.

2.7 -1.5 kJ.

2.8 85 MJ.

2.9 $30\ J\ K^{-1}\ mol^{-1}$, $22\ J\ K^{-1}\ mol^{-1}$.

2.10 640 kJ, 640 s.

2.11 $80\ J\ K^{-1}$, -1.2 kJ, -1.2 kJ.

2.12 $+2.2$ kJ, $+2.2$ kJ, $+1.6$ kJ.

2.13 -1.0 kJ, $+13$ kJ, $+12$ kJ.

2.14 $-4.56\ MJ\ mol^{-1}$.

2.15 $-126\ kJ\ mol^{-1}$.

2.16 $-432\ kJ\ mol^{-1}$.

2.17 $+79\ kJ\ mol^{-1}$.

2.18 $641\ J\ K^{-1}$.

2.19 $1.58\ kJ\ K^{-1}$, 2.05 K.

2.20 (a) $-2.80\ MJ\ mol^{-1}$; (b) $-2.80\ MJ\ mol^{-1}$; (c) $-1.28\ MJ\ mol^{-1}$.

2.21 $+65.49\ kJ\ mol^{-1}$.

2.22 $-383\ kJ\ mol^{-1}$.

2.23 $1.90\ kJ\ mol^{-1}$.

2.24 -25 kJ, 9.8 m.

2.25 (a) $-2205\ kJ\ mol^{-1}$; (b) $-2200\ kJ\ mol^{-1}$.

2.26 (a) exothermic; (b, c) endothermic.

2.27 (a) $-57.20\ kJ\ mol^{-1}$; (b) $-176.01\ kJ\ mol^{-1}$; (c) $-32.88\ kJ\ mol^{-1}$; (d) $-55.84\ kJ\ mol^{-1}$.

2.28 $+11.3\ kJ\ mol^{-1}$.

2.29 $-56.98\ kJ\ mol^{-1}$.

3.1 54 kJ, $-1.9 \times 10^2\ J\ K^{-1}$.

3.2 $+26\ J\ K^{-1}$.

3.3 6.6 L.

3.4 $3\ J\ K^{-1}$.

3.5 $+87.8\ J\ K^{-1}\ mol^{-1}$, $-87.8\ J\ K^{-1}\ mol^{-1}$.

3.6 (a) $-386.1\ J\ K^{-1}$; (b) $+92.6\ J\ K^{-1}$; (c) $-153.1\ J\ K^{-1}$; (d) $-21.0\ J\ K^{-1}$; (e) $+512.0\ J\ K^{-1}$.

3.7 (a) -521.5 kJ; (b) $+25.8$ kJ; (c) -178.7 kJ; (d) -212.40 kJ; (e) -5798 kJ.

3.8 (a) -522.1 kJ; (b) $+25.78$ kJ; (c) -178.6 kJ; (d) -212.55 kJ; (e) -5798 kJ.

3.9 $-49.7\ kJ\ mol^{-1}$.

3.10 $\Delta S = n(C_V - R) \ln 2$.

3.11 $817.9\ kJ\ mol^{-1}$.

3.12 $+0.95\ J\ K^{-1}\ mol^{-1}$.

3.13 $-2.42\ kJ\ mol^{-1}$.

3.14 3.01.

3.15 1500 K.

3.16 $\Delta H^{\ominus} = +2.77$ kJ; $\Delta S^{\ominus} = -16.5\ J\ K^{-1}\ mol^{-1}$.

3.17 $+12.3\ kJ\ mol^{-1}$.

3.18 $-41.0\ kJ\ mol^{-1}$.

3.19 0.90, 0.10.

3.20 (a, c, and e).

3.21 (b and d).

3.22 (a) $+53\ kJ\ mol^{-1}$; (b) $-53\ kJ\ mol^{-1}$.

3.23 $-14.38\ kJ\ mol^{-1}$.

3.24 (a) 9.24; (b) $-12.9\ kJ\ mol^{-1}$; (c) $+161\ kJ\ mol^{-1}$; (d) $+248\ J\ K^{-1}\ mol^{-1}$.

3.25 (a) 1110 K; (b) 397 K.

4.1 (a) 1.7 kg; (b) 31 kg; (c) 1.4 g.

4.2 Yes; 3 Torr or more.

4.6 $886.8\ cm^3$.

4.7 6.4 MPa.

4.8 $x_A = 0.920$, $x_B = 0.080$; $y_A = 0.968$, $y_B = 0.032$.

4.9 82 g mol^{-1}.

4.10 381 g mol^{-1}.

4.11 $-0.09\,°C$.

4.12 (a) 3.4 mmol kg^{-1}; (b) 34 mmol kg^{-1}.

4.13 0.51 mmol kg^{-1} N$_2$, 0.27 mmol kg^{-1} O$_2$.

4.14 0.17 mol kg^{-1}.

4.15 $-0.16\,°C$.

4.16 87 kg mol^{-1}.

4.17 14 kg mol^{-1}.

4.18 (a) $y_T = 0.36$; (b) $y_T = 0.82$.

4.23 (a) 80 per cent by mass.

5.1 (a) $a(COCl)a(Cl)/a(CO)a(Cl_2)$;
(b) $a(SO_3)^2/a(SO_2)^2a(O_2)$;
(c) $a(HBr)_2/a(H_2)a(Br_2)$;
(d) $a(O_2)^3/a(O_3)^2$.

5.2 5.4×10^{-4}.

5.3 2.6×10^{-4} bar.

5.4 (a) 1.0×10^{-2} mol L^{-1} PCl$_3$; (b) 52 per cent.

5.5 -41.1 kJ mol^{-1}.

5.6 x(borneol) = 0.90, x(isoborneol) = 0.10.

5.7 (a) 9.24; (b) -12.9 kJ mol^{-1}; (c) $+161$ kJ mol^{-1}; (d) $+248$ J K^{-1} mol^{-1}.

5.8 0.010 atm N$_2$, 0.010 atm H$_2$; 1.9×10^{-5} atm NH$_3$.

5.9 $K = 4\alpha^2 p/(1 - \alpha^2) \approx 4\alpha^2 p$.

5.10 (a) H$_2$SO$_4$(A$_1$) + H$_2$O(B$_2$) $\rightleftharpoons$ H$_3$O$^+$(A$_2$) + HSO$_4^-$(B$_1$);
(b) HF(A$_1$) + H$_2$O(B$_2$) $\rightleftharpoons$ H$_3$O$^+$(A$_2$) + F$^-$(B$_1$);
(c) C$_6$H$_5$NH$_3^+$(A$_1$) + H$_2$O(B$_2$) $\rightleftharpoons$ H$_3$O$^+$(A$_2$) + C$_6$H$_5$NH$_2$(B$_1$);
(d) H$_2$PO$_4^-$(A$_1$) + H$_2$O(B$_2$) $\rightleftharpoons$ H$_3$O$^+$(A$_2$) + HPO$_4^{2-}$(B$_1$);
(e) HCOOH(A$_2$) + H$_2$O(B$_2$) $\rightleftharpoons$ H$_3$O$^+$(A$_2$) + HCO$_2^-$(B$_1$);
(f) NH$_2$NH$_3^+$(A$_1$) + H$_2$O(B$_2$) $\rightleftharpoons$ H$_3$O$^+$(A$_2$) + N$_2$H$_4$(B$_1$).

5.11 (a) 1.6×10^{-7} mol L^{-1}, 6.80; (b) 1.6×10^{-7} mol L^{-1}, 6.80.

5.12 (a) D$_2$O + D$_2$O $\rightleftharpoons$ D$_3$O$^+$ + OD$^-$;
(b) 14.870;
(c) 3.67×10^{-8} mol L^{-1};
(d) 7.435, 7.435;
(e) pD + pOD = pK_w.

5.13 (a) 4.70, 9.30; (b) 0.00, 14.00; (c) 13.30, 0.70; (d) 4.30, 9.70.

5.14 (a) 0.050 mol L^{-1}, 1.30; (b) 0.050 mol L^{-1}, 12.70.

5.15 (a) <7, NH$_4^+$ + H$_2$O $\rightleftharpoons$ H$_3$O$^+$ + NH$_3$;
(b) >7, CO$_3^{2-}$ + H$_2$O $\rightleftharpoons$ HCO$_3^-$ + OH$^-$;
(c) >7, H$_2$O + F$^-$ $\rightleftharpoons$ HF + OH$^-$; (d) 7;
(e) <7, [Al(OH$_2$)$_6$]$^{3+}$ + H$_2$O $\rightleftharpoons$ H$_3$O$^+$ + [Al(OH$_2$)$_5$OH]$^{2+}$;
(f) <7, [Co(OH$_2$)$_6$]$^{2+}$ + H$_2$O $\rightleftharpoons$ H$_3$O$^+$ + [Co(OH$_2$)$_5$OH]$^+$.

5.16 (a) 9.18, (b) 4.74, (c) 0, strong acid.

5.17 (a) 3.08, 8.3×10^{-4}; (c) 2.8.

5.19 (a) 1.8 per cent; (b) 0.29 per cent; (c) 1.8 per cent.

5.20 (a) 1.9, 12.1, 6.5 per cent; (b) 5.0, 9.0, 98.8 per cent; (c) 1.1, 12.9, 73 per cent.

5.21 (a) 6.5; (b) 2.1; (c) 1.5.

5.22 (a) 0.047, 0.053, 6.5×10^{-5}, 0.053, 1.9×10^{-13} mol L^{-1};
(b) 0.050, 8.1×10^{-5}, 7.1×10^{-15}, 8.1×10^{-5}, 1.2×10^{-10} mol L^{-1}.

5.23 (a) 2.87; (b) 4.56; (c) 12.5 mL; (d) 4.74; (e) 25.0 mL; (f) 8.72.

5.24 (a) 4.74; (b) 5.01, $+0.27$; (c) 4.14, -0.60.

5.25 (a) 2 to 4; (b) 3 to 5; (c) 6 to 8.

5.26 (a) 4.0×10^{-6}, 5.40; 3.6.

5.27 (a) 5.1; (b) 8.9; (c) 2.88.

5.28 7.9.

5.30 (a) H$_3$PO$_4$/NaH$_2$PO$_4$; (b) NaH$_2$PO$_4$/Na$_2$HPO$_4$.

5.31 (a) $a(Ag^+)a(Br^-)$;
(b) $a(Ag^+)^2a(S^{2-})$;
(c) $a(Ca^{2+})a(OH^-)^2$;
(d) $a(Ag^+)^2a(CrO_4^{2-})$.

5.32 (a) 1.3×10^{-4} mol L^{-1}; (b) 1.3×10^{-4} mol L^{-1}; (c) 1.6×10^{-5} mol L^{-1}; (d) 6.2×10^{-4} mol L^{-1}.

5.33 Yes.

5.34 (a) 7.7×10^{-10} mol L^{-1}; (b) 3.1×10^{-3} mol L^{-1}; (c) 1.6×10^{-7} mol L^{-1}; (d) 2.1×10^{-7} mol L^{-1}.

6.1 1.25×10^{-5} mol L^{-1}.

6.2 34.2 mV.

6.3 Mn | MnCl$_2$(aq) | Cl$_2$(g) | Pt, -1.18 V.

6.4 (a) 2Ag$^+$ + Zn $\rightarrow$ 2Ag + Zn^{2+}, $+1.56$ V;
(b) Cd + 2H$^+$ $\rightarrow$ Cd^{2+} + H$_2$, $+0.40$ V;
(c) Cr^{3+} + 3[Fe(CN)$_6$]$^{4-}$ $\rightarrow$ Cr + 3[Fe(CN)$_6$]$^{3-}$, -1.10 V;

(d) Ag$_2$CrO$_4$ + 2Cl$^-$ $\rightarrow$ 2Ag + CrO$_4^{2-}$ + Cl$_2$, -0.91 V;
(e) Sn^{4+} + 2Fe^{2+} $\rightarrow$ Sn^{2+} + 2Fe^{3+}, -0.62 V;
(f) Cu + MnO$_2$ + 4H$^+$ $\rightarrow$ Cu^{2+} + Mn^{2+} + 2H$_2$O, $+0.89$ V.

6.5 (a) Zn |ZnSO$_4$| |CuSO$_4$| Cu, $+1.10$ V;
(b) Pt |H$_2$| HCl |AgCl| Ag, $+0.22$ V;
(c) Pt |H$_2$| H$^+$, H$_2$O |O$_2$| Pt, $+1.23$ V;
(d) Pt |H$_2$| HI |I$_2$| Pt, $+0.54$ V.

6.6 See above.

6.7 (a) -1.20 V; (b) -1.18 V.

6.8 (a) -363 kJ mol^{-1};
(b) -405 kJ mol^{-1};
(c) -291 kJ mol^{-1};
(d) $+122$ kJ mol^{-1}.

6.9 (a) $+0.324$ V; (b) $+0.45$ V.

6.10 -0.62 V.

6.11 -1.23 V, $+233$ kJ, $+300$ kJ; $+232$ kJ.

6.12 (a) 1.6×10^{-8} mol L^{-1}; (b) $+0.12$ V.

6.13 (a) 6.5×10^9; (b) 1.5×10^{12}; (c) 2.8×10^{-16}; (d) 1.7×10^{16}; (e) 8.2×10^{-7}.

6.14 1.8×10^{-10}, 9.04×10^{-7}.

6.15 $E = E^\ominus - (RT/6F)\ln Q$, $Q = a(Cr^{3+})^2/a(Cr_2O_7^{2-})a(H^+)^{14}$.

6.16 0.86.

6.17 0.

6.18 (a) 9.19×10^{-9} mol L^{-1}; (b) 8.5×10^{-17}.

6.19 Pb |PbSO$_4$| H$_2$SO$_4$ |Hg$_2$SO$_4$| Hg or Pb |PbSO$_4$| H$_2$SO$_4$ |H$_2$SO$_4$| Hg$_2$SO$_4$| Hg; 0.95 V.

6.20 (a) $+1.23$ V; (b) $+2.85$ V.

6.21 $+0.85$ V.

6.22 3.6×10^{-8}.

6.23 $+0.22$ V.

7.1 1.0(A), 0.5(B), 1.5(D); mol L^{-1} s^{-1}.

7.2 L^2 mol^{-2} s^{-1}.

7.3 1.03×10^4 s; (a) 501 Torr; (b) 530 Torr.

7.4 (a) L mol^{-1} s^{-1}, L^2 mol^{-2} s^{-1};
(b) atm^{-1} s^{-1}, atm^{-2} s^{-1}.

7.5 2.7×10^3 y.

7.6 (a) 0.64 μg; (b) 0.18 μg.

7.7 (a) 0.05 mol L^{-1}, 0.095 mol L^{-1}; (b) 0.001 mol L^{-1}, 0.051 mol L^{-1}.

7.8 1.2×10^5 s.

7.9 first order, 2.8×10^{-4} s^{-1}.

7.10 64.9 kJ mol^{-1}, 4.32×10^8 mol L^{-1} s^{-1}.

7.11 $H_2O(1)$, $Br^-(1)$, overall (2).

7.12 rate $= k_2 K^{1/2}[A_2]^{1/2}[B]$.

7.13 rate $= k_1 K[A][B]$, $k = k_1 k_2/k_2'$.

7.14 $1.52 \text{ mmol L}^{-1} \text{ s}^{-1}$.

7.15 $[S] = K_M$.

7.16 rate $=$
$-k_1[R_2] - k_2(k_1/k_4)^{1/2}[R_2]^{3/2}$.

7.17 0.28 to 2.2 kPa, 0.14 to 8.9 kPa;
>0.11 kPa.

7.18 3.3×10^{18}.

7.19 0.521.

7.20 rate $= k_1 k_3[A][AH][B]/$
$\{k_2[BH^+] + k_3[A]\}$.

7.21 rate $= k_3 K K_a[HA][BH^+]$.

7.22 rate $= k'[AH]$,
$k' = k_a + k_a k_c/2kk_d$, with
$$k = \left(\frac{k_a}{4k_b}\right)\left\{1 + \sqrt{1 + \frac{8k_b k_c}{k_a k_d}}\right\}$$

7.23 3.6 mmol L^{-1}, $5.0 \text{ } \mu\text{mol L}^{-1} \text{ s}^{-1}$;
0.33 s^{-1}.

7.24 noncompetitive.

8.1 0.17 kW.

8.2 5.2×10^{16} Hz.

8.3 $1.6 \times 10^6 \text{ m s}^{-1}$.

8.4 (a) $8.83 \times 10^{-28} \text{ kg m s}^{-1}$;
(b) $9.5 \times 10^{-24} \text{ kg m s}^{-1}$;
(c) $3.5 \times 10^{-35} \text{ kg m s}^{-1}$.

8.5 50.6 nm.

8.6 70 nm.

8.7 6.09 nm.

8.8 $0.17 L$ or $0.83 L$.

8.9 5.9×10^{-24} J.

8.10 (a) 3.31×10^{-19} J, 199 kJ mol^{-1};
(b) 3.61×10^{-19} J, 218 kJ mol^{-1};
(c) 4.97×10^{-19} J, 299 kJ mol^{-1};
(d) 9.93×10^{-19} J, 598 kJ mol^{-1};
(e) 1.32×10^{-15} J,
$7.98 \times 10^5 \text{ kJ mol}^{-1}$;
(f) 1.99×10^{-23} J, $0.012 \text{ kJ mol}^{-1}$.

8.11 (a) $1.10 \times 10^{-27} \text{ kg m s}^{-1}$;
(b) $1.20 \times 10^{-27} \text{ kg m s}^{-1}$;
(c) $1.66 \times 10^{-27} \text{ kg m s}^{-1}$;
(d) $3.31 \times 10^{-27} \text{ kg m s}^{-1}$;
(e) $4.42 \times 10^{-24} \text{ kg m s}^{-1}$;
(f) $6.63 \times 10^{-32} \text{ kg m s}^{-1}$.

8.12 (a) $2.8 \times 10^{18} \text{ s}^{-1}$;
(b) $2.8 \times 10^{20} \text{ s}^{-1}$.

8.13 6000 K.

8.14 (a) no ejection;
(b) 3.19×10^{-19} J.

8.15 no.

8.16 (a) 7×10^{-19} J; (b) 40 kJ mol^{-1};
(c) $4 \times 10^{-13} \text{ kJ mol}^{-1}$.

8.17 (a) 6.6×10^{-29} m;

8.18 (b) 6.6×10^{-36} m; (c) 99.7 pm.

8.18 (a) 123 pm; (b) 39 pm;
(c) 3.88 pm.

8.19 (a) $1.1 \times 10^{-28} \text{ m s}^{-1}$;
(b) 1×10^{-27} m.

8.20 $5 \times 10^{-25} \text{ kg m s}^{-1}$, $5 \times 10^5 \text{ m s}^{-1}$.

8.21 1.12 fJ.

8.22 (a) $1.6 \times 10^{-33} \text{ J m}^{-3}$;
(b) 0.25 mJ m^{-3}.

8.23 6.52×10^{-34} J s.

8.24 486.3 nm.

8.25 6.

8.26 6842 cm^{-1}.

8.27 14.0 eV.

8.28 101 pm, 376 pm.

8.29 (a) 0; (b) 0; (c) $\sqrt{6}\hbar$; (d) $\sqrt{2}\hbar$; (e)
$\sqrt{2}\hbar$; angular nodes $= l$; radial
nodes $= n - l - 1$.

8.30 (a) 1; (b) 9; (c) 25.

8.31 $0.35a_0$.

8.32 (a) 110 pm; (b) 86 pm.

8.33 (a) F; (b) A; (c) A; (d) F; (e) A.

8.34 (a) 2; (b) 6; (c) 10; (d) 18.

8.35 H $1s^1$; He $1s^2$; Li [He]$2s^1$;
Be [He]$2s^2$; B [He]$2s^2 2p^1$;
C [He]$2s^2 2p^2$; N [He]$2s^2 2p^3$;
O [He]$2s^2 2p^4$; F [He]$2s^2 2p^5$;
Ne [He]$2s^2 2p^6$; Na [Ne]$3s^1$;
Mg [Ne]$3s^2$; Al [Ne]$3s^2 3p^1$;
Si [Ne]$3s^2 3p^2$; P [Ne]$3s^2 3p^3$;
Si [Ne]$3s^2 3p^4$; Cl [Ne]$3s^2 3p^5$;
Ar [Ne]$3s^2 3p^6$.

8.36 $n_2 \rightarrow 6$.

8.37 397.13 nm, 3.40 eV.

8.38 $987\,663 \text{ cm}^{-1}$; lines at
$137\,175 \text{ cm}^{-1}$, $185\,187 \text{ cm}^{-1}$, ...;
122.5 eV.

9.1 (a) $1\sigma_g^2$, $b = 1$; (b) $1\sigma_g^2 2\sigma_u^2$, $b = 0$;
(c) $1\sigma_g^2 2\sigma_u^2 2\pi_u^4$, $b = 2$.

9.2 (a) $1\sigma_g^2 2\sigma_u^2$; (b) $\dots 1\pi_u^4 2\sigma_g^2$;
(c) $1\pi_g^1 1\pi_g^1$.

9.3 (a) $\dots \sigma^2$; (b) $\dots \pi^{*2}$; (c) $\dots \sigma^2$.

9.4 C_2.

9.5 (a) C_2, CN; (b) NO, O_2, F_2.

9.6 $R(XeF^+) < R(XeF)$.

9.7 (a) g; (b) not relevant; (c) g;
(d) u.

9.8 $a_2(g)$, $e_1(g)$, $e_2(u)$, $b_2(g)$.

9.9 N_2.

9.11 $N = 1/(1 + 2\lambda S + \lambda^2)^{1/2}$.

9.12 (a, c).

9.13 (a) angular; (b) octahedral;
(c) square planar; (d) see-saw.

9.15 (a) $6\alpha + 8\beta$; (b) $5\alpha + 7\beta$.

9.16 4.3×10^{-19} J (460 nm).

10.1 $3.9 \times 10^3 \text{ kJ mol}^{-1}$.

10.2 ratio $= 1.13$.

10.3 $2 ax + 1$ eq.

10.4 (a) ortho: 0.7 D; (b) meta: 0.4 D;
(c) para: 0 (certain).

10.5 1.4 D.

10.6 $V = -\mu q/4\pi\varepsilon_0 R^2$.

10.7 $V = 2\mu^2/4\pi\varepsilon_0 R^3$.

10.8 $R = 2^{1/6}\sigma$.

10.12 $a = (2\pi/3)N_A\sigma^3$,
$b = (2\pi/3)N_A^2\varepsilon(\sigma'^3 - \sigma^3)$.

10.14 (a) $3.3 \times 10^{-9} \text{ m}^2 \text{ s}^{-1}$;
(b) $8.3 \times 10^{-10} \text{ m}^2 \text{ s}^{-1}$.

10.15 (a) $d = \lambda(t/\tau)^{1/2}$; (b) 10^6.

10.16 (a) 0.9069; (b) 0.5236.

10.19 (326), (111), (122), ($3\bar{2}\bar{2}$).

10.22 249 pm, 176 pm, 432 pm.

10.23 70.7 pm.

10.24 0.21 cm.

10.25 $3.96 \times 10^{-28} \text{ m}^3$.

10.26 4, 4.01 g cm^{-3}.

10.27 190 pm.

10.28 bcc.

10.29 8.97 g cm^{-3}.

11.1 (b, c, d, e, f, g).

11.2 328.7 N m^{-1}.

11.3 $0.999\,999\,925 \times \lambda$,
$6.36 \times 10^7 \text{ m s}^{-1}$.

11.4 $2.4 \times 10^4 \text{ km s}^{-1}$; 8.4×10^5 K.

11.5 (a) 53 ps; (b) 5 ps; (c) 2 ns.

11.6 (a) 50 cm^{-1}; (b) 0.5 cm^{-1}.

11.7 HF: 967.1 N m^{-1},
HCl: 515.6 N m^{-1},
HBr: 411.8 N m^{-1},
HI: 314.2 N m^{-1}.

11.8 DF: 3002 cm^{-1}, DCl: 2144 cm^{-1},
DBr: 1886 cm^{-1}, DI: 1640 cm^{-1}.

11.9 (a) yes; (b) no.

11.10 80 per cent.

11.11 $7.9 \times 10^5 \text{ cm}^2 \text{ mol}^{-1}$.

11.12 1.5 mmol L^{-1}.

11.14 $450 \text{ mol L}^{-1} \text{ cm}^{-1}$.

11.15 $160 \text{ mol L}^{-1} \text{ cm}^{-1}$, 23 per cent.

11.16 (a) 0.9 m; (b) 3 m.

11.18 1:6:15:20:15:6:1 heptet.

11.19 (a) 1:2:3:2:1 quintet.
(b) 1:3:6:7:6:3:1 heptet.

11.20 $-1.625 \times 10^{-26} \text{ J} \times m_I$.

11.21 6.116×10^{-26} J, 92.3 MHz.

11.22 3.523 T.

11.23 (a) $11 \text{ } \mu\text{T}$; (b) $53 \text{ } \mu\text{T}$.

11.24 (a) 460 Hz; (b) 2.66 kHz.

Index

T after a page number refers to a Table in the text.